Design of Concrete Buildings for Earthquake and Wind Forces

SECOND EDITION

by S. K. Ghosh, August W. Domel, Jr., and David A. Fanella

5420 Old Orchard Road, Skokie, Illinois 60077-1083
Phone: 708/966-6200

© 1995 Portland Cement Association

All rights reserved

Printed in U.S.A.

Library of Congress catalog card number 91-46033

ISBN 0-89312-098-7

About the authors: S. K. Ghosh is Director, August W. Domel, Jr. is former Senior Structural Engineer, and David A. Fanella is Coordinating Structural Engineer, Engineering Services, Codes and Standards, Portland Cement Association, Skokie, Illinois.

Preface to the First Edition

The publication in 1961 of the landmark volume, "Design of Multistory Concrete Buildings for Earthquake Motions, " by Blume, Newmark and Corning, gave major impetus to the construction of concrete buildings in regions of high seismicity. In the three decades since, significant strides have been made in the earthquake resistant design and construction of reinforced concrete buildings. The treatment of reinforced concrete in U.S. seismic codes is now more favorable than it has been in the past. However, a comprehensive guide to aid the designer in the detailed seismic design of concrete buildings has not been available so far. This publication seeks to fill that void. It is hoped that the availability of this guide will give further impetus to the construction of concrete buildings in regions of significant seismicity.

The detailed design of reinforced concrete buildings utilizing the various structural systems recognized in U.S. seismic codes is illustrated. All designs are according to the provisions of the 1991 edition of the Uniform Building Code (UBC). Design of the same building is carried out for regions of high, moderate, and low seismicity, and for wind, so that this publication would be of use and interest in all parts of the country.

This manual will hopefully assist the engineer and designer in the proper application of the wind and seismic design provisions of the 1991 UBC which has adopted, with modifications, the seismic detailing requirements of the 1989 edition of "Building Code Requirements for Reinforced Concrete (ACI 318-89, Revised 1992)." The emphasis, except in Chapter 1, is placed on "how-to-use" the Code. For background information on the design and detailing provisions, beyond what is contained in Chapter 1, the reader is referred to the commentary portion of "Recommended Lateral Force Requirements" by the Seismology Committee of the Structural Engineers Association of California, and to the "Commentary on Building Code Requirements for Reinforced Concrete (ACI 318R-89)."

This manual will also be a valuable aid to students, educators, architects, building code authorities, and others involved in the design, construction, and regulation of concrete buildings.

Although every attempt has been made to impart editorial consistency to the seven chapters, some inconsistencies probably still remain. Since this is the very first edition of a fairly extensive volume, some errors, almost certainly, are also to be found. PCA would be grateful to any reader who would bring such errors and inconsistencies to our attention. Other suggestions for improvement are also more than welcome.

Acknowledgments

James P. Barris, former Director of Codes and Standards, initiated this project during his tenure at the Portland Cement Association. The writer would like to express his gratitude to him.

Mohammad Kleit, Tom Behringer, and Francisco Javier Perez-Caballero deserve sincere thanks for contributing to Chapters 2 and 5, Chapter 3 and Chapter 4, respectively. All three were the writer's graduate students at the University of Illinois at Chicago when they worked on the project. Mohammad Kleit and Francisco Perez are now members of the Computer Programs Marketing staff at PCA. The latter's many suggestions concerning all the chapters have been especially useful.

Aphrodite Lisa Lehocky, also of PCA's Computer Programs Marketing staff, was responsible for the layout, design and word-processing of this entire complex manuscript including its many tables. She also drafted all the figures. Her dedicated help was a major factor in the completion of this project.

A thorough review of the entire manuscript by Gerald B. Neville of the International Conference of Building Officials is much appreciated.

Finally, the writer would like to express his deep appreciation of the efforts of his co-author, August W. Domel, Jr. of PCA's Engineered Structures and Codes staff. Dedicated professionalism and much hard work on his part finally brought this project to fruition.

Skokie, Illinois   S. K. Ghosh
February, 1992

Preface to the Second Edition

This second edition updates this publication to the 1994 edition of the Uniform Building Code. An Appendix B has been added to Chapter 1, which lists and briefly discusses the significant changes from the 1991 to the 1994 edition of the Uniform Building Code. All the examples have been redone to conform to the 1994 edition of the Uniform Building Code.

One major change from the 1991 to the 1994 edition of the UBC is a new procedure for the design of reinforced concrete shearwalls in combined bending and axial compression. This new design procedure is not part of the "Building Code Requirements for Reinforced Concrete (ACI 318-89, Revised 1992)" as yet. Design of shearwalls by 1991 UBC provisions has been retained in this edition. This is because in many jurisdictions, those provisions will probably be retained for some time to come. Also, the contrast between the 1991 and the 1994 designs would probably be of interest to most readers. In Chapters 4, 5 and 6, the shearwall configuration in the core has been changed, to make it more practical, and to increase the torsional resistance of the buildings considered.

Three-dimensional structural analysis by the Portland Cement Association computer program *PCA-Frame* has been introduced in this edition. Direct comparisons have been provided in Chapters 2 and 3 between the results of the 3-D and the 2-D analyses of the prior edition, where accidental torsion had to be accounted for separately. Encouraging agreement was observed between comparable sets of results.

As always, every attempt has been made to be as accurate as possible in the design examples. Also, an attempt has been made to significantly improve the figures. Readers are urged to bring any errors or misprints they may find to our attention. Any suggestions for improvement would also be gratefully received.

Acknowledgments

David A. Fanella has been largely responsible for the update of this publication to the 1994 edition of the Uniform Building Code. His dedication, professionalism and thoroughness are much appreciated.

Deborah Terrill and Cami Lobb have done the word processing, the graphics and the layout of this second edition. They deserve much credit and gratitude for the high quality of production of this volume.

Skokie, Illinois
September, 1995

S. K. Ghosh

Table of Contents

Chapter 1

Introduction

This chapter was written for the first edition of this publication issued in 1992. A new appendix 1-B has been added to update the chapter to 1994 UBC. The scope of the second edition of this publication remains essentially the same as for the first edition (see Sect. 1.10).

1.1   Evolution Of U.S. Seismic Codes

Over the past few decades, a pattern of American seismic building code development had emerged. Provisions were first proposed by the Structural Engineers Association of California (SEAOC) in its *Recommended Lateral Force Requirements* (commonly referred to as the Blue Book) [1.1], then adopted by the International Conference of Building Officials (ICBO) in the *Uniform Building Code* (UBC) [1.2] and finally (often with modifications) by the American National Standards Institute (ANSI) Standard A58.1 [1.3] and the other two model codes-*The BOCA National Building Code* (BOCA/NBC) published by the Building Officials and Code Administrators International [1.4] and the *Standard Building Code* (SBC) published by the Southern Building Code Congress International [1.5].

A departure from the above pattern occurred when in 1972 the National Science Foundation and the National Bureau of Standards initiated a Cooperative Program in Building Practices for Disaster Mitigation. Under that program, the Applied Technology Council (ATC) developed a document entitled *Tentative Provisions for the Development of Seismic Regulations for Buildings* [1.6]. This document, commonly referred to as ATC 3-06, underwent thorough review by the building community in ensuing years. Trial designs were conducted to establish the technical validity of the new provisions and to assess their impact. All of this subsequent effort culminated in the publication in 1985 of the NEHRP (National Earthquake Hazards Reduction Program) Recommended Provisions for the Development of Seismic Regulations for New Buildings [1.7]. 1988 and 1991 updates of this document have been issued.

In 1980, the SEAOC Seismology Committee undertook the task of developing an ATC-based revision of their Blue Book. This extensive effort resulted in the latest edition of the SEAOC Blue Book [1.1]. It has been adopted into the 1988 edition of the Uniform Building Code [1.2]. Changes from the 1988 to the 1991 edition of the UBC have not been major, except for a significant change in format. Earthquake regulations of Section 2312 of 1988 UBC are now in Sections 2330 through 2339 of 1991 UBC.

The BOCA National Building Code [1.4] and the *Standard Building Code* [1.5] have traditionally adopted the seismic provisions of the ANSI Standard [1.3]. However, that situation has now changed. Both have adopted the 1988 NEHRP Provisions [1.7], and are in the process of updating to the 1991 NEHRP Provisions. The model code adoption procedure has gone ahead of the standard writing procedure, and it is now up to ASCE Committee 7 (formerly ANSI Committee A58.1) to catch up. A 1988 edition of the ASCE/ANSI Standard was published in 1990, without updated seismic provisions [1.3]. A decision has been made to adopt NEHRP 1991 Provisions into ASCE 7 (formerly ANSI A58.1). However, the publication of this updated Standard is still in the future.

Since 1991 UBC [1.2] represents the latest seismic design practice in the most seismically active regions of this country, and since 1985 UBC represents that practice before the influence of ATC/NEHRP [1.6, 1.7], a comparison of the seismic design provisions of the 1985 and 1991 Uniform Building Codes is provided in this chapter.

1.2   Code Design Criteria

The procedures and limitations for the design of structures by the UBC are determined considering zoning, site characteristics, occupancy, configuration, structural system and height. Two of the major parameters in the selection of design criteria are occupancy and structural configuration.

Four categories of occupancy are defined in Table 23-K of UBC-91: Essential, Hazardous, Special and Standard. The "Hazardous Facilities" category was added to the UBC in 1988.

Structural configuration is addressed by defining two categories of structural irregularities in Tables 23-M (Vertical Structural Irregularities) and 23-N (Plan Structural Irregularities). Five different types of vertical structural irregularities are defined in Table 23-M (Fig. 1-1): Stiffness Irregularity—Soft Story, Weight (Mass) Irregularity, Vertical Geometric Irregularity, In-plane Discontinuity in Vertical Lateral Force-Resisting Elements, and Discontinuity in Capacity—Weak Story. An exception is provided to the definition of Stiffness Irregularity and Mass Irregularity. Where no story drift ratio under design lateral forces is greater than 1.3 times the story drift ratio of the story above, a structure may be deemed to not have Stiffness or Mass Irregularity. Five different types of plan irregularity are defined in Table 23-N (Fig. 1-2): Torsional Irregularity (to be considered when diaphragms are not flexible), Reentrant Corners, Diaphragm Discontinuity, Out-of-plane Offsets, and Nonparallel Lateral Force-Resisting Systems. Regular structures are defined as having no significant physical discontinuities in plan or vertical configuration or in their lateral force-resisting systems such as those identified for irregular structures.

Two different analysis procedures are recognized in UBC-91 for the determination of seismic effects on structures. The Dynamic Lateral Force Procedure of Section 2335 is always acceptable for design. The Static Lateral Force Procedure of Section 2334 is allowed only under certain conditions of regularity, occupancy, and height.

The applicability of the Static Lateral Force Procedure of Section 2334 is detailed below:

ZONE 1

ALL STRUCTURES
*   Any occupancy
*   Any height
*   Regular or irregular

ZONE 2 (A&B)

STANDARD OCCUPANCY
*   Any height
*   Regular or irregular

ESSENTIAL, HAZARDOUS, SPECIAL OCCUPANCY
*   Regular structures < 240 ft, except those having period > 0.7 sec. and located on Soil Profile Type S_4
*   Irregular structures ≤ stories or 65 ft
*   Structures having flexible upper portion and rigid lower portion....each portion regular, within certain limitations

ZONES 3, 4
*   Regular structures < 240 ft, except those having period > 0.7 sec. and located on Soil Profile Type S4
*   Irregular structures ≤ 5 stories or 65 ft
*   Structures having flexible upper portion and rigid lower portion...within certain limitations

Although any structure may be designed using the Dynamic Lateral Force Procedure of Section 2335, it is required to be used in situations detailed below:

ZONE 2 (A&B)

ESSENTIAL, HAZARDOUS, SPECIAL OCCUPANCY
*   Regular structures > 240 ft
*   Irregular structures > 5 stories or 65 ft with stiffness, weight or geometric vertical irregularity or with irregular features not defined in Tables 23-M, 23-N (except as permitted by Section 2334(c) 2)
*   Regular or irregular structures having period > 0.7 sec. and located on Soil Profile Type S4

ZONES 3,4
*   Regular structures > 240 ft
*   Irregular structures > 5 stories or 65 ft with stiffness, weight or geometric vertical irregularity or with irregular features not defined in Tables 23-M, 23-N (except as permitted by Section 2334(c) 2)
*   Structures > 5 stories or 65 ft with mixed vertical structural systems (except as permitted by Section 2334(c) 2)
*   Regular or irregular structures having period > 0.7 sec. and located on Soil Profile Type S4

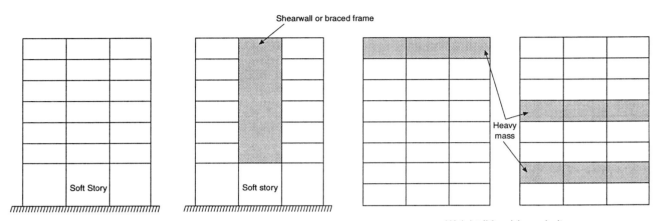

Vertical Stiffness Irregularity—Soft Story
"Soft Story" Stiffness < 70% Story Stiffness Above

Weight (Mass) Irregularity
Story Mass > 150% Adjacent Story Mass

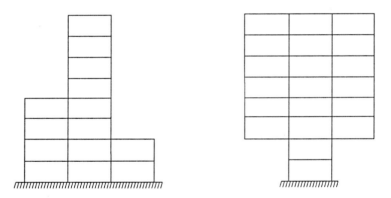

Vertical Geometric Irregularity
Story Dimension > 130% Adjacent Story Dimension

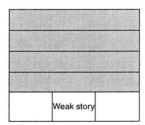

Vertical Strength Irregularity-Weak Story
"Weak Story" Strength < 80% Story Strength Above

Figure 1-1 Vertical Structural Irregularities

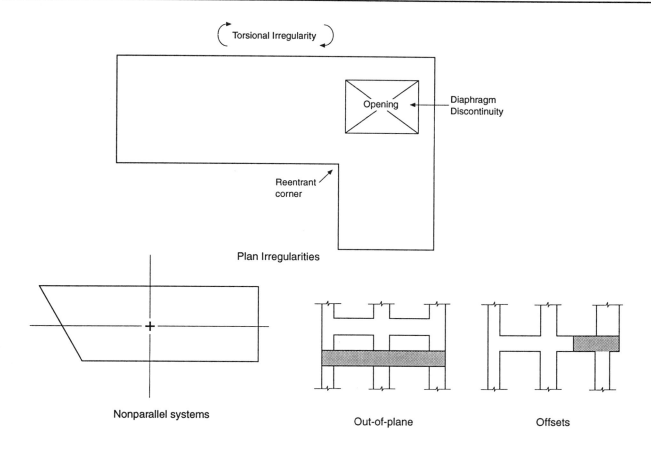

Figure 1-2 Plan Structural Irregularities

Structures with a vertical discontinuity in capacity (weak story) are not permitted to be over two stories or 30 ft in height where the weak story has a calculated strength of less than 65% of the story above, except where the weak story is capable of resisting a total lateral seismic force of $3 R_w/8$ times the design force prescribed in Section 2334.

Irregular structures are beyond the scope of this publication. The Dynamic Lateral Force Procedure is used and illustrated in Chapter 5, "Dual System." The Static Lateral Force Procedure is used everywhere else. The remaining discussion in this chapter is centered primarily around the Static Lateral Force Procedure.

1.3   Structural Systems

When a structure responds elastically to ground motions during a severe earthquake, the maximum response accelerations may be several times the maximum ground acceleration, and depend on the mass and stiffness of the structure and the magnitude of the damping. It is generally uneconomical and also unnecessary to design a structure to respond in the elastic range to the maximum likely earthquake-induced inertia forces. Thus, the design seismic horizontal force recommended by codes are generally much less than the elastic response inertia forces induced by a major earthquake.

Experience has shown that structures designed to the level of seismic horizontal forces recommended by codes can survive major earthquake shaking. This is due to the ability of well-designed structures to dissipate seismic energy by inelastic deformations in certain localized regions of certain members. Decrease in structural stiffness caused by accumulating damage and soil-structure interaction also help at times. It should be evident that use of the level of seismic design loads recommended by codes implies that the critical regions of elastically deforming members should have sufficient inelastic deformability to enable the structure to survive without collapse when subjected to several cycles of loading well into the inelastic range. This means avoiding all forms of brittle failure and achieving adequate inelastic deformability by flexural yielding of members.

The elastic (acceleration) response spectrum implicit in the applicable code yields the earthquake force induced in a structure by an earthquake of intensity as specified for the given seismic zone, corresponding to fully elastic, rather than inelastic, structural response. The code-specified design earthquake force is derived from the above force by suitable reduction to account for inelastic deformations and damping. The reduction is dependent on the structural system used. The structural systems for concrete buildings that are defined in 1991 UBC are as follows (Fig. 1-3):

Moment-Resisting Frame System. A structural system with an essentially complete space frame providing support for gravity loads. Moment-resisting frames provide resistance to lateral loads primarily by flexural action of members. In Seismic Zones 0 and 1, the moment-resisting frames can be Ordinary Moment-Resisting Frames (OMRF) proportioned to satisfy the requirements of Sections 2601 through 2618. In Seismic Zone 2 (A&B), reinforced concrete frames resisting forces induced by earthquake motions must be, at a minimum, Intermediate Moment-Resisting Frames (IMRF) proportioned to satisfy Section 2625(k) in addition to the requirements of Sections 2601 through 2618. In Seismic Zones 3 and 4, reinforced concrete frames resisting forces induced by earthquake motions must be Special Moment-Resisting Frames (SMRF) proportioned to satisfy Sections 2625(c), (d), (e), (g) and (h) in addition to the requirements of Sections 2601 through 2617.

Dual System. A structural system with three essential features: (i) An essentially complete space frame provides support for gravity loads, (ii) Resistance to lateral loads is provided by moment-resisting frames capable of resisting at least 25% of the design base shear, and by shearwalls, (iii) The two systems (moment frames and shearwalls) are designed to resist the design base shear in proportion to their relative rigidities. In Seismic Zones 3 and 4, the moment frames resisting lateral loads must be Special Moment-Resisting Frames (SMRF) as defined earlier, and the shearwalls must also be specially detailed to satisfy Sections 2625(c), (f) and (h) in addition to the applicable requirements of Sections 2601 through 2617. In Seismic Zone 2 (A&B), the moment frames resisting lateral loads must be Intermediate Moment-Resisting Frames (IMRF) as defined earlier; the shearwalls do not require any special detailing. In Seismic Zones 1 and 0, the shearwalls do not require any special detailing, and the moment frames can be Ordinary Moment-Resisting Frames (OMRF). It should be noted that the concept of the Dual System really loses its validity in Seismic Zones 1 and 0, because it is questionable whether the OMRF can act as a backup to the ordinarily detailed shearwall, the inelastic deformabilities of both systems being comparable. It is common practice in the lower seismic zones to design structures containing shearwalls and frames satisfying only criterion (iii) above. These structural systems should be called shearwall-frame interactive systems, and not Dual Systems.

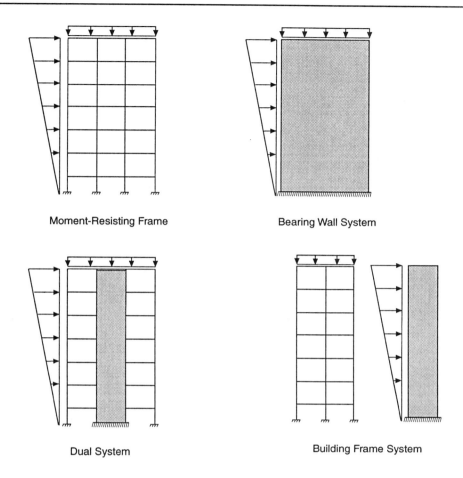

Figure 1-3 Earthquake-Resisting Structural Systems of Reinforced Concrete

In UBC-85, there was a fourth criterion, in addition to the three enumerated above, that had to be satisfied for a shearwall-frame system to qualify as a "Dual System." "The shear walls or braced frames acting independently of the ductile moment-resisting portions of the space frame shall resist the total required lateral forces" (Table 23-I of UBC-85). This requirement was at best counterproductive. It took incentive away from the design engineer who would have liked to design his shearwall-frame system as a dual system. If the shearwalls had to resist 100 percent of the design lateral forces, what then was the benefit in making the moment-resisting frames part of the lateral-force resisting system? It is true that the design force level went down by 20% when this was done (K decreased from 1.0 to 0.8). However, the moment-resisting frames then had to be detailed by the special requirements of 2625(c), (d), (e), (g) and (h). In most cases, the cost of the increased detailing requirements more than outweighed any savings resulting from the reduced design force level. Also, there were many instances where there simply were not sufficient shearwalls in a frame-wall building for the shearwalls to carry 100% of the design lateral forces. The fourth criterion, in addition to being counterproductive, was also not conservative. The unnecessarily overdesigned shearwalls attracted more shear forces to themselves in the event of an actual earthquake, which obviously is not a desirable phenomenon. Fortunately, the undesirable fourth criterion has been dropped from UBC-88/91. For a shearwall-frame system to qualify as a Dual System, only the three criteria listed at the beginning of the previous paragraph need to be satisfied per the current code. This, in addition to a decrease in the design force level for the Dual System, discussed later, will make this system much more attractive to the designer. A properly designed Dual System consisting of specially detailed shearwalls and SMRF is in fact the best structural system that can be designed in reinforced concrete for taller buildings. Because of the lateral rigidity contributed by shearwalls, this system provides a high degree of damage control in addition to life safety.

Building Frame System. A structural system with an essentially complete space frame providing support for gravity loads. Resistance to lateral loads is provided by shearwalls. In Seismic Zones 3 and 4, the shearwalls must be specially detailed to satisfy Sections 2625 (c), (f), and (h) in addition to the applicable requirements of Sections 2601 through 2617. In lower seismic zones, there are no special detailing requirements for the shearwalls. It should be noted that the concept of the Building Frame System also becomes somewhat meaningless in Seismic Zones 1 and 0. In those regions, making the frames part of the lateral force-resisting system does not involve any special ductility details for the frame members. Thus, to design the shearwalls to resist 100% of the design base shear does not make economic or practical sense. As mentioned earlier, in Seismic Zones 1 and 0, structures containing shearwalls and frames are and should be designed as shearwall-frame interactive systems.

The seismic safety of the building frame system is totally dependent on good-faith satisfaction of the so-called deformation compatibility requirements. These recognize that when the designated lateral force resisting system of a structure deforms laterally under an earthquake of intensity anticipated by the code, the subsystems that have been arbitrarily designated to be outside of the lateral force resisting system will have no choice but to deform together, because they are connected at every floor level through the floor slabs. If in the course of that earthquake-induced lateral displacement, the subsystems designed for gravity loads only are unable to sustain their gravity load carrying capacity, then life-safety is compromised. It is thus a specific requirement of all seismic codes that structural elements or subsystems designated not to be part of the lateral force resisting system be able to sustain their gravity load carrying capacity at a lateral displacement equal to a multiple times the computed elastic displacement of the lateral force resisting system under code-prescribed design seismic forces. The amplified elastic displacement of the lateral force resisting system is supposed to provide an estimate of the actual displacement of the entire structure caused by an earthquake of intensity anticipated by the code. If under the estimated earthquake-induced displacements, the gravity loads would cause inelasticity (exceedance of design moment capacity) in any structural element initially designed for gravity only, that structural element must also be detailed for inelastic deformability. If satisfaction of the deformation compatibility requirement would require that ductility details be provided in structural members originally designed for gravity only, then the engineer ought to review his original decision, and consider making such structural elements or subsystems part of the lateral force resisting system. In other words, he should consider an alternative structural system for his building.

Bearing Wall System. A structural system without an essentially complete space frame providing support for gravity loads. Bearing walls provide support for all or most gravity loads. Resistance to lateral load is provided by the same bearing walls acting as shearwalls. In Seismic Zones 3 and 4, the shearwalls must be specially detailed to satisfy Sections 2625 (c), (f), and (h) in addition to the applicable requirements of Sections 2601 through 2617. In lower seismic zones, there are no special detailing requirements for shearwalls.

The equation for design base shear in UBC-91 contains a parameter R_w which effects the reduction from elastic response seismic force to design seismic force levels. $R_w = 3$ would correspond more or less to elastic response seismic forces. By contrast, the Moment-Resisting Frame System with SMRF, the Dual System with specially detailed shearwalls and SMRF, the Building Frame System, and the Bearing Wall System are designed for R_w values of 12, 12, 9 and 6, respectively. Thus, the Bearing Wall System, the Building Frame System, and the remaining two systems are designed for one-half, one-third, and one-quarter of the elastic response force level, respectively.

UBC-91 restricts the Bearing Wall System to a maximum height of 160 ft, and the Building Frame System to a maximum height of 240 ft.

Tables 1-1, 1-2, and 1-3 summarize the choices of reinforced concrete structural systems in Seismic Zones 3 and 4, Seismic Zone 2 (A and B), and Seismic Zone 1, respectively.

Table 1-1 Reinforced Concrete Structural Systems for Seismic Zones 3 and 4

Structural System	R_w	Height Limit
Moment Resisting Frame System Beam-column frames (SMRF)	12	NL
Dual System Specially detailed shearwalls with SMRF	12	NL
Building Frame System Specially detailed shearwalls with OMRF	8	240
Bearing Wall System Specially detailed shearwalls	6	160

NL = No Limit

Table 1-2 Reinforced Concrete Structural Systems for Seismic Zone 2

Structural System	R_w
Dual System Shearwalls with IMRF	9
Building Frame System Shearwalls with OMRF	8

Moment Resisting Frame System
    Beam-column frames (IMRF)
    Slab-column frames (IMRF)
	8

Bearing Wall System Shearwalls	6

Table 1-3   Reinforced Concrete Structural Systems for Seismic Zone 1

Structural System	R_w
Shearwall-Frame Interactive System Shearwalls with OMRF	8
Bearing Wall System Shearwalls	6

Moment Resisting Frame System
    Beam-column frames (OMRF)
    Slab-column frames (OMRF)
	5

1.4   Design Base Shear

Table 1-4 summarizes the provisions concerning design base shear in the 1985 and 1991 editions of the Uniform Building Code [1.2]. V in Table 1-4 is the total lateral force or shear at the base for which a building in a seismic zone is to be designed. W is the total seismic dead load.

1.4.1   Total Seismic Dead Load, W

In UBC-85, W was the total dead load as defined in Section 2302, including a partition loading specified in Section 2304(d) where applicable. In Section 2302, dead load was the vertical load due to the weight of all permanent structural and nonstructural components of a building, such as walls, floors, roofs, and fixed service equipment. Section 2304(d) required that floors in office buildings and in other buildings where partition locations are subject to change must be designed to support, in addition to all other loads, a uniformly distributed dead load equal to 20 psf. Access floor systems might be designed to support, in addition to all other loads, a uniformly distributed dead load equal to 10 psf.

UBC-85 also required that W equal the total dead load plus 25% of the floor live load in storage and warehouse occupancies. "Where the design snow load is 30 psf or less, no part need be included in the value of W. Where the snow load is greater than 30 psf, the snow load shall be included; however, where the snow load duration warrants, the building official may allow the snow load to be reduced up to 75%."

In UBC-91, seismic dead load, W, is the total dead load and applicable portions of other loads listed below:

1. In storage and warehouse occupancies, a minimum of 25% of the floor live load shall be applicable.

2. Where a partition load is used in the floor design, a load of not less than 10 psf shall be included.

3. Where the snow load is greater than 30 psf, the snow load shall be included. Where considerations of siting, configuration and load duration warrant, the snow load may be reduced up to 75% when approved by the building official.

4. Total weight of permanent equipment shall be included.

Table 1-4 Design Base Shear Forces

UBC-85

V = $ZIKCSW$ (1)

Z = 1 in Zone 4, 3/4 in Zone 3, 3/8 in Zone 2, and 3/16 in Zone 1

I = 1.5 for essential facilities, 1.25 for assembly buildings, 1.0 for other occupancies

K = See Tables 1-7 — 1-10

C = $1/15\sqrt{T} \leq 0.12$ (2) See Table 1-13 for T

S = $1.0 + T/T_s - 0.5(T/T_s)^2$ for $T/T_s \leq 1.0$ $\qquad T \geq 0.3$ sec.

 = $1.2 + 0.6 T/T_s - 0.3 (T/T_s)^2$ for $T/T_s > 1.0$ $\qquad 0.5 \leq T_s \leq 2.5$ sec.

 $\geq$ 1

 = 1.5 when T_s is not properly established, OR

S = 1.0 for rock of any characteristic, either shale-like or crystalline in nature (such material may be characterized by a shear wave velocity greater than 2500 ft/sec.); or stiff soil conditions where the soil depth is less than 200 ft and the soil types overlying rock are stable deposits of sands, gravels, or stiff clays (Soil Profile Type S_1).

 = 1.2 for deep cohesionless or stiff clay soil conditions, including sites where the soil depth exceeds 200 ft and the soil types overlying rock are stable deposits of sands, gravels, or stiff clays (Soil Profile Type S_2).

 = 1.5 for soft to medium-stiff clays and sands, characterized by 30 ft or more of soft to medium-stiff clay with or without intervening layers of sand or other cohesionless soils (Soil Profile Type S_3).

In locations where the soil properties are not known in sufficient detail to determine the soil profile type or where the profile does not fit any of the three types, Soil Profile Type S_3 should be used.

CS $\leq$ 0.14

UBC-91

V = $\dfrac{ZIC}{R_w} W$ (4)

Z = 0.4 in Zone 4, 0.3 in Zone 3, 0.2 in Zone 2B, 0.15 in Zone 2A, and 0.05 in Zone 1

I = 1.25 for essential facilities and hazardous facilities, 1.0 for special occupancy structures and standard occupancy structures

C = $\dfrac{1.25S}{T^{2/3}} \geq 0.075 R_w$ (5) See Table 1-13 for T

 $\leq$ 2.75

S = 1.0 for soil profile with either (a) a rock-like material characterized by a shear-wave velocity greater than 2500 ft/sec. or by other suitable means of classification, or (b) stiff or dense soil condition where the soil depth is less than 200 ft (Soil Profile Type S_1).

 = 1.2 for a soil profile with dense or stiff soil conditions, where the soil depth exceeds 200 ft (Soil Profile Type S_2).

 = 1.5 for a soil profile 70 ft or more in depth and containing more than 20 ft of soft to medium stiff clay but not more than 40 ft of soft clay (Soil Profile Type S_3).

 = 2.0 for a soil profile containing more than 40 ft of soft clay characterized by a shear-wave velocity less than 500 ft/sec. (Soil Profile Type S_4).

In locations where the soil properties are not known in sufficient detail to determine the soil profile, Soil Profile Type S_3 should be used.

R_w = See Tables 1-7 — 1-10.

1.4.2   Seismic Zone Factor, Z

Appendix 1-A to this chapter provides background to the seismic zoning map in UBC-88/91, which is different from the seismic zoning map in UBC-85. In the 1988/91 seismic zoning map, the former Seismic Zone 2 has been divided into Seismic Zones 2B (areas in Zone 2 that are west of the Continental Divide) and 2A (Zone 2 areas east of the Continental Divide).

Table 1-5 compares the seismic zone factors Z of the 1985 Code with those of the 1988/91 Codes. It should be noted that in 1985, the zone factor Z for each of the four seismic zones was a value normalized to Seismic Zone 4 without any physical significance other than a higher design base shear for zones of higher seismicity. Thus, Z for Seismic Zone 4 through 1 were 1, 3/4, 3/8, and 3/16, respectively. Starting in 1988, Z reflects the effective horizontal peak ground acceleration (as a percent of gravity) expected in a particular seismic zone. The current Z values indicate that the effective peak ground accelerations are 0.4g, 0.3g, 0.2g, 0.15g, and 0.075g in Seismic Zones 4, 3, 2B, 2A, and 1, respectively.

Table 1-5 Seismic Zone Factor, Z

Zone	UBC-85	UBC-91
4	1	0.4
3	3/4	0.3
2B	3/8	0.2
2A	3/8	0.15
1	3/16	0.075

1.4.3   Importance Factor, I

As mentioned earlier, four categories of occupancy are defined in Table 23-K of UBC-91: Essential, Hazardous, Special and Standard. The "Hazardous Facilities" category was added to the UBC in 1988.

Table 1-6 shows that Essential and Hazardous occupancies are assigned an important Factor I of 1.25, whereas Special and Standard occupancies are permitted the use of an I factor of 1.0. Essential, Hazardous, and Special occupancies require construction observation (Section 307) in addition to compliance with Sections 305 and 306, whereas Standard occupancies require compliance with UBC Sections 305 and 306 only. It should be noted that the importance factor was higher for essential facilities and public assembly buildings in 1985; however, the other special requirement (of structural observation) did not apply.

Table 1-6 Importance Factor, I

Occupancy	UBC-85	UBC-91
Essential Facilities	1.5	1.25
Hazardous Facilities		1.25
Public Assembly	1.25	1.0
Other Occupancies	1.0	1.0

1.4.4   Horizontal Force Factor, K/Response Modification Factor, R_w

The horizontal force factor, K, of UBC-85, and the response modification factor, R_w, of UBC-88/91 for the structural systems described earlier—the Moment-Resisting Frame System, the Dual System, the Building Frame System, and the Bearing Wall System—are listed in Tables 1-7, 1-8, 1-9, and 1-10, respectively. The height limits imposed on the various systems by UBC-88 in Seismic Zones 3 and 4 are shown in the tables. Also listed in the tables are values of 1/K, 3/K and $3R_w/8$. Elastic lateral displacements under code-prescribed seismic forces, when multiplied by 3/K of UBC-85, were supposed to produce realistic estimates of actual lateral displacements caused by an earthquake of intensity anticipated by the code. $3R_w/8$ serves the same role in UBC-88/91. 1/K was used in UBC-85 for purposes of control of interstory drift.

Table 1-7 Horizontal Force Factors and Response Modification Factors for Moment-Resisting Frame Systems

UBC-85	K	1/K ($\geq$ 1)	3/K
Buildings with a ductile moment-resisting space frame designed in accordance with the following criteria: The ductile moment-resisting space frame shall have the capacity to resist the total required lateral force.	0.67	1.50	4.50
Buildings with a moment-resisting space frame where the above criterion is not met or where special design and detailing requirements for a ductile moment-resisting frame are not satisfied.	1.00	1.00	3.00

UBC-91		R_w	$3R_w/8$	Height Limit*
Moment-resisting frame system. An essentially complete frame provides support for gravity loads. Resistance to lateral load is provided primarily by flexural action of members.	1) Special moment-resisting frame (SMRF) of concrete	12	4.5	NL
	2) Intermediate moment-resisting frame (IMRF) of concrete [1]	8	3.0	
	3) Ordinary moment-resisting frame (OMRF) of concrete [2]	5	1.875	

* Height limits are applicable in Zones 3 and 4. Maximum heights may be increased by up to 50% for regular structures which are not occupied by nor accessible to the general public. NL = No Limit.
[1] Prohibited in Seismic Zones 3 and 4, except as permitted in Section 2338(b).
[2] Prohibited in Seismic Zones 2, 3, and 4.

The horizontal force factor, K, of UBC-85 was intended to account for differences in the inelastic deformability or energy dissipation capacity of various structural systems. Only three different structural systems for buildings were formally recognized in UBC-85: the box system (now called the Bearing Wall System), the dual system with ductile moment-resisting frames, and the ductile moment-resisting frame system (now called the Special Moment-Resisting Frame System). All other structural systems for buildings fell in an unclassified category, and were assigned a K-value of 1.0. The Intermediate Moment-Resisting Frame, the Ordinary Moment-Resisting Frame, the Building Frame System, as well as the Dual System with ordinary shearwalls and Intermediate Moment-Resisting Frames would all fall under that unclassified category of UBC-85.

Table 1-8 Horizontal Force Factors and Response Modification Factors for Dual Systems

UBC-85	K	1/K (≥ 1)	3/K
Buildings with a dual bracing system consisting of a ductile moment-resisting space frame and shearwalls or braced frames using the following design criteria: The frames and shearwalls or braced frames shall resist the total lateral force in accordance with their relative rigidities considering the interaction of the shearwalls and frames. The shearwalls or braced frames acting independently of the ductile moment-resisting portions of the space frame shall resist the total required lateral force. The ductile moment-resisting space frame shall have the capacity to resist not less than 25% of the required lateral force.	0.80	1.25	3.75

UBC-91	R_w	$3R_w/8$	Height Limit*
Dual System: An essentially complete space frame provides support for gravity loads. Resistance to lateral load is provided by shearwalls or braced frames and moment-resisting frames (SMRF, IMRF, or OMRF). The moment resisting frames shall be designed to independently resist at least 25% of the design base shear. The two systems shall be designed to resist the total design base shear in proportion to their relative rigidities considering the interaction of the dual system at all levels			
Concrete Shearwalls with SMRF	12	4.50	NL
Concrete Shearwalls with IMRF[1]	9	3.375	

Height limits are applicable in Zones 3 and 4. Maximum heights may be increased by up to 50% for regular structures which are not occupied by nor accessible to the general public. NL = No Limit.
[1] *Prohibited in Seismic Zones 3 and 4, except as permitted in Section 2338(b).*

One of the most significant changes from the 1985 to the 1988 edition of the UBC is the replacement of the horizontal force factor, K, with the reciprocal of the response modification factor, $1/R_w$. The response modification factor, R_w, accounts for the reduction in structural response caused by damping and inelasticity. It essentially represents the ratio of forces which would develop in a structure under the specific ground motion if it behaved in an entirely linear elastic manner to the prescribed design forces which anticipate significant yielding. It serves the same purpose as the coefficient K did in the UBC-85 base shear formula. However, division by R_w makes it clearer than before that the code-prescribed seismic design force for a particular structural system represents a reduction from the design force level that would have been required for elastic response to an earthquake of intensity anticipated by the code. The switch from K to R_w also enables a finer distinction to be made among the seismic capabilities of various structural systems. While only three broad categories of structural systems used to be specifically recognized in UBC-85, there are now many more subclasses of structural systems listed in Table 23-O, each with its own assigned R_w value. As should be evident from Table 1-7, and as mentioned earlier under "Structural Systems," three categories of moment-resisting frames of concrete are specifically recognized in UBC-88/91. It should be noted that the Intermediate Moment-Resisting Frame (IMRF) is prohibited in Seismic Zones 3 and 4, except as permitted by Section 2338(b). According to that section, IMRF may be used in Seismic Zones 3 and 4 for nonbuilding structures in Occupancy Categories III and IV (special and standard occupancies) if the structure is less than 50 ft in height and if an $R_w = 4.0$ is used for design. The Ordinary Moment-Resisting Frame (OMRF) is prohibited in Seismic Zones 2, 3 and 4. Table 1-8 shows, as was also mentioned under "Structural Systems," that two categories of Dual Systems are recognized in UBC-88/91: one that combines an SMRF with a specially detailed shearwall, and another that combines an IMRF with an ordinarily detailed shearwall. The latter system is prohibited in Seismic Zones 3 and 4. While the Building Frame System (Table 1-9) and the Bearing Wall System (Table 1-10) do not have different categories shown, it was pointed out earlier under "Structural Systems" that special detailing requirements of Section 2625(c), (f) and (h) apply to the shearwalls of these systems in Seismic Zones 3 and 4, while no such requirements apply in the lower seismic zones.

Table 1-9 Horizontal Force Factors and Response Modification Factors for Building Frame Systems

UBC-85	K	1/K (≥ 1)	3/K
Buildings with a dual bracing system consisting of a moment-resisting space frame and shearwalls or braced frames where the three criteria for a dual system with a K = 0.8 (Table 1-8) are not met or where the space frame is not designed and detailed to satisfy the special requirements for a ductile moment-resisting space frame.	1.00	1.00	3.00

UBC-91	R_w	$3R_w/8$	Height Limit*
Building frame system with concrete shearwalls. An essentially complete frame provides support for gravity loads. Resistance to lateral loads is provided by shearwalls or braced frames	8	3.00	240 ft

* Height limits are applicable in Zones 3 and 4. Maximum heights may be increased by up to 50% for regular structures which are not occupied by nor accessible to the general public. NL = No Limit.

Roughly speaking, if $R_w = 8/K$, status quo between the 1985 and 1988/91 codes is maintained, and the base shear value is not affected by the switch from K to $1/R_w$. Table 1-11 shows that the status quo is maintained for the Bearing Wall System, the Building Frame System, the SMRF and the IMRF. The R_w value is liberalized for both categories of Dual Systems, in recognition of the superior performance of this system in past earthquakes. The R_w value is made more conservative than the corresponding K value in UBC-85 for the OMRF. It should be noted that the R_w value for the IMRF was 7 in UBC-88, and has been increased to 8 in UBC-91. More detailed comparisons between the design force levels of UBC-85 and UBC-91 are provided later in this chapter.

Table 1-10 Horizontal Force Factors and Response Modification Factors for Bearing Wall Systems

UBC-85	K	1/K (≥ 1)	3/K
Buildings with a box system, i.e., a structural system without a complete vertical load-carrying space frame. The required forces are resisted by shearwalls or braced frames. EXCEPTION: Buildings not more than three stories in height with stud wall framing and using plywood horizontal diaphragms and plywood vertical shear panels for the lateral force system may use K = 1.0.	1.33	1.00	2.25

UBC-91	R_w	$3R_w/8$	Height Limit*
Bearing wall system: A structural system without a complete vertical load-carrying frame. Bearing walls or bracing systems provide support for all or most gravity loads. Resistance to lateral loads is provided by shearwalls or braced frames.			
Concrete Shearwalls	6	2.25	160 ft

* Height limits are applicable in Zones 3 and 4. Maximum heights may be increased by up to 50% for regular structures which are not occupied by nor accessible to the general public. NL = No Limit

1.4.5   The Coefficient, C

In UBC-85, C was a coefficient related to the fundamental period of vibration of the structure, which in turn is a function of the mass and the stiffness of the structure. C in UBC-85 was proportional to $1/\sqrt{T}$.

In UBC-88/91, C incorporates the site-structure resonance factor, S. Also, it is now proportional to $1/T^{2/3}$, which indicates a change in the shape of the design spectrum particularly at longer periods.

According to the Commentary to the SEAOC Blue Book [1.1], "The ZC portion of the base shear equation is an engineering representation of the site dependent earthquake ground motion having a 10% probability of exceedance in 50 years. It may be considered as a multi-mode, effective acceleration response spectrum envelope. As such, it represents the forces and displacements induced in a linear elastic structure when this structure is subjected to the maximum expected ground motion." In comparison with the response spectrum shapes given in Fig. 23-3 of UBC-91, which vary with the reciprocal of T, the formula for C (Eq. 34-2 of UBC-91) uses the 2/3 power of T to provide a multi-mode response envelope.

Table 1-11 Response Modification Factors For Reinforced Concrete Structural Systems

Structural System	Special Requirements or Restrictions	UBC-85 K Value (Equivalent R_w Value)*	UBC-91 R_w Value	Height Limit (ft) **
Bearing Wall System	Special detailing in Zones 3, 4	1.33 (6)	6	160
	Ordinary detailing in Zones 1, 2			
Building Frame System	Special detailing of shearwalls and deformation compatibility requirements for frames in Zones 3, 4	1.00 (8)	8	240
	Ordinary detailing of shearwalls and deformation compatibility requirements for frames in Zones 1, 2			
Moment-Resisting Frame System	Special Moment Frames	0.67 (12)	12	No Limit
	Intermediate Moment Frames restricted to Zone 2 and lower	1.00 (8)	8	
	Ordinary Moment Frames restricted to Zone 1 and lower	1.00 (8)	5	
Dual System	Special Moment Frames + Specially detailed shearwalls	0.80 (10)	12	No Limit
	Intermediate Moment Frames + Ordinarily detailed shearwalls restricted to Zone 2 and lower	1.00 (8)	9	

* Equivalency is based on $R_w = 8/K$.
** Height limits are applicable in Zones 3 and 4 only.

A comparison of the spectral ratios for the three spectra shown in Fig. 23-3 with the S factor values given in Table 23-J will show that those portions of the spectra that are different for each soil type do not have the same ratios. In a design analysis that uses a response spectrum, the effects of individual modes are computed separately and then combined to give the total building response. The more complex buildings are usually taller and as a consequence have longer fundamental response periods. A simple method of introducing a more conservative design approach for long-period structures is achieved by reducing the exponent which is used with the period of the structure T in the formulation of the factor C. The reduction of the exponent from a value of 1 to a value of 2/3 results in a slower attenuation of the lateral force coefficient with increasing period than is obtained with the corresponding response spectrum. The slower attenuation effect increases with increasing period. Whereas the values of S in Table 23-J range from 1.0 to 1.5 for the first three soil types, the ratios of the separate spectral values range between 1.0 and 2.0 for the same three soil types. A similar comparison for soil type 4 cannot be made because of the wide variety of response spectra that could be produced by soils in this group.

A maximum limit on C = 2.75 for any structure and soil site condition is given to provide a simple seismic load evaluation for design projects where it is not practical to evaluate the site soil conditions and the structural period.

A lower limit of $C/R_w = 0.075$ has also been prescribed. In consideration of the present uncertainty in forecasting ground motion and structural response, it has been judged prudent to establish this minimum design requirement until more experience and knowledge are available for long-period structures subjected to strong levels of ground motion. Table 1-12 lists the maximum and the minimum design base shears for moment-resisting frame systems in various seismic zones.

Table 1-12 Limiting Design Base Shears For Moment-Resisting Frame Systems

Seismic Zone	R_w	Design Base Shear — V/W (%)	
		Minimum	Maximum
4	12	3.00	9.17
3	12	2.25	6.88
2B	8	1.50	6.88
2A	8	1.13	5.16

Table 1-13 Fundamental Period of Vibration of Structure

UBC-85	UBC-91
Moment-resisting frames, deforming freely, provide the entire lateral resistance	
$T_a = 0.1\,N$ (6)	Steel: $T_a = 0.035 h_n^{3/4}$ (7a) Concrete: $T_a = 0.030 h_n^{3/4}$ (7b)
Other structural systems	
$T_a = 0.5 h_n / \sqrt{D}$ (8)	$T_a = C_t h_n^{3/4},\ C_t = 0.020$ or $0.1/\sqrt{A_c}$ (9a) $A_c = \Sigma A_e [0.2 + (D_e/h_n)^2]$ (9b) $D_e/h_n \le 0.9$ (9c)
All structural systems	
$T = 2\pi \sqrt{\left(\sum_{i=1}^{n} w_i \delta_i^2 \right) \Big/ \left(g \sum_{i=1}^{n} f_i \delta_i \right)}$ (10)	$T = 2\pi \sqrt{\left(\sum_{i=1}^{n} w_i \delta_i^2 \right) \Big/ \left(g \sum_{i=1}^{n} f_i \delta_i \right)}$ (10) The value of C, obtained using T from Eq. 10, shall not be less than 80% of C_a, based on T_a calculated from Eqs. 7 or 9.

T = *fundamental period of vibration of structure in seconds in direction of analysis*
T_a = *approximate value of T*
N = *total number of stories above the base to level n which is uppermost in the main portion of the structure*
h_n = *building height in ft above base*
D = *plan dimension in ft in direction of analysis*
A_c = *combined effective area, in sft, of shearwalls in first story of structure*
A_e = *minimum cross-sectional shear area in any horizontal plane in the first story, in sft, of a shearwall*
D_e = *length in ft of a shearwall in the first story in direction parallel to the applied forces*
w_i = *that portion of w which is located at or assigned to level i*
f_i = *distributed portion of total lateral force at level i; lateral fore should be distributed approximately in accordance with the principles of Table 1-14, or in some other rational manner*
δ_i = *diflection at level i relative to the base, due to applied lateral forces f $_i$*

1.4.6   Fundamental Period of Vibration, T

The fundamental periods of vibration of structures, as given in UBC-85 and UBC-91, are summarized in Table 1-13. The analytical formula for period, Eq. 10 of Table 1-13, gives a more accurate rationally based estimate of the fundamental period than do Eqs. 6 to 9, and is the same in both editions of the UBC, except that there are different restrictions attached to it. However, Eq. 10 cannot be used until a design for the building under consideration, at least a preliminary one, is available. Simple formulas (6) to (9), involving only a general description of the building type and overall dimensions, are necessary to estimate the period of vibration in order to calculate an initial base shear and proceed with a preliminary design.

Equation 6, applicable to moment-resisting frame systems, was changed to Equation 7 in UBC-88/91. Equation 7 is intended to produce values for T_a generally lower than the true fundamental periods of vibration of moment frame buildings. Equations 7a and 7b represent judgmental lower-bound fits to fundamental periods of vibration as computed from accelerograms recorded in the upper stories of 17 steel frame buildings (Eq. 7a) and 14 concrete frame buildings (Eq. 7b) during the 1971 San Fernando earthquake [1.6, 1.7—Commentary]. Equation 7b has been carefully studied in recent times, and modifications to it have been proposed [1.8].

Figure 1-4 show comparisons between Eq. 6 and Eqs. 7, for steel and concrete frame buildings of various story heights. It may be stated that for the story heights typical of concrete buildings, the UBC-91 and UBC-85 formulas given approximate fundamental periods that are comparable, at least up to the 20-story height range studied.

For non-frame structures, Eq. 9a from UBC-88/91 usually gives a longer period than the previous period formula, Eq. (8) of UBC-85. This can be seen from Fig. 1-5, if it is understood that the plan dimension of a building usually increases as its height increases.

Equation 10 usually yields periods considerably longer than those given by the approximate formulas. A stipulation that was not in UBC-85 was added to UBC-88 to the effect that if the base shear coefficient C, based on period T as given by Eq. 10, falls below 80% of the C_a value computed using T_a as given by the approximate formulas, then $0.8 C_a$ rather than C is to be used in computations of base shear for buildings. This restriction has been put in place to prevent abuse of rational period computation. Obviously, improperly long periods can be computed through unreasonable stiffness assumptions, resulting in lower than reasonable design base shears.

1.4.7   Site Coefficient, S

An S coefficient was introduced into the base shear formula to account for the variability of soil conditions in the Blue Book of 1974. Prior to that time there had been no variation in the base shear coefficients for different site conditions, although the Blue Book Commentary prior to 1974 had pointed out that "The absence of a soil factor...should not be interpreted as meaning that the effect of soil conditions on building response is not important." [1.1—Commentary]

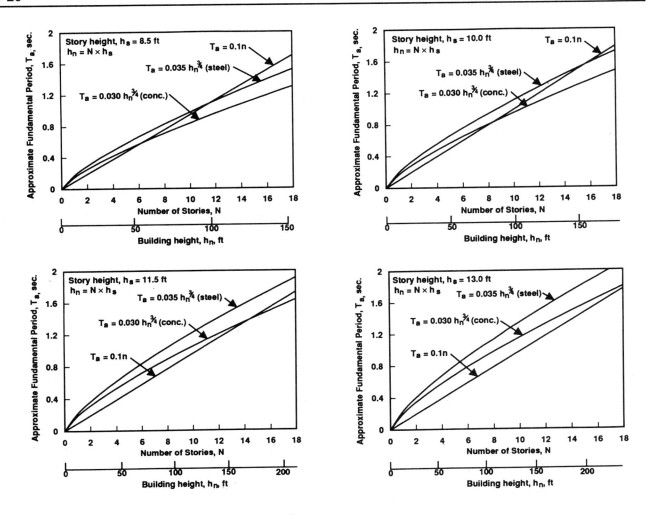

Figure 1-4 Approximate Fundamental Period of Moment-Resisting Frame Buildings

The approach used in the 1974 Blue Book, consistent with the state of the art at that time, was to consider the possibility of resonance between the structure and the soil by comparing the natural period of vibration of the structure, T, and the fundamental period of the site, T_s. If some resonant interaction was possible, then the design base shear was increased by use of an S factor whose value was greater than unity. The value of S was limited between 1.0 and 1.5, with the maximum value occurring when the structure and the site periods were equal. Simple parabolic functions were used to obtain values of S when the two periods were not equal. Acceptable values of T_s were between the limits of 0.5 and 2.5 seconds. The range recognized that on shallow soil profiles the simple concept of a fundamental period was not valid. The structural period that could be used for computing the S factor was restricted to be not less than 0.3 second.

Shortly after the issuance of the 1974 Blue Book, several studies appeared with some statistical treatment of strong motion response spectra recorded on different types of soil deposits. By using these results, it was possible to develop a more direct approach for the inclusion of site effects in a design code formulation. Such an approach was implemented in the ATC 3-06 document. Three sets of site conditions were established, each with its assigned S value. UBC-85 gave the designer a choice between the old approach and the new ATC-based approach to the determination of site coefficient, S. Only the newer approach was retained in UBC-88.

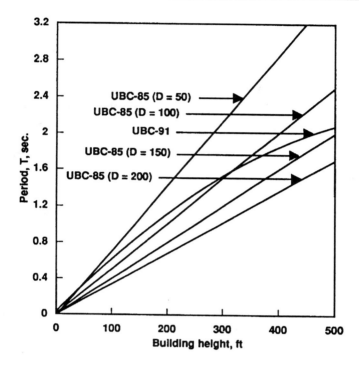

D = Plan dimension in ft in direction of analysis

Figure 1-5 Approximate Fundamental Period of Non-Frame Buildings

Following the extensive damage in Mexico City during the Mexican Earthquake of September, 1985, an additional soil profile (Type S_4) was added to Table 23-J of UBC-88, with an assigned S value of 2.0, to represent the anticipated response of deep deposits of soft clays.

It should again be noted that at short response periods, the value of C, which includes the site coefficient S, and is computed by using Eq. 34.2 of UBC-91, need not exceed a value of 2.75.

It is strongly recommended in the latest Blue Book [1.1] that for structures located on soft soil profiles, a site-specific study should be performed to develop seismic design criteria. The value of the site coefficient for soils of the S_4 classification was really included in the Blue Book for guidance purposes, in recognition of the fact that a site-specific study may not be warranted for some structures.

1.4.8   Comparison of 1985 and 1991 Levels of Design Base Shear

The changes from 1985 to 1991 UBC outlined in the preceding subsections produce design base shears for various reinforced concrete structural systems that are compared below.

1.4.8.1   Special Moment-Resisting Frame System

Figure 1-6 shows the comparison between 1985 and 1991 levels of design base shear for a Special Moment-Resisting Frame System of concrete in Seismic Zone 4. Values of I and S are assumed to be equal to unity. It is obvious that the base shears are comparable except for very low buildings which are governed by the upper bound limit on base shear which used to be C ≤ 0.12 and CS ≤ 0.14 in 1985, and is now C ≤ 2.75. The upper bound limits are clearly different. There is some discrepancy between the 1985 and 1991 design force levels for relatively tall buildings also, where the new lower bound limit of C ≥ 0.075 R_w governs. This limit was not there in the 1985 Code. In the intermediate height range, the design force levels are comparable because the K and R_w values for the SMRF are equivalent, as mentioned earlier, and because the 1985 and 1991 approximate period formulas, though different, yield estimated periods that are comparable.

The comparison shown in Fig. 1-6 also applies exactly to the SMRF in Seismic Zone 3 where all the forces are proportionately lower by a factor of 0.75.

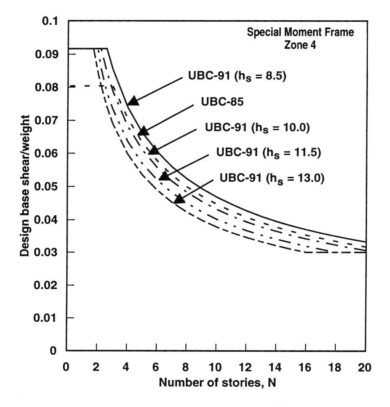

UBC-85: *Z = 1.0, I = 1, K = 0.67, C = 1/15 √T̄, S = 1, T = 0.1N*

UBC-91: *Z = 0.40, I = 1, C = 1.25S/T²ᐟ³, R_w = 12, S = 1, T = 0.03h_n³ᐟ⁴*

h_s = Floor height in feet

Figure 1-6 Design Base Shear for Special Moment Frames in Zone 4

1.4.8.2   Intermediate Moment-Resisting Frame System

The design base shears according to UBC-85 and those according to UBC-91 for the Intermediate Moment-Resisting Frame System of concrete in Seismic Zone 2B are compared in Fig. 1-7. Both I and S are once again assumed to be equal to unity. Since the K and R_w factors for this structural system are also equivalent, and since the same 1985 and 1991 approximate period formulas apply to the IMRF as to the SMRF, the comparison would have been the same as shown in Fig. 1-6, except for an effective increase in the Z factor for Seismic Zone 2B over the corresponding 1985 Z value for Seismic Zone 2 (see Table 1-5). Because of this increase, however, the 1991 design force levels are significantly higher than the corresponding 1985 design force levels.

The comparison of design force levels shown in Fig. 1-6 for the SMRF in Seismic Zone 4 also applies exactly to the IMRF in Seismic Zone 2A; all the forces in the latter case are simply proportionally lower by a factor of 0.375.

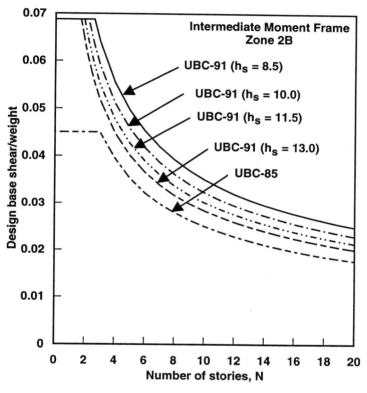

UBC-85: $Z = 3/8, I = 1, K = 1.00, C = 1/15\sqrt{T}, S = 1, T = 0.1N$

UBC-91: $Z = 0.20, I = 1, C = 1.25S/T^{2/3}, R_w = 8, S = 1, T = 0.03h_n^{3/4}$

h_s = *Floor height in feet*

Figure 1-7 Design Base Shear for Intermediate Moment Frames in Zone 2B

1.4.8.3   Ordinary Moment-Resisting Frame System

Figure 1-8 shows the comparison between 1985 and 1991 design base shears for an Ordinary Moment-Resisting Frame System of concrete in Seismic Zone 4. I and S are both assumed equal to one. It has been mentioned earlier that the 1985 and 1991 approximate period formulas yield comparable period estimates for this structural system. However, since the 1991 R_w value of 5 is substantially lower than the R_w that would have been equivalent to the 1985 K value of 1.0, the 1991 design force levels are substantially higher than the corresponding 1985 design force levels, irrespective of height.

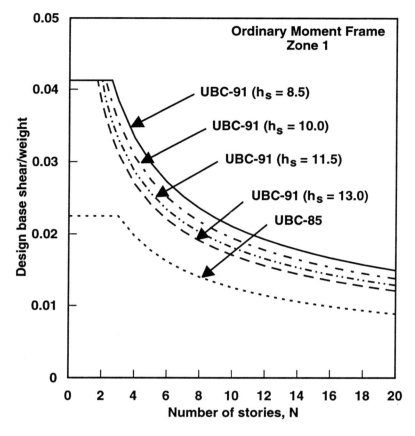

$UBC\text{-}85$: $Z = 3/16$, $I = 1$, $K = 1.00$, $C = 1/15\sqrt{T}$, $S = 1$, $T = 0.1N$

$UBC\text{-}91$: $Z = 0.10$, $I = 1$, $C = 1.25S/T^{2/3}$, $R_w = 5$, $S = 1$, $T = 0.03h_n^{3/4}$

h_s = Floor height in feet

Figure 1-8 Design Base Shear for Ordinary Moment Frames in Zone 1

1.4.8.4 Bearing Wall System

Design base shears for a Bearing Wall System of reinforced concrete in Seismic Zone 4 from the 1985 and 1991 Codes are compared in Fig. 1-9. Values of I and S are assumed to be unity. The base shears are mostly lower by the new code, because the K and R_w values for this system are equivalent, and because the 1991 approximate period formula for non-frame structures yields periods that are longer than those given by the corresponding 1985 formula. An exception occurs for short buildings which must be designed for higher forces according to the 1991 Code. This is because the upper bound limit on design base shear is different in the 1991 Code (C ≥ 2.75) from that in the 1985 Code (C ≥ 0.12, CS ≥ 0.14).

The comparison shown in Fig. 1-9 is also applicable in Zone 3, Zone 2A and Zone 1. It, however, is not applicable in Zone 2B, where an effective increase in the Z value from 1985 to 1988 and 1991 causes the comparison to be as shown in Fig. 1-10.

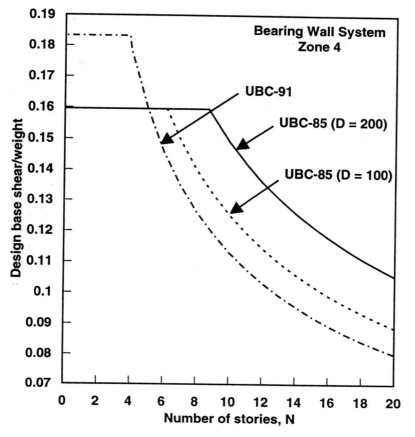

$$UBC\text{-}85: \ Z = 1.0, I = 1, K = 1.33, C = 1/15\sqrt{T}, S = 1, T = 0.05h_n/\sqrt{D}$$

$$UBC\text{-}91: \ Z = 0.40, I = 1, C = 1.25S/T^{2/3}, R_w = 6, S = 1, T = 0.02h_n^{3/4}$$

D = Plan dimension in ft in direction of analysis

Floor height = 10 ft 0 in.

Figure 1-9 Design Base Shear for Bearing Wall System in Zone 4

1.4.8.5   Building Frame System

The design base shears according to UBC-85 and those according to UBC-91 for a Building Frame System of reinforced concrete in Seismic Zone 4 are compared in Fig. 1-11. Values of I and S are again assumed to be equal to one. The comparison is very much like that shown in Fig. 1-9 for the Bearing Wall System. The R_w and K values for this system are comparable. The 1991 period formula yields longer periods for this system than the corresponding 1985 formula. Also, the 1991 upper bound design base shear is higher than the corresponding 1985 level.

The comparison illustrated in Fig. 1-11 is also valid in Zone 3, Zone 2A and Zone 1. It is not valid in Zone 2B where an effective increase in the seismic zone factor, Z, from 1985 to 1991 causes the comparison to change to that shown in Fig. 1-12.

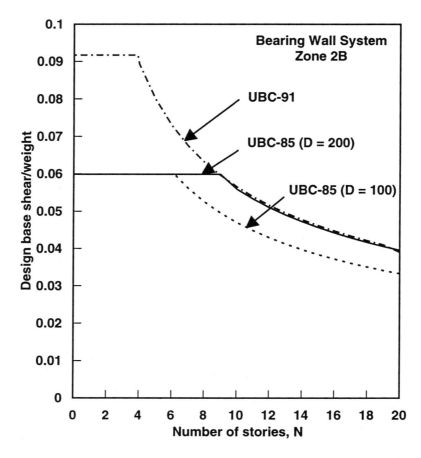

UBC-85:* $Z = 3/8, I = 1, K = 1.33, C = 1/15\sqrt{T}, S = 1, T = 0.05h_n/\sqrt{D}$

UBC-91: $Z = 0.20, I = 1, C = 1.25S/T^{2/3}, R_w = 6, S = 1, T = 0.02h_n^{3/4}$

D = Plan dimension in ft in direction of analysis

Floor height = 10 ft 0 in.

Figure 1-10 Design Base Shear for Bearing Wall System in Zone 2B

1.4.8.6   Dual System with SMRF

The design base shear levels from the 1985 and 1991 Codes for a Dual System with specially detailed shearwalls and SMRF in Seismic Zone 4 are compared in Fig. 1-13. I and S are both assumed to be equal to unity. It can been seen that the 1991 design force levels are lower than the 1985 levels because the 1991 approximate period formula yields longer periods for this system than the corresponding 1985 formula, and because the R_w value of 12 is higher than the R_w corresponding to the old K value of 0.8. The comparison of Fig. 1-13 is also valid in Seismic Zone 3.

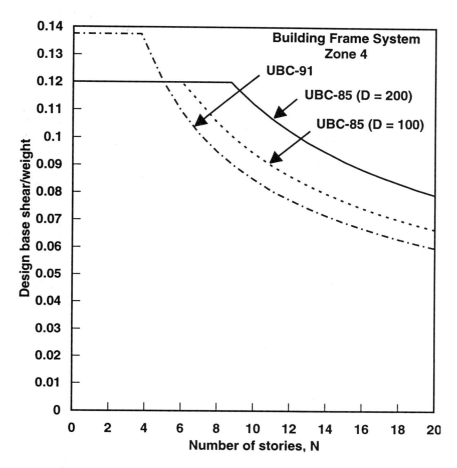

UBC-85: $Z = 1.0, I = 1, K = 1.00, C = 1/15\sqrt{T}, S = 1, T = 0.05h_n/\sqrt{D}$

UBC-91: $Z = 0.40, I = 1, C = 1.25S/T^{2/3}, R_w = 8, S = 1, T = 0.02h_n^{3/4}$

D = *Plan dimension in ft in direction of analysis*

Floor height = 10 ft 0 in.

Figure 1-11 Design Base Shear for Building Frame System in Zone 4

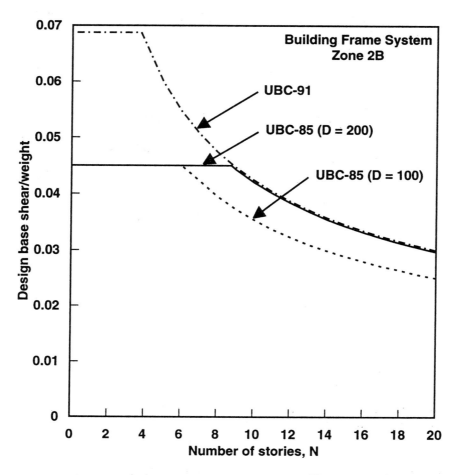

UBC-85: $Z = 3/8, I = 1, K = 1.00, C = 1/15\sqrt{T}, S = 1, T = 0.05h_n/\sqrt{D}$

UBC-91: $Z = 0.20, I = 1, C = 1.25S/T^{2/3}, R_w = 8, S = 1, T = 0.02h_n^{3/4}$

D = Plan dimension in ft in direction of analysis

Floor height = 10 ft 0 in.

Figure 1-12 Design Base Shear for Building Frame System in Zone 2B

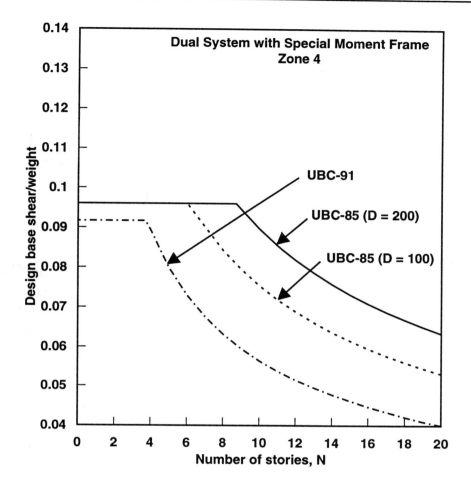

$$UBC\text{-}85: \ Z = 1.0, I = 1, K = 0.80, C = 1/15\sqrt{T}, S = 1, T = 0.05h_n/\sqrt{D}$$

$$UBC\text{-}91: \ Z = 0.40, I = 1, C = 1.25S/T^{2/3}, R_w = 12, S = 1, T = 0.02h_n^{3/4}$$

D = *Plan dimension in ft in direction of analysis*

Floor height = *10 ft 0 in.*

Figure 1-13 Design Base Shear for Dual System with Special Moment Frame in Zone 4

1.4.8.7   Dual System with IMRF

The 1985 and 1991 design force levels for a Dual System with ordinarily detailed shearwalls and IMRF in Seismic Zone 2A are compared in Fig. 1-14. I and S are once again both equal to one. The comparison is not unlike that shown in Fig. 1-13, because the R_w value is higher than the R_w corresponding to the K value of the 1985 Code, and because the 1991 approximate period formula yields longer periods for this structural system, as compared with the corresponding 1985 formula. The comparison, however, is different for Zone 2B, as shown in Fig. 1-15. The 1991 design force levels are generally higher, because of an increase in Z-factor over that prescribed in the 1985 Code, as noted earlier.

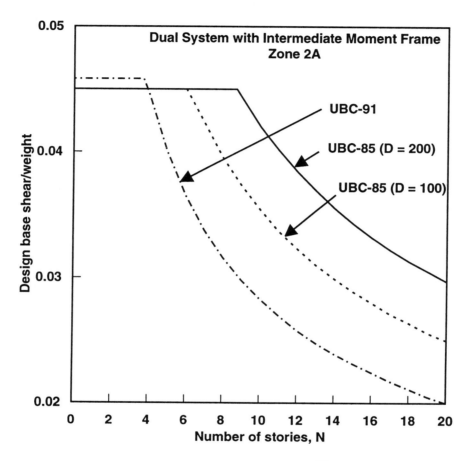

$$UBC\text{-}85:\ Z = 3/8,\ I = 1,\ K = 1.00,\ C = 1/15\sqrt{T},\ S = 1,\ T = 0.05h_n/\sqrt{D}$$
$$UBC\text{-}91:\ Z = 0.15,\ I = 1,\ C = 1.25S/T^{2/3},\ R_w = 9,\ S = 1,\ T = 0.02h_n^{3/4}$$
D = Plan dimension in ft in direction of analysis

Floor height = 10 ft 0 in.

Figure 1-14 Design Base Shear for Dual System with Intermediate Moment Frame in Zone 2A

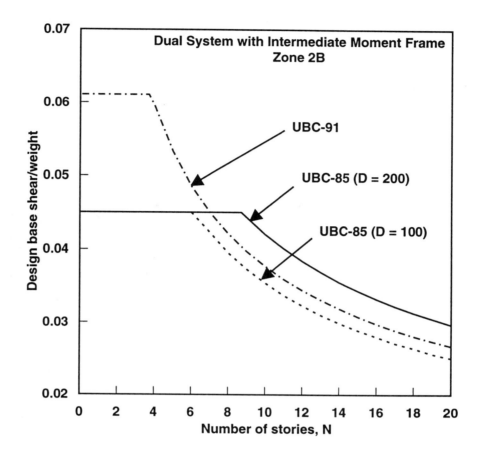

UBC-85: $Z = 3/8, I = 1, K = 1.00, C = 1/15\sqrt{T}, S = 1, T = 0.05h_n/\sqrt{D}$

UBC-91: $Z = 0.20, I = 1, C = 1.25S/T^{2/3}, R_w = 9, S = 1, T = 0.02h_n^{3/4}$

$D = Plan\ dimension\ in\ ft\ in\ direction\ of\ analysis$

$Floor\ height = 10\ ft\ 0\ in.$

Figure 1-15 Design Base Shear for Dual System with Intermediate Moment Frame in Zone 2B

1.5   Distribution Of Lateral Forces Along Building Height

Table 1-14 summarizes the distribution of design lateral forces along the height of a building, as prescribed in the 1985, 1988, and 1991 editions of the UBC. A portion of the design base shear is concentrated at the top of flexible buildings (T > 0.7 sec.), to account for higher mode response. The rest of the design base shear is distributed linearly, varying from a maximum value at the top to a minimum at the bottom, in correspondence with fundamental mode response.

Table 1-14 Distribution of Design Lateral Forces Along Building Height

UBC-85, UBC-91
$V = F_t + \sum\limits_{i=1}^{n} F_i$
$F_t \leq 0.25V$
$\quad = 0.07TV \qquad$ for $T > 0.7$ sec.
$\quad = 0 \qquad\qquad$ for $T \leq 0.7$ sec.
$F_x = \dfrac{(V - F_t)w_x h_x}{\sum\limits_{i=1}^{n} w_i h_i}$

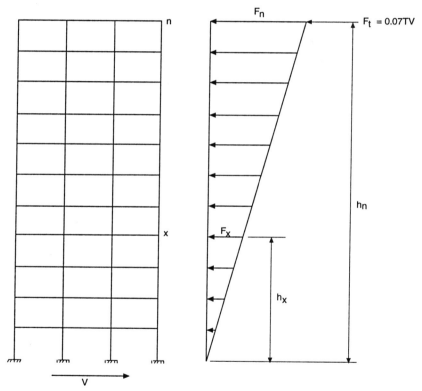

Vertical Force Distribution

$F_i, F_n, F_x \quad = \quad$ *lateral force applied to level i, n or x, respectively*
$F_t \qquad\quad = \quad$ *portion of V considered concentrated at top of structure in addition to F_n*
$h_i, h_x \qquad = \quad$ *height in ft above base to level i, or x, respectively*
$w_i, w_x \qquad = \quad$ *that portion of w which is located at or assigned to level i or x, respectively*
$W \qquad\quad = \quad$ *total seismic dead load*
$T \qquad\quad = \quad$ *fundamental period of vibration of structure in seconds in direction of analysis*
$V \qquad\quad = \quad$ *total lateral force or shear at the base for which a building in a seismic zone is to be designed*

1.6 Drift Limitation

UBC-85 required that: "Lateral deflections or drift of a story relative to its adjacent stories shall not exceed 0.005 times the story height...The displacement calculated from the application of the required lateral forces shall be multiplied by (1.0/K) to obtain the drift. The ratio (1.0/K) shall not be less than 1.0." This requirement can be summarized as follows:

$$K < 1 \qquad \Delta < 0.005 \, K \, h_s$$
$$K \geq 1 \qquad \Delta < 0.005 \, h_s$$

where h_s = story height, and $\Delta = \delta_i - \delta_{i-1}$, δ_i being the horizontal displacement at level i relative to the base due to applied lateral forces given by Eqs. 12-6 through 12-8 of UBC-85 which are the same as Eqs. 34-6 through 34-8 of UBC-91.

In UBC-91, the allowable story drift is adjusted to $0.04/R_w$ times or 0.005 times the story height, whichever is less, for buildings having fundamental periods below 0.7 sec. The allowable drift is $0.03/R_w$ or 0.004 times the story height, whichever is less, for buildings having fundamental periods equal to or greater than 0.7 sec. The new provisions may be summarized as follows:

$$T < 0.7 \text{ sec.} \qquad \Delta \leq \frac{0.04h_s}{R_w}$$

$$\leq 0.005h_s$$

$$T \geq 0.7 \text{ sec.} \qquad \Delta \leq \frac{0.03h_s}{R_w}$$

$$\leq 0.004h_s$$

It may be noted that the drift limits that now apply to buildings with T < 0.7 sec., applied in UBC-88 to buildings less than 65 ft in height. The drift limits that now apply to buildings with T ≥ 0.7 sec., applied in UBC-88 to buildings 65 ft or taller in height.

The story drift limitations of UBC-85 and UBC-91 are compared in Table 1-15 for reinforced concrete structural systems permitted in Seismic Zones 3 and 4. It may be noted that the 1985 and 1991 drift limits are comparable for shorter buildings (T < 0.7 sec.), while the 1991 drift limits are tighter than the 1985 drift limits for taller buildings (T ≥ 0.7 sec.).

Table 1-15 Story Drift Limitations

Structural System	UBC-85	UBC-91	
		T < 0.7 sec.	T ≥ 0.7 sec.
Special Moment-Resisting Frame System	$0.0034h_s$	$0.0033h_s$	$0.0025h_s$
Dual System with SMRF	$0.004h_s$	$0.0033h_s$	$0.0025h_s$
Building Frame System	$0.005h_s$	$0.005h_s$	$0.0038h_s$
Bearing Wall System	$0.005h_s$	$0.005h_s$	$0.004h_s$

1.7   Miscellaneous Changes

There are significant changes from the 1985 to the 1988/91 editions of the UBC that are difficult to categorize, and are discussed in this section.

1.7.1   Combinations of Structural Systems

UBC-88 added provisions for mixed structural systems that have been retained intact in UBC-91.

It is specifically required that the value of R_w used in the design of any story shall be less than or equal to the value of R_w used in the given direction for the story above (2334(c)2; also see Fig. 1-16). This requirement need not apply to a story where the dead weight above that story is less than 10% of the total dead weight of the structure.

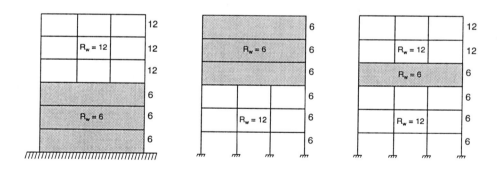

R_w (any story) ≤ R_w (story above)

Figure 1-16 Mixed Vertical Structural Systems

In Seismic Zones 3 and 4, where a structure has a bearing wall system in only one direction, the value of R_w used for design in the orthogonal direction shall not be greater than that used for the bearing wall system (2334(c)3).

Only combinations of Dual Systems and Special Moment-Resisting Frames may be used to resist seismic forces in structures exceeding 160 ft in height in Seismic Zones 3 and 4.

1.7.2 Horizontal Distribution of Shear and Horizontal Torsional Moments

UBC-85 required the following:

The total lateral force shall be distributed to the various vertical elements of the lateral force-resisting system in proportion to their rigidities considering the rigidity of the horizontal bracing system or diaphragm. Rigid elements that are assumed not to be part of the lateral force-resisting system may be incorporated into buildings, provided that their effect on the action of the system is considered and provided for in the design (2303(6)1).

Provisions shall be made for the increased forces induced on resisting elements of the structural system resulting from torsion due to eccentricity between the center of application of the lateral forces and the center of rigidity of the lateral force-resisting system. Forces shall not be decreased due to torsional effects (2303(b)2).

In addition, where the vertical resisting elements depend on diaphragm action for shear distribution at any level, the shear-resisting elements shall be capable of resisting a torsional moment assumed to be equivalent to the story shear acting with an eccentricity of not less than 5% of the maximum building dimension at that level (2312(e)4).

The first of the above requirements is the same, and is stated slightly more precisely in UBC-88/91: "The design story shear...in any story...shall be distributed to the various elements of the vertical lateral force-resisting system in proportion to their rigidities, considering the rigidity of the diaphragm" (2334(e)).

UBC-88/91 clarifies that the second requirement applies only when diaphragms are not flexible. Diaphragms are considered flexible when the maximum lateral deformation of the diaphragm is more than two times the average story drift of the associated story. This may be determined by comparing the computed midpoint in-plane deflection of the diaphragm under lateral load with the story drift of adjoining vertical resisting elements under equivalent tributary lateral load (2334(f)).

UBC-88/91 states much more clearly than before that the torsional design moment at a given story shall be the moment resulting from eccentricities between applied design lateral forces at levels above that story and the center of rigidity of the vertical resisting elements in that story plus an accidental torsion. To compute the accidental torsion, the mass at each level shall be assumed to be displaced from the calculated center of mass in each direction a distance equal to 5% of the building dimension at that level perpendicular to the direction of the force under consideration. Note that this is substantially different from the corresponding 1985 provision (2334(f)).

UBC-88/91 further requires that where a torsional irregularity exists, as defined in Table 23-N, the effects shall be accounted for by increasing the accidental torsion at each level by an amplification factor, A_x, given by Eq. 34-9 (2334(f)).

The Tri-Services Manual [1.9] provides a simple way of accounting for accidental torsion in seismic design.

1.7.3 Discontinuous Lateral Force-Resisting Elements

Overturning moments on discontinuous shear resisting elements are to be carried as loads to the foundation. In Seismic Zones 3 and 4, columns supporting the overturning forces of discontinuous shearwalls are to be designed to resist the axial forces resulting from the dead load, 0.8 time the live load, and $3R_w/8$ times the earthquake force (when the effects of gravity and earthquake forces are in the same direction), and 0.85 time the dead load and $3R_w/8$ times the earthquake force (when the effects of gravity and earthquake forces are in opposite directions). The columns carrying these factored loads may be designed to their axial load strength, and must meet special detailing requirements outlined later (2334(g)2).

The one-third stress increase allowed by Section 2303(d) may be exceeded for soils for combinations including earthquake effects, when substantiated by geotechnical data (2910(b)).

The above provisions were added for the first time to UBC-88.

1.7.4   P-Δ Effects

Unlike UBC-85, UBC-88/91 specifically requires that member forces and story drifts induced by P-Δ effects shall be considered in the evaluation of overall structural frame stability. P-Δ need not be considered when the ratio of secondary moment to primary moment does not exceed 0.10. The ratio may be evaluated for any story as the product of the dead, floor live load and snow load (as required in 2303(f)) above the story, times the seismic drift in that story, divided by the product of the seismic shear in that story and the height of that story. UBC-88 allowed that in Zones 3 and 4, "designs conforming to the drift limitations in Section 2312(e)8 may be deemed to satisfy this requirement." This relaxation has been replaced in UBC-91 by: "In Seismic Zones Nos. 3 and 4, P-Δ need not be considered where the story drift ratio does not exceed $0.02/R_w$" (2334(i)).

1.7.5   Vertical Component of Earthquake Force

Unlike UBC-85, UBC-88/91 contains the following Zone 3, 4 requirement:

A horizontal cantilever component shall be designed for a net upward force equal to 0.2 times the weight of the component (2334(i)).

In addition to all other applicable load combinations, horizontal prestressed components shall be designed using no more than 50% of the dead load for gravity load, alone or in combination with the lateral force effects (2334(j)).

Design for dead load in members will usually assure against problems resulting from downward accelerations, while the typical load factors will provide assurance against failure resulting from upward accelerations. Since cantilevers do not have continuity, it was felt necessary to provide some additional assurance [1.1—Commentary]. Also, both simply supported and continuous prestressed beams should be checked for the reduced vertical load combination.

1.7.6   Lateral Force on Elements of Structures and Non-Structural Components Supported by Structures

The provisions of the 1985 UBC for the anchorage of structural elements and non-structural components were revised and added to in UBC-88. Consideration was given for the first time to whether these components are rigid or non-rigid, and whether located at ground level or higher in the structure where amplification can be significant. The 1988 requirements have been further refined and added to in UBC-91, including modifications to Table 23-P (2336).

1.7.7 Uplift Effects

UBC-88/91 contains the following specific requirement not included in UBC-85: "Consideration shall be given to design for uplift effects caused by seismic loads. For materials which use working stress procedures, dead loads shall be multiplied by 0.85 when used to reduce uplift" (2337(a)).

1.7.8 Orthogonal Effects

UBC-88/91, unlike UBC-85, requires that in Seismic Zones 2, 3 and 4, provision shall be made for the effects of earthquake forces acting in a direction other than the principal axes in each of the following circumstances:

1) The structure has nonparallel lateral force-resisting systems (e.g. triangular shaped building).
2) The structure has torsional irregularity as defined in Table 23-N.
3) A column of a structure forms part of two or more intersecting lateral force-resisting systems.

If the axial load in a column due to seismic forces acting along either principal axis is less than 20% of the column allowable axial load, then the above provision need not apply.

The requirement that orthogonal effects be considered may be satisfied by designing an element for 100% of the prescribed seismic forces in one direction plus 30% of the prescribed forces in the perpendicular direction. The combination requiring the greater component strength must be used for design (2337(a)).

Alternatively, the effects of the prescribed seismic forces along two orthogonal directions may be combined on a square root of the sum of the squares (SRSS) basis. Because the SRSS methodology loses the signs of the terms, use of its results requires great care and caution.

1.7.9 Connections

UBC-88 added a requirement, continued in UBC-91, that connections which resist seismic forces be designed and detailed on the drawings.

1.7.10 Exterior Elements

Section 2312(j)3C of UBC-85 on Exterior Elements was expanded in Section 2312(h)2D(iii) of UBC-88. Section 2337(b)4B of UBC-91 contains the same expanded provisions.

1.7.11 Building Separation

UBC-85 Section 2312(h) used to require that "All portions of structures shall be designed and constructed to act as an integral unit in resisting horizontal forces unless separated structurally by a distance sufficient to avoid contact under deflection from seismic action or wind forces." UBC-88 required that all structures must be separated from adjoining structures, and that separations must allow for $3R_w/8$ times the displacement due to seismic forces. UBC-91 contains the same requirement. However, the following exception has been added: "Smaller separations may be permitted when justified by rational analyses based on maximum expected ground motions. As a minimum, building separations shall not be less than $(R_w/8) \geq 1$ times the sum of displacements due to code-specified seismic forces" (2337(b)11).

1.7.12   Ties and Continuity

UBC-88 added provisions to ensure that all parts of a structure will act together under seismic excitation, without localized separation, loss of support, or collapse. These provisions, unchanged, are to be found in Section 2337(b)5 of UBC-91.

1.7.13   Diaphragms

UBC-88/91, unlike UBC-85, requires that connections of diaphragms to the vertical elements and to collectors and connections of collectors to the vertical elements in structures in Seismic Zones 3 and 4, having torsional irregularity, reentrant corners, diaphragm discontinuity or out-of-plane offsets, as defined in Table 23-N (Plan Structural Irregularities), shall be designed without considering the one-third increase usually permitted in allowable stresses for elements resisting earthquake forces (2337(b)9E).

Also, in structures in Seismic Zones 3 and 4 having reentrant corners, diaphragm chord and drag members shall be designed considering independent movement of the projecting wings of the structure. Each of these diaphragm elements must be designed for the more severe of two assumptions: motion of the projecting wings in the same direction, motion of the projecting wings in opposing directions. This requirement may be deemed satisfied if the Dynamic Lateral Force Procedures of Section 2335 in conjunction with a three-dimensional model are used to determine the lateral seismic forces for design (2337(b)9F).

1.7.14   Perimeter Frames

UBC-85 Section 2312(j)1C required that in Seismic Zones 2, 3 and 4, all concrete frames located in the perimeter line of vertical support had to be ductile (special) moment-resisting space frames. An exception was provided for frames in the perimeter line of vertical support of buildings designed with shearwalls taking 100% of the design lateral forces.

The above requirement has been dropped from UBC-88/91, as being without proper justification, providing the design engineer with more flexibility.

1.8   Proportioning of Members-Load and Strength Reduction Factors

UBC-91, as in previous editions of the UBC, requires (2625(c)4) that for earthquake loading in Seismic Zones 3 and 4, the load factors given by Eqs. 9-2 and 9-3 be modified to:

$$U = 1.4 (D + L + E)$$
$$U = 0.9 D \pm 1.4 E$$

where

U = required strength to resist factored loads or related internal moments and forces,

D = dead loads, or related internal moments and forces,

L = live loads, or related internal moments and forces,

E = load effects of earthquake, or related internal moments and forces.

Section 2609(d)4 of UBC-91 specifies that in Seismic Zones 3 and 4, strength reduction factors shall be as given in the rest of Section 2609(d) except for the following:

The shear strength reduction factor shall be 0.6 for the design of walls, topping slabs used as diaphragms over precast concrete members and structural framing members, with the exception of joints, if their nominal strength is less than the shear corresponding to development of their nominal flexural strength. The shear strength reduction factor for joints is 0.85.

Section 2625(c)3 further requires that the strength reduction factor for axial compression and flexure shall be 0.5 for all frame members with factored axial compressive force exceeding $A_g f'_c/10$, if the transverse reinforcement does not conform to 2625(e)4. A_g = gross area of section, and f'_c = specified compressive strength of concrete.

1.9 Seismic Detailing Requirements

The seismic detailing requirements of Section 2625 of UBC-85 were based on Appendix A to ACI 318-83, "Building Code Requirements for Reinforced Concrete"[1.10]. Some of the changes from ACI 318-83 Appendix A to ACI 318-89 Chapter 21 [1.10] were incorporated into UBC-88. Section 2625 of UBC-91 is the same as Chapter 21 of ACI 318-89, with a few exceptions. The following is a list (not meant to be exhaustive) of major changes in Section 2625 of UBC from the 1985 to the 1991 edition. The few remaining differences between the seismic detailing requirements of UBC-91 Section 2625 and those of ACI 318-89 Chapter 21 are also discussed. It should be noted that when UBC-91 Chapter 26 text is different from the corresponding ACI 318-89 text, such differing text is printed in italics in UBC-91. Also, vertical lines along the margins of UBC-91 indicate changes from the corresponding provisions of UBC-88.

1. A new term, "seismic hook," was introduced in UBC-88, and is defined the same way in UBC-91 (2625(b)). It is defined as a 135-degree bend with a six-bar-diameter, but not less than 3-in., extension that engages the longitudinal reinforcement and projects into the interior of the stirrup or hoop. Paradoxically this term was not used in the definitions of "hoop" and "crosstie" in UBC-88, as it should have been. These definitions remained as they were in UBC-85; a six-bar-diameter extension (without a 3-in. minimum) was required beyond a 135-degree bend. The definitions of "hoop" and "crosstie" have been revised in UBC-91 to the same definitions as in ACI 318-89. The six-bar-diameter extension beyond the 135-degree bend now has a 3 in. minimum limit on it. Also, the 90-degree hooks of two successive crossties engaging the same longitudinal bars are now required to be alternated end for end. However, the term "seismic hook" is still not used in these definitions. Indeed, this term is not defined in ACI 318-89.

2. The shear strength reduction factor for walls, for topping slabs used as diaphragms over precast concrete members and for structural members, except joints, where their nominal shear strength is less than the shear corresponding to the development of their nominal flexural strength is specified to be 0.6 in UBC-91 (2609(d)4) as well as in UBC-88 (2625(e)3A). The corresponding provision in ACI 318-89 (Section 9.3.4.1) talks in terms of "any structural member," rather than "walls, topping slabs used as diaphragms over precast concrete members and structural framing members." ACI 318-89 further adds that the nominal flexural strength shall be determined considering the most critical factored axial loads including earthquake effects. The corresponding UBC-85 (2625(c)3A) and ACI 318-83 provision was for "any structural member if its nominal shear strength is less than the shear corresponding to development of its nominal flexural strength <u>for the factored-load combinations, including earthquake effect</u>." The underlined part of the provision was less than clear.

3. The compressive strength of lightweight concrete is in no case to exceed 6000 psi (2625(c)5B). This provision was new in UBC-88, is unchanged in UBC-91, was not included in ACI 318-83, and is not in ACI 318-89.

4. Section 2625(c)6 of UBC-85/88/91 specifies that welding of stirrups, ties, inserts or other similar elements to longitudinal reinforcing bars shall not be permitted. This requirement was not in ACI 318-83, but has been added to ACI 318-89, with one significant modification. ACI 318-89 Section 21.2.6 prohibits the welding of stirrups, ties, inserts, or other similar elements to longitudinal reinforcement required by design. Thus welding of stirrups, etc., to longitudinal construction bars is permitted.

5. The "scope" of Section 2625(d) of UBC-91, "Flexural Members of Frames," is editionally revised from that in UBC-88 and UBC-85. The new "scope" is identical to that in ACI 318-89, and avoids ambiguity that existed before.

6. The "scope" of Section 2625(e) of UBC-91, "Frame Members Subjected to Bending and Axial Load," is also editionally revised from that in UBC-88 and UBC-85. The new "scope" is again identical to that in ACI 318-89, and avoids ambiguity that existed before.

7. In frame members subjected to bending and axial load (columns), lap splices are permitted only within the center half of the member length. According to UBC-85 as well as ACI 318-83, these "shall be proportioned as tension splices." According to UBC-88/91, these shall be proportioned as Class A tension splices (Section 2625(e)3B). This change has not been made in ACI 318-89.

8. The coefficient 0.12 in Eq. 25-4 of UBC-85/88 has been reduced to 0.09 in Eq. 25-4 of ACI 318-91 (2625(e)4A(ii)). The same change has been made in Eq. 21-4 of ACI 318-89. ACI-ASCE Committee 352, Joints and Connections in Monolithic Concrete Structures, recommended this change based on the observed behavior of tied columns which have properly detailed hoops and crossties. [1.11]

9. Any area of a column which, for architectural purposes, extends more than 4 in. beyond the confined core is required to have minimum reinforcement as for nonseismic columns (2625(e)4A(v)). This provision was new in UBC-88, continues into UBC-91, was not part of ACI 318-83, and has not been included in ACI 318-89.

10. When the calculated point of contraflexure is not within the middle half of the column clear height, special transverse reinforcement as specified in 2625(e)4A(i) must be provided over the full height of the member (2625(e)4A(vi)). This provision was new in UBC-88, continues into UBC-91, was not included in ACI 318-83, and is not part of ACI 318-89.

11. Over the end regions of columns in Seismic Zones 3,4 where special transverse reinforcement is required, the maximum spacing of such reinforcement is specified in ACI 318-83/89 to be one-quarter of the minimum member dimension, or 4 in., whichever is less. The maximum spacing is 4 in., without the one-quarter minimum member dimension option, in UBC-85/88/91 (2625(e)4B).

12. Where columns support reactions from discontinued stiff members, the special transverse column reinforcement must be placed above the discontinuity for at least the development length of the largest longitudinal column reinforcement. Where the column is supported on a wall, the special transverse reinforcement must be placed below the discontinuity for the same development length. Where the column terminates on a footing or mat, UBC-88 required that special transverse reinforcement must be placed below the top of the footing or mat for the same development length or the lead length of a standard hook. This was a modification and clarification of the corresponding 1985 provision. ACI 318-83 did not contain any similar require-

ments. ACI 318-89 adopted the UBC-88 provision, except that where the column terminates on a footing or mat, special transverse reinforcement is required to extend at least 12 in. into the footing or mat. The modified ACI 318-89 provision has been adopted into UBC-91 (2625(e)4E).

13. A UBC-85/88/91 provision not included in ACI 318-83 or 318-89 requires that at any section where the design strength ϕP_n of the column is less than the sum of the shears V_e computed in accordance with Section 2625(h)1 for all the beams framing into the column above the level under consideration, special transverse reinforcement as specified in 2625(e)4A through C must be provided. For beams framing into opposite sides of the column, the moment components may be assumed to be of opposite sign. For the determination of the design strength, ϕP_n, of the column, these moments may be assumed to result from the deformation of the frame along any one principal axis (2624(e)4F).

14. A new ACI 318-89 and UBC-91 provision requires that where transverse reinforcement as specified in 2625(e)4A through C is not provided throughout the full length of a column, the remainder of the column length must contain spiral or hoop reinforcement with center-to-center spacing not exceeding the smaller of six times the diameter of the longitudinal column bars or 6 in. Until this provision was instituted, transverse reinforcement in a column in Seismic Zones 3 and 4, outside the end regions containing special transverse reinforcement, used to be governed by 2607(k). The minimum amount and maximum spacing specified in that section were simply insufficient. Column failures occurring just outside of the end regions containing special transverse reinforcement have been observed in actual earthquakes. The new provision of 2625(e)4G is thus a welcome addition to the code, although it increases the cost of reinforced concrete columns in regions of high seismicity.

15. Splices in horizontal shearwall reinforcement must be staggered. Splices in two curtains, where used, shall not occur in the same location (2625(f)2D). This provision was new in UBC-88, continues into UBC-91, was not included in ACI 318-83, and is not part of ACI 318-89.

16. An ACI 318-89 change adopted into UBC-91 requires that boundary members of structural diaphragms shall be proportioned to resist the sum of the factored force acting in the plane of the diaphragm and the force obtained from dividing the factored moment at the section by the distance between the edges of the diaphragm at that section (2625(f)3D).

17. Another ACI 318-89 change adopted into UBC-91 requires that except when V_u in the plane of the wall is less than $A_{cv}\sqrt{f'_c}$, transverse reinforcement terminating at the edges of shearwalls without boundary elements must have a standard hook engaging the edge reinforcement, or the edge reinforcement must be enclosed in U stirrups having the same size and spacing as, and spliced to, the transverse reinforcement (2625(f)3E).

18. Structural steel members conforming to Chapter 27 and encased monolithically in the walls at the edges may be used as boundary members, provided adequate shear transfer is provided between the steel and the concrete (2625(f)3F). This provision was new in UBC-88, has continued into UBC-91, was not part of ACI 318-83, and is not included in ACI 318-89.

19. The provisions of Section 2625(f)5 of UBC-85/88/91 concerning coupling beams are not included in ACI 318-83 or 318-89.

20. A cast-in-place topping on a precast floor system may serve as the diaphragm, provided the cast-in-place topping acting alone is proportioned and detailed to resist the design forces (2625(f)7). This provision was new in UBC-88, continues into UBC-91, was not part of ACI 318-83, and is not included in ACI 318-89.

21. Diaphragms used to resist prescribed lateral forces must not be less than 2 in. thick. Topping slabs placed over precast floor and roof elements must not be less than 2 1/2 in. thick (2625(f)8). This provision was new in UBC-88, continues into UBC-91, was not included in ACI 318-83, and is not part of ACI 318-89.

22. UBC-91 has added new provisions for wall piers that are not in ACI 318-89 (2624(f)9).

23. Where longitudinal beam reinforcing bars extend through a joint, the column depth in the direction of loading must not be less than 20 times the diameter of the largest longitudinal beam bar (2625(g)1D). This provision was new in UBC-88, continues into UBC-91, is intended to minimize slippage of beam bars at interior beam-column joints, and is in accordance with recommendations from ACI-ASCE Committee 352 [1.11]. This provision was not included in ACI 318-83, and is not part of ACI 318-89.

24. When a beam-column joint is confined by structural members as specified in 2625(g)2B, within the depth of the shallowest framing member, the 4 in. spacing of transverse reinforcement specified in 2625(e)4B may be increased, but shall not exceed 6 in. This is a new ACI 318-89 requirement adopted into UBC-91.

25. Effective cross-sectional area within a joint (in a plane parallel to plane of reinforcement generating shear in the joint) is joint depth times effective joint width, where joint depth is overall column depth, and effective joint width is as shown in Fig. 1-17. This revises the definition of A_j in UBC-85, which was the same as in ACI 318-83, and makes it consistent with the areas used in test reports from which the joint shear strength formulas were developed. UBC-91 and ACI 318-89 contain the same definition of A_j as in UBC-88 (2625(g)3).

26. UBC-85, as in ACI 318-83, specified nominal shear strength values of $20\sqrt{f'_c}\,A_j$ for confined joints, and $15\sqrt{f'_c}\,A_j$ for other joints. UBC-91, as in ACI 318-89, specifies three levels of nominal shear strength: $20\sqrt{f'_c}\,A_j$ for joints confined on all four faces, $15\sqrt{f'_c}\,A_j$ for joints confined on three faces or on two opposite faces, and $12\sqrt{f'_c}\,A_j$ for other joints. The three-level shear strength provision is based on the recommendation of ACI-ASCE Committee 352 [1.11]. Test data reviewed by that committee indicated that the value of $15\sqrt{f'_c}\,A_j$ is unconservative when applied to corner joints.

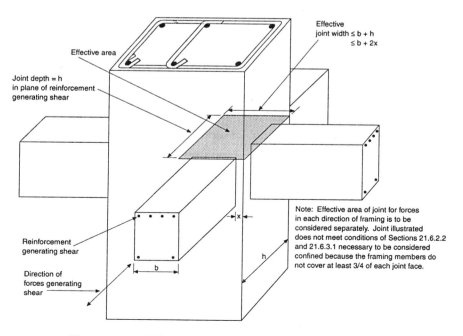

Figure 1-17 Effective Area Within Beam-Column Joint

27. For frame members subjected primarily to bending (beams), the design shear force, V_e, is to be calculated considering the member loaded with the factored tributary gravity load along its span according to ACI 318-83 and 318-89. In UBC-85/88/91, only unfactored tributary gravity load is considered (2625(h)1). The use of factored tributary gravity load was felt to be unnecessarily conservative by SEAOC [1.1].

28. Computation of design shear force, V_e, based on probable flexural strength should apply to frame members subjected to combined bending and axial load (columns) as well as to frame members subjected primarily to bending (beams). Also, there is no need to calculate shear based on the full column moment capacity when the upper limit on moment is controlled by flexural members framing into the column. Section 2625(h)1B was rewritten in UBC-88 to point these out. The provisions of UBC-85 and ACI 318-83 were similar in this regard. The revised provision of UBC-88/91 and ACI 318-89 are similar once again.

29. Section 2625(h)2 of UBC-85 and UBC-88 mandated that for determining the required transverse reinforcement in frame members, the quantity V_c must be taken as zero if the factored axial compressive force including earthquake effects is less than $(A_g f_c')$. According to ACI 318-83, 318-89, and now UBC-91, for V_c to be neglected, the earthquake-induced shear force must additionally represent one-half or more of the total design shear.

30. UBC-85/88/91 contains an inspection clause (2625(j)) that is not included in ACI 318-83 or 318-89.

31. The deformation compatibility requirement of ACI 318-89 Section 21.8 is quite different from and more comprehensive than that of UBC-91 Section (2625(i).

The detailing requirements for frames in regions of moderate seismic risk, applicable to Intermediate Moment-Resisting (beam-column as well as slab-column) Frames of concrete in Zones 2A and B, as contained in Section 2625(k), have not changed from UBC-85 to UBC-91 or from ACI 318-83 to ACI 318-89.

PCA's "Notes on ACI 318-89," [1.12] contains a good discussion of changes in seismic detailing requirements from ACI 318-83 to ACI 318-89. For guidance on standard non-seismic detailing the reader may refer to [1.13, 1.14].

1.10   Scope of Publication

Design of the various structural systems discussed in this chapter in different seismic zones, and in non-seismic areas for gravity and wind only, is illustrated in this publication.

A 14-story flat plate-column frame building is designed in Chapter 2 for Seismic Zones 0 (70 mph wind), 1 and 2A. It is designed as an Ordinary Moment-Resisting Frame ($R_w = 5$) in Seismic Zone 1, and as an Intermediate Moment-Resisting Frame ($R_w = 8$) in Seismic Zone 2A. It is concluded that an attempt to design the particular example building for Seismic Zone 2B would result in impractical slab thicknesses and slab reinforcements. It has been noted that a flat plate-column frame cannot be made part of the lateral force-resisting system of a structure in Seismic Zone 3 or 4.

Chapter 3 is devoted to the design of a 12-story beam-column frame in Seismic Zones 4, 2B, 1 and 0 (70 mph wind). It is designed as an SMRF ($R_w = 12$) in Seismic Zone 4, as an IMRF ($R_w = 8$) in Seismic Zone 2B, and as an OMRF ($R_w = 5$) in Seismic Zone 1. Seismic Zones 3 and 2A designs would have been duplications of Seismic Zones 4 and 2B designs, respectively, for scaled-down forces; thus, these are omitted.

A 10-story Building Frame System, with shearwalls providing resistance to 100% of the design seismic forces, and with an essentially complete space frame providing support for gravity loads, is designed in Chapter 4 for Seismic Zones 4 and 2B. It has been mentioned earlier that in Seismic Zones 1 and 0, the concept of the Building Frame System loses its validity. Zone 3 and 2A designs are not illustrated, to avoid essential repetition of Zone 4 and 2B designs, respectively. R_w is 8 for both the designs for Seismic Zones 4 and 2B. The shearwalls are specially detailed in Seismic Zone 4, and are not so detailed in Seismic Zone 2B. A comparison of the results for the shearwall design using UBC-91 and UBC-94 provisions is given for Seismic Zone 4. The deformation compatibility between the lateral force-resisting shearwalls and the gravity-carrying frame system is paid special attention.

A 20-story Dual System consisting of shearwalls and beam-column frames is designed in Chapter 5 for Seismic Zones 4 and 2B. Since the building height exceeds 240 ft, the Dynamic Lateral Force Procedure is employed in Seismic Zone 4. The shearwalls are specially detailed and the moment frames are SMRF in Seismic Zone 4; R_w is equal to 12. As in Chapter 4, a comparison between UBC-91 and UBC-94 is given for the design of shearwalls. The shearwalls are ordinarily detailed and the moment frames are IMRF in Seismic Zone 2B; R_w is equal to 9. It has been pointed out earlier that the concept of the Dual System is not valid in Seismic Zones 1 and 0. Zone 3 and 2A designs are not included, as being essentially repetitive of Zone 4 and 2B designs, respectively.

A shearwall-frame interactive system is designed for Seismic Zones 1 and 0 (70 mph wind) in Chapter 6. This system is not specifically recognized as such in UBC-94, and no specific R_w value is assigned. The value of R_w used is 8, the same as that for a Building Frame System. This system is definitely not inferior to the Building Frame System from the point of view of seismic resistance.

Finally, a Bearing Wall System, with shearwalls resisting gravity as well as lateral loads, is designed for Seismic Zones 4, 2B and 0 (70 mph wind) in Chapter 7. The value of R_w is 6 in Seismic Zones 4 and 2B. The shearwalls are specially detailed in Seismic Zone 4, and are not so detailed in Seismic Zone 2B. Designs for other seismic zones are omitted as being essentially repetitive of those illustrated.

The applicable sections of the Uniform Building Code (1994 edition) are referenced in italics in the right hand column throughout the text.

References

1.1 Seismology Committee, Structural Engineers Association of California, *Recommended Lateral Force Requirements and Commentary*, San Francisco, California, 1988.

1.2 International Conference of Building Officials, *Uniform Building Code*, Whittier, California, 1985, 1988, 1991.

1.3 American Society of Civil Engineers, *ASCE Standard Minimum Design Loads for Buildings and Other Structures*, ASCE 7-88 (formerly ANSI A58.1), New York, New York, 1990.

1.4 Building Officials and Code Administrators International, *The BOCA National Building Code, Country Club Hills*, Illinois, 1990.

1.5 Southern Building Code Congress International, *Standard Building Code*, Birmingham, Alabama, 1991.

1.6 Applied Technology Council, *Tentative Provisions for the Development of Seismic Regulations for Buildings*, ATC Publication ATC 3-06, NBS Special Publication 510, NSF Publication 78-8, U.S. Government Printing Office, Washington, D.C., 1978.

1.7 Building Seismic Safety Council, *NEHRP (National Earthquake Hazards Reduction Program) Recommended Provisions for the Development of Seismic Regulations for New Buildings*, Washington, D.C., 1985, 1988, 1991.

1.8 Bendimerad, F.M., Shah, H. C., and Hoskins, T., "Extension of Study on Fundamental Period of Reinforced Concrete Moment-Resisting Frame Structures," The John A. Blume Earthquake Engineering Center, Stanford University, Stanford, California, January 1991.

1.9 Departments of the Army, the Navy, and the Air Force, *Technical Manual: Seismic Design for Buildings*, (Army TM 5-809-10, Navy NAVFAC P-355, Air Force AFM 88-3, Chap. 13), Washington, D.C., February 1982.

1.10 Committee 318, American Concrete Institute, *Building Code Requirements for Reinforced Concrete (ACI 318-89) and Commentary (ACI 318R-89)*, Detroit, Michigan, 1989.

1.11 Committee 352, American Concrete Institute—American Society of Civil Engineers, "Recommendations for Design of Beam-Column Joints in Monolithic Reinforced Concrete Structures," (ACI 352R-76) (Reaffirmed 1981), Detroit, Michigan, 1976. Also, ACI Manual of Concrete Practice, Part 3.

1.12 Portland Cement Association, "Notes on ACI 318-89," S. K. Ghosh and B. G. Rabbat, editors, Skokie, Illinois, 1990.

1.13 Committee on Reinforcing Bar Detailing, the Concrete Reinforcing Steel Institute, *Reinforcing Bar Detailing*, 3rd Edition, Schaumburg, Illinois, 1988.

1.14 Committee 315, American Concrete Institute, *ACI Detailing Manual* (ACI SP-66 (88)), Detroit, Michigan, 1988.

Appendix 1-A

Background to Seismic Zoning Map in 1988 and 1991 UBC

The following information is compiled from readily available sources: a 1969 paper by Algermissen [A.1], the commentary portion of ASCE Standard A7 (formerly ANSI A58.1) [1.3], the commentary portion of ATC 3-06, "Tentative Provisions for the Development of Seismic Regulations for Buildings" [1.6], and the commentary portion of the SEAOC Blue Book [1.1]. It is presented here in one concise package for the convenience of the reader.

A seismic probability map of the United States was prepared in 1948 by F. P. Ulrich with the advice of seismologists throughout the United States [A.1] and was issued by the Coast and Geodetic Survey in 1948. In 1949, the seismic probability map was revised such that the Charleston, South Carolina area was changed from Zone 3 to Zone 2 and a Zone 3 was set up for the Puget Sound region of Washington which had formerly been included in Zone 2 [A.2]. The revised map was adopted by the Pacific Coast Building Officials Conference (later ICBO) for inclusion in the 1952 edition of the Uniform Building Code. Subsequent editions of the UBC, up to and including the 1967 edition, included this map with no changes.

The seismic zone map in UBC-85 evolved from the work of Algermissen [A.3] during the 1960s and is based on the maximum recorded intensity of shaking without regard to the frequency with which such shaking might occur. As originally published, this map, adopted into the 1970 and 1973 editions of the UBC, had four Zones (0, 1, 2, and 3), so that several areas in the Eastern United States were in the same zone as California. During the 1970s the map was modified and eventually adopted into the 1976 edition of the UBC with a Zone 4 in parts of California and Nevada, and with the Z factor for the remaining Zone 3 areas reduced. The boundary between Zones 3 and 4 was determined by proximity to the major fault systems. The boundary was set at 25 miles from a fault considered capable of generating an earthquake of magnitude 7.0 or greater and 15 miles from a fault that could generate an earthquake of magnitude between 6.0 and 7.0. The distances were based on the Schnabel-Seed attenuation curve [A.4] and a value of 0.3g for the peak acceleration at the boundary on a rock site.

Algermissen and Perkins subsequently published a new contour map for peak ground acceleration on rock, based on a uniform probability of occurrence throughout the 48 contiguous states [A.5]. The probability that the contoured peak acceleration would not be exceeded was given as 90% in 50 years. In developing this map, the first step was to delineate zones within which earthquakes may occur and establish for each such zone the frequency of earthquakes with different magnitudes. Attenuation equations were selected for both the eastern and the western parts of the country. These several parameters served as input to a computer program that computed the frequency for different peak accelerations at all points of a gridwork covering the 48 states.

The Seismic Risk Committee of ATC 3 modified the Algermissen-Perkins map to make it more suitable for use in a document resembling a model code. The concept of Effective Peak Acceleration (EPA), related to the damageability of ground shaking, was introduced, and certain small zones of a very high peak acceleration were eliminated, partly on the basis that their retention would constitute microzoning, which the Committee had been instructed to avoid. The contours of the Algermissen-Perkins map were smoothed, so as to avoid the appearance of great precision, and in some locales the contours were shifted on the basis of more recent knowledge. The result was a map that retained the basic principles and trends of the Algermissen-Perkins map, but lacked the internal consistency of that map. It was estimated that the probability of not exceeding the contoured values of effective peak acceleration within 50 years was 80% to 95%. Maps of EPA for Alaska, Hawaii, Puerto Rico, and several territories were drawn using the best available information from various sources as guidance [1.6].

It was decided during the initial stages of development of ATC 3 that ground-shaking regionalization maps should take into account the distance from anticipated earthquake sources. This decision reflects the observation that the higher frequencies in ground motion attenuate more rapidly with distance than the lower frequencies. Thus, at distances of 100 km or more from a major earthquake, flexible buildings may be more seriously affected than stiff buildings. To accomplish the objectives of the decision, it proved necessary to use two separate ground motion parameters and, therefore, to prepare a separate map for each.

The two parameters used to characterize the intensity of design ground shaking were the Effective Peak Acceleration (EPA), and the Effective Peak Velocity (EPV). To best understand the meaning of EPA and EPV, they should be considered as normalizing factors for construction of smoothed elastic response spectra for ground motions of normal duration. The EPA is proportional to spectral ordinates for periods in the range of 0.1 to 0.5 seconds, while the EPV is proportional to spectral ordinates at a period of about 1 second (Fig. A-1). The constant of proportionality (for a spectrum corresponding to 5% of critical damping) is set at a standard value of 2.5 in both cases.

The EPA and EPV are related to peak ground acceleration and peak ground velocity, but are not necessarily the same as or even proportional to peak acceleration and velocity. When very high frequencies are present in the ground motion, the EPA may be significantly less than the peak acceleration. This is consistent with the observation that chopping off the highest peak in an acceleration time history has very little effect upon the response spectrum computed from that motion, except at periods much shorter than those of interest in ordinary building practice. Furthermore, a rigid foundation tends to screen out very high frequencies in the free field motion. On the other hand, the EPV is generally greater than the peak velocity at large distances from a major earthquake. Ground motions increase in duration and become more periodic with distance. These factors tend to produce proportionately larger increases in that portion of the response spectrum represented by the EPV.

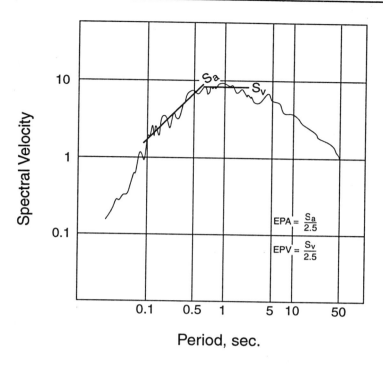

Figure A-1 Schematic Representation of Effective Peak Acceleration and Effective Peak Velocity

If an earthquake is of very short or very long duration, then it may be necessary to correct the EPA and EPV values to more closely represent the event. It is well documented that two motions having different durations but similar response spectra cause different degrees of damage—the damage being less for the shorter duration. In particular, there have been numerous instances where motions with very large accelerations and short durations have caused very little or even no damage. Thus when expressing the significance of a ground motion to design, it may be appropriate to decrease the EPA and EPV obtained from the elastic spectrum for a motion of short duration. On the other hand, for a motion of very long duration it may be appropriate to increase the EPA and EPV.

For ease in developing base shear formulas, it proved desirable to also express EPV by a dimensionless parameter (A_v) which is an acceleration coefficient. This parameter is referred to as the Effective Peak Velocity-Related Acceleration Coefficient. The relationship between EPV and A_v is given in Table A-1.

Table A-1 Relationship Between EPV and A_v

Effective Peak Velocity (in./sec.)	Velocity-Related Acceleration Coefficient, A_v
12	0.4
6	0.2
3	0.1
1.5	0.05

The ATC map for EPV was constructed by modifying the map for EPA. At all locations where a contour gives the highest EPA in a region, the EPV along that contour was set at the corresponding value of EPV. For example, the EPV was set at 3 in./sec. along the contour for EPA = 0.1g in the vicinity of the Appalachian Mountains and in South Carolina. In California, EPV was set at 12 in./sec. along the contours for EPA = 0.4g. A study by McGuire [A.6], based upon strong motion records from California, indicated that the distance required for EPV to decrease by a factor of 2 is about 80 miles. Thus, in the western part of the country, the contours for EPV = 6 in./sec. were located about 80 miles outside the contours for EPV = 12 in./sec. Similarly, in Washington and Utah where the highest contour is at 0.2g, corresponding to EPV = 6 in./sec., the net contour for EPV = 3 in./sec. was located about 80 miles away. The strong motion data available to McGuire were inadequate beyond a distance of about 100 miles. To estimate the attenuation of EPV beyond this distance, it was assumed that at large distances from an earthquake the logarithm of EPV would be linearly proportional to Modified Mecalli Intensity (MMI). Data from large earthquakes in California suggested that MMI decreased roughly linearly with distance, which would translate into EPV continuing to halve at equal increments of distance. Thus subsequent contours were also spaced at about 80 miles. For the midwest and east, reliance was placed entirely on information about the attenuation of MMI. Available studies indicated that MMI decays logarithmically with distance and that for the first 100 miles from a large earthquake the attenuation is roughly the same as in the west. The implication is that the distance required for EPV to halve increases with distance. Thus, starting from the contour for EPV = 6 in./sec. centered on southeastern Missouri, the contour for EPV = 3 in./sec. would be about 80 miles away, and the contour for EPV = 1.5 in./sec. would be 160 miles beyond that. It was also stipulated that a contour for EPV should never fall inside the corresponding contour for EPA. After these various rules were applied to produce a set of contours for EPV, considerable smoothing was done and contours were joined where they fell close together. The EPV map has neither the detailed theoretical basis nor the internal consistency of the Algermissen-Perkins map for peak acceleration.

The ATC 3-06 map for Effective Peak Velocity-Related Acceleration Coefficient was selected as the basis for the zoning map in ANSI A58.1-1982. The contour map was converted into a zoning map as follows:

A_v	Zone
≥ 0.4	4
0.2 to 0.4	3
0.1 to 0.2	2
0.05 to 0.1	1
≤ 0.05	0

There were several reasons for the choice of the map for EPV over that for EPA. First, the contours for EPA are very closely spaced in some regions. This is especially true in California and Nevada; use of the EPA map or the Algermissen-Perkins map would have meant that both states would be subdivided into five zones. Second, since only one map was to be used for the sake of simplicity, it seemed appropriate to use the more conservative map. Finally, use of the map for EPV meant a more modest change in the zonation of the country from that given in the 1979 UBC (the zonation in the 1985 UBC is the same).

The seismic zoning map in ANSI A58.1-1982, when compared with that in UBC-85, is seen to reduce the Z factor in many areas, especially in the eastern U.S.

The 1987 SEAOC Seismology Committee collaborated with a number of other organizations and individuals to produce a zone map of the United States for use in the 1988 UBC. There is only one zone map, but the philosophy embodied in the two ATC maps was followed. In drawing the zone boundaries, both acceleration and velocity-related maps were consulted, and, if they disagreed, the one indicating the higher zone prevailed. The design base shear equation was modified so that values of Z would correspond to the estimated values of effective peak acceleration. It was intended that the boundary between Zone 3 and Zone 4 correspond to a peak acceleration of 0.3g (or equivalent velocity) and the boundary between Zone 2 and Zone 3 correspond to 0.2g. The map is based principally on the ATC maps with modifications reflecting other sources of information, including the maps of acceleration and velocity published by Algermissen et al. in 1982 [A.7]. The ATC maps and the maps by Algermissen et al. are based primarily on the historical record of earthquakes, which in California is short relative to the repeat time of major earthquakes even on the most active faults. To avoid the danger of missing areas that are potentially active but simply by chance have not shown activity during the short time of the historical record, geologic data were consulted, particularly data on fault slip rates. On the basis of geologic data, Zones 3 and 4 were extended in Southern California to encompass the Garlock fault. The modification was made with the intent to be consistent with the philosophy that the Z coefficient should correspond to ground motion values with a 10% probability of being exceeded in 50 years. The criterion was abandoned, however, in making many other modifications. It was abandoned in drawing the boundary between Zone 2 and Zone 3 in California and Oregon. The boundary was drawn so as to exclude Zone 2 from California in order to accommodate the desire to retain the structural detailing requirements of Zones 3 and 4 for the state. The criterion was also abandoned in the southwestern part of California and elsewhere to accommodate local opinion.

The 1988 seismic zoning map has remained unchanged in the 1991 edition of the UBC.

References

[A.1] Roberts, E. B., and Ulrich, F. P., "Seismological Activities of the U.S. Coast and Geodetic Survey in 1948," Bulletin of the Seismological Society of America, Vol. 40, 1950, pp. 195-216.

[A.2] Roberts, E. B., and Ulrich, F. P., "Seismological Activities of the U.S. Coast and Geodetic Survey in 1949," Bulletin of the Seismological Society of America, Vol. 41, 1951, pp. 205-220.

[A.3] Algermissen, S. T., "Seismic Risk Studies in the United States," Proceedings of the Fourth World Conference on Earthquake Engineering, Santiago, Chile, 1969, V. I, pp. A-1.14—A-1.27.

[A.4] Schnabel, P. B., and Seed, H. B., "Acceleration in Rock for Earthquakes in the Western United States," Bulletin of the Seismological Society of America, Vol. 63, 1973, pp. 501-516.

[A.5] Algermissen, S. T., and Perkins, D. M., *A Probalistic Estimate of Maximum Acceleration in Rock in the Contiguous United States*, U. S. Geological Survey Open File Report 76-416, 1976, 45 pp.

[A.6] McGuire, R. K.,*Seismic Structural Response Risk Analysis, Incorporating Peak Response Regressions on Earthquake Magnitude and Distance*, Research Report R 74-51, Department of Civil Engineering, Massachusetts Institute of Technology, 1974, 371 pp.

[A.7] Algermissen, S. T., Perkins, D. M., Thenhaus, P. C., Hansen, S. L., and Bender, B. L., *Probalistic Estimates of Maximum Acceleration and Velocity in Rock in the Contiguous United States*, U. S. Geological Survey Open File Report 82-1033, 1982, 99 pp.

Appendix 1-B

1994 UBC Update

The most major changes from the 1991 edition to the 1994 edition of the Uniform Building Code, which are of interest to the readers of this publication, are discussed in this appendix. These include changes in format, changes in earthquake design requirements, and changes in concrete related provisions.

1-B.1   Changes in Format

Significant changes in format have been made from the 1991 edition to the 1994 edition of the Uniform Building Code, as described below.

1-B.1.1   Marginal Markings

Solid vertical lines in the margins within the body of the code indicate a change from the requirements of the 1991 edition except where an entire chapter was revised, a new chapter was added, or a change was minor. Where an entire chapter was revised, (for example, Chapter 21, Masonry) or a new chapter was added, a notation appears at the beginning of that chapter. Deletion indicators (➜) are provided in the margin where a paragraph or item listing has been deleted if the deletion resulted in a change of requirements.

1-B.1.2   Common Code Format

The provisions of the 1994 edition of the Uniform Building Code have been reformatted into the common code format established by the Council of American Building Officials. The new format establishes a common format of chapter designations for the three model building codes published in the United States. For instance, concrete related provisions are now in Chapter 19 in all three model codes. Apart from those changes approved by the ICBO membership, the reformatting has not changed the technical content of the code. The two chapters of interest in this publication have changed as follows: Chapter 23 of the 1991 edition is now Chapter 16, Structural Forces, of the 1994 edition, and Chapter 26 of the 1991 edition is now chapter 19, Concrete, of the 1994 edition.

1-B.1.3   Three-Volume Set

Provisions of the Uniform Building Code and UBC Standards have been divided into a three-volume set. Volume 1 accommodates administrative, fire- and life-safety, and field inspection provisions. Chapter 1 through 15 and Chapters 24 through 35 are printed in Volume 1 in their entirety. Any appendices associated with these chapters are printed in their entirety at the end of Volume 1. Excerpts from certain chapters from Volume 2 are reprinted in Volume 1 to provide greater usability.

Volume 2 accommodates structural engineering design provisions, and specifically contains Chapters 16 through 23 printed in their entirety. Included in this volume are design standards previously published in the UBC Standards. Design standards have been added to their respective chapters as divisions of the chapters. Any appendices associated with these chapters are printed in their entirety at the end of Volume 2. Excerpts of certain chapters from Volume 1 are reprinted in Volume 2 to provide greater usability.

Volume 3 contains material, testing, and installation standards.

1-B.1.4   Metrication

The Uniform Building Code has been metricated for the 1994 edition. The metric conversions are provided in parentheses following the conventional U.S. units. Where industry has made metric conversions available, the conversions conform to current industry standards.

Formulas also are provided with metric equivalents. Each metric equivalent formula immediately following the corresponding formula in conventional U. S. units is denoted by "For SI" preceding the metric equivalent. Some formulas are independent of units and, thus, are not provided with metric equivalents. Multiplying conversion factors have been provided for formulas where metric forms were unavailable. Tables are provided with multiplying conversion factors in subheadings for each tabulated unit of measurement.

1-B.1.5   New Decimal Numbering System

Each section and subsection of the 1994 code is numbered following a decimal system. For example, Section 2337(b)4A of the 1991 code is now Section 1637.2.4.1 of the 1994 code.

1-B.2   Changes in Earthquake Design Requirements

The following changes from 1991 UBC are significant:

1. 1991 UBC Section 2334(b)2B required that the value of coefficient $C = 1.25 \ S/T^{2/3}$ based on fundamental period T calculated by Method B (Rayleigh formula or other rational analysis) must not be less than 80 percent of the value obtained by using T from Method A (approximate formula). 1994 UBC Section 1628.2.2 Item 2 requires that the value of T from Method B shall not be over 30 percent greater than the value of T obtained from Method A in Seismic Zone 4 and 40 percent in Seismic Zone 1, 2, and 3. This is not a major change, as should be apparent below.

1991 UBC: $V_{(\text{Method B})} \geq 0.80 \, V_{(\text{Method A})}$

1994 UBC: $T_{(\text{Method B})} \leq 1.3 \, T_{(\text{Method A})}$ or
$V_{(\text{Method B})} \geq 0.84 \, V_{(\text{Method A})}$ in Zone 4

$T_{(\text{Method B})} \leq 1.4 \, T_{(\text{Method A})}$ or
$V_{(\text{Method B})} \geq 0.80 \, V_{(\text{Method A})}$ in Zones 1, 2, 3

The design lateral forces used to determine the calculated drift (Section 1628.8.3) may be derived from a value of C based on rationally determined period neglecting the above limitations and also neglecting the lower-bound ratio for C/R_w of 0.075 (as in 1991 UBC Section 2334(h)4).

2. While 1991 UBC Section 2334(c) addressed vertical combinations of structural systems and combinations of structural systems along different axes, combinations of different structural systems along the same axis were not specifically considered. 1994 UBC Section 1628.3.4 requires that, for other than dual systems, where a combination of different structural systems is utilized to resist lateral forces in the same direction, the value of R_w used in that direction shall not be greater than the least value for any of the systems utilized in that same direction.

3. A new provision in 1994 UBC (Section 1628.8.2 Exception 2) is that there shall be no drift limit in single-story steel-framed industrial buildings, warehouses, garages, aircraft hangers, and similar structures. Structures on which this exception is used shall not have the equipment attached to the structural frame or shall have such equipment detailed to accommodate the additional drift. Walls which are laterally supported by the frame shall be designed to accommodate the drift.

4. Another new provision in 1994 UBC (Section 1631.2.9 Item 3) is that design forces for flexible diaphragms and their connections, providing lateral supports for walls or frames of masonry or concrete, shall be calculated using an R_w not to exceed 6.

5. The building separation provisions of Section 1631.2.11 now include a property line setback provision. When a structure adjoins a property line not common to a public way, that structure shall also be set back from the property line by at least $3R_w/8$ times the displacement of that structure. Similar exceptions apply to building separation as well as to property line setback requirements.

6. A new section on nonbuilding structures (1632) has been created out of existing text. Original intent has been clarified. Lateral force procedures for nonbuilding structures with structural systems similar to those of buildings (as listed in Table 16-N) shall be selected in accordance with the provisions of Section 1627. Rigid nonbuilding structures (T < 0.06 sec.), including their anchorages, shall be designed for the lateral force $V = 0.5ZIW$. V shall be distributed according to the distribution of mass and shall be assumed to act in any horizontal direction. Other nonbuilding structures shall be designed to resist minimum seismic lateral forces not less than those given by Section 1628, using R_w from Table 16-P, and with $C/R_w \geq 0.4$ (the corresponding minimum value of C/R_w was 0.5 in 1991 UBC). The vertical distribution of the lateral seismic forces may be determined by using the provisions of Section 1628.4 or by using the procedures of Section 1629.

7. In Table 16-J, new or revised descriptions of soil profiles S_1, S_2, and S_3 have been provided, with the intent that commonly encountered soil profiles would fit better into one of the listed categories.

8. In Table 16-K, occupancy categories have been redefined according to code-defined occupancies in Chapter 3. Three importance factors for each occupancy category have been listed: seismic importance factor I for

the entire structure, seismic importance factor I_p for elements, components and equipment, and wind importance factor I_w.

9. The underlining in Table B-1 indicates changes in UBC Table 16-N, Structural Systems, from the 1991 to the 1994 edition of the code. It should also be noted that four systems have been redefined or newly defined in Section 1625: Dual System, Moment-Resisting Wall Frame (masonry), Ordinary Braced Frame (steel), and Special Eccentrically Braced Frame (steel).

10. The 1994 UBC seismic zone map is somewhat different from the 1991 map. The notable changes from 1991 are as follows:

   a. Parts of southern Arizona, including the city of Tucson, previously in Zone 1, are now in Zone 2A.

   b. Parts of the island of Oahu, including the city of Honolulu, previously in Zone 1, are now in Zone 2A.

   c. Western parts of the state of Oregon, previously in Zone 2B, are now in Zone 3.

   d. Western parts of the state of Washington, previously in Zone 2B, are now in Zone 3.

   e. The city of San Diego, previously in Zone 3, is now in Zone 4.

The impacts of c and d above on the design and construction of concrete structures are particularly severe, because of stringent high seismic zone detailing requirements.

Table B-1 Structural Systems
(adapted from Table 16-N of UBC-94, underlining indicates changes from UBC-91)

Basic Structural System	Lateral-Force-Resisting System		R_w	H (ft)
1. Bearing wall system	1.	Light-framed walls with shear panels		
		a. Wood structural panel walls for structures 3 stories or less	8	65
2. Building frame system	2.	Light-framed walls with shear panels		
		a. Wood structural panel walls for structures 3 stories or less	9	65
	4.	Ordinary braced frames		
		a. Steel	8	160
		b. Concrete	8	—
		c. Heavy timber	8	65
	5.	Special concentrically braced frames		
		a. Steel	9	240
3. Moment-resisting frame system	2.	Masonry moment-resisting wall frame	9	160
	4.	Ordinary moment-resisting frames (OMRF)		
		a. Steel[6]	6	160
		b. Concrete[7,8]	5	—
4. Dual System	3.	Ordinary braced frames		
		a. Steel with steel SMRF	10	NL
		b. Steel with steel OMRF	6	160
		c. Concrete with concrete SMRF	9	—
		d. Concrete with concrete IMRF	6	—
	4.	Special concentrically braced frames		
		a. Steel with steel SMRF	11	NL
		b. Steel with steel OMRF	6	160

[6]Ordinary moment-resisting frames in Seismic Zone 1 meeting the requirements of Section 2211.6 may use $R_w = 12$.
[7]Prohibited in Seismic Zones 2, 3, and 4
[8]Prohibited in Seismic Zones 2A, 2B, 3, and 4. See Section 1631.2.7

1-B.3 Changes in Concrete Related Provisions

The following changes from 1991 UBC are significant:

1. The entire Chapter 26 of UBC-91 has been reviewed line-by-line with the source document, ACI 318-89 (Revised 1992), and revised editorially to make the same code provisions as identical as practicable. Where Chapter 26 differed technically from the ACI 318 Standard, or where subsequent changes have been made through the code change process, the differing provisions remain in UBC-94 Chapter 19, and are shown in italics. For easy cross-reference, UBC-94 Chapter 19 has been reformatted, so that section numbering corresponds to that of ACI 318-89 (Revised 1992). For example, Section 1912.3.1 of UBC-94 corresponds to Section 12.3.1 of ACI 318-89 (Revised 1992).

2. A new section 1903.10 in UBC-94 requires that the welding of reinforcing steel, metal inserts and connections in reinforced concrete conform to UBC Standard 19-2. The Standard requires that, subject to certain amendments made in the Standard, the welding of concrete reinforcing steel for splices (prestressing steel excepted), steel connection devices, inserts, anchors, and anchorage details, as well as any other welding required in reinforced concrete construction, be in accordance with the *Structural Welding Code-Reinforcing Steel*, ANSI/AWS D1.4-92, published by the American Welding Society.

3. Several changes have been made in Section 1907.10.5, requirements for ties in compression members. Section 1907.10.5.1 is identical to UBC-91 Section 2607(k)3A. Section 1907.10.5.2 is much briefer than Section 2607(k)3B because Seismic Zone 2, 3, 4 requirements (which more properly belong in Section 1921) have been deleted. The Zone 3, 4 requirements were less stringent than those in Section 1921.4.4, and never governed. The Zone 2 requirements were the same as those in Section 1921.8.5.1. Section 1907.10.5.3 is the same as previous Section 2607(k)3C. Section 2607(k)3D has been deleted. The seismic hook requirement for Zones 3 and 4 is covered by cross-referencing in Section 1907.1.3 Item 4. Requirements for additional ties around bolts for Zones 3, 4 and for Zone 2 are to be found in Sections 1921.4.4.2 and 1921.8.5.5, respectively. Sections 1907.10.5.4 and 1907.10.5.5 are identical to Sections 2607(k)3E and 2607(k)3F, respectively. Section 1907.10.5.6 is the same as the first sentence of previous (deleted) section 2607(k)3D.

4. Section 1912.5.3.6 prescribes that the required development length of a hooked epoxy-coated bar must be 1.2 times that of a hooked uncoated bar. This is a feature of ACI 318-95 that has already been incorporated into UBC-94. A similar provision for straight epoxy-coated bars is contained in UBC-91 and ACI 318-89 (Revised 1992)

5. A new Section 1912.14.3.6 requires that welded splices and mechanical connections shall maintain the clearance and cover requirements of Sections 1907.6 and 1907.7. Another new Section 1912.15.6 states that mechanical connections need not be staggered as required...provided the clearance and cover requirements of Sections 1907.6 and 1907.7 are maintained and, "at 90 percent of the yield stress, the strain measured over the full length of the connector does not exceed 50 percent of the strain of an unspliced bar when the maximum computed design load stress does not exceed 50 percent of the yield stress." The strain control intended by the requirement within quotes is unclear, making this provision difficult to implement until further clarification is made in the code. It should be noted that staggering of welded and mechanical splices is still required by Section 1921.2.6.1 for Zone 3 and 4 applications.

6. Section 2615(i)3B of UBC-91 required that "Connection between precast wall and supporting member shall have a tensile strength not less than 50 A_g, in pounds, where A_g is cross-sectional area of wall." This requirement was felt to be excessive and was controversial. It has now been replaced with Section 1915.8.3.2

of UBC-94, which requires that precast wall element base connections shall have a tensile strength of not less than calculated forces. In tilt-up construction, this connection may be to an adjacent floor slab, designed for the calculated forces but not less than 200 pounds per linear foot of wall.

7. Section 1921.2.1.4 of UBC-94 clearly spells out that the requirements of 1921.2 through 1921.7 (should be 1921.6) apply to structural reinforced concrete members <u>that are part of the lateral force-resisting system</u> in Seismic Zones 3 and 4. 1921.7 is a misprint and needs to be corrected. As stated in Section 1921.2.2.4, Section 1921.7 applies to structural members assumed not to be part of the lateral force-resisting system.

8. In Section 1921.2.4.1, a new exception has been included. Footings of buildings three stories or less may have concrete with f'_c less than 3000 psi, but not less than 2500 psi.

9. Section 2625(f)2D of UBC-91 contained the requirement: "Splices in horizontal reinforcement shall be staggered. Splices in two curtains where used shall not occur in the same location." In UBC-94, staggering of splices of horizontal shearwall reinforcement is required only "Where necessary to avoid congestion." This is a significant change brought about by the recognition that earthquake-induced cracking of shearwalls is invariably diagonal, rather than vertical, so that concerns over a vertical plane of weakness created by non-staggered splices need not exist.

10. Design provisions for Zone 3 and 4 shearwalls subject to combined bending and axial loads have been made much more sensible and rational in UBC-94 (Section 1921.6.5). According to UBC-91 provisions (Section 2625(f)3), adopted from ACI 318, a shearwall must be provided with boundary elements at the ends, if the extreme compression fiber stress caused by the entire factored axial load tributary to the wall and the entire factored overturning moment at the base of the wall exceeds 0.2 f'_c. The stresses are calculated using a linearly elastic model of the shearwall and gross section properties. Boundary elements, where required, must be designed to carry the entire factored axial load tributary to the wall, and the entire factored moment at the base of the wall by tension and compression in the boundary elements. In other words, the boundary elements are required to be designed completely ignoring the fact that the shearwall web even exists. This gives rise to oversized boundary elements typically containing a large number of longitudinal reinforcing bars. The code further requires that these bars be tied and cross-tied at a maximum spacing of 4 in. (100 mm), with every other bar laterally supported by a corner of a tie or crosstie. All these requirements result in boundary elements that are not only unnecessarily uneconomical, but are very difficult to construct. The result has been that buildings that could benefit from inclusion of shearwalls are being constructed without them. Also, the oversized boundary elements make the shearwalls over-strong in flexure, thereby potentially attracting larger shear forces to them in actual earthquakes than what the walls are designed for, making premature shear failure more likely.

Wood [B.1], in research carried out at the University of Illinois, examined the effects of boundary element transverse reinforcement amounts lower and higher than that required by ACI 318 on the deformability of shearwalls. It was found that higher confinement translated into higher deformability as long as a wall failed in flexure. However, the deformability of a wall failing in shear was independent of such confinement.

Wood [B.2], and later Wallace and Moehle [B.3], also determined that the ACI/UBC boundary element confinement requirements are excessive, except when the ratio of shearwall cross-sectional area to floor plan area is very low (on the order of half a percent).

The new provisions have the following primary features:

1. A shearwall is designed for flexure and axial load considering the entire cross-section, including web(s), to be effective, as in a short column. Shear resistance is still provided by the web, without any contribution from the overhanging flanges.

2. Wall is screened to eliminate cases where special boundary zone detailing is not required. Walls having $P_u \leq 0.10 A_g f'_c$ and either $M_u/V_u \ell_w \leq 1.0$ or $V_u \leq 3 \ell_w h \sqrt{f'_c}$ and $M_u/V_u \ell_w \leq 3.0$ are exempt. Walls with $P_u > 0.35 P_o$ are not permitted to resist earthquake-induced forces.

3. Two options are provided for cases where boundary elements with special details are needed: (a) Conservative approach: provide boundary elements over $0.25 \ell_w$ at each end; (b) Alternatively, determine compressive strains at wall edges when wall is subject to design earthquake displacements, using cracked section properties. Provide confinement wherever compressive strain exceeds 0.003.

The detailing requirements for confinement reinforcement, when needed, are much less stringent than in ACI 318-89 (Revised 1992) and UBC-91. The maximum spacing is 6 in. (150 mm) or 6 times the longitudinal bar diameter, whichever is smaller, rather than 4 in. (100 mm).

11. The deformation compatibility requirement of Section 1921.7 for frame members that are not part of the lateral force resisting system has been rewritten to bring it in line with the corresponding requirement in ACI 318-89 (Revised 1992). The requirements of Section 1921.7.1 are summarized in Fig. B-1(a). Frame members that are designated not to be part of the lateral force resisting system must be detailed according to Section 1921.7.1.1 or 1921.7.1.2, depending on the magnitude of moments induced in those members when subjected to $3(R_W/8)$ times the displacements resulting from the specified lateral forces, in conformance with the requirements of Section 1631.2.4. It should be noted that these elastically computed moments must be combined with unfactored gravity load effects before comparisons are made with design moment strengths. If the design moment strength of a member is not exceeded, the behavior of that non-lateral-force-resisting member is elastic under the estimated design earthquake displacements, and then only the minimum and maximum longitudinal reinforcement requirements of Section 1921.3.2.1 apply (this section applies to beams or flexural members only, i.e. to members with $P_u \leq A_g f'_c/10$). If the design moment strength is exceeded, the non-lateral-force-resisting member behavior under the estimated design earthquake displacements is inelastic; then, depending on whether the member is a beam ($P_u \leq A_g f'_c/10$) or a column ($P_u > A_g f'_c/10$), different detailing requirements apply. Beams must conform with Section 1921.3.2.1 mentioned above, and also with Section 1921.3.4.1 which requires a beam to be designed for a shear force that is computed assuming that moments of opposite sign corresponding to probable strength M_{pr} act at the joint faces and that the beam is loaded with tributary gravity load along its span. Columns must conform with Section 1921.4.4 (transverse reinforcement) and also with Sections 1921.4.5.1 and 1921.4.5.2. Section 1921.4.5.1 is comparable to Section 1921.3.4.1, and requires that a column be designed for the maximum shear force that can reasonably be induced in it. Section 1921.4.5.2 is meaningless because it applies to members with $P_u \leq A_g f'_c/20$, and columns by definition are subject to $P_u > A_g f'_c/10$.

Section 1921.7.2 requires that all columns that are designated not to be part of the lateral force resisting system conform to the transverse reinforcement requirements of Sections 1921.7.2.1 through 1921.7.2.4, unless they conform to the more stringent transverse reinforcement requirements of Section 1921.4.4. Figure B-1(b) presents a summary of the combined deformation compatibility requirements of Section 1921.7.1 and 1921.7.2.

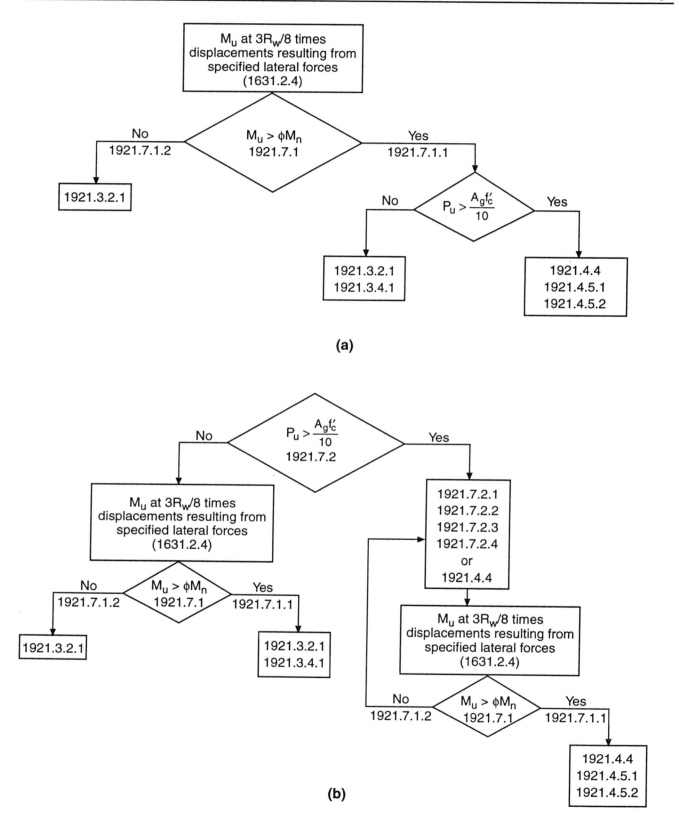

Figure B-1 Provisions for Frame Members Not Part of the Lateral Force Resisting System

12. Section 1923 of UBC-94 contains new design provisions for structural plain concrete that are patterned after ACI 318.1-89, *Building Code Requirements for Plain Concrete.* The limitations on the use of plain concrete, as given in Section 1923.1.2, should be noted. Also to be noted are the seismic requirements of Section 1923.10. Structural plain concrete members are not permitted in buildings in Seismic Zones 2, 3, and 4, with the exception of: (1) footings of dwellings (Group R, Division 3 occupancy) and private garages (Group U, Division 1 occupancy), and (2) nonstructural slabs supported directly on the ground or by approved structural systems.

13. Significant changes have been made in Table 19-E, which lists allowable service loads on embedded bolts, from the corresponding Table 26-E of UBC-91. The table is part of Section 1925.1, Service Load Design for Anchorage to Concrete (Section 2624(a) of UBC-91). The allowable service loads have been reviewed, and revised in some cases, based on new test data. Some allowable values have been reduced. Some values, for larger bolt sizes and higher concrete strengths, have been increased. Allowable values in both tension and shear are now specifically given for three different concrete strengths: 2000, 3000, and 4000 psi. Also, minimum edge distance and maximum spacing requirements have been added.

14. Sections 2624(b) and (c) of UBC-91 have been editorially revised and rearranged into Sections 1925.3.2, 1925.3.3, and 1925.3.4 of UBC-94, to clarify strength design of anchorage to concrete. Sections 1925.3.2, 1925.3.3, and 1925.3.4 address design strengths in (1) tension, (2) shear, and (3) combined tension and shear, respectively. Section 1925.3.4 now restricts the design strength in combined tension and shear to be no greater than the design strength in direct tension or the design strength in direct shear (see Fig. B-2).

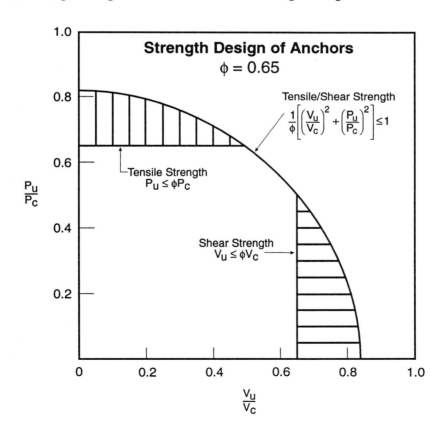

Figure B-2 Strength Interaction Diagram for Anchor Bolts Subjected to Combined Tension and Shear

References

[B.1]   Wood, S. L. 1991a. "Observed Behavior of Slender Reinforced Concrete Walls Subjected to Cyclic Loading." Publication SP-127. American Concrete Institute, Detroit, Michigan. 453-477.

[B.2]   Wood, S. L. 1991b. "Performance of Reinforced Concrete Buildings During the 1985 Chile Earthquake." Implications for the Design of Structural Walls. *Earthquake Spectra*. EERI. Volume 7, No. 4, 607-638.

[B.3]   Wallace, J. W., Moehle, J. P. 1992. "Ductility and Detailing Requirements of Bearing Wall Buildings." *Journal of Structural Engineering*. ASCE. Vol. 118, No. 6, 1625-1644.

Chapter 2

Flat Plate-Column Frame

2.1   Introduction

2.1.1   General

Reinforced concrete slabs provide a versatile and economical means of supporting gravity loads and form an integral portion of the structural frame to resist lateral loads. Slabs supported by isolated columns arranged in regular rows that permit the slab to deflect in two orthogonal directions are known as two-way slab systems. A two-way slab system may have beams along all column lines, beams along discontinuous edges only, drop panels, column capitals, or, none of the mentioned elements, in which case it is referred to as a flat plate.

The conventionally reinforced flat plate without post-tensioning is generally 5 in. to 11 in. thick, with spans ranging from 15 ft to 30 ft. It has been the most widely used slab system in reinforced concrete since the early 1950s with the majority of its applications being in residential multi-story construction. Flat plates are economical because they can be constructed with minimum field labor, due to the use of simple formwork and reinforcement arrangements. Additionally, they require the least story height for a specified headroom, thus cutting the cost of exterior cladding, wiring, and HVAC ducts. In many cases, if a ceiling finish is used, it can be applied directly to the slab soffit, thereby, eliminating costly false ceilings.

2.1.2   Design Criteria

The design of a 14-story flat plate building subjected to gravity, wind, and seismic loads will be illustrated. A typical plan and elevation of the structure are shown in Figs. 2-1 and 2-2. The column size and the slab thickness are assumed to be the same throughout the building. Also, the columns at the base of the building are assumed to be fixed.

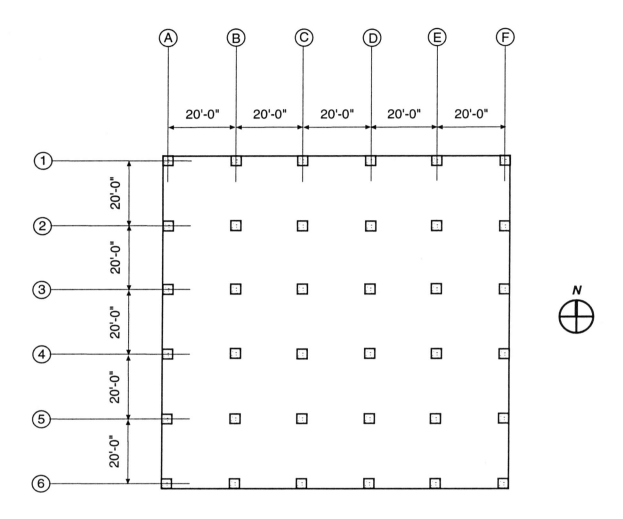

Figure 2-1 Typical Floor Plan of Example Building

The bottom story height is 12.5 ft and all other stories are 9.5 ft high. Other design data are as follows:

Service loads:
    Live load: 40 psf
    Superimposed dead load: 20 psf

Material properties:
    Concrete: f'_c = 4 ksi (5 ksi for columns below the 4th floor)
                         w_c = 150 pcf
    Reinforcement: f_y = 60 ksi

Preliminary Sizes:
    Slab: 8.5 in.
    Columns: 18 × 18 in.

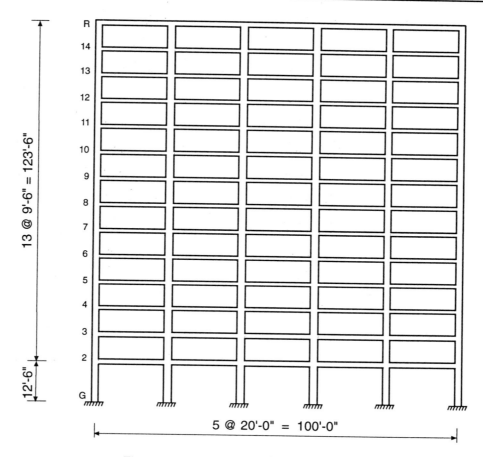

Figure 2-2 Elevation of Example Building

The slab and columns will be designed so the structure will carry, in addition to its own weight, all live loads, superimposed dead loads, and lateral loads. Both wind loads and seismic loads will be imposed on the structure. Loads will be determined in accordance with the 1994 edition of the Uniform Building Code (UBC). Code references from the UBC will be given. Corresponding sections in ACI 318-89 (Revised 1992) can be easily determined from the UBC sections; for example, ACI 318 Sect. 9.5.3 corresponds to UBC Sect. 1909.5.3.

2.1.3 Outline of Solution

Considering the least permissible slab thickness prescribed by the UBC, which is based on deflection control and other serviceability requirements, a minimum slab thickness is determined. Preliminary design under gravity loads addressing shear strength requirements is then carried out to determine the adequacy of the chosen thickness. Lateral loads, due to either wind or seismic excitation, are then imposed on the structure, and design under combined gravity and lateral loads is performed.

When subjected to lateral loads, flat plate structures behave similarly to rigid frames (i.e., frames comprising of beams and columns). Structures consisting of flat plates and columns are thus analyzed as frames by considering slab strips spanning between the columns as beams. The effective slab width is a function of the relative dimensions of the column section with respect to the longitudinal and transverse spans of the slab and the distribution of reinforcement in the slab. ACI 318 Commentary Sect. R13.3.1.2 suggests using one-half to one-fourth of the slab panel width (between adjoining bay center lines) as the effective slab width. The structural systems of the example building are identical in the N-S and E-W directions. Therefore, only the N-S direction is considered here. A three-dimensional analysis of the building is performed for the lateral load cases using *PCA-Frame* [2.1], which is a general purpose structural analysis program for two- and three-dimensional structures subjected to static loads. In the model, rigid diaphragms are assigned at each floor level, and rigid-end offsets are assigned at the ends of each horizontal member so that results are automatically obtained at the faces of each support. An effective width of one-third the bay width for an interior frame and one-sixth the bay width for an exterior frame were used. It will be shown that the internal forces in the columns and slabs do not vary much as the effective width of an interior slab varies from one-fourth to one-half of the bay width.

For comparison purposes, a two-dimensional analysis of the building is also performed. The model consists of two different frames, F-1 and F-2, coupled together by hinged rigid links at all floor levels, which impose equal horizontal displacements on the two frames (see Fig. 2-3). Frame F-1 represents the lumping of exterior slab-column frames, and F-2 represents the lumping of interior slab-column frames. Lumping is achieved for the columns by adding their areas and moments of inertia at the same column line; for the slabs, the moments of inertia are added at the same floor level. In the 2-D analysis, the same effective slab width was used as in the 3-D analysis.

The following sections present the design of the example 14-story reinforced concrete flat plate building subjected to gravity and lateral (wind and seismic) loads. In all cases, the results from the 3-D analysis with an effective slab width equal to one-third the bay width were used for the design of the members.

2.2   Design for Gravity and Wind Loads

2.2.1   Preliminary Slab Thickness

Assuming 18-in. square columns, a minimum slab thickness is determined. The minimum overall thickness, h, for slab systems without edge and interior beams is governed by the provisions of UBC Table 19-C-2:

$$h = \frac{\ell_n}{30} \hspace{4cm} \textit{1909.5.3.2}$$

where ℓ_n is the longest clear span length in inches.

For the building shown in Fig. 2-1,

$$\ell_n = (20 \times 12) - 18 = 222 \text{ in.}$$

Therefore, minimum h = 222/30 = 7.4 in. > 5.0 in.   O.K.   *1909.5.3.2*

However, preliminary calculations show that in order to satisfy strength requirements for combined loading, an 8.5 in. thickness is required.

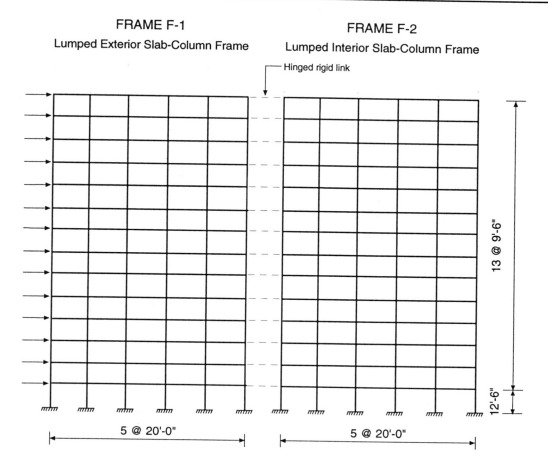

Figure 2-3 Model of the 2-D Lateral Load Resisting System

2.2.2   Shear Strength of Slab Subjected to Gravity Loads

For a slab thickness of 8.5 in., the effective depth d is taken as 7.125 in. (3/4 in. cover and #5 bars). The design loads are:

Factored slab weight = 1.4 × 150 × (8.5/12) = 149 psf   *1909.2.1*

Factored superimposed dead load = 1.4 × 20 = 28 psf

Factored live load = 1.7 × 40 = 68 psf

Total factored load w_u = 149 + 28 + 68 = 245 psf = 0.245 ksf

Shear caused by wide beam action must be checked. Investigation of wide beam action is made at a distance d from the face of the support in either direction, as shown in Fig. 2-4.

$$V_u = 0.245 \times \left(10 - \frac{9}{12} - \frac{7.125}{12}\right) \times 20 = 42.4 \text{ kips}$$ *1911.1.3.1*

$$V_c \quad = 2\sqrt{f'_c} \; b_w d = 2\sqrt{4000} \times 240 \times 7.125/1000 = 216.3 \text{ kips}$$

1911.3.1.1

$$\phi V_c = 0.85 \times 216.3 = 183.9 \text{ kips} \; > V_u \quad \text{O.K.}$$

1909.3

Two-way shear must also be checked. Since there are no shear forces at the centerlines of adjacent panels, the factored shear force at a distance d/2 around the support is (see Fig. 2-4):

$$V_u \quad = 0.245 \times \left[20^2 - \frac{(18+7.125)^2}{144} \right] = 96.9 \text{ kips}$$

1911.12.1.2

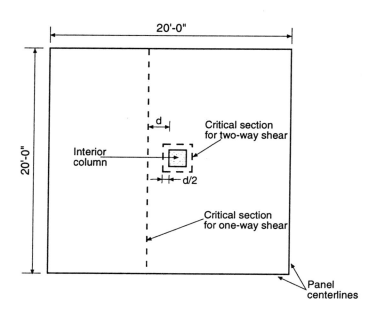

Figure 2-4 Critical Sections for Shear

For square columns:

$$V_c \quad = 4\sqrt{f'_c} \; b_o d$$

1911.12.2.1

$$= 4\sqrt{4000} \times 4 \times (18+7.125) \times 7.125/1000 = 181.2 \text{ kips}$$

$$\phi V_c = 0.85 \times 181.2 = 154.0 \text{ kips} \; > V_u \quad \text{O.K.}$$

1909.3

Therefore, preliminary design indicates that an 8.5-in. thick slab is adequate for control of deflection and shear strength. It should be noted that the check for shear strength is only preliminary at this stage. The shear stress will be higher than that indicated by the above calculations because of added shear caused by unbalanced moment transfer from the slab to the columns. A more refined check for adequacy of shear strength is made at a later stage.

2.2.3   Applicability of Direct Design Method for Gravity Load Analysis

Slab systems can be designed by using relatively exact analyses such as finite element computer programs, or, approximate analyses, such as the Direct Design Method (UBC 1913.6.1). The Direct Design Method, which is utilized in this publication, is an approximate design procedure based on theoretical analysis, construction procedures, and performance history of slab systems. This method can be used only if certain geometric and loading criteria are met. All the required criteria were met, as indicated below:

- There shall be a minimum of three continuous spans.
  Structure has five continuous spans.

- Panels shall be rectangular with a longer span to shorter span length ratio less than 2.0.
  Bays for this structure are square and have a long-to-short span ratio of 1.0, which is less than 2.0.

- Successive span lengths shall not differ by more than one-third the longer span.
  Span lengths for this structure are all equal.

- Columns may be offset a maximum of 10% of the span from either axis between centerlines of successive columns.
  Columns for this structure are all evenly spaced in both directions.

- All loads shall be due to gravity only and uniformly distributed over an entire panel. Live loads shall not exceed three times the dead load.
  This criterion is also satisfied.

Ratio of service dead-to-live load is $\beta_a = 126/40 = 3.2 > 2.0$; therefore, pattern loading effects may be neglected (UBC 1913.6.10).

Thus, the Direct Design Method can be used for gravity load analysis, and pattern loading effects may be neglected.

2.2.4   Factored Loads and Moments due to Gravity

A design strip centered on an interior column line, and spanning in the N-S direction (the E-W direction is identical), as shown in Fig. 2-5, is considered. The 20-ft wide design strip is divided into a column strip and two-half middle strips. With equal span lengths in the direction of analysis as well as in the orthogonal direction, the column strip has a width equal to:

$$\ell_1/2 = \ell_2/2 = 20/2 = 10 \text{ ft}$$

1913.2

The middle strip, bounded by two column strips, is also 10 ft wide.

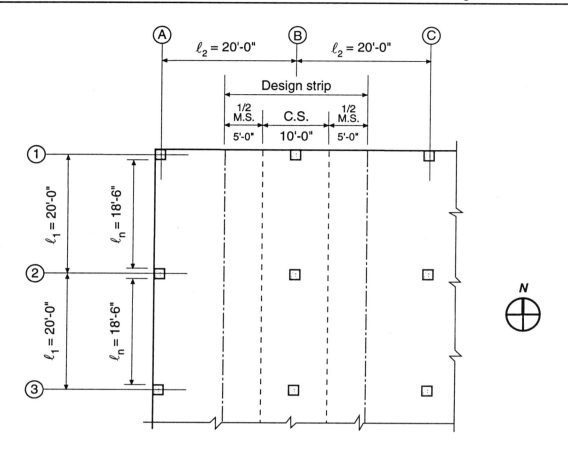

Figure 2-5 Interior Design Strip

2.2.4.1   Factored Moments in Slab

Total factored static moment:

$$M_o = \frac{w_u \ell_2 \ell_n^2}{8} \qquad\qquad 1913.6.2.2$$

$$= \frac{0.245 \times 20 \times 18.5^2}{8} = 209.6 \text{ ft-kips}$$

This moment must be divided into negative and positive moments, and then into column and middle strip moments according to the distribution factors given in UBC 1913.6.3, 1913.6.4, and 1913.6.6. The distributed moments for flat plates are given in Table 2-1.

Table 2-2 summarizes the moments in the column and the middle strips for an end and an interior span of the design strip considered.

Table 2-1 Design Moments for Flat Plate or Flat Slab Supported Directly on Columns

Slab Moments	END SPAN			INTERIOR SPAN	
	1 Exterior Negative	2 Positive	3 First Interior Negative	4 Positive	5 Interior Negative
Total Moment	$0.26\,M_o$	$0.52\,M_o$	$0.70\,M_o$	$0.35\,M_o$	$0.65\,M_o$
Column Strip	$0.26\,M_o$	$0.31\,M_o$	$0.53\,M_o$	$0.21\,M_o$	$0.49\,M_o$
Middle Strip	0	$0.21\,M_o$	$0.17\,M_o$	$0.14\,M_o$	$0.16\,M_o$

Note: All negative moments are at face of support.

Table 2-2 Factored Slab Moments due to Gravity Loads

Span Location		M_u (ft-kips)
END SPAN		
Column Strip	Exterior Negative	$0.26\,M_o = \quad 54.5$
	Positive	$0.31\,M_o = \quad 65.0$
	Interior Negative	$0.53\,M_o = \ 111.1$
Middle Strip	Exterior Negative	0.0
	Positive	$0.21\,M_o = \quad 44.0$
	Interior Negative	$0.17\,M_o = \quad 35.6$
INTERIOR SPAN		
Column Strip	Positive	$0.21\,M_o = \quad 44.0$
	Negative	$0.49\,M_o = \ 102.7$
Middle Strip	Positive	$0.14\,M_o = \quad 29.3$
	Negative	$0.16\,M_o = \quad 33.5$

2.2.4.2   Factored Axial Loads and Moments in Columns

Design axial loads due to gravity on an interior column and an exterior column of a typical interior frame are shown in Tables 2-3 and 2-4, respectively. The live load reduction factor is in accordance with Eq. (6-1) in UBC 1606.

The factored moment due to gravity to be transferred from the slab (with equal spans of adjacent slab panels in the direction of analysis as well as in the orthogonal direction) to the interior columns above and below the slab is given by:

$$M_i = 0.07 \times (0.5 \times w_L \times \ell_2 \times \ell_n^2) \qquad \textit{1913.6.9.2}$$

$$= 0.07 \times (0.5 \times 1.7 \times 0.04 \times 20 \times 18.5^2) = 16.3 \text{ ft-kips}$$

Each interior column at the roof level is subjected to this entire moment at the top of the column. In general, the 16.3 ft-kip moment will be distributed to the columns inversely proportional to their lengths above and below the floor level under consideration. With the same column size and length above and below the slab:

$$M_u = 16.3/2 = 8.2 \text{ ft-kips}$$

Table 2-3 Axial Loads due to Gravity on an 18-in. Square Interior Column of an Interior Frame

Floor	Dead Load (psf)	Live Load (psf)	Area, A (ft²)	Reduction Factor (1-R/100)*	Reduced Live Load (psf)	Factored Load (kips)**	Cumulative Factored Load (kips)
R	126	40	400	0.80	32.0	97	97
14	126	40	800	0.48	19.2	89	186
13	126	40	1200	0.40	16.0	86	272
12	126	40	1600	0.40	16.0	86	358
11	126	40	2000	0.40	16.0	86	444
10	126	40	2400	0.40	16.0	86	530
9	126	40	2800	0.40	16.0	86	616
8	126	40	3200	0.40	16.0	86	702
7	126	40	3600	0.40	16.0	86	788
6	126	40	4000	0.40	16.0	86	874
5	126	40	4400	0.40	16.0	86	960
4	126	40	4800	0.40	16.0	86	1046
3	126	40	5200	0.40	16.0	86	1132
2	126	40	5600	0.40	16.0	88	1220

* $R = r(A - 150)$
** Factored load = [1.4(Dead Load) + 1.7(Reduced Live Load)] $\times$ Tributary Area. Factored load includes
  1.4 $\times$ column weight = 0.5 kips/ft

Table 2-4 Axial Loads due to Gravity on an 18-in. Square Exterior Column of an Interior Frame

Floor	Dead Load (psf)	Live Load (psf)	Area, A (ft²)	Reduction Factor (1-R/100)*	Reduced Live Load (psf)	Factored Load (kips)**	Cumulative Factored Load (kips)
R	126	40	200	0.96	38.4	53	53
14	126	40	400	0.80	32.0	51	104
13	126	40	600	0.64	25.6	49	153
12	126	40	800	0.48	19.2	47	200
11	126	40	1000	0.40	16.0	46	246
10	126	40	1200	0.40	16.0	46	292
9	126	40	1400	0.40	16.0	46	338
8	126	40	1600	0.40	16.0	46	384
7	126	40	1800	0.40	16.0	46	430
6	126	40	2000	0.40	16.0	46	476
5	126	40	2200	0.40	16.0	46	522
4	126	40	2400	0.40	16.0	46	568
3	126	40	2600	0.40	16.0	46	614
2	126	40	2800	0.40	16.0	47	661

* $R = r(A - 150)$
** Factored load = [1.4(Dead Load) + 1.7(Reduced Live Load)] $\times$ Tributary Area. Factored load includes
  1.4 $\times$ column weight = 0.5 kips/ft

For the columns below the 2nd floor level,

$$M_u = \frac{9.5}{9.5 + 12.5}(16.3) = 7.0 \text{ ft-kips}$$

The columns above this level will be subjected to 16.3 - 7.0 = 9.3 ft-kips.

The total exterior negative moment $M_u = 54.5$ ft-kips (see Table 2-2) from the slab must be transferred directly to the exterior column. Similar to the interior columns, the exterior columns will carry their portion of this moment inversely proportional to their lengths.

2.2.5 Wind Load Analysis

Table 2-5 shows the wind loads in the N-S direction computed according to UBC 1618 with the basic wind speed = 70 mph, Exposure B, and an importance factor = 1.0. Three-dimensional elastic analyses of the lateral load resisting system under wind loads, considering different effective slab widths of one-half, one-third, and one-fourth the slab panel width (between adjoining bay center lines) were carried out. The resulting internal forces at the faces of the supports (bending moments, shear forces, and axial forces) in the structural members of an interior frame are shown in Figs. 2-6 and 2-7.

Table 2-5 Wind Forces on Example Building

Floor Level	Height (ft)	Tributary Height (ft)	Windward			Leeward			Total Force (kips)
			C_e	Pressure (psf)	Force (kips)	C_e	Pressure (psf)	Force (kips)	
R	136.0	4.75	1.24	12.5	6.0	1.24	7.8	3.7	9.7
14	126.5	9.50	1.22	12.3	11.7	1.24	7.8	7.4	19.1
13	117.0	9.50	1.19	12.0	11.4	1.24	7.8	7.4	18.8
12	107.5	9.50	1.16	11.7	11.1	1.24	7.8	7.4	18.5
11	98.0	9.50	1.12	11.3	10.7	1.24	7.8	7.4	18.1
10	88.5	9.50	1.08	10.9	10.3	1.24	7.8	7.4	17.7
9	79.0	9.50	1.04	10.5	10.0	1.24	7.8	7.4	17.4
8	69.5	9.50	0.99	10.0	9.5	1.24	7.8	7.4	16.9
7	60.0	9.50	0.95	9.6	9.1	1.24	7.8	7.4	16.5
6	50.5	9.50	0.90	9.1	8.6	1.24	7.8	7.4	16.0
5	41.0	9.50	0.84	8.5	8.0	1.24	7.8	7.4	15.4
4	31.5	9.50	0.77	7.8	7.4	1.24	7.8	7.4	14.8
3	22.0	9.50	0.69	7.0	6.6	1.24	7.8	7.4	14.0
2	12.5	11.00	0.62	6.3	6.9	1.24	7.8	8.6	15.5

The internal forces in the slabs and columns do not vary much with the change of the effective slab width within the range considered. This indicates that an average effective slab width of one-third the slab panel width (between adjoining bay center lines) is reasonable and satisfactory for interior frames, with one-half of this being the effective width for exterior frames.

For comparison purposes, a two-dimensional analysis of the building was performed based on the criteria given in Section 2.1.3. Figures 2-8 and 2-9 give the ratios of the 2-D to 3-D results for the columns and slabs in an interior frame with an effective slab width equal to one-third of the bay width. It can be seen from the figures that the 2-D analysis results are consistently larger than the 3-D results, but not by much.

Effective Slab Width/Bay Width

	$\frac{1}{2}$	$\frac{1}{3}$	$\frac{1}{4}$	$\frac{1}{2}$	$\frac{1}{3}$	$\frac{1}{4}$	$\frac{1}{2}$	$\frac{1}{3}$	$\frac{1}{4}$
R	6	11	15	28	35	40	45	48	51
	-14	-18	-22	7	3	-1	20	13	8
14	41	50	57	67	75	83	77	82	88
	4	-2	-8	37	31	24	47	38	30
13	60	68	77	104	113	121	114	120	127
	22	15	8	72	65	57	82	72	63
12	79	88	96	138	147	156	147	154	161
	42	34	26	107	100	92	116	106	97
11	97	105	113	173	181	190	181	187	195
	60	53	45	142	134	127	149	140	131
10	114	122	130	206	215	223	213	220	227
	79	71	63	176	169	161	182	174	165
9	132	139	147	238	247	255	244	251	259
	97	89	81	209	202	195	214	206	198
8	149	155	162	270	278	286	274	282	289
	115	107	99	242	235	228	246	238	231
7	165	171	178	301	309	317	304	311	318
	133	125	117	274	267	261	276	269	262
6	181	186	192	331	338	345	332	339	346
	151	142	135	305	299	293	306	300	294
5	198	201	205	359	365	370	358	365	369
	171	164	159	334	330	326	333	329	325
4	209	208	206	387	388	386	385	387	385
	191	189	191	362	362	363	361	361	362
3	212	200	185	407	391	371	402	388	369
	189	207	231	413	424	438	404	418	434
2	231	174	123	360	296	239	354	292	236
	597	660	716	661	721	774	658	719	773

Bending Moments (in.-kips)

Figure 2-6 Results of 3-D Analysis under Wind Forces for Columns of Interior Frame

Effective Slab Width/Bay Width

R	1/2	1/3	1/4	1/2	1/3	1/4	1/2	1/3	1/4
R	-0.1	-0.1	-0.1	0.3	0.3	0.4	0.6	0.6	0.5
14	0.4	0.4	0.4	0.9	0.9	0.9	1.1	1.1	1.0
13	0.7	0.7	0.7	1.5	1.6	1.6	1.7	1.7	1.7
12	1.1	1.1	1.1	2.2	2.2	2.2	2.3	2.3	2.3
11	1.4	1.4	1.4	2.8	2.8	2.8	2.9	2.9	2.9
10	1.7	1.7	1.7	3.4	3.4	3.4	3.5	3.5	3.5
9	2.0	2.0	2.0	3.9	4.0	4.0	4.0	4.0	4.0
8	2.3	2.3	2.3	4.5	4.5	4.5	4.6	4.6	4.6
7	2.6	2.6	2.6	5.0	5.1	5.1	5.1	5.1	5.1
6	2.9	2.9	2.9	5.6	5.6	5.6	5.6	5.6	5.6
5	3.2	3.2	3.2	6.1	6.1	6.1	6.1	6.1	6.1
4	3.5	3.5	3.5	6.6	6.6	6.6	6.5	6.6	6.6
3	3.5	3.6	3.7	7.2	7.2	7.1	7.1	7.1	7.0
2	5.5	5.6	5.6	6.8	6.8	6.8	6.7	6.7	6.7

Shear Forces (kips)

Effective Slab Width/Bay Width

R	1/2	1/3	1/4	1/2	1/3	1/4	1/2	1/3	1/4
R	0.1	0.1	0.1	0.2	0.1	0.1	0.0	0.0	0.0
14	0.1	0.3	0.4	0.5	0.3	0.3	0.0	0.0	0.0
13	0.7	0.9	1.0	0.6	0.5	0.4	0.0	0.0	0.0
12	1.6	1.8	2.0	0.8	0.6	0.5	0.0	0.0	0.0
11	2.8	3.1	3.3	0.9	0.7	0.6	0.0	0.0	0.0
10	4.5	4.8	5.0	1.0	0.8	0.6	0.0	0.0	0.0
9	6.4	6.8	7.0	1.1	0.9	0.7	0.0	0.0	0.0
8	8.8	9.1	9.4	1.2	0.9	0.7	0.0	0.0	0.0
7	11.4	11.8	12.1	1.2	0.9	0.7	0.0	0.0	0.0
6	14.4	14.8	15.1	1.2	0.9	0.7	0.0	0.0	0.0
5	17.8	18.2	18.4	1.1	0.8	0.7	0.0	0.0	0.0
4	21.4	21.8	22.0	1.0	0.8	0.6	0.0	0.0	0.0
3	25.3	25.6	25.7	0.8	0.7	0.6	0.0	0.0	0.0
2	29.4	29.3	29.2	0.5	0.5	0.4	0.0	0.0	0.0

Axial Forces (kips)

Figure 2-6 (continued)

Effective Slab Width/Bay Width

	$\frac{1}{2}$	$\frac{1}{3}$	$\frac{1}{4}$	$\frac{1}{4}$	$\frac{1}{3}$	$\frac{1}{2}$	$\frac{1}{2}$	$\frac{1}{3}$	$\frac{1}{4}$	$\frac{1}{4}$	$\frac{1}{3}$	$\frac{1}{2}$	$\frac{1}{2}$	$\frac{1}{3}$	$\frac{1}{4}$
R	-4	-11	-16	-14	-9	-1	-27	-28	-29	-28	-28	-25	-25	-28	-28
14	-28	-34	-38	-37	-32	-26	-54	-53	-53	-53	-53	-53	-54	-53	-53
13	-69	-72	-75	-72	-69	-64	-90	-88	-87	-87	-88	-88	-90	-88	-87
12	-110	-112	-114	-111	-108	-103	-127	-125	-124	-124	-125	-126	-127	-125	-125
11	-151	-153	-154	-149	-147	-143	-165	-163	-162	-162	-163	-164	-165	-163	-162
10	-192	-192	-193	-188	-185	-182	-201	-200	-199	-199	-200	-200	-201	-200	-199
9	-232	-232	-232	-225	-223	-220	-237	-236	-235	-235	-236	-236	-237	-236	-235
8	-271	-270	-269	-262	-261	-257	-271	-271	-270	-270	-271	-271	-272	-271	-271
7	-310	-308	-307	-299	-297	-294	-305	-305	-305	-305	-305	-305	-305	-305	-305
6	-348	-345	-342	-334	-333	-331	-337	-338	-337	-337	-338	-337	-338	-338	-338
5	-385	-380	-377	-367	-367	-366	-368	-369	-368	-368	-369	-368	-369	-369	368
4	-420	-412	-405	-395	-399	-401	-398	-397	-394	-394	-397	-398	-399	-397	-394
3	-447	-431	-417	-408	-419	-429	-423	-415	-406	-406	-415	-423	-424	-415	-406
2	-466	-423	-392	-380	-406	-439	-418	-393	-371	-371	-393	-420	-421	-394	-372

Bending Moments (in.-kips)

Effective Slab Width/Bay Width

	$\frac{1}{2}$	$\frac{1}{3}$	$\frac{1}{4}$	$\frac{1}{4}$	$\frac{1}{3}$	$\frac{1}{2}$	$\frac{1}{2}$	$\frac{1}{3}$	$\frac{1}{4}$	$\frac{1}{4}$	$\frac{1}{3}$	$\frac{1}{2}$	$\frac{1}{2}$	$\frac{1}{3}$	$\frac{1}{4}$
R	0.0	-0.1	-0.1	0.1	0.1	0.0	-0.2	-0.2	-0.3	0.3	0.2	0.2	-0.2	-0.2	-0.3
14	-0.2	-0.3	-0.3	0.3	0.3	0.2	-0.5	-0.5	-0.5	0.5	0.5	0.5	-0.5	-0.5	-0.5
13	-0.6	-0.6	-0.7	0.7	0.6	0.6	-0.8	-0.8	-0.8	0.8	0.8	0.8	-0.8	-0.8	-0.8
12	-1.0	-1.0	-1.0	1.0	1.0	1.0	-1.1	-1.1	-1.1	1.1	1.1	1.1	-1.1	-1.1	-1.1
11	-1.3	-1.4	-1.4	1.4	1.4	1.3	-1.5	-1.5	-1.5	1.5	1.5	1.5	-1.5	-1.5	-1.5
10	-1.7	-1.7	-1.7	1.7	1.7	1.7	-1.8	-1.8	-1.8	1.8	1.8	1.8	-1.8	-1.8	-1.8
9	-2.0	-2.1	-2.1	2.1	2.1	2.0	-2.1	-2.1	-2.1	2.1	2.1	2.1	-2.1	-2.1	-2.1
8	-2.4	-2.4	-2.4	2.4	2.4	2.4	-2.4	-2.4	-2.4	2.4	2.4	2.4	-2.4	-2.4	-2.4
7	-2.7	-2.7	-2.7	2.7	2.7	2.7	-2.8	-2.8	-2.8	2.8	2.8	2.8	-2.8	-2.8	-2.8
6	-3.1	-3.1	-3.1	3.1	3.1	3.1	-3.0	-3.0	-3.0	3.0	3.0	3.0	-3.0	-3.0	-3.0
5	-3.4	-3.4	-3.4	3.4	3.4	3.4	-3.3	-3.3	-3.3	3.3	3.3	3.3	-3.3	-3.3	-3.3
4	-3.7	-3.7	-3.6	3.6	3.7	3.7	-3.6	-3.6	-3.6	3.6	3.6	3.6	-3.6	-3.6	-3.6
3	-3.9	-3.8	-3.7	3.7	3.8	3.9	-3.8	-3.7	-3.7	3.7	3.7	3.8	-3.8	-3.7	-3.7
2	-4.1	-3.7	-3.5	3.5	3.7	4.1	-3.8	-3.5	-3.3	3.3	3.5	3.8	-3.8	-3.6	-3.4

Shear Forces (kips)

Figure 2-7 Results of 3-D Analysis under Wind Forces for Slabs of Interior Frame

Level	Col 1	Col 2	Col 3
R	1.18 / 1.11	1.20 / 1.52	1.25 / 1.38
14	1.16 / 1.50	1.19 / 1.19	1.20 / 1.21
13	1.19 / 1.20	1.19 / 1.23	1.21 / 1.22
12	1.19 / 1.24	1.20 / 1.22	1.21 / 1.23
11	1.18 / 1.21	1.19 / 1.21	1.20 / 1.21
10	1.19 / 1.20	1.19 / 1.20	1.20 / 1.20
9	1.19 / 1.20	1.19 / 1.20	1.20 / 1.20
8	1.20 / 1.20	1.21 / 1.20	1.21 / 1.21
7	1.21 / 1.20	1.21 / 1.21	1.21 / 1.21
6	1.22 / 1.21	1.21 / 1.21	1.22 / 1.21
5	1.21 / 1.19	1.22 / 1.22	1.22 / 1.22
4	1.24 / 1.20	1.21 / 1.20	1.20 / 1.20
3	1.25 / 1.20	1.25 / 1.22	1.24 / 1.22
2	1.26 / 1.11	1.25 / 1.12	1.24 / 1.12

Bending Moments

Level	Col 1	Col 2	Col 3
R	1.00	1.33	1.17
14	1.25	1.33	1.27
13	1.29	1.25	1.29
12	1.27	1.27	1.30
11	1.29	1.29	1.28
10	1.29	1.26	1.29
9	1.30	1.28	1.30
8	1.30	1.31	1.30
7	1.31	1.29	1.31
6	1.31	1.30	1.30
5	1.31	1.31	1.31
4	1.31	1.29	1.29
3	1.31	1.33	1.32
2	1.16	1.19	1.19

Shear Forces

Level	Col 1	Col 2	Col 3
R	1.00	1.00	1.00
14	1.00	1.33	1.00
13	1.22	1.00	1.00
12	1.22	1.17	1.00
11	1.19	1.14	1.00
10	1.15	1.13	1.00
9	1.13	1.11	1.00
8	1.13	1.22	1.00
7	1.13	1.22	1.00
6	1.12	1.22	1.00
5	1.12	1.25	1.00
4	1.11	1.13	1.00
3	1.12	1.14	1.00
2	1.12	1.20	1.00

Axial Forces

Figure 2-8 Ratio of 2-D to 3-D Results under Wind Forces for Columns of Interior Frame

R	1.09	1.07	1.10	1.04	1.03
14	1.09	1.10	1.10	1.09	1.10
13	1.09	1.09	1.09	1.08	1.09
12	1.10	1.09	1.10	1.09	1.10
11	1.09	1.09	1.09	1.09	1.09
10	1.09	1.09	1.08	1.08	1.08
9	1.08	1.08	1.08	1.08	1.08
8	1.09	1.08	1.08	1.08	1.08
7	1.09	1.09	1.09	1.09	1.09
6	1.09	1.09	1.09	1.09	1.09
5	1.10	1.09	1.09	1.09	1.09
4	1.10	1.09	1.06	1.09	1.09
3	1.11	1.11	1.10	1.10	1.10
2	1.13	1.12	1.11	1.11	1.11

Bending Moments

R	1.00	1.00	1.50	1.50	1.50
14	1.00	1.00	1.00	1.00	1.00
13	1.00	1.00	1.13	1.13	1.13
12	1.10	1.10	1.09	1.09	1.09
11	1.07	1.07	1.07	1.07	1.07
10	1.06	1.06	1.11	1.11	1.11
9	1.05	1.05	1.10	1.10	1.10
8	1.08	1.08	1.08	1.08	1.13
7	1.11	1.11	1.07	1.07	1.07
6	1.06	1.06	1.10	1.10	1.10
5	1.09	1.09	1.09	1.09	1.09
4	1.08	1.08	1.08	1.08	1.08
3	1.11	1.11	1.11	1.11	1.11
2	1.14	1.14	1.11	1.11	1.11

Shear Forces

Figure 2-9 Ratio of 2-D to 3-D Results under Wind Forces for Slabs of Interior Frame

2.2.6   Design for Combined Loading

2.2.6.1   Slab Design

The required slab reinforcement for an interior design strip at floor level 3 due to the combined effects of gravity and wind loads is summarized in Table 2-6.

Table 2-6 Required Slab Reinforcement at Floor Level 3 due to Combined Gravity and Wind Loads

Span Location		M_u (ft-kips)	b (in.)	d (in.)	A_s (in.^{2})	$A_{s(min)}$ (in.^{2})	No. of #4 bars	No. of #5 bars
END SPAN								
Column Strip	Ext. Negative	86.7	120	7.125	2.79	1.84	14	9
		-21.5	120	7.125	0.68	—	4**	3**
	Positive	65.0	120	7.125	2.11	1.84	11	8*
	Int. Negative	115.9	120	7.125	3.76	1.84	19	13
Middle Strip	Ext. Negative	0.0	120	7.125	0.0	1.84	10	8*
	Positive	44.0	120	7.125	1.43	1.84	10	8*
	Int. Negative	38.6	120	7.125	1.25	1.84	10	8*
INTERIOR SPAN								
Column Strip	Positive	44.0	120	7.125	1.43	1.84	10	8*
	Negative	109.7	120	7.125	3.59	1.84	18	12
Middle Strip	Positive	29.3	120	7.125	0.91	1.84	10	8*
	Negative	36.6	120	7.125	1.20	1.84	10	8*

*s_{max} = 2h = 17" < 18"; 120/17 = 7.1 spaces, say 8 bars
**No. of bottom bars to be anchored into the exterior column

Applicable load combinations were taken from UBC 1909.2. For the interior design strip, the column strip width and the middle strip width are 120 in., as shown in Fig. 2-5. The required flexural reinforcement was determined using the design aid given in Table 2-7.

Also, the minimum area of flexural reinforcement = 0.0018bh (UBC 1913.4.1). Note that the design for other floor levels would be similar to what follows for floor level 3.

Sample calculations for the interior negative moment in the column strip, end span:

- M_u = $1.4 M_D + 1.7 M_L$ *1909.2.1*

  M_u = $0.53 M_o = 111.1$ ft-kips   (see Table 2-2)

- M_u = $0.75 (1.4 M_D + 1.7 M_L \pm 1.7 M_W)$ *1909.2.2*

Total factored moment in the design strip (see Table 2-1 and Fig. 2-7):

M_u = $0.75 (0.7 M_o \pm 1.7 M_W)$ *1913.6.3.3*

     = $0.75 [(0.7 \times 209.6) \pm (1.7 \times 419/12)]$

     = 154.6 ft-kips, 65.5 ft-kips

75% of this moment is distributed to the column strip:

M_u = $0.75 \times 154.6 = 115.9$ ft-kips (governs)   *1913.6.4*

- M_u = $0.9 M_D \pm 1.3 M_W$ *1909.2.2*

  M_D = $0.70 \times 0.126 \times 20 \times 18.5^2/8 = 75.5$ ft-kips   *1913.6.3.3*

  M_u = $(0.9 \times 75.5) \pm (1.3 \times 419/12) = 113.3$ ft-kips, 22.6 ft-kips

75% of this moment is distributed to the column strip:

M_u = $0.75 \times 113.3 = 85.0$ ft-kips

Required area of flexural reinforcement, A_s, using Table 2-7:

$M_u/\phi f'_c bd^2 = (115.9 \times 12)/(0.9 \times 4 \times 120 \times 7.125^2) = 0.0634$

From design aid, $\omega = 0.066$

ρ = $\omega f'_c/f_y = (0.066 \times 4)/60 = 0.0044$

A_s = $0.0044 \times 120 \times 7.125 = 3.76$ in.^{2}

$A_{s(min)} = 0.0018 \times 120 \times 8.5 = 1.84$ in.$^2 < 3.76$ in.^{2} O.K.

No. of #4 bars = 3.76/0.20 = 18.8, use 19 bars

No. of #5 bars = 3.76/0.31 = 12.1, use 13 bars

Table 2-7 Design Aid: Flexural Strength of Rectangular Beam Sections with Tension Reinforcement Only

ω	.000	.001	.002	.003	.004	.005	.006	.007	.008	.009
0.0	0	.0010	.0020	.0030	.0040	.0050	.0060	.0070	.0080	.0090
0.01	.0099	.0109	.0119	.0129	.0139	.0149	.0159	.0168	.0178	.0188
0.02	.0197	.0207	.0217	.0226	.0236	.0246	.0256	.0266	.0275	.0285
0.03	.0295	.0304	.0314	.0324	.0333	.0343	.0352	.0362	.0372	.0381
0.04	.0391	.0400	.0410	.0420	.0429	.0438	.0448	.0457	.0467	.0476
0.05	.0485	.0495	.0504	.0513	.0523	.0532	.0541	.0551	.0560	.0569
0.06	.0579	.0588	.0597	.0607	.0616	.0625	.0634	.0643	.0653	.0662
0.07	.0671	.0680	.0689	.0699	.0708	.0717	.0726	.0735	.0744	.0753
0.08	.0762	.0771	.0780	.0789	.0798	.0807	.0816	.0825	.0834	.0843
0.09	.0852	.0861	.0870	.0879	.0888	.0897	.0906	.0915	.0923	.0932
0.10	.0941	.0950	.0959	.0967	.0976	.0985	.0994	.1002	.1011	.1020
0.11	.1029	.1037	.1046	.1055	.1063	.1072	.1081	.1089	.1098	.1106
0.12	.1115	.1124	.1133	.1141	.1149	.1158	.1166	.1175	.1183	.1192
0.13	.1200	.1209	.1217	.1226	.1234	.1243	.1251	.1259	.1268	.1276
0.14	.1284	.1293	.1301	.1309	.1318	.1326	.1334	.1342	.1351	.1359
0.15	.1367	.1375	.1384	.1392	.1400	.1408	.1416	.1425	.1433	.1441
0.16	.1449	.1457	.1465	.1473	.1481	.1489	.1497	.1506	.1514	.1522
0.17	.1529	.1537	.1545	.1553	.1561	.1569	.1577	.1585	.1593	.1601
0.18	.1609	.1617	.1624	.1632	.1640	.1648	.1656	.1664	.1671	.1679
0.19	.1687	.1695	.1703	.1710	.1718	.1726	.1733	.1741	.1749	.1756
0.20	.1764	.1772	.1779	.1787	.1794	.1802	.1810	.1817	.1825	.1832
0.21	.1840	.1847	.1855	.1862	.1870	.1877	.1885	.1892	.1900	.1907
0.22	.1914	.1922	.1929	.1937	.1944	.1951	.1959	.1966	.1973	.1981
0.23	.1988	.1985	.2002	.2010	.2017	.2024	.2031	.2039	.2046	.2053
0.24	.2060	.2067	.2075	.2082	.2089	.2096	.2103	.2110	.2117	.2124
0.25	.2131	.2138	.2145	.2152	.2159	.2166	.2173	.2180	.2187	.2194
0.26	.2201	.2208	.2215	.2222	.2229	.2236	.2243	.2249	.2256	.2263
0.27	.2270	.2277	.2284	.2290	.2297	.2304	.2311	.2317	.2324	.2331
0.28	.2337	.2344	.2351	.2357	.2364	.2371	.2377	.2384	.2391	.2397
0.29	.2404	.2410	.2417	.2423	.2430	.2437	.2443	.2450	.2456	.2463
0.30	.2469	.2475	.2482	.2488	.2495	.2501	.2508	.2514	.2520	.2527
0.31	.2533	.2539	.2546	.2552	.2558	.2565	.2571	.2577	.2583	.2590
0.32	.2596	.2602	.2608	.2614	.2621	.2627	.2633	.2639	.2645	.2651
0.33	.2657	.2664	.2670	.2676	.2682	.2688	.2694	.2670	.2706	.2712
0.34	.2718	.2724	.2730	.2736	.2742	.2748	.2754	.2760	.2766	.2771
0.35	.2777	.2783	.2789	.2795	.2801	.2807	.2812	.2818	.2824	.2830
0.36	.2835	.2841	.2847	.2853	.2858	.2864	.2870	.2875	.2881	.2887
0.37	.2892	.2898	.2904	.2909	.2915	.2920	.2926	.2931	.2937	.2943
0.38	.2948	.2954	.2959	.2965	.2970	.2975	.2981	.2986	.2992	.2997
0.39	.3003	.3008	.3013	.3019	.3024	.3029	.3035	.3040	.3045	.3051

Design: Using factored moment M_u, enter table with $M_u / \phi \, f'_c \, bd^2$; find ω and compute steel percentage ρ from $\rho = \omega \, f'_c / f_y$.

Investigation: Enter table with ω from $\omega = \rho \, f_y / f'_c$; find value of $M_n / f'_c \, bd^2$ and solve for nominal strength M_n.

According to UBC 1913.3.3.2 and 1911.12.6.1, the unbalanced moment at each column is transferred by flexure and by eccentricity of shear. The amount of slab reinforcement and the shear strength of the slab needs to be checked at the end support, the first interior support, and the second interior support.

- ***End Support—Additional Flexural Reinforcement Required for Moment Transfer***

The unbalanced moment at this location is equal to the maximum factored moment in the column strip:

— Gravity:

$$M_u = 0.26\,M_o = 54.5 \text{ ft-kips (see Table 2-2)}$$

— Gravity + wind (see Fig. 2-7 for bending moment due to wind):

$$M_u = 0.75\,(0.26\,M_o + 1.7\,M_W)$$

$$= 0.75\,(54.5 + 1.7 \times 431/12) = 86.7 \text{ ft-kips (governs)}$$

$$M_u = 0.9\,M_D + 1.3\,M_W$$

$$= (0.9 \times 0.26 \times 0.126 \times 20 \times 18.5^2/8) + (1.3 \times 431/12) = 71.9 \text{ ft-kips}$$

A fraction of the unbalanced moment must be transferred over an effective width b_e:

$$b_e = c_2 + 3h \qquad\qquad\qquad 1913.3.3.2$$

$$= 18 + (3 \times 8.5) = 43.5 \text{ in.}$$

The fraction of unbalanced moment transferred by flexure is:

$$\gamma_f = \cfrac{1}{1 + \dfrac{2}{3}\sqrt{b_1/b_2}} \qquad\qquad 1913.3.3.2$$

where $b_1 = 18 + (7.125/2) = 21.56$ in.

$b_2 = 18 + 7.125 = 25.13$ in.

Therefore,

$$\gamma_f = \cfrac{1}{1 + \dfrac{2}{3}\sqrt{21.56/25.13}} = 0.62$$

Unbalanced moment transferred by flexure:

$$\gamma_f M_u = 0.62 \times 86.7 = 53.8 \text{ ft-kips}$$

$$\gamma_f M_u / \phi\, f'_c b d^2 = (53.8 \times 12)/(0.9 \times 4 \times 43.5 \times 7.125^2) = 0.0812$$

From Table 2-7,

$$\omega = 0.086$$

$$\rho = \omega\, f'_c/f_y = 0.086 \times 4/60 = 0.0057$$

$$A_s = 0.0057 \times 43.5 \times 7.125 = 1.78 \text{ in.}^2$$

No. of #4 bars = 1.78/0.20 = 8.9, use 9 bars

Provide the required 9-#4 bars by concentrating 9 of the column strip bars (14-#4) within the 43.5 in. slab width over the column. For symmetry, add one column strip bar to the remaining 5 bars so that 3 bars will be on each side of the 43.5 in. strip. Check bar spacing:

For 9-#4 within 43.5 in. width: 43.5/9 = 4.8 in. < 17.0 in.   O.K.

For 6-#4 within (120.0 - 43.5) = 76.5 in. width: 76.5/6 = 12.8 in. < 17.0 in.   O.K.

No additional bars are required for moment transfer. Figure 2-10 shows the reinforcement detail for the top bars at the exterior columns.

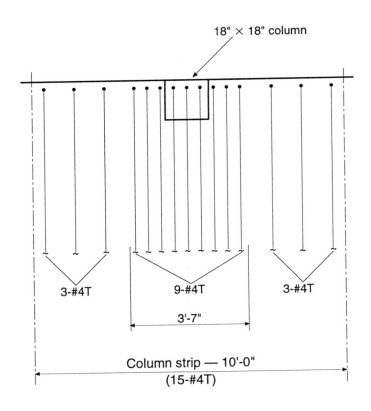

Figure 2-10 Reinforcement Detail for 15-#4 Top Bars at Exterior Columns

- **End Support—Check for Shear Strength**

The combined shear stress at the inside face of the critical transfer section must also be checked. The properties of the critical section for the exterior column bending perpendicular to the edge can be determined from Case C in Fig. 2-11:

$$A_c = (2b_1 + b_2)d = [(2 \times 21.56) + 25.13] \times 7.125 = 486.3 \text{ in.}^2$$

$$J/c = [2b_1^2 d (b_1 + 2b_2) + d^3 (2b_1 + b_2)]/6b_1 = 3868.4 \text{ in.}^3$$

— Gravity:

$$V_u = w_u (0.5 \ \ell_1 \ \ell_2 - b_1 b_2) - (M_1 - M_2)/\ell_n$$

where M_1 and M_2 are the negative moments at the right (interior) end and the left (exterior) end of the span, respectively:

$$M_1 = 0.70 M_o = 146.7 \text{ ft-kips}$$

1913.6.3.3

$$M_2 = 0.26 M_o = 54.5 \text{ ft-kips}$$

Therefore,

$$V_u = 0.245 [0.5 \times 20^2 - (21.56 \times 25.13/144)] - (146.7 - 54.5)/18.5$$

$$= 48.1 - 5.0 = 43.1 \text{ kips}$$

For the gravity load case, UBC 1913.6.3.6 requires that the unbalanced moment transferred by eccentricity of shear must be the full column strip nominal flexural strength, M_n, provided at the exterior support section.

For 15-#4 bars,

$$A_s = 15 \times 0.20 = 3.0 \text{ in.}^2$$

$$\rho = 3.0/(120 \times 7.125) = 0.0035$$

$$\omega = \rho f_y / f_c' = 0.0035 \times 60/4 = 0.053$$

From Table 2-7,

$$M_n/ f_c' bd^2 = 0.0513$$

$$M_n = 0.0513 \times 4 \times 120 \times 7.125^2 = 1250 \text{ in.-kips}$$

Combined shear stress at face of critical section:

$$v_u = \frac{V_u}{A_c} + \frac{\gamma_v M_n c}{J}$$

$$= \frac{43,100}{486.3} + \frac{(1 - 0.62)(1250 \times 1000)}{3868.4}$$

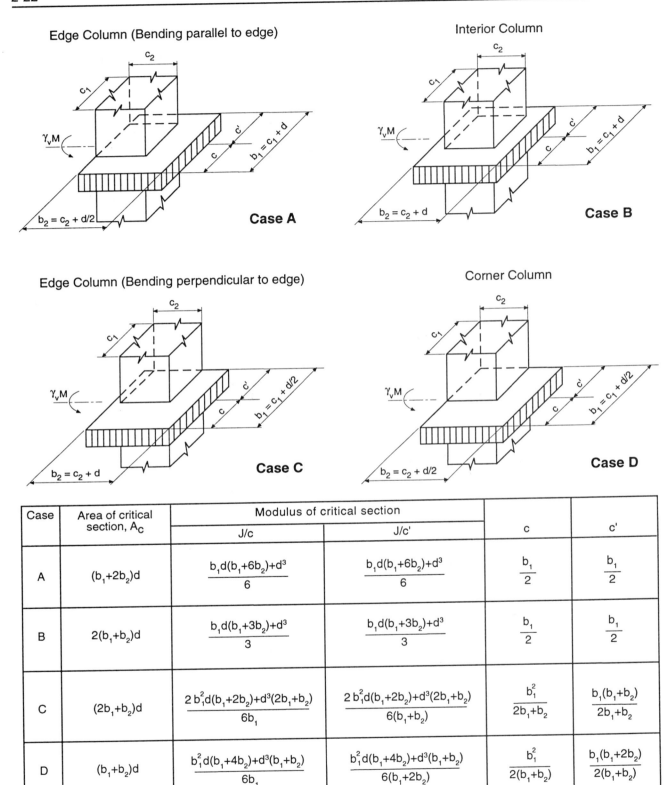

Case	Area of critical section, A_C	Modulus of critical section		c	c'
		J/c	J/c'		
A	$(b_1+2b_2)d$	$\dfrac{b_1 d(b_1+6b_2)+d^3}{6}$	$\dfrac{b_1 d(b_1+6b_2)+d^3}{6}$	$\dfrac{b_1}{2}$	$\dfrac{b_1}{2}$
B	$2(b_1+b_2)d$	$\dfrac{b_1 d(b_1+3b_2)+d^3}{3}$	$\dfrac{b_1 d(b_1+3b_2)+d^3}{3}$	$\dfrac{b_1}{2}$	$\dfrac{b_1}{2}$
C	$(2b_1+b_2)d$	$\dfrac{2 b_1^2 d(b_1+2b_2)+d^3(2b_1+b_2)}{6b_1}$	$\dfrac{2 b_1^2 d(b_1+2b_2)+d^3(2b_1+b_2)}{6(b_1+b_2)}$	$\dfrac{b_1^2}{2b_1+b_2}$	$\dfrac{b_1(b_1+b_2)}{2b_1+b_2}$
D	$(b_1+b_2)d$	$\dfrac{b_1^2 d(b_1+4b_2)+d^3(b_1+b_2)}{6b_1}$	$\dfrac{b_1^2 d(b_1+4b_2)+d^3(b_1+b_2)}{6(b_1+2b_2)}$	$\dfrac{b_1^2}{2(b_1+b_2)}$	$\dfrac{b_1(b_1+2b_2)}{2(b_1+b_2)}$

Figure 2-11 Critical Section Properties for Shear Stress Computations

$$= 88.6 + 122.8 = 211.4 \text{ psi}$$

$$\phi v_c = \phi 4 \sqrt{f'_c}$$

1911.12.2.1

$$= 0.85 \times 4 \sqrt{4000} = 215.0 \text{ psi} > 211.4 \text{ psi} \quad \text{O.K.}$$

— Gravity + wind:

$$V_u = 0.75 (1.4 V_D + 1.7 V_L + 1.7 V_W)$$

From previous calculations, $1.4 V_D + 1.7 V_L = 43.1$ kips. Also, from Fig. 2-7, $V_W = 3.8$ kips. Therefore,

$$V_u = 0.75 \times [43.1 + (1.7 \times 3.8)] = 37.2 \text{ kips}$$

When wind is considered, shear stress computations can be based on the actual unbalanced moment, rather than on the nominal moment strength of the column strip. The actual unbalanced moment at an exterior slab section is 86.7 ft-kips (see Table 2-6). The combined shear stress is:

$$v_u = \frac{37,200}{486.3} + \frac{(1 - 0.62)(86.7 \times 12,000)}{3868.4}$$

$$= 76.5 + 102.2 = 178.7 \text{ psi} < 215.0 \text{ psi} \quad \text{O.K.}$$

• **End Support—Design for Positive Moment**

The end support must be checked for positive moment. The controlling loading condition will occur when gravity loads are at a minimum with maximum lateral load effects (see Table 2-6):

$$M_D = M_{floorweight} + M_{superimposed}$$

$$= 0.26 \, w_D \, \ell_2 \, \ell_n^2 / 8$$

$$= 0.26 \times (0.150 \times 8.5/12 + 0.020) \times 20 \times 18.5^2 / 8 = 28.1 \text{ ft-kips}$$

$$M_W = 431 \text{ in.-kips (see Fig. 2-7)}$$

$$= 35.9 \text{ ft-kips}$$

$$M_u = 0.9 M_D - 1.3 M_W$$

$$= (0.9 \times 28.1) - (1.3 \times 35.9) = -21.5 \text{ ft-kips}$$

The required area of steel to carry this moment is 0.68 in.² For the 43.5 in. slab width which must resist 62% of this moment, the required area of steel is 0.42 in.² A total of 4-#4 bars in the column strip, with two of these bars located in the 43.5 in. effective slab width, are required to satisfy positive moment requirements at the exterior column. Note that the integrity steel provisions in UBC 1913.4.8 require that at least two of the bottom column strip bars be anchored into the column.

• *First Interior Support—Additional Flexural Reinforcement Required for Moment Transfer*

In determining the total factored negative moment at the exterior face of the first interior support, the following are taken into account:

1) Maximum factored moment due to gravity at the first interior support equals $0.70 \, M_o = 146.7$ ft-kips (see Table 2-1), and

2) Maximum unfactored moment due to wind at the same support equals 34.9 ft-kips (419 in.-kips, see Fig. 2-7).

Therefore, the total factored moment in the design strip at the first interior support is:

$$M_u = 0.75 \times [146.7 + (1.7 \times 34.9)]$$

$$= 154.5 \text{ ft-kips} > 146.7 \text{ ft-kips}$$

This moment is to be distributed to the column strip (75%) and the two-half middle strips (25%).

Therefore, for the column strip:

$$M_u = 0.75 \times 154.5 = 115.9 \text{ ft-kips}$$

The fraction of the column strip moment to be resisted by the 43.5-in. wide strip is:

$$\gamma_f M_u = 0.6 \times 115.9 = 69.5 \text{ ft-kips}$$

where $\quad \gamma_f \quad = 0.6$ for $b_1 = b_2 = 25.13$ in.

Using Table 2-7, the required area of steel $A_s = 2.33$ in.^{2}

For #5 bars, the total number of bars required $= 2.33/0.31 = 7.5$, use 8 bars.

Provide the required 8-#5 bars by concentrating 8 of the column strip bars (13-#5) within the 43.5 in. width over the column. For symmetry, add one column strip bar to the remaining 5 bars so that 3 bars will be on each side of the 43.5 in. strip. Check bar spacing:

For 8-#5 within 43.5 in. width: $43.5/8 = 5.4$ in. < 17.0 in.   O.K.

For 6-#5 within 76.5 in. width: $76.5/6 = 12.8$ in. < 17.0 in.   O.K.

No additional bars are required for moment transfer.

• *First Interior Support—Check for Shear Strength*

Combined shear stress at the face of the critical transfer section must be checked. Properties of the critical shear transfer section at an interior column are (see Case B in Fig. 2-11): $A_c = 716.1$ in.^{2} and $J/c = 6117.6$ in.^{3}

— Gravity:

$$V_u = w_u \, (\ell_1 \ell_2 - b_1 b_2) + (M_1 - M_2)/\ell_n$$

$$= 0.245 \, (20^2 - 25.13^2/144) + (146.7 - 54.5)/18.5$$

$$= 96.9 + 5.0 = 101.9 \text{ kips}$$

The difference between the slab moments acting on the two faces of the interior support needs to be transferred by shear to the first interior column. From Table 2-1, the exterior negative moment at the face of the support is $0.70 \, M_o = 146.7$ ft-kips, and the interior negative moment at the face of the support is $0.65 \, M_o = 136.2$ ft-kips. Therefore, the unbalanced moment $= 146.7 - 136.2 = 10.5$ ft-kips. The combined shear stress is:

$$v_u = \frac{101,900}{716.1} + \frac{0.4 \times 10.5 \times 12,000}{6117.6}$$

$$= 142.3 + 8.2 = 150.5 \text{ psi} < 215.0 \text{ psi} \quad \text{O.K.}$$

— Gravity + wind:

$$V_u = 0.75 \, (1.4 \, V_D + 1.7 \, V_L + 1.7 \, V_W)$$

where $1.4 \, V_D + 1.7 \, V_L = 101.9$ kips and $V_W = 3.8$ kips (see Fig. 2-7). Therefore,

$$V_u = 0.75 \, [101.9 + (1.7 \times 3.8)] = 81.3 \text{ kips}$$

The unbalanced moment at the first interior support due to gravity loads is the difference between the moments acting on the two sides of the support, while the unbalanced moment due to wind loads is the sum of the moments acting on the two sides of that same support (see Fig. 2-7):

$$M_u = 0.75 \times [(146.7 - 136.2) + 1.7 \, (34.9 + 34.6)]$$

$$= 0.75 \times (10.5 + 118.2) = 96.5 \text{ ft-kips}$$

The combined shear stress is:

$$v_u = \frac{81,300}{716.1} + \frac{0.4 \times 96.5 \times 12,000}{6117.6}$$

$$= 113.5 + 75.7 = 189.2 \text{ psi} < 215.0 \text{ psi} \quad \text{O.K.}$$

- ***First Interior Support—Design for Positive Moment***

The first interior span must be checked for positive moment at the support. The controlling loading condition will occur when gravity loads are at a minimum with maximum lateral load effects:

$$M_D = M_{\text{floorweight}} + M_{\text{superimposed}}$$

$$= 0.70 w_D \, \ell_2 \ell_n^2 / 8$$

$$= 0.70 \times (0.150 \times 8.5/12 + 0.020) \times 20 \times 18.5^2/8 = 75.5 \text{ ft-kips}$$

$$M_W \quad = 419 \text{ in.-kips (Fig. 2-7)}$$

$$= 34.9 \text{ ft-kips}$$

$$M_u \quad = 0.9 \, M_D - 1.3 \, M_W$$

$$= (0.9 \times 75.5) - (1.3 \times 34.9) = 22.6 \text{ ft-kips}$$

The slab will not require positive moment reinforcement at the first interior support, except for the structural integrity reinforcement given in UBC 1913.4.8.

- ***Second Interior Support—Additional Flexural Reinforcement Required for Moment Transfer***

The factored negative moment due to gravity on either side of the second interior support is $0.65 \, M_o = 136.2$ ft-kips (see Table 2-1). At the same support, the negative moment due to wind loads is 34.6 ft-kips (see Fig. 2-7). The total factored negative moment on either side of the second interior support is then:

$$M_u \quad = 0.75 \times [136.2 + (1.7 \times 34.6)]$$

$$= 146.2 \text{ ft-kips} > 136.2 \text{ ft-kips}$$

This is less than 154.5 ft-kips which is the maximum factored negative moment due to combined gravity and wind loads at the first interior support. Therefore, the first interior support reinforcement will be provided at the second and subsequent interior support sections.

- ***Second Interior Support—Check for Shear Strength***

The combined shear stress at the face of the critical transfer section must be checked. The unbalanced moment at the second interior support due to gravity loads is zero (moments on both faces of the support are equal); due to wind loads, it is the sum of the moments acting on both sides of that support.

Therefore, the total unbalanced moment is:

$$M_u \quad = 0.75 \times [0.0 + 1.7 \times (34.6 + 34.6)]$$

$$= 0.75 \times 117.6 = 88.2 \text{ ft-kips}$$

This unbalanced moment is less than 96.5 ft-kips which is the unbalanced moment at the first interior support. Considering that the properties of the critical shear transfer section and the factored shear force due to gravity and wind loads are the same at both supports, the shear stress at the faces of the critical section around the second interior support is less than the shear stress at the first interior support, and consequently, it is less than the permissible shear stress.

- ***Midspan Region of Each Span***

Wind loads cause negligible moments in the midspan regions. Reinforcement provided for gravity loads is sufficient.

• *Reinforcement Details*

Suggested reinforcement details for the slab designed in this section are shown in Fig. 2-12. Bar lengths were obtained from UBC Fig. 19-1. A summary of the slab reinforcement is given in Table 2-8.

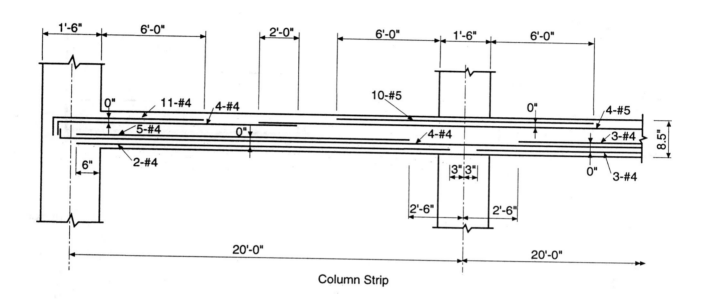

Column Strip

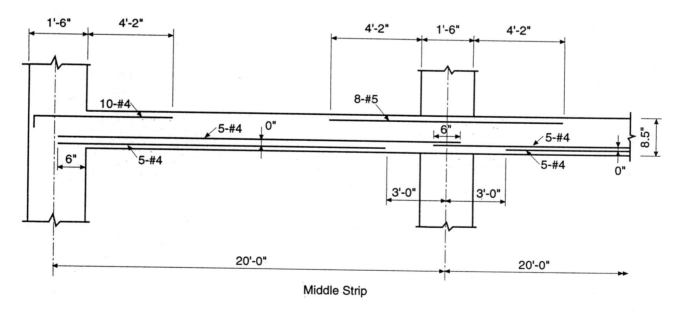

Middle Strip

Figure 2-12 Reinforcement Details for Interior Design Strip, Floor Level 3, Combined Gravity and Wind Loads

Table 2-8 Summary of Slab Reinforcement for Interior Design Strip, Floor Level 3, Combined Gravity and Wind Loads (8.5-in. Slab, 18-in. Square Columns)

Span Location		M_u (ft-kips)	No. of #4 bars	No. of #5 bars
END SPAN				
Column Strip	Ext. Negative	86.7	15	—
	Positive	65.0	11	—
	Int. Negative	115.9	—	14
Middle Strip	Ext. Negative	0.0	10	—
	Positive	44.0	10	—
	Int. Negative	38.6	—	8
INTERIOR SPAN				
Column Strip	Positive	44.0	10	—
	Negative	109.7	—	14
Middle Strip	Positive	29.3	10	—
	Negative	36.6	—	8

2.2.6.2   Column Design

Table 2-9 summarizes the factored axial loads, bending moments, and shear forces for an interior and an exterior column located above the 4th floor level in an interior frame. The points representing these combinations (including slenderness effects) are shown in Fig. 2-13 for an interior column and Fig. 2-14 for an exterior column. The interaction diagrams were obtained using *PCACOL* [2.2] with 8-#11 bars ($\rho_g = 3.9\%$) for an interior column and 4-#11 bars ($\rho_g = 1.9\%$) for an exterior column. At this time, the designer may choose to increase the size of the interior columns so as to decrease the amount of reinforcement. This, obviously, would require recomputing the member forces, which is not done here.

Table 2-9 Summary of Factored Axial Loads, Bending Moments, and Shear Forces for a Typical 18-in. Square Column above the 4th Floor Level, Combined Gravity and Wind Loads

(1) U $= 1.4D + 1.7L$	Eq. (9-1), UBC 1909.2.1	
(2) U $= 0.75 (1.4D + 1.7L \pm 1.7W)$	Eq. (9-2), UBC 1909.2.2	
(3) U $= 0.9D \pm 1.3W$	Eq. (9-3), UBC 1909.2.2	

Load Combination	INTERIOR COLUMN				EXTERIOR COLUMN			
	Axial Load (kips)	Bending Moment (ft-kips)		Shear Force (kips)	Axial Load (kips)	Bending Moment (ft-kips)		Shear Force (kips)
		Top	Bottom			Top	Bottom	
1	960	8.2	-8.2	1.9	522	-27.3	27.3	-6.2
2: sidesway right	719	44.9	-41.2	9.2	368	-0.9	3.1	-0.6
sidesway left	721	-32.6	28.9	-6.4	415	-41.8	37.9	-8.7
3: sidesway right	531	39.5	-35.8	8.6	258	9.1	-5.1	1.6
sidesway left	533	-39.5	35.8	-8.6	306	-34.5	30.5	-7.4

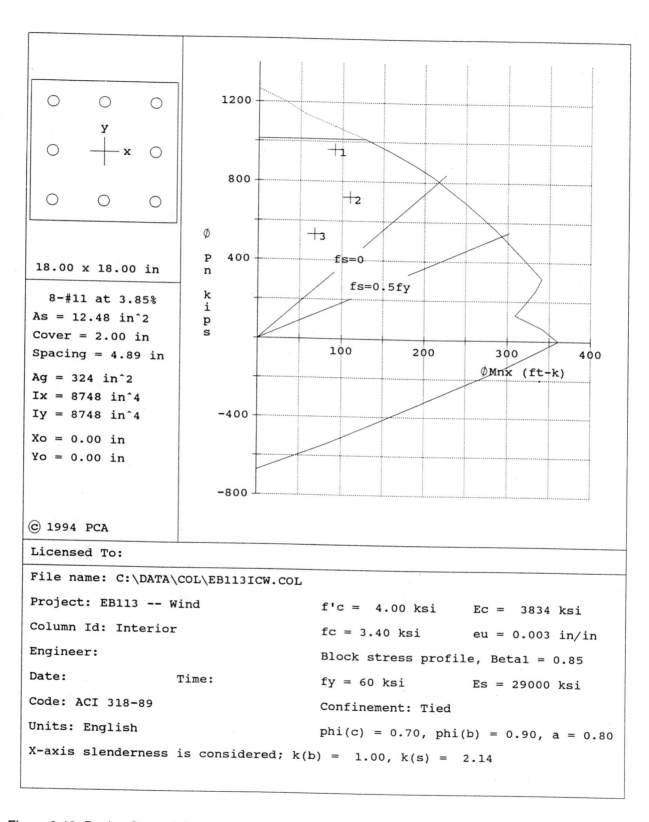

18.00 x 18.00 in

8-#11 at 3.85%
As = 12.48 in^2
Cover = 2.00 in
Spacing = 4.89 in

Ag = 324 in^2
Ix = 8748 in^4
Iy = 8748 in^4

Xo = 0.00 in
Yo = 0.00 in

© 1994 PCA

Licensed To:

File name: C:\DATA\COL\EB113ICW.COL

Project: EB113 -- Wind

Column Id: Interior

Engineer:

Date: Time:

Code: ACI 318-89

Units: English

X-axis slenderness is considered; k(b) = 1.00, k(s) = 2.14

f'c = 4.00 ksi   Ec = 3834 ksi

fc = 3.40 ksi   eu = 0.003 in/in

Block stress profile, Beta1 = 0.85

fy = 60 ksi   Es = 29000 ksi

Confinement: Tied

phi(c) = 0.70, phi(b) = 0.90, a = 0.80

Figure 2-13 Design Strength Interaction Diagram for an 18-in. Square Interior Column above the 4th Floor Level, Interior Frame

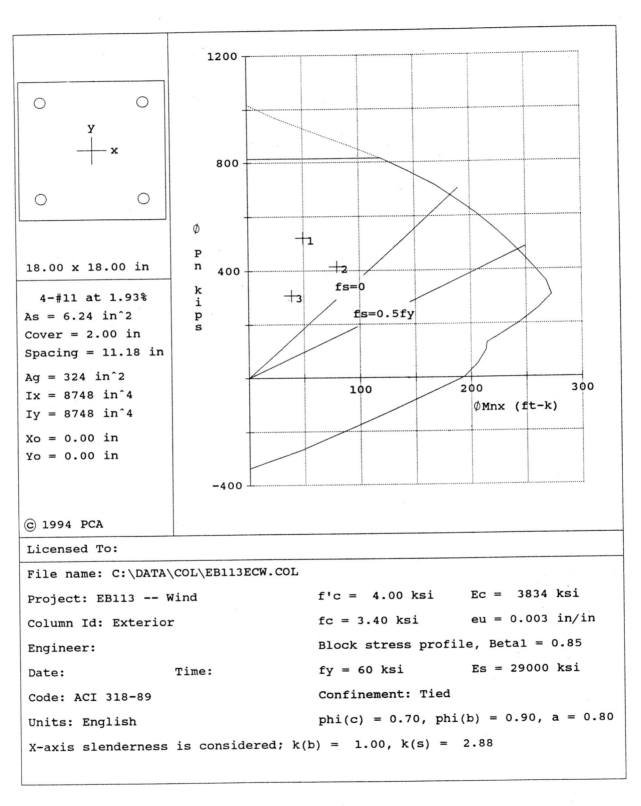

18.00 x 18.00 in

4-#11 at 1.93%
As = 6.24 in^2
Cover = 2.00 in
Spacing = 11.18 in

Ag = 324 in^2
Ix = 8748 in^4
Iy = 8748 in^4

Xo = 0.00 in
Yo = 0.00 in

© 1994 PCA

Licensed To:

File name: C:\DATA\COL\EB113ECW.COL

Project: EB113 -- Wind   f'c = 4.00 ksi   Ec = 3834 ksi

Column Id: Exterior   fc = 3.40 ksi   eu = 0.003 in/in

Engineer: Block stress profile, Beta1 = 0.85

Date: Time: fy = 60 ksi   Es = 29000 ksi

Code: ACI 318-89   Confinement: Tied

Units: English   phi(c) = 0.70, phi(b) = 0.90, a = 0.80

X-axis slenderness is considered; k(b) = 1.00, k(s) = 2.88

Figure 2-14 Design Strength Interaction Diagram for an 18-in. Square Exterior Column above the 4th Floor
Level, Interior Frame

Transverse reinforcement must be provided along the length of the column to satisfy both shear and confinement requirements. From Table 2-9, the maximum shear force is $V_u = 9.2$ kips which occurs in an interior column subjected to both gravity and wind loads. Column shear strength without shear reinforcement is:

$$V_c = 2 \left(1 + \frac{N_u}{2000 A_g} \right) \sqrt{f'_c}\ b_w d \qquad\qquad 1911.3.1.2$$

where N_u is the factored axial load in the column which in this case is 719 kips (see Table 2-9). Therefore,

$$V_c = 2 \left(1 + \frac{719 \times 1000}{2000 \times 18^2} \right) \sqrt{4000} \times 18 \times 15.3/1000$$

$$= 73.5 \text{ kips}$$

$$\phi V_c = 0.85 \times 73.5 = 62.5 \text{ kips} > V_u \quad \text{O.K.}$$

Although no shear reinforcement is required, #4 ties must be provided with vertical spacing not to exceed the smallest of the values given in UBC 1907.10.5:

16 longitudinal bar diameters = $16 \times 1.41 = 22.6$ in.

48 tie bar diameters = $48 \times 0.5 = 24.0$ in.

least column dimension = 18.0 in. (governs)

All of the column bars from the floor below are to be lap spliced with the column bars of the floor above. In this situation, these splices are usually made just above the construction joint at the floor level. It can be seen from Figs. 2-13 and 2-14 that all load combinations fall within the compression zone of the interaction diagrams; thus, compression splices will be used in all columns (UBC 1912.17.2.1). According to UBC 1912.16.1, the required length of the compression lap splice is:

$$0.0005\ f_y\ d_b = 0.0005 \times 60,000 \times 1.41 = 42.3 \text{ in.}$$

A lap splice of 43.0 in. will be provided, as shown in Fig. 2-15 for an interior column. A similar detail can be obtained for an exterior column.

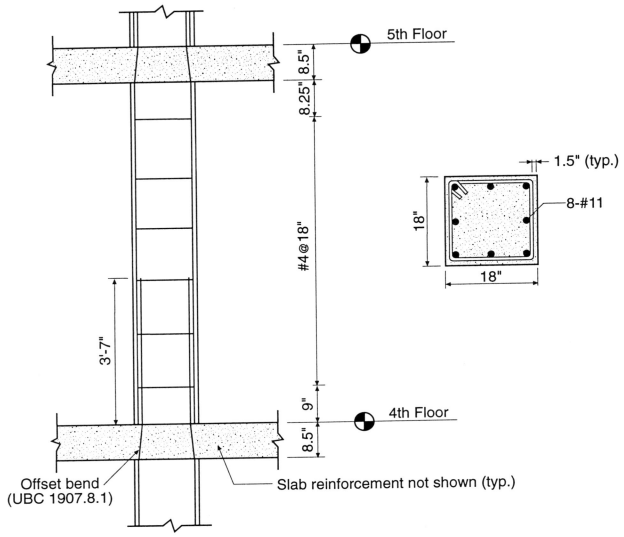

Figure 2-15 *Reinforcement Details for an 18-in. Square Interior Column above the 4th Floor Level, Interior Frame*

2.2.7   Overturning and Serviceability

According to UBC 1619.1, the base overturning moment for the entire structure shall not exceed two-thirds of the resisting moment due to the dead loads. The total overturning moment M_{ov} caused by the wind loads in this case is (see Table 2-5):

$$M_{ov} = (9.7 \times 136.0) + (19.1 \times 126.5) + (18.8 \times 117.0) + (18.5 \times 107.5) + (18.1 \times 98.0)$$

$$+ (17.7 \times 88.5) + (17.4 \times 79.0) + (16.9 \times 69.5) + (16.5 \times 60.0) + (16.0 \times 50.5)$$

$$+ (15.4 \times 41.0) + (14.8 \times 31.5) + (14.0 \times 22.0) + (15.5 \times 12.5)$$

$$= 17,211 \text{ ft-kips}$$

In order to determine the resisting moment M_r, the total dead load of the structure must be computed:

$$DL = 14 \times \left[\left(\frac{8.5}{12} \times 0.150 \times 101.5^2 \right) + \left(0.02 \times 101.5^2 \right) \right]$$

$$+ 36 \times \left(\frac{18^2}{144} \times 0.150 \times 126.1 \right) = 18,209 + 1532$$

$$= 19,741 \text{ kips}$$

Knowing that this dead load acts through the center of mass, the total resisting moment is:

$$M_r = 19,741 \times \frac{101.5}{2} = 1,001,856 \text{ ft-kips}$$

Check overturning criterion:

$$\frac{2}{3} M_r = 667,904 \text{ ft-kips} > 17,211 \text{ ft-kips} \quad \text{O.K.}$$

No provisions are made in the UBC pertaining to the overall serviceability of a structure subjected to wind loads. Traditionally, designers have limited the deflection at the top of the structure to the height of the building divided by some constant to satisfy overall serviceability requirements. The constant that is used most often is 500. For this 14-story building, the deflection at the roof level is 1.2 in. which is less than $(136 \times 12)/500 = 3.3$ in. Consequently, no overall serviceability problems should be evident for this building subjected to wind loads.

2.3   Design for Seismic Zone 1

2.3.1   General

An 8.5-in. thick slab supported on 18-in. columns is sufficient to carry the combined gravity and wind loads. This structure will now be designed for combined gravity and Seismic Zone 1 forces. Preliminary calculations show that the 8.5-in. thick slab will be insufficient to resist punching shear at interior columns. Therefore, the slab thickness will be increased to 9 in. for this case. Likewise, column sizes are increased to 20-in. square.

The total lateral force (or shear) at the base, V, is computed according to Eq. (28-1) of UBC-94. For this example, the importance factor, I, and the soil factor, S, have been assigned the values of unity (standard occupancy structure; see UBC Table 16-K) and 1.2 (soil profile type S_2; see UBC Table 16-J), respectively. The period of the structure is calculated from Eq. (28-3) of UBC-94, with the coefficient C_t equal to 0.03 (reinforced concrete moment-resisting frame). The response coefficient R_w is equal to 5 (Ordinary Moment-Resisting Frame, OMRF, of concrete; see UBC 1627.6 and Table 16-N). The seismic forces resulting from the vertical distribution of the base shear, V, according to Eqs. (28-6), (28-7), and (28-8) of UBC-94 are listed in Table 2-10.

Table 2-10 Seismic Forces for Zone 1

Floor Level	Weight, w_x (kips)	Height, h_x (ft)	$w_x h_x$ (ft-kips)	Lateral Force (kips)
R	1435	136.0	195,160	83.3
14	1501	126.5	189,877	47.0
13	1501	117.0	175,617	43.5
12	1501	107.5	161,358	39.9
11	1501	98.0	147,098	36.4
10	1501	88.5	132,839	32.9
9	1501	79.0	118,579	29.4
8	1501	69.5	104,320	25.8
7	1501	60.0	90,060	22.3
6	1501	50.5	75,801	18.8
5	1501	41.0	61,541	15.2
4	1501	31.5	47,282	11.7
3	1501	22.0	33,022	8.2
2	1526	12.5	19,075	4.7
Σ	20,973		1,551,629	419.1

Building height	h_n	$= 136\ ft$
Building weight	W	$= 20,973\ kips$
Fundamental period	T	$= 0.030\ (h_n)^{3/4} = 1.2\ sec.$
Seismic zone factor	Z	$= 0.075$
Importance factor	I	$= 1.0$
Response coefficient	R_w	$= 5$
Soil factor	S	$= 1.2$
Coefficient	C	$= 1.25\ S/T^{2/3} = 1.33$
Base shear	V	$= ZICW/R_w = 419.1\ kips$
Top level force	F_t	$= 0.07TV = 35.0\ kips$

2.3.2 Torsional Effects

Since the structural layout of the building is symmetrical, there is no actual eccentricity between the center of mass and the center of rigidity. However, UBC 1628.5 requires that "Where diaphragms are not flexible, the mass of each level shall be assumed to be displaced from the calculated center of mass in each direction a distance equal to 5 percent of the building dimension at that level perpendicular to the direction of the force under consideration." Additionally, the story shear V_x shall be distributed to the various elements of the vertical lateral-force-resisting system in proportion to their rigidities, considering the rigidity of the diaphragm.

The results from the 3-D analysis for the first interior frame are given in Figs. 2-16 and 2-17 for the columns and slabs, respectively. Similar to the analysis for the wind loads, an effective width of one-third the bay width was used for an interior frame and one-sixth the bay width for an exterior frame. Also, rigid diaphragms were specified at each floor level. In order to comply with UBC 1628.5, the seismic forces were displaced a distance of $0.05 \times 100 = 5$ ft from the center of mass at each floor level.

For comparison purposes, a 2-D analysis was also performed. The story shears for the interior and exterior frames, accounting for accidental torsion, were required for the analysis. Based on the above UBC requirements, the story shear due to accidental torsion in any frame at a particular level is given by:

R	9.6 / -1.1	20.4 / 7.9	23.3 / 10.1
14	18.1 / 4.2	27.1 / 15.0	28.7 / 16.7
13	20.8 / 8.3	34.4 / 22.7	36.0 / 24.3
12	24.2 / 12.5	40.5 / 29.7	41.9 / 31.2
11	26.9 / 16.3	46.0 / 36.3	47.4 / 37.6
10	29.4 / 19.9	51.0 / 42.3	52.2 / 43.5
9	31.5 / 23.2	55.4 / 47.7	56.4 / 48.8
8	33.3 / 26.2	59.2 / 52.6	60.1 / 53.5
7	34.8 / 29.0	62.3 / 56.9	63.0 / 57.5
6	35.9 / 31.4	64.8 / 60.8	61.2 / 65.2
5	36.6 / 34.6	66.3 / 64.0	66.5 / 64.2
4	35.4 / 38.5	66.2 / 67.7	66.3 / 67.9
3	30.9 / 42.8	61.5 / 76.4	61.3 / 75.9
2	22.0 / 114.0	40.5 / 124.0	40.1 / 123.0

Bending Moments (ft-kips)

R	0.9	3.0	3.5
14	2.3	4.4	4.8
13	3.1	6.0	6.3
12	3.9	7.4	7.7
11	4.6	8.7	9.0
10	5.2	9.8	10.1
9	5.8	10.9	11.1
8	6.3	11.8	11.9
7	6.7	12.5	12.7
6	7.1	13.2	13.3
5	7.5	13.7	13.7
4	7.8	14.1	14.1
3	7.8	14.5	14.5
2	10.9	13.1	13.1

Shear Forces (kips)

R	1.0	0.3	0.0
14	2.5	0.8	0.0
13	5.0	1.1	0.0
12	8.3	1.5	0.0
11	12.3	1.8	0.1
10	17.0	2.0	0.1
9	22.4	2.2	0.1
8	28.3	2.4	0.1
7	34.7	2.5	0.1
6	41.5	2.6	0.1
5	48.7	2.6	0.1
4	56.1	2.5	0.1
3	63.5	2.4	0.1
2	70.1	2.2	0.1

Axial Forces (kips)

Figure 2-16 Results of 3-D Analysis under Seismic Zone 1 Forces for Columns of First Interior Frame

R	-9.6	-8.8	-12.7	-12.5	-12.5
14	-16.5	-16.0	-20.4	-20.4	-20.6
13	-24.5	-23.8	-27.8	-27.7	-27.9
12	-32.0	-31.1	-34.9	-34.8	-35.0
11	-39.0	-37.9	-41.4	-41.4	-41.5
10	-45.4	-44.1	-47.4	-47.3	-47.5
9	-51.2	-49.8	-52.7	-52.6	-52.8
8	-56.4	-54.8	-57.3	-57.3	-57.4
7	-61.0	-59.3	-61.4	-61.3	-61.4
6	-64.9	-63.1	-64.7	-64.7	-64.8
5	-68.1	-66.2	-67.2	-67.2	-67.3
4	-70.0	-68.2	-68.6	-68.6	-68.7
3	-69.3	-67.8	-67.9	-67.8	-67.9
2	-63.0	-60.9	-59.8	-59.8	-59.9

Bending Moments (ft-kips)

R	-1.0	-1.4	-1.4
14	-1.8	-2.2	-2.2
13	-2.6	-3.0	-3.0
12	-3.4	-3.8	-3.8
11	-4.2	-4.5	-4.5
10	-4.8	-5.1	-5.1
9	-5.5	-5.7	-5.7
8	-6.0	-6.2	-6.2
7	-6.5	-6.6	-6.6
6	-6.9	-7.0	-7.0
5	-7.3	-7.3	-7.3
4	-7.5	-7.4	-7.4
3	-7.4	-7.3	-7.3
2	-6.7	-6.5	-6.5

Shear Forces (kips)

Figure 2-17 Results of 3-D Analysis under Seismic Zone 1 Forces for Slabs of First Interior Frame

$$V_{Txi} = \frac{0.05 \times DIM_x \times V_x}{D_{xi}} \times \frac{R_{xi}D_{xi}^2}{\Sigma R_{xi}D_{xi}^2}$$

where

DIM_x = building dimension perpendicular to the direction of analysis, at floor level x.

R_{xi} = relative rigidity of frame i at floor level x.

V_x = story shear, in the direction of analysis, at floor level x.

D_{xi} = distance, perpendicular to the direction of analysis, from the center of rigidity to the centerline of frame i, at floor level x.

The values of DIM_x and D_{xi} are obtained from the building geometry. The relative rigidity of a frame at a particular level is determined in a rational manner, as indicated below.

In all of the 2-D lateral load analyses carried out for this example, a rigid diaphragm has been assumed, represented by the hinged rigid links that impose equal displacements on the various frames at every floor level (see Fig. 2-3). The story shear at any level is thus automatically distributed to the various frames in proportion to their relative lateral rigidities. Therefore, the story shear carried by a frame is a measure of its relative rigidity.

The structure is analyzed along each principal plan axis under the distributed lateral forces that add up to the base shear V. Since the building is doubly symmetrical about the two principal plan axes, the analyses in the two directions are identical. In either direction, the building is analyzed as two plane frames connected by hinged links (see Sect. 2.1.3 and Fig. 2-3). The first of these frames consists of four interior frames lumped together. The second plane frame represents the lumping of the two exterior frames.

Analysis indicates that the base shear of 419.1 kips (in either direction), which is also the story shear in the first story (supporting level 2), is distributed to the two exterior and the four interior frames according to the percentages shown in Table 2-11. The table also shows how the story shears carried by the frames are distributed to the columns. This latter information is for interest only, and is not necessary for the purposes of computation of torsional shears.

Table 2-11 Distribution of Story Shear Forces to Columns and Frames (First Story)

Column Line	Column Shears (kips)						Frame Shear V_{xi} (kips)	Relative Rigidity R_{xi}*
	A	B	C	D	E	F		
1	9.43	10.73	10.70	10.70	10.73	9.43	61.72	0.1473
2	10.61	13.21	13.13	13.13	13.21	10.61	73.91	0.1764
3	10.61	13.21	13.13	13.13	13.21	10.61	73.91	0.1764
4	10.61	13.21	13.13	13.13	13.21	10.61	73.91	0.1764
5	10.61	13.21	13.13	13.13	13.21	10.61	73.91	0.1764
6	9.43	10.73	10.70	10.70	10.73	9.43	61.72	0.1473
						Σ	419.08	1.0002

*R_{xi} = Frame shear V_{xi} / Story shear V_x

The relative rigidity from Table 2-11 is listed in the second column of Table 2-12. Note that this relative rigidity equals the frame shear V_{xi} when the story shear V_x is taken equal to one. The third, fourth, fifth, and sixth columns of this table should be self-explanatory; it should be pointed out that $\Sigma R_{xi} \times D_{xi}^2$ refers to all the frames in both orthogonal directions of the building. The seventh column lists the shear per frame due to accidental torsion caused by a unit story shear. The calculation of the quantities listed in this column is shown below the table. The last column lists the total unit story shear per frame, including effects of accidental torsion. The quantities in the last column, when multiplied by the actual story shear, V_x, would yield the frame shears at that particular story level, including effects of accidental torsion.

The torsional shear calculations shown in Table 2-12 must be repeated for each story. The results of such analyses for the doubly symmetrical example building are given in Table 2-13.

Table 2-12 Torsional Shear Calculations for the First Story

E-W Direction* $DIM_x = 100 \text{ ft}$

Frame	$R_{xi} = V_{xi}$ ($V_x =1$)	D_{xi} (ft)	D_{xi}^2 (ft^2)	$R_{xi} \times D_{xi}^2$ (ft^2)	$\dfrac{R_{xi} \times D_{xi}^2}{\Sigma R_{xi} \times D_{xi}^2}$	V_{Txi} ($V_x=1$)	$V_{xi} + V_{Txi}$
1	0.1473	50	2500	368.3	0.169	0.0169	0.1642
2	0.1764	30	900	158.8	0.073	0.0122	0.1886
3	0.1764	10	100	17.6	0.008	0.0040	0.1804
4	0.1764	10	100	17.6	0.008	-0.0040	0.1724
5	0.1764	30	900	158.8	0.073	-0.0122	0.1642
6	0.1473	50	2500	368.3	0.169	-0.0169	0.1304
Σ	1.0002			1089.4			

$\Sigma R_{xi} D_{xi}^2 = 2 \times 1089.4 = 2178.8$ (considering frames in N-S direction)

V_{xi} $(V_x = 1)$ = Unit story shear per frame, not including effects of accidental torsion (same as R_{xi})

V_{Txi} $(V_x = 1)$ $= \dfrac{R_{xi} \times D_{xi}^2}{\Sigma R_{xi} \times D_{xi}^2} \times \dfrac{0.05 \times DIM_x \times 1}{D_{xi}}$

= Shear per frame due to accidental torsion caused by a unit story shear

$V_{xi} + V_{Txi}$ = Total unit story shear per frame, including effects of accidental torsion

* N-S direction is the same

The revised percentages given in columns three and five of Table 2-13, when multiplied by the appropriate story shear, yield revised frame shears, including the accidental torsional effects, at the various story levels. From these revised story shears, the revised lateral forces including accidental torsional effects at the various floor levels of each frame can be easily determined. The various frames then need to be analyzed separately under the revised lateral forces, including accidental torsional effects, applicable to them. The ratio of the 2-D to 3-D results are given in Figs. 2-18 and 2-19 for the columns and slabs of the first interior frame, respectively.

Table 2-13 Distribution of Story Shear to Interior and Exterior Frames, Seismic Zone 1

Level	% of story shear carried by one exterior frame	Revised % of story shear accounting for torsion, carried by an exterior frame	% of story shear carried by first interior frame	Revised % of story shear, accounting for torsion, carried by first interior frame
R	9.4	10.7	20.5	22.3
14	10.8	12.5	19.6	21.2
13	10.8	12.5	19.6	21.2
12	10.8	12.5	19.6	21.2
11	10.8	12.5	19.6	21.2
10	10.8	12.5	19.6	21.2
9	10.8	12.5	19.6	21.2
8	10.8	12.5	19.6	21.2
7	10.8	12.5	19.6	21.2
6	10.8	12.5	19.6	21.2
5	10.7	12.2	19.7	21.3
4	10.8	12.4	19.6	21.2
3	10.7	12.2	19.7	21.3
2	14.7	16.4	17.6	18.9

Bending Moments

Level			
R	1.13	1.17	1.19
	1.10	1.13	1.17
14	1.08	1.14	1.15
	1.10	1.14	1.16
13	1.09	1.13	1.14
	1.02	1.14	1.15
12	1.08	1.13	1.14
	1.04	1.13	1.14
11	1.08	1.13	1.13
	1.06	1.13	1.13
10	1.08	1.22	1.22
	1.06	1.13	1.13
9	1.07	1.12	1.13
	1.06	1.12	1.13
8	1.07	1.12	1.13
	1.06	1.12	1.13
7	1.07	1.12	1.12
	1.06	1.12	1.13
6	1.07	1.12	1.12
	1.07	1.12	1.12
5	1.07	1.13	1.13
	1.03	1.12	1.12
4	1.11	1.13	1.13
	1.06	1.11	1.11
3	1.11	1.17	1.17
	0.99	1.10	1.10
2	1.21	1.23	1.23
	1.09	1.11	1.10

Shear Forces

Level			
R	1.11	1.23	1.29
14	1.17	1.31	1.26
13	1.17	1.23	1.24
12	1.18	1.22	1.30
11	1.18	1.23	1.24
10	1.15	1.21	1.23
9	1.16	1.22	1.20
8	1.16	1.22	1.22
7	1.15	1.22	1.22
6	1.15	1.21	1.22
5	1.13	1.22	1.22
4	1.18	1.21	1.21
3	1.13	1.22	1.22
2	1.18	1.21	1.22

Axial Forces

Level			
R	1.00	1.33	1.00
14	1.20	1.38	1.00
13	1.16	1.18	1.00
12	1.16	1.13	1.00
11	1.15	1.11	1.00
10	1.14	1.15	1.00
9	1.13	1.18	1.00
8	1.12	1.17	1.00
7	1.11	1.16	1.00
6	1.11	1.15	1.00
5	1.11	1.15	1.00
4	1.10	1.16	1.00
3	1.10	1.16	1.00
2	1.02	1.14	1.00

Figure 2-18 Ratio of 2-D to 3-D Results under Seismic Zone 1 Forces for Columns of an Interior Frame

R	1.08	1.08	1.08	1.07	1.06
14	1.09	1.09	1.14	1.09	1.09
13	1.09	1.08	1.09	1.09	1.08
12	1.04	1.11	1.07	1.08	1.08
11	1.07	1.07	1.07	1.07	1.07
10	1.07	1.06	1.07	1.07	1.07
9	1.07	1.07	1.07	1.07	1.07
8	1.07	1.07	1.13	1.05	1.07
7	1.06	1.06	1.06	1.06	1.06
6	1.06	1.06	1.06	1.06	1.06
5	1.06	1.06	1.06	1.06	1.06
4	1.07	1.07	1.07	1.07	1.07
3	1.08	1.08	1.08	1.08	1.08
2	1.12	1.12	1.11	1.11	1.11

R	1.40	1.71	1.71
14	1.56	1.50	1.50
13	1.42	1.37	1.37
12	1.32	1.26	1.26
11	1.27	1.22	1.22
10	1.23	1.20	1.20
9	1.19	1.16	1.16
8	1.15	1.15	1.15
7	1.14	1.14	1.14
6	1.13	1.11	1.11
5	1.11	1.11	1.10
4	1.21	1.22	1.22
3	1.09	1.10	1.10
2	1.12	1.12	1.11

Bending Moments   *Shear Forces*

Figure 2-19 Ratio of 2-D to 3-D Results under Seismic Zone 1 Forces for Slabs of an Interior Frame

It should be noted that the calculation procedure outlined above for the 2-D analysis is applicable only to structures with approximately the same stiffness in both orthogonal directions. If the stiffnesses in the two directions are dissimilar, as can be readily determined from the elastic lateral deflection caused in those directions by the code-prescribed seismic forces, appropriate adjustments must be made to the relative stiffnesses of frames spanning in the two orthogonal directions, so that both sets of values correspond to essentially the same lateral deflections under similar lateral forces (see Sect. 3.2.1).

2.3.3   Factored Loads and Moments due to Gravity

A 9-in. slab thickness has an effective depth of 7.625 in. (3/4 in. cover and #5 bars). The design loads are:

Factored slab weight = $1.4 \times 150 \times 9/12 = 157.5$ psf

Factored superimposed dead load = $1.4 \times 20 = 28$ psf

Factored live load = $1.7 \times 40 = 68$ psf

Total factored load $w_u = 157.5 + 28 + 68 = 253.5$ psf $= 0.254$ ksf

With 20-in. square columns, the clear span length $\ell_n = 20 - 20/12 = 18.33$ ft.

2.3.3.1   Factored Moments in Slab

Total factored static moment:

$$M_o = \frac{w_u \ell_2 \ell_n^2}{8}$$

$$= 0.254 \times 20 \times \frac{18.33^2}{8} = 213.4 \text{ ft-kips}$$

Table 2-14 summarizes the moments in the column and the middle strips for an interior design strip.

Table 2-14 Factored Slab Moments due to Gravity Loads

Span Location		M_u (ft-kips)
END SPAN		
Column Strip	Exterior Negative Positive Interior Negative	$0.26\,M_o = 55.5$ $0.31\,M_o = 66.2$ $0.53\,M_o = 113.1$
Middle Strip	Exterior Negative Positive Interior Negative	0.0 $0.21\,M_o = 44.8$ $0.17\,M_o = 36.3$
INTERIOR SPAN		
Column Strip	Positive Negative	$0.21\,M_o = 44.8$ $0.49\,M_o = 104.6$
Middle Strip	Positive Negative	$0.14\,M_o = 29.9$ $0.16\,M_o = 34.1$

2.3.3.2   Factored Axial Loads and Moments in Columns

Design axial loads due to gravity on an interior column and an exterior column of a typical interior frame are shown in Tables 2-15 and 2-16, respectively. The live load reduction factor is in accordance with Eq. (6-1) in UBC 1606.

Table 2-15 Axial Loads due to Gravity on a 20-in. Square Interior Column of an Interior Frame

Floor	Dead Load (psf)	Live Load (psf)	Area, A (ft^2)	Reduction Factor (1-R/100)*	Reduced Live Load (psf)	Factored Load (kips)**	Cumulative Factored Load (kips)
R	132.5	40	400	0.80	32.0	102	102
14	132.5	40	800	0.48	19.2	93	195
13	132.5	40	1200	0.40	16.0	91	285
12	132.5	40	1600	0.40	16.0	91	376
11	132.5	40	2000	0.40	16.0	91	467
10	132.5	40	2400	0.40	16.0	91	558
9	132.5	40	2800	0.40	16.0	91	649
8	132.5	40	3200	0.40	16.0	91	739
7	132.5	40	3600	0.40	16.0	91	830
6	132.5	40	4000	0.40	16.0	91	921
5	132.5	40	4400	0.40	16.0	91	1012
4	132.5	40	4800	0.40	16.0	91	1102
3	132.5	40	5200	0.40	16.0	91	1193
2	132.5	40	5600	0.40	16.0	93	1286

* $R = r(A - 150)$
** Factored load = [1.4(Dead Load) + 1.7(Reduced Live Load)] $\times$ Tributary Area. Factored load includes
   1.4 $\times$ column weight = 0.6 kips/ft

Table 2-16 Axial Loads due to Gravity on a 20-in. Square Exterior Column of an Interior Frame

Floor	Dead Load (psf)	Live Load (psf)	Area, A (ft²)	Reduction Factor (1-R/100)*	Reduced Live Load (psf)	Factored Load (kips)**	Cumulative Factored Load (kips)
R	132.5	40	200	0.96	38.4	56	56
14	132.5	40	400	0.80	32.0	54	110
13	132.5	40	600	0.64	25.6	52	161
12	132.5	40	800	0.48	19.2	49	210
11	132.5	40	1000	0.40	16.0	48	259
10	132.5	40	1200	0.40	16.0	48	307
9	132.5	40	1400	0.40	16.0	48	355
8	132.5	40	1600	0.40	16.0	48	403
7	132.5	40	1800	0.40	16.0	48	452
6	132.5	40	2000	0.40	16.0	48	500
5	132.5	40	2200	0.40	16.0	48	548
4	132.5	40	2400	0.40	16.0	48	596
3	132.5	40	2600	0.40	16.0	48	645
2	132.5	40	2800	0.40	16.0	50	695

* $R = r(A - 150)$
** Factored load = [1.4(Dead Load) + 1.7(Reduced Live Load)] × Tributary Area. Factored load includes 1.4 × column weight = 0.6 kips/ft

The factored gravity moment to be transferred from the slab to the interior columns above and below the slab is:

$$M_i = 0.07 \times (0.5 \times 1.7 \times 0.04 \times 20 \times 18.33^2) = 16.0 \text{ ft-kips}$$

As noted previously, this moment will be distributed to the columns inversely proportional to their lengths above and below the floor slab under consideration. The total exterior negative moments from the slab must be transferred directly to the exterior columns.

2.3.4   Design for Combined Loading

2.3.4.1   Slab Design

The required slab reinforcement for an interior design strip at floor level 4 due to the combined effects of gravity and Seismic Zone 1 forces is summarized in Table 2-17. Applicable load combinations were taken from UBC 1909.2 with 1.1E substituted for W in Eqs. (9-2) and (9-3).

Sample calculations for the interior negative moment in the column strip, end span:

- $M_u = 1.4 M_D + 1.7 M_L$ 1909.2.1

  $M_u = 0.53 M_o = 0.53 \times 213.4 = 113.1 \text{ ft-kips}$

- $M_u = 0.75 (1.4 M_D + 1.7 M_L \pm 1.87 M_E)$ 1909.2.2
  1909.2.3

Table 2-17 Required Slab Reinforcement at Floor Level 4 due to Combined Gravity and Seismic Zone 1 Loads

Span Location		M_u (ft-kips)	b (in.)	d (in.)	A_s (in.^{2})	$A_{s(min)}$ (in.^{2})	No. of #4 bars	No. of #5 bars
END SPAN								
Column Strip	Ext. Negative	139.8	120	7.625	4.25	1.94	22	14
		-74.1	120	7.625	2.21	—	12**	8**
	Positive	66.2	120	7.625	1.97	1.94	10	8*
	Int. Negative	155.8	120	7.625	4.76	1.94	24	16
		-20.6	120	7.625	0.60	—	3**	2**
Middle Strip	Ext. Negative	0.0	120	7.625	0.00	1.94	10	8*
	Positive	44.8	120	7.625	1.32	1.94	10	8*
	Int. Negative	51.9	120	7.625	1.54	1.94	10	8*
		-6.9	120	7.625	0.20	—	1**	1**
INTERIOR SPAN								
Column Strip	Positive	44.8	120	7.625	1.32	1.94	10	8*
	Negative	150.3	120	7.625	4.58	1.94	23	15
		-24.9	120	7.625	0.73	—	4**	3**
Middle Strip	Positive	29.9	120	7.625	0.88	1.94	10	8*
	Negative	50.1	120	7.625	1.48	1.94	10	8*
		-8.3	120	7.625	0.24	—	2**	1**

$^*s_{max}$ = 2h = 18"; 120/18 = 6.7 spaces, say 8 bars
***No. of bars required to be anchored or continuous at support*

Total factored moment in the design strip (see Table 2-1 and Fig. 2-17):

M_u = 0.75 [(0.7 × 213.4) ± (1.87 × 68.2)] = 207.7 ft-kips, 16.4 ft-kips   *1913.6.3.3*

75% of this moment is distributed to the column strip:

M_u = 0.75 × 207.7 = 155.8 ft-kips (governs)   *1913.6.4*

• M_u = 0.9 M_D ± 1.43 M_E *1909.2.2*
          1909.2.3

M_D = 0.70 × 0.1325 × 20 × 18.33^2/8 = 77.9 ft-kips   *1913.6.3.3*

M_u = (0.9 × 77.9) ± (1.43 × 68.2) = 98.4 ft-kips, -27.4 ft-kips

75% of this moment is distributed to the column strip; therefore,

M_u = 73.8 ft-kips, -20.6 ft-kips

Required area of flexural reinforcement, A_s, using Table 2-7:

$$M_u/\phi f_c' bd^2 = (155.8 \times 12)/(0.9 \times 4 \times 120 \times 7.625^2) = 0.0744$$

From design aid, $\omega = 0.078$

$$A_s = 0.078 \times 120 \times 7.625 \times 4/60 = 4.76 \text{ in.}^2$$

$$A_{s(min)} = 0.0018 \times 120 \times 9 = 1.94 \text{ in.}^2 < 4.76 \text{ in.}^2 \quad \text{O.K.}$$

No. of #4 bars = 4.76/0.2 = 23.8, use 24 bars

No. of #5 bars = 4.76/0.31 = 15.4, use 16 bars

The amount of slab reinforcement and the shear strength of the slab need to be checked at the end support, the first interior support, and the second interior support (UBC 1913.3.3.2 and 1911.12.6.1).

- **End Support—Additional Flexural Reinforcement Required for Moment Transfer**

At this location, the unbalanced moment is equal to the maximum factored moment in the column strip:

— Gravity:

$$M_u = 0.26 M_o = 55.5 \text{ ft-kips} \quad \text{(see Table 2-14)}$$

— Gravity + seismic (see Fig. 2-17 for bending moment due to Seismic Zone 1 forces):

$$M_u = 0.75 (0.26 M_o + 1.87 M_E)$$

$$= 0.75 (55.5 + 1.87 \times 70.0) = 139.8 \text{ ft-kips (governs)}$$

$$M_u = 0.9 M_D + 1.43 M_E$$

$$= (0.9 \times 0.26 \times 0.1325 \times 20 \times 18.33^2/8) + (1.43 \times 70.0) = 126.1 \text{ ft-kips}$$

A fraction of the unbalanced moment must be transferred over an effective width b_e:

$$b_e = c_2 + 3h \hspace{4cm} \textit{1913.3.3.2}$$

$$= 20 + (3 \times 9) = 47.0 \text{ in.}$$

The fraction of unbalanced moment transferred by flexure is:

$$\gamma_f = \cfrac{1}{1 + \cfrac{2}{3}\sqrt{b_1/b_2}} \hspace{3cm} \textit{1913.3.3.2}$$

where $\quad b_1 = 20 + (7.625/2) = 23.81 \text{ in.}$

$\qquad\qquad b_2 = 20 + 7.625 = 27.63 \text{ in.}$

Therefore, $\gamma_f = 0.62$. Thus, the unbalanced moment transferred by flexure is:

$$\gamma_f M_u = 0.62 \times 139.8 = 86.7 \text{ ft-kips}$$

Required flexural reinforcement in 47.0 in. width:

$$\gamma_f M_u / \phi \, f'_c bd^2 = (86.7 \times 12)/(0.9 \times 4 \times 47.0 \times 7.625^2) = 0.1058$$

$$\omega \quad = 0.114 \text{ (from Table 2-7)}$$

$$A_s \quad = 0.114 \times 47.0 \times 7.625 \times 4/60 = 2.72 \text{ in.}^2$$

No. of #4 bars = 2.72/0.2 = 13.6, use 14 bars

Provide the required 14-#4 bars by concentrating 14 of the column strip bars (22-#4) within the 47.0 in. width over the column. Check bar spacing:

For 14-#4 within 47.0 in. width: 47.0/14 = 3.4 in. < 18.0 in. O.K.

For 8-#4 within (120.0 - 47.0) = 73.0 in. width: 73.0/8 = 9.1 in. < 18.0 in. O.K.

No additional bars are required for moment transfer. Figure 2-20 shows the reinforcement detail for the top bars at the exterior columns.

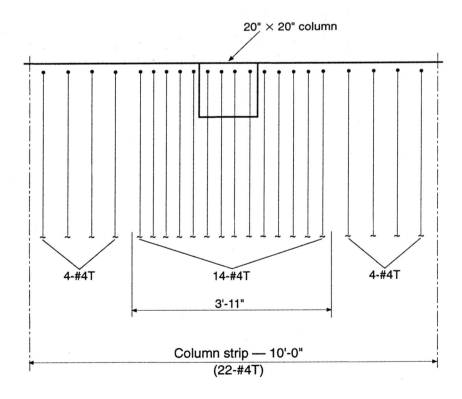

Figure 2-20 Reinforcement Detail for 22-#4 Top Bars at Exterior Columns

- ### *End Support—Check for Shear Strength*

The combined shear stress at the inside face of the critical transfer section must also be checked. The properties of the critical section for these exterior columns bending perpendicular to the edge can be determined from Case C in Fig. 2-11:

$$A_c = (2b_1 + b_2)d = [(2 \times 23.81) + 27.63] \times 7.625 = 573.8 \text{ in.}^2$$

$$J/c = [2b_1^2 d (b_1 + 2b_2) + d^3 (2b_1 + b_2)]/6b_1 = 5018.6 \text{ in.}^3$$

— Gravity:

$$V_u = w_u (0.5 \, \ell_1 \ell_2 - b_1 b_2) - (M_1 - M_2)/\ell_n$$

where
$$M_1 = 0.70 \, M_o = 149.4 \text{ ft-kips}$$

$$M_2 = 0.26 \, M_o = 55.5 \text{ ft-kips}$$

$$V_u = 0.254 [0.5 \times 20^2 - (23.81 \times 27.63/144)] - (149.4 - 55.5)/18.33 = 44.5 \text{ kips}$$

For the gravity load case, the unbalanced moment transferred by eccentricity of shear must be the full column strip nominal flexural strength, M_n, provided at the exterior support section (UBC 1913.6.3.6).

For 22-#4 bars,

$$A_s = 4.40 \text{ in.}^2$$

$$\rho = 4.40/(120 \times 7.625) = 0.0048$$

$$\omega = 0.0048 \times 60/4 = 0.072$$

From Table 2-7,

$$M_n/ f_c' bd^2 = 0.0689$$

$$M_n = 0.0689 \times 4 \times 120 \times 7.625^2 = 1923 \text{ in.-kips}$$

Combined shear stress at face of critical section:

$$v_u = \frac{V_u}{A_c} + \frac{\gamma_v M_n c}{J}$$

$$= \frac{44,500}{573.8} + \frac{0.38 \times 1923 \times 1000}{5018.6}$$

$$= 77.6 + 145.6 = 223.2 \text{ psi}$$

$$\phi v_c = 0.85 \times 4 \sqrt{4000} = 215.0 \text{ psi} \approx v_u, \text{ say O.K. } (\approx 4\% \text{ overstress})$$

— Gravity + seismic:

$$V_u = 0.75 (1.4 V_D + 1.7 V_L + 1.87 V_E)$$

where $1.4 V_D + 1.7 V_L = 44.5$ kips and $V_E = 7.5$ kips (see Fig. 2-17). Therefore,

$$V_u = 0.75 \times [44.5 + (1.87 \times 7.5)] = 43.9 \text{ kips}$$

When seismic loads are considered, shear stress computations can be based on the actual unbalanced moment, rather than on the nominal moment strength of the column strip. The actual unbalanced moment at an exterior slab section is 139.8 ft-kips (see Table 2-17). Therefore, the combined shear stress is:

$$v_u = \frac{43,900}{573.8} + \frac{0.38 \times 139.8 \times 12,000}{5018.6}$$

$$= 76.5 + 127.0 = 203.5 \text{ psi} < 215.0 \text{ psi} \quad \text{O.K.}$$

- *End Support—Design for Positive Moment*

The end support must be checked for positive moment. The controlling load combination will occur when gravity loads are at a minimum with seismic loads at a maximum:

$$M_D = M_{floorweight} + M_{superimposed}$$

$$= 0.26 \, w_D \, \ell_2 \, \ell_n^2 / 8$$

$$= 0.26 \times 0.1325 \times 20 \times 18.33^2 / 8 = 28.9 \text{ ft-kips}$$

$$M_E = 70 \text{ ft-kips (see Fig. 2-17)}$$

$$M_u = 0.9 \, M_D - 1.43 \, M_E$$

$$= (0.9 \times 28.9) - (1.43 \times 70) = -74.1 \text{ ft-kips}$$

The required area of steel to carry this moment is 2.21 in.^{2} (see Table 2-17). The 47.0 in. slab width must resist 62% of this moment; thus, the required area of steel is 1.39 in.^{2} A total of 12-#4 bars in the column strip, with 7 of these bars located in the 47.0 in. effective slab width, are required to satisfy positive moment requirements at the exterior column. Also, the integrity steel provisions in UBC 1913.4.8 require that at least two of the bottom column strip bars be anchored into the column.

- *First Interior Support—Additional Flexural Reinforcement Required for Moment Transfer*

At this location:

1) Maximum factored moment due to gravity loads equals $0.70 \, M_o = 149.4$ ft-kips

2) Maximum unfactored moment due to Seismic Zone 1 loads is 68.2 ft-kips (see Fig. 2-17)

Therefore, the total factored moment in the design strip is:

$$M_u = 0.75 \times [149.4 + (1.87 \times 68.2)] = 207.7 \text{ ft-kips} > 149.4 \text{ ft-kips}$$

In the column strip:

$$M_u = 0.75 \times 207.7 = 155.8 \text{ ft-kips}$$

In the middle strip:

$$M_u = 0.25 \times 207.7 = 51.9 \text{ ft-kips}$$

The fraction of the column strip moment to be resisted by the 47.0 in. wide strip is:

$$\gamma_f M_u = 0.6 \times 155.8 = 93.5 \text{ ft-kips}$$

where $\gamma_f = 0.6$ for $b_1 = b_2 = 27.63$ in.

Using Table 2-7, the required area of steel $A_s = 2.94$ in.^{2}

Total number of #5 bars = 2.94/0.31 = 9.5, use 10 bars.

Provide the required 10-#5 bars by concentrating 10 of the column strip bars (16-#5) within the 47.0 in. width over the column. Check bar spacing:

For 10-#5 within 47.0 in. width: 47.0/10 = 4.7 in. < 18.0 in. O.K.

For 6-#5 within 73.0 in. width: 73.0/6 = 12.2 in. < 18.0 in. O.K.

No additional bars are required for moment transfer.

- ***First Interior Support—Check for Shear Strength***

Combined shear stress at the face of the critical transfer section must be checked. Properties of the critical transfer section at an interior column are (see Case B in Fig. 2-11): $A_c = 842.6$ in.^{2} and $J/c = 7906.4$ in.^{3}

— Gravity:

$$V_u = w_u (\ell_1 \ell_2 - b_1 b_2) + (M_1 - M_2)/\ell_n$$

$$= 0.254 [20^2 - (27.63^2/144)] + (149.4 - 55.5)/18.33$$

$$= 100.3 + 5.1 = 105.4 \text{ kips}$$

The unbalanced moment is the difference between the negative moments on the faces of the column: $0.70 M_0 - 0.65 M_0 = 0.05 \times 213.4 = 10.7$ ft-kips.

The combined shear stress is:

$$v_u = \frac{105,400}{842.6} + \frac{0.4 \times 10.7 \times 12,000}{7906.4}$$

$$= 125.1 + 6.5 = 131.6 \text{ psi} < 215.0 \text{ psi} \quad \text{O.K.}$$

— Gravity + seismic:

$$V_u = 0.75 \, (1.4 \, V_D + 1.7 \, V_L + 1.87 \, V_E)$$

where $1.4 \, V_D + 1.7 \, V_L = 105.4$ kips and $V_E = 7.4$ kips (see Fig. 2-17). Therefore,

$$V_u = 0.75 \, [105.4 + (1.87 \times 7.4)] = 89.4 \text{ kips}$$

The unbalanced moment at the first interior support due to gravity loads is the difference between the moments acting on the two sides of the support, while the unbalanced moment due to seismic loads is the sum of the moments acting on the two sides of that same support:

$$M_u = 0.75 \times [(149.4 - 138.7) + 1.87 \, (68.2 + 68.6)]$$

$$= 0.75 \times (10.7 + 255.8) = 199.9 \text{ ft-kips}$$

The combined shear stress is:

$$v_u = \frac{89,400}{842.6} + \frac{0.4 \times 199.9 \times 12,000}{7906.4}$$

$$= 106.1 + 121.4 = 227.5 \text{ psi} \approx 215.0 \text{ psi, say O.K. } (\approx 6\% \text{ overstress})$$

- ***First Interior Support—Design for Positive Moment***

The first interior support must be designed for positive moment:

$$M_D = M_{floorweight} + M_{superimposed}$$

$$= 0.70 \times 0.1325 \times 20 \times 18.33^2/8 = 77.9 \text{ ft-kips}$$

$$M_E = 68.2 \text{ ft-kips}$$

$$M_u = (0.9 \times 77.9) - (1.43 \times 68.2) = -27.4 \text{ ft-kips}$$

The column strip resists 75% of this moment, with the remainder to be resisted by the middle strip. This requires that 3-#4 bars in the column strip and 1-#4 in the middle strip be fully developed at the support (see Table 2-17).

- ***Second Interior Support—Additional Flexural Reinforcement Required for Moment Transfer***

The factored negative moment due to gravity on either side of the second interior support is $0.65 \, M_o = 138.7$ ft-kips. At the same support, the negative moment due to the seismic forces is 68.7 ft-kips (see Fig. 2-17). The total factored moment at the second interior support is then:

$$M_u = 0.75 \times [138.7 + (1.87 \times 68.7)] = 200.4 \text{ ft-kips} > 138.7 \text{ ft-kips}$$

This moment is less than 207.7 ft-kips which is the maximum factored negative moment due to combined gravity and seismic loads at the first interior support. Therefore, the first interior support reinforcement will be provided at the second and subsequent interior supports.

- *Second Interior Support—Check for Shear Strength*

The combined shear stress at the face of the critical transfer section must be checked. The unbalanced moment due to gravity loads at this location is zero; due to seismic loads, it is the sum of the moments acting on both sides of that support. Therefore,

$$M_u = 0.75 \times [0.0 + 1.87 \times (68.6 + 68.7)]$$

$$= 0.75 \times 256.8 = 192.6 \text{ ft-kips}$$

This unbalanced moment is less than 199.9 ft-kips which is the unbalanced moment at the first interior support. Since the properties of the critical shear transfer section and the factored shear force due to gravity and seismic loads are the same at both supports, the shear stress at the second interior support is less than the shear stress at the first interior support, and consequently, it is less than the permissible shear stress.

- *Second Interior Support—Design for Positive Moment*

Table 2-17 shows the number of bars that must be fully developed at the support in the column and middle strips.

- *Midspan Region of Each Span*

Seismic loads cause negligible moments in the midspan regions. Reinforcement provided for gravity loads is sufficient.

- *Reinforcement Details*

Reinforcement details for the slab designed in this section are shown in Fig. 2-21. Bar lengths were determined from UBC Fig. 19-1. A summary of the slab reinforcement is given in Table 2-18.

Table 2-18 *Summary of Slab Reinforcement for Interior Design Strip, Floor Level 4, Combined Gravity and Seismic Zone 1 Loads (9-in. Slab, 20-in. Square Columns)*

Span Location		M_u (ft-kips)	No. of #4 bars	No. of #5 bars
END SPAN				
Column Strip	Ext. Negative	139.8	22	—
	Positive	66.2	12	—
	Int. Negative	155.8	—	16
Middle Strip	Ext. Negative	0.0	10	—
	Positive	44.8	10	—
	Int. Negative	51.9	—	8
INTERIOR SPAN				
Column Strip	Positive	44.8	10	—
	Negative	150.3	—	16
Middle Strip	Positive	29.9	10	—
	Negative	50.1	—	8

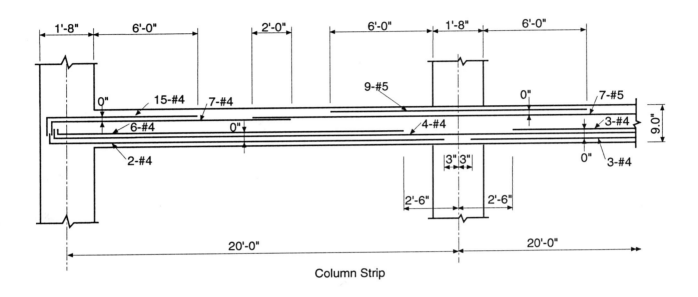

Column Strip

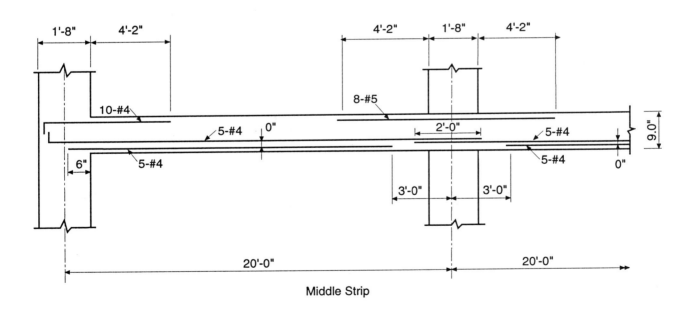

Middle Strip

Figure 2-21 Reinforcement Details for Interior Design Strip, Floor Level 4, Combined Gravity and Seismic
Zone 1 Loads

2.3.4.2   Column Design

Table 2-19 summarizes the factored axial loads, bending moments, and shear forces for the first interior column and an exterior column located above the 2nd floor level in the first interior frame. Figures 2-22 and 2-23 show the interaction diagrams for the interior column and exterior column, respectively, including all of the applicable load combinations.

Table 2-19 Summary of Factored Axial Loads, Bending Moments, and Shear Forces for a Typical 20-in. Square Column above the 2nd Floor Level, Combined Gravity and Seismic Zone 1 Loads

(1) $U = 1.4D + 1.7L$	Eq. (9-1), UBC 1909.2.1
(2) $U = 0.75 (1.4D + 1.7L \pm 1.87E)$	Eq. (9-2), UBC 1909.2.3
(3) $U = 0.9D \pm 1.43E$	Eq. (9-3), UBC 1909.2.3

Load Combination	INTERIOR COLUMN				EXTERIOR COLUMN			
	Axial Load (kips)	Bending Moment (ft-kips)		Shear Force (kips)	Axial Load (kips)	Bending Moment (ft-kips)		Shear Force (kips)
		Top	Bottom			Top	Bottom	
1	1194	8.0	-8.0	1.8	645	-27.7	27.7	-6.3
2: sidesway right	892	92.0	-113.0	21.7	395	22.6	-39.3	6.2
sidesway left	899	-80.3	101.2	-19.0	573	-64.1	80.8	-15.7
3: sidesway right	665	88.0	-109.3	22.5	267	31.1	-48.2	9.1
sidesway left	671	-88.0	109.3	-22.5	449	-57.2	74.3	-15.0

It can be seen from the figures that 8-#11 bars ($\rho_g = 3.12\%$) and 4-#11 ($\rho_g = 1.56\%$) bars are adequate for the interior and the exterior column, respectively.

Transverse reinforcement must be provided along the length of the column to satisfy both shear and confinement requirements. The maximum factored shear force $V_u = 22.5$ kips occurs in an interior column subjected to gravity and seismic loads (see Table 2-19). Column shear strength without shear reinforcement is:

$$V_c = 2 \left(1 + \frac{N_u}{2000 A_g} \right) \sqrt{f_c'} \, b_w d \qquad\qquad 1911.3.1.2$$

In this case, the factored axial load $N_u = 665$ kips (see Table 2-19). Therefore,

$$V_c = 2 \left(1 + \frac{665 \times 1000}{2000 \times 20^2} \right) \sqrt{5000} \times 20 \times 17.3/1000$$

$$= 89.6 \text{ kips}$$

$$\phi V_c = 0.85 \times 89.6 = 76.2 \text{ kips} > V_u \quad \text{O.K.}$$

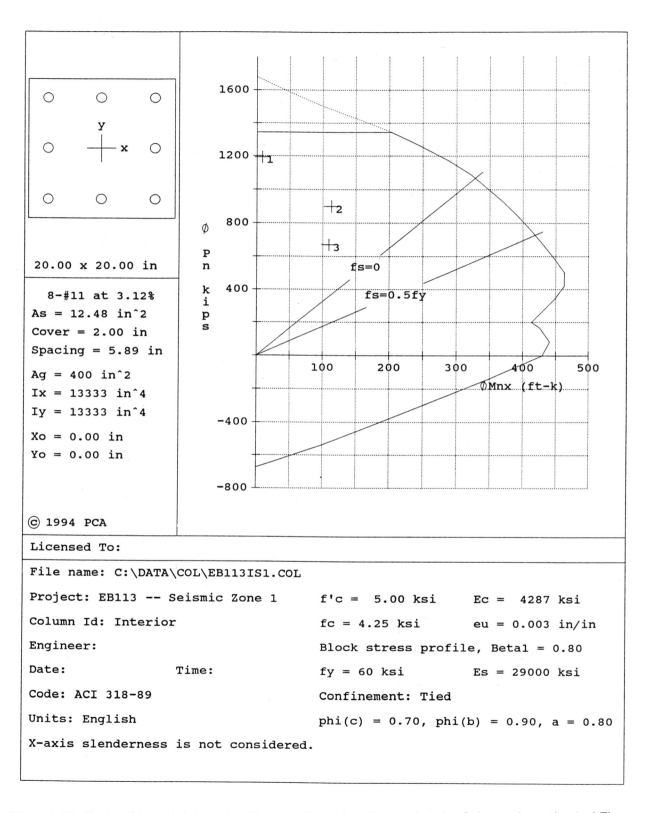

20.00 x 20.00 in

8-#11 at 3.12%
As = 12.48 in^2
Cover = 2.00 in
Spacing = 5.89 in

Ag = 400 in^2
Ix = 13333 in^4
Iy = 13333 in^4

Xo = 0.00 in
Yo = 0.00 in

© 1994 PCA

Licensed To:

File name: C:\DATA\COL\EB113IS1.COL

Project: EB113 -- Seismic Zone 1

Column Id: Interior

Engineer:

Date: Time:

Code: ACI 318-89

Units: English

X-axis slenderness is not considered.

f'c = 5.00 ksi   Ec = 4287 ksi

fc = 4.25 ksi   eu = 0.003 in/in

Block stress profile, Beta1 = 0.80

fy = 60 ksi   Es = 29000 ksi

Confinement: Tied

phi(c) = 0.70, phi(b) = 0.90, a = 0.80

Figure 2-22 Design Strength Interaction Diagram for a 20-in. Square Interior Column above the 2nd Floor Level, First Interior Frame

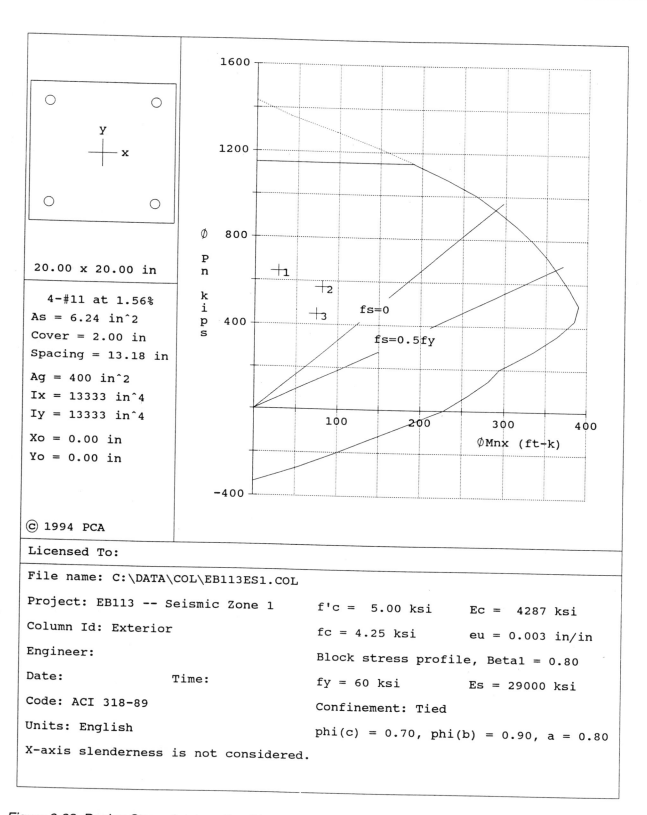

20.00 x 20.00 in

4-#11 at 1.56%
As = 6.24 in^2
Cover = 2.00 in
Spacing = 13.18 in

Ag = 400 in^2
Ix = 13333 in^4
Iy = 13333 in^4

Xo = 0.00 in
Yo = 0.00 in

© 1994 PCA

Licensed To:

File name: C:\DATA\COL\EB113ES1.COL

Project: EB113 -- Seismic Zone 1

Column Id: Exterior

Engineer:

Date: Time:

Code: ACI 318-89

Units: English

X-axis slenderness is not considered.

f'c = 5.00 ksi   Ec = 4287 ksi

fc = 4.25 ksi   eu = 0.003 in/in

Block stress profile, Beta1 = 0.80

fy = 60 ksi   Es = 29000 ksi

Confinement: Tied

phi(c) = 0.70, phi(b) = 0.90, a = 0.80

Figure 2-23 Design Strength Interaction Diagram for a 20-in. Square Exterior Column above the 2nd Floor Level, First Interior Frame

For confinement purposes, #4 ties must be provided along the length of the column with vertical spacing not to exceed the smallest of the following values (UBC 1907.10.5):

      16 longitudinal bar diameters = $16 \times 1.41 = 22.6$ in.

      48 tie bar diameters = $48 \times 0.5 = 24.0$ in.

      least column dimension = 20.0 in. (governs)

The column bars from the floor below are to be lap spliced with the column bars of the floor above. It can be seen from Figs. 2-22 and 2-23 that all column bars are in compression for all load combinations; therefore, compression splices will be used in all columns. The required length of the compression lap splice is (UBC 1912.16.1):

      $0.0005 \, f_y \, d_b = 0.0005 \times 60{,}000 \times 1.41 = 42.3$ in.

A lap splice of 43.0 in. will be provided for an interior column as shown in Fig. 2-24. A similar detail can be obtained for an exterior column.

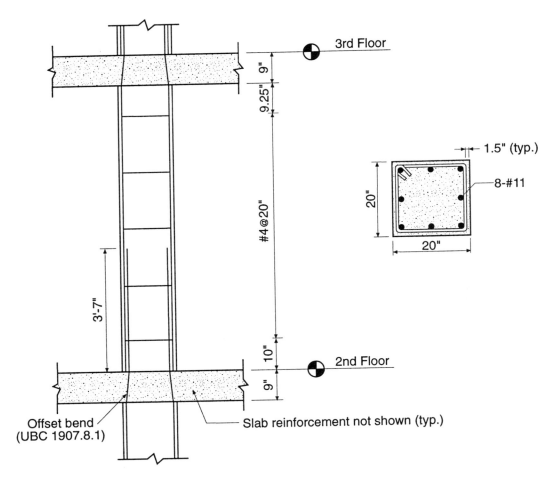

Figure 2-24 Reinforcement Details for a 20-in. Square Interior Column above the 2nd Floor Level, First Interior Frame

2.3.4.3 Story Drift Limitation and P-Δ Effects

According to UBC 1628.8.2, the story drift to story height ratio shall not exceed $0.03/R_w$ nor 0.004 for structures with a fundamental period greater than 0.7 seconds. With $R_w = 5$, the second criterion controls and the story drift limit is equal to $0.004 \times 9.5 \times 12 = 0.46$ in. The lateral displacements at the different floor levels as well as the interstory drifts for Seismic Zone 1 forces are shown in Table 2-20. The maximum allowable drift is never exceeded.

UBC 1628.9 allows the P-Δ effects to be ignored when the ratio of secondary moment to primary moment does not exceed 0.1. The ratio may be evaluated for any story as the product of the total dead and live load above the story times the seismic drift in that story, divided by the product of the seismic shear in that story times the height of that story. The ratio $\Sigma w_i \Delta_i / V_i h_{si}$ is calculated at all story levels for Seismic Zone 1 in Table 2-21. Based on these results, P-Δ effects need not be considered.

2.4 Design for Seismic Zone 2A

2.4.1 General

Preliminary calculations show that in order to satisfy Seismic Zone 2A strength requirements, the slab thickness must be increased to 10 in., and the column size increased to 22-in. square.

The estimated floor weights and the seismic forces are summarized in Table 2-22. The results of the 3-D analysis for the first interior frame are given in Figs. 2-25 and 2-26 for the columns and the slabs, respectively.

For the 2-D analysis, the torsional effects were calculated in the same manner as previously described for Seismic Zone 1. An interior frame was reanalyzed using the increased lateral forces caused by torsional effects. The comparison of the 2-D to the 3-D results are shown in Figs. 2-27 and 2-28.

Seismic Zone 2 has special design and detailing requirements in addition to those for Seismic Zone 1. The special detailing requirements for columns and two-way slabs without beams are given in UBC 1921.8. These special detailing requirements are:

For columns (UBC 1921.8.5):

*   Maximum tie spacing shall not exceed s_o over a length ℓ_o measured from the joint face. Spacing s_o shall not exceed (1) eight times the diameter of the smallest longitudinal bar enclosed, (2) twenty-four times the diameter of the tie bar, (3) one-half the smallest cross-sectional dimension of the frame member, and (4) twelve inches. Length ℓ_o shall not be less than (1) one-sixth the clear span of the member, (2) maximum cross-sectional dimension of the member, and (3) eighteen inches.

*   The first tie shall be located at no more than $s_o/2$ from the joint face.

*   Joint reinforcement shall conform to UBC 1911.11.2.

*   Tie spacing shall not exceed twice the spacing s_o anywhere along the column height.

Table 2-20 Lateral Displacements of Example Building under Seismic Zone 1 Forces

Floor Level	Total Lateral Displacement (in.)	Interstory Drift (in.)
R	2.26	0.05
14	2.21	0.08
13	2.13	0.10
12	2.03	0.13
11	1.90	0.14
10	1.76	0.16
9	1.60	0.18
8	1.42	0.19
7	1.23	0.20
6	1.03	0.21
5	0.82	0.22
4	0.60	0.22
3	0.38	0.20
2	0.18	0.18

Table 2-21 P-Δ Check under Seismic Zone 1 Forces

Floor Level	Gravity Loads * Σw_i (kips)	Story Drift Δ_i (in.)	$\Sigma w_i \Delta_i$ (in.-kips)	Story Shear V_i (kips)	Story Height h_{si} (in.)	$V_i h_{si}$ (in.-kips)	$\Sigma w_i \Delta_i / V_i h_{si}$
R	1683	0.05	84.2	83.3	114	9496.2	0.009
14	3349	0.08	267.9	130.3	114	14,854.2	0.018
13	5015	0.10	501.5	173.8	114	19,813.2	0.025
12	6681	0.13	868.5	213.7	114	24,361.8	0.036
11	8347	0.14	1168.6	250.1	114	28,511.4	0.041
10	10,013	0.16	1602.1	283.0	114	32,262.0	0.050
9	11,679	0.18	2102.2	312.4	114	35,613.6	0.059
8	13,345	0.19	2535.6	338.2	114	38,554.8	0.066
7	15,011	0.20	3002.2	360.5	114	41,097.0	0.073
6	16,677	0.21	3502.2	379.3	114	43,240.2	0.081
5	18,343	0.22	4035.5	394.5	114	44,973.0	0.090
4	20,009	0.22	4402.0	406.2	114	46,306.8	0.095
3	21,675	0.20	4335.0	414.4	114	47,241.6	0.092
2	23,366	0.18	4205.9	419.1	150	62,865.0	0.067

* Includes floor weight, superimposed dead load, and reduced roof and floor live loads

For two-way slabs without beams (UBC 1921.8.6):

• Factored slab moment at support related to earthquake effects shall be determined for load combinations defined by Eqs. (9-2) and (9-3). All reinforcement provided to resist M_s, the portion of slab moment balanced by support moment, shall be placed within the column strip defined in UBC 1913.2.1.

• The fraction, defined by Eq. (13-1), of moment M_s shall be resisted by reinforcement placed within the effective slab width specified in UBC 1913.3.2.

• Not less than one-half of the reinforcement in the column strip at the support shall be placed within the effective slab width specified in UBC 1913.3.2.

- Not less than one-fourth of the top reinforcement at the support in the column strip shall be continuous throughout the span.

- Continuous bottom reinforcement in the column strip shall not be less than one-third of the top reinforcement at the support.

- Not less than one-half of all bottom reinforcement at midspan shall be continuous and shall develop its yield strength at face of support as defined in UBC 1913.6.2.5.

- At discontinuous edges of the slab, all top and bottom reinforcement shall be developed at the face of the support as defined in UBC 1913.6.2.5.

Table 2-22 Seismic Forces for Zone 2A

Floor Level	Weight, w_x (kips)	Height, h_x (ft)	$w_x h_x$ (ft-kips)	Lateral Force (kips)
R	1582	136.0	215,152	115.1
14	1661	126.5	210,117	65.0
13	1661	117.0	194,337	60.1
12	1661	107.5	178,558	55.3
11	1661	98.0	162,778	50.4
10	1661	88.5	146,999	45.5
9	1661	79.0	131,219	40.6
8	1661	69.5	115,440	35.7
7	1661	60.0	99,660	30.8
6	1661	50.5	83,881	26.0
5	1661	41.0	68,101	21.1
4	1661	31.5	52,322	16.2
3	1661	22.0	36,542	11.3
2	1692	12.5	21,150	6.5
Σ	23,206		1,716,253	579.6

Building height h_n $= 136\ ft$
Building weight W $= 23,206\ kips$
Fundamental period T $= 0.030\ (h_n)^{3/4} = 1.2\ sec.$
Seismic zone factor Z $= 0.15$
Importance factor I $= 1.0$
Response coefficient R_w $= 8$
Soil factor S $= 1.2$
Coefficient C $= 1.25\ S/T^{2/3} = 1.33$
Base shear V $= ZICW/R_w = 579.6\ kips$
Top level force F_t $= 0.07TV = 48.5\ kips$

The special requirement of UBC 1921.8.3 is for shear design of beams, columns, and two-way slabs and is as follows:

Design shear strength of beams, columns, and two-way slabs resisting earthquake effects shall not be less than either (1) the sum of the shear associated with development of nominal moment strengths of the member at both restrained ends of the clear span and the shear calculated for gravity loads, or (2) the maximum shear obtained from design load combinations which include earthquake effect E, with E assumed to be twice that prescribed in UBC 1613. The second criterion is used for the shear design of the slab in this section.

R	12.9 / -1.9	28.1 / 10.7	32.8 / 14.2
14	24.9 / 5.4	37.5 / 20.6	40.0 / 23.2
13	28.6 / 11.1	47.5 / 31.2	50.2 / 33.8
12	33.2 / 16.9	55.9 / 41.0	58.4 / 43.4
11	37.0 / 22.3	63.6 / 50.1	66.0 / 52.3
10	40.4 / 27.2	70.5 / 58.4	72.6 / 60.4
9	43.4 / 31.8	76.5 / 65.9	78.4 / 67.7
8	45.9 / 36.0	81.8 / 72.7	83.4 / 74.2
7	48.0 / 39.8	86.1 / 78.6	87.4 / 79.8
6	49.5 / 43.2	89.6 / 84.0	90.6 / 84.0
5	50.4 / 47.7	91.6 / 88.6	92.1 / 89.0
4	48.7 / 53.3	91.5 / 93.8	91.8 / 94.1
3	42.4 / 59.5	84.8 / 106.0	84.6 / 105.0
2	29.8 / 159.0	55.0 / 172.0	54.9 / 171.0

Bending Moments (ft-kips)

R	1.2	4.1	5.0
14	3.2	6.1	6.7
13	4.2	8.3	8.8
12	5.3	10.2	10.7
11	6.2	12.0	12.5
10	7.1	13.6	14.0
9	7.9	15.0	15.4
8	8.6	16.3	16.6
7	9.2	17.3	17.6
6	9.8	18.3	18.5
5	10.3	19.0	19.1
4	10.7	19.5	19.6
3	10.7	20.1	20.0
2	15.1	18.2	18.1

Shear Forces (kips)

R	1.1	0.6	0.0
14	3.3	1.2	0.0
13	6.6	1.8	0.1
12	11.1	2.4	0.1
11	16.6	2.9	0.1
10	23.0	3.3	0.2
9	30.3	3.7	0.2
8	38.4	4.0	0.2
7	47.1	4.3	0.2
6	56.5	4.4	0.3
5	66.4	4.5	0.3
4	76.6	4.4	0.3
3	86.7	4.4	0.3
2	95.9	4.1	0.3

Axial Forces (kips)

Figure 2-25 Results of 3-D Analysis Under Seismic Zone 2A Forces for Columns of the First Interior Frame

R	-12.4	-11.4	-17.6	-17.2	-17.4
14	-21.7	-21.1	-28.1	-27.9	-28.4
13	-32.7	-31.6	-38.0	-37.9	-38.3
12	-42.9	-41.6	-47.7	-47.6	-48.0
11	-52.4	-50.9	-56.6	-56.4	-56.8
10	-61.1	-59.3	-64.6	-64.5	-64.9
9	-68.9	-67.0	-71.8	-71.7	-72.0
8	-76.0	-73.9	-78.2	-78.1	-78.4
7	-82.2	-80.0	-83.6	-83.5	-83.8
6	-87.6	-85.3	-88.1	-88.1	-88.3
5	-92.0	-89.5	-91.5	-91.5	-91.6
4	-94.7	-92.3	-93.4	-93.4	-93.5
3	-93.7	-91.7	-92.2	-92.2	-92.3
2	-85.1	-82.4	-81.1	-81.2	-81.3

Bending Moments (ft-kips)

R	-1.3	-1.9	-1.9
14	-2.4	-3.1	-3.1
13	-3.5	-4.2	-4.2
12	-4.7	-5.2	-5.3
11	-5.7	-6.2	-6.3
10	-6.6	-7.1	-7.1
9	-7.5	-7.9	-7.9
8	-8.3	-8.6	-8.6
7	-8.9	-9.2	-9.2
6	-9.5	-9.7	-9.7
5	-10.0	-10.1	-10.1
4	-10.3	-10.3	-10.3
3	-10.2	-10.1	-10.2
2	-9.2	-8.9	-8.9

Shear Forces (kips)

Figure 2-26 Results of 3-D Analysis under Seismic Zone 2A Forces for Slabs of the First Interior Frame

Bending Moments

R	1.08	1.13	1.15
	1.28	1.13	1.18
14	1.04	1.08	1.10
	1.02	1.08	1.11
13	1.05	1.09	1.11
	0.96	1.09	1.11
12	1.05	1.09	1.10
	0.99	1.10	1.10
11	1.05	1.09	1.10
	1.01	1.10	1.10
10	1.05	1.10	1.10
	1.02	1.10	1.10
9	1.05	1.10	1.10
	1.03	1.10	1.10
8	1.05	1.10	1.10
	1.03	1.10	1.10
7	1.05	1.10	1.10
	1.03	1.10	1.10
6	1.05	1.10	1.11
	1.04	1.11	1.10
5	1.10	1.10	1.11
	0.98	1.10	1.11
4	1.08	1.10	1.10
	1.03	1.08	1.08
3	1.09	1.14	1.14
	1.01	1.10	1.09
2	1.05	1.12	1.12
	1.00	1.01	1.01

Shear Forces

R	1.14	1.24	1.27
14	1.09	1.19	1.21
13	1.13	1.19	1.21
12	1.13	1.20	1.20
11	1.13	1.20	1.21
10	1.14	1.20	1.21
9	1.14	1.20	1.21
8	1.14	1.20	1.21
7	1.14	1.21	1.21
6	1.15	1.21	1.21
5	1.13	1.21	1.22
4	1.15	1.20	1.20
3	1.14	1.22	1.22
2	1.03	1.07	1.07

Axial Forces

R	1.22	1.18	1.00
14	1.18	1.17	1.00
13	1.15	1.18	1.00
12	1.14	1.18	1.00
11	1.12	1.18	1.00
10	1.11	1.18	1.00
9	1.10	1.18	1.00
8	1.10	1.18	1.00
7	1.09	1.18	1.00
6	1.09	1.18	1.00
5	1.09	1.19	1.00
4	1.09	1.19	1.00
3	1.09	1.18	1.00
2	1.09	1.17	1.00

Figure 2-27 Ratio of 2-D to 3-D Results under Seismic Zone 2A Forces for Columns of an Interior Frame

Bending Moments

R	1.04	1.03	1.05	1.04	1.04
14	1.05	1.04	1.05	1.05	1.05
13	1.04	1.04	1.05	1.04	1.04
12	1.04	1.04	1.04	1.04	1.04
11	1.04	1.04	1.04	1.04	1.04
10	1.04	1.04	1.04	1.04	1.04
9	1.04	1.04	1.04	1.04	1.04
8	1.04	1.04	1.04	1.04	1.04
7	1.04	1.04	1.04	1.04	1.04
6	1.05	1.04	1.04	1.04	1.05
5	1.04	1.04	1.05	1.05	1.05
4	1.05	1.05	1.05	1.05	1.05
3	1.06	1.06	1.05	1.06	1.05
2	1.09	1.08	1.07	1.07	1.07

Shear Forces

R	1.05	1.06	1.06
14	1.06	1.07	1.07
13	1.06	1.06	1.06
12	1.06	1.06	1.06
11	1.06	1.06	1.07
10	1.06	1.06	1.06
9	1.06	1.06	1.06
8	1.06	1.06	1.06
7	1.06	1.06	1.06
6	1.06	1.06	1.06
5	1.06	1.06	1.06
4	1.07	1.07	1.07
3	1.08	1.08	0.99
2	1.10	1.09	1.09

Figure 2-28 Ratio of 2-D to 3-D Results under Seismic Zone 2A Forces for Slabs of an Interior Frame

2.4.2   Factored Loads and Moments due to Gravity

The slab with a thickness of 10 in. has an effective depth of 8.625 in. (3/4 in. cover and #5 rebar). The design loads are:

        Factored slab weight = $1.4 \times 150 \times 10/12 = 175$ psf

        Factored superimposed dead load = $1.4 \times 20 = 28$ psf

        Factored live load = $1.7 \times 40 = 68$ psf

        Total factored load $w_u = 175 + 28 + 68 = 271$ psf $= 0.271$ ksf

With 22-in. square columns, the clear span length $\ell_n = 20 - (22/12) = 18.17$ ft.

2.4.2.1   Factored Moments in Slab

Total factored static moment:

$$M_o = \frac{w_u \ell_2 \ell_n^2}{8} \qquad\qquad\qquad 1913.6.2.2$$

$$= \frac{0.271 \times 20 \times 18.17^2}{8} = 223.7 \text{ ft-kips}$$

Table 2-23 summarizes the moments in the column and the middle strips for an interior design strip.

Table 2-23 Factored Slab Moments due to Gravity Loads

Span Location		M_u (ft-kips)
END SPAN		
Column Strip	Exterior Negative Positive Interior Negative	$0.26\,M_o = $ 58.2 $0.31\,M_o = $ 69.3 $0.53\,M_o = $ 118.6
Middle Strip	Exterior Negative Positive Interior Negative	0.0 $0.21\,M_o = $ 47.0 $0.17\,M_o = $ 38.0
INTERIOR SPAN		
Column Strip	Positive Negative	$0.21\,M_o = $ 47.0 $0.49\,M_o = $ 109.6
Middle Strip	Positive Negative	$0.14\,M_o = $ 31.3 $0.16\,M_o = $ 35.8

2.4.2.2 Factored Axial Loads and Moments in Columns

Design axial loads due to gravity on an interior column and an exterior column of an interior frame are shown in Tables 2-24 and 2-25, respectively. The live load reduction factor is in accordance with Eq. (6-1) in UBC 1606.

The factored gravity moment to be transferred from the slab to the interior column above and below the slab is:

$$M_i = 0.07 \times (0.5 \times 1.7 \times 0.04 \times 20 \times 18.17^2) = 15.7 \text{ ft-kips}$$

This moment will be distributed to the columns inversely proportional to their lengths above and below the floor slab under consideration. The total exterior negative moments from the slab must be transferred directly to the exterior columns.

Table 2-24 Axial Loads due to Gravity on a 22-in. Square Interior Column of an Interior Frame

Floor	Dead Load (psf)	Live Load (psf)	Area, A (ft^2)	Reduction Factor (1-R/100)*	Reduced Live Load (psf)	Factored Load (kips)**	Cumulative Factored Load (kips)
R	145	40	400	0.80	32.0	110	110
14	145	40	800	0.48	19.2	101	211
13	145	40	1200	0.40	16.0	99	309
12	145	40	1600	0.40	16.0	99	408
11	145	40	2000	0.40	16.0	99	507
10	145	40	2400	0.40	16.0	99	605
9	145	40	2800	0.40	16.0	99	704
8	145	40	3200	0.40	16.0	99	803
7	145	40	3600	0.40	16.0	99	902
6	145	40	4000	0.40	16.0	99	1000
5	145	40	4400	0.40	16.0	99	1099
4	145	40	4800	0.40	16.0	99	1198
3	145	40	5200	0.40	16.0	99	1297
2	145	40	5600	0.40	16.0	101	1398

* $R = r(A - 150)$

** *Factored load = [1.4(Dead Load) + 1.7(Reduced Live Load)] × Tributary Area. Factored load includes 1.4 × column weight = 0.7 kips/ft*

Table 2-25 Axial Loads due to Gravity on a 22-in. Square Exterior Column of an Interior Frame

Floor	Dead Load (psf)	Live Load (psf)	Area, A (ft²)	Reduction Factor (1-R/100)*	Reduced Live Load (psf)	Factored Load (kips)**	Cumulative Factored Load (kips)
R	145	40	200	0.96	38.4	60	60
14	145	40	400	0.80	32.0	58	118
13	145	40	600	0.64	25.6	56	174
12	145	40	800	0.48	19.2	54	228
11	145	40	1000	0.40	16.0	53	281
10	145	40	1200	0.40	16.0	53	334
9	145	40	1400	0.40	16.0	53	386
8	145	40	1600	0.40	16.0	53	439
7	145	40	1800	0.40	16.0	53	492
6	145	40	2000	0.40	16.0	53	544
5	145	40	2200	0.40	16.0	53	597
4	145	40	2400	0.40	16.0	53	650
3	145	40	2600	0.40	16.0	53	702
2	145	40	2800	0.40	16.0	55	757

* $R = r(A - 150)$
** Factored load = [1.4(Dead Load) + 1.7(Reduced Live Load)] $\times$ Tributary Area. Factored load includes 1.4 $\times$ column weight = 0.7 kips/ft

2.4.3   Design for Combined Loading

2.4.3.1   Slab Design

The required slab reinforcement for an interior design strip at floor level 4 due to the combined effects of gravity and Seismic Zone 2A forces is summarized in Table 2-26.

The amount of slab reinforcement and the shear strength of the slab need to be checked at the end support, the first interior support, and the second interior support.

• *End Support—Additional Flexural Reinforcement Required for Moment Transfer*

At this location, the unbalanced moment is equal to 176.5 ft-kips which is the maximum factored moment in the column strip (see Table 2-26). A fraction of this moment must be transferred over an effective width b_e:

$$b_e = c_2 + 3h$$

$$= 22 + (3 \times 10) = 52.0 \text{ in.}$$

The unbalanced moment transferred by flexure is:

$$\gamma_f M_u = 0.62 \times 176.5 = 109.4 \text{ ft kips}$$

The required area of steel in the 52.0 in. width to resist this moment is $A_s = 3.0$ in.^{2} which is equivalent to 10 - #5 bars.

Table 2-26 Required Slab Reinforcement at Floor Level 4 due to Combined Gravity and Seismic Zone 2A Loads

Span Location		M_u (ft-kips)	b (in.)	d (in.)	A_s (in.^{2})	$A_{s(min)}$ (in.^{2})	No. of #4 bars	No. of #5 bars
END SPAN								
Column Strip	Ext. Negative	176.5	120	8.625	4.74	2.16	24	16
		-107.4	120	8.625	2.84	—	15**	10**
	Positive	69.3	120	8.625	1.81	2.16	11	8*
	Int. Negative	186.3	120	8.625	4.98	2.16	25	17
		-43.7	120	8.625	1.14	—	6**	4**
Middle Strip	Ext. Negative	0.0	120	8.625	0.0	2.16	11	8*
	Positive	47.0	120	8.625	1.22	2.16	11	8*
	Int. Negative	62.1	120	8.625	1.61	2.16	11	8*
		-14.6	120	8.625	0.38	—	2**	2**
INTERIOR SPAN								
Column Strip	Positive	47.0	120	8.625	1.22	2.16	11	8*
	Negative	180.1	120	8.625	4.84	2.16	25	16
		-47.9	120	8.625	1.25	—	7**	4**
Middle Strip	Positive	31.3	120	8.625	0.81	2.16	11	8*
	Negative	60.0	120	8.625	1.57	2.16	11	8*
		-16.0	120	8.625	0.41	—	2**	2**

*s_{max} = 18"; 120/18 = 6.7 spaces, say 8 bars
**No. of bars required to be anchored or continuous at support

Provide the required 10-#5 bars by concentrating 10 of the column strip bars (16-#5) within the 52.0 in. width over the column. Check bar spacing:

For 10-#5 within 52.0 in. width: 52.0/10 = 5.2 in. < 18.0 in. O.K.

For 6-#5 within 68.0 in. width: 68.0/6 = 11.3 in. < 18.0 in. O.K.

No additional bars are required for moment transfer. Figure 2-29 shows the reinforcement detail for the top bars at the exterior columns.

• **End Support—Check for Shear Strength**

The combined shear stress at the inside face of the critical transfer section must also be checked. The properties of the critical section are (see Case C in Fig. 2-11): A_c = 718.0 in.^{2} and J/c = 6962.3 in.^{3}

— Gravity:

$$V_u = w_u (0.5 \, \ell_1 \ell_2 - b_1 b_2) - (M_1 - M_2)/\ell_n$$

where b_1 = 22 + (8.625/2) = 26.31 in.

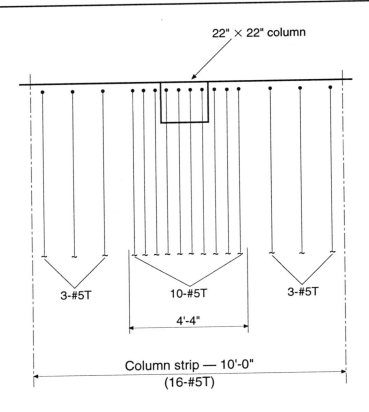

Figure 2-29 *Reinforcement Detail for 16-#5 Top Bars at Exterior Columns*

b_2 $= 22 + 8.625 = 30.63$ in.

M_1 $= 0.70M_o = 156.6$ ft-kips

M_2 $= 0.26M_o = 58.2$ ft-kips

V_u $= 0.271 [0.5 \times 20^2 - (26.31 \times 30.63/144)] - (156.6 - 58.2)/18.17$

$= 52.7 - 5.4 = 47.3$ kips

For the gravity load case, the unbalanced moment transferred by eccentricity of shear must be the full column strip nominal flexural strength, M_n, provided at the exterior support section (UBC 1913.6.3.6). With 16-#5 bars at this location, $M_n = 204.8$ ft-kips. The combined shear stress at the face of the critical section is:

$$v_u = \frac{V_u}{A_c} + \frac{\gamma_v M_n c}{J}$$

$$= \frac{47,300}{718.0} + \frac{0.38 \times 204.8 \times 12,000}{6962.3}$$

$$= 65.9 + 134.2 = 200.0 \text{ psi}$$

$$\phi v_c = 0.85 \times 4\sqrt{4000} = 215.0 \text{ psi} > v_u \quad \text{O.K.}$$

— Gravity + seismic:

The maximum factored shear force due to gravity and seismic forces will be determined using the second provision in UBC 1921.8.3, which requires that the earthquake effect E in the design load combination be twice that prescribed in UBC 1613:

$$V_u = 0.75 [(1.4 V_D + 1.7 V_L) + (2 \times 1.87 V_E)]$$

where $1.4 V_D + 1.7 V_L = 47.3$ kips and $V_E = 10.3$ kips (see Fig. 2-26). Therefore,

$$V_u = 0.75 \times [47.3 + (2 \times 1.87 \times 10.3)] = 64.4 \text{ kips}$$

When seismic loads are considered, shear stress computations will be based on the actual unbalanced moment, rather than the nominal moment strength of the column strip. The actual unbalanced moment at an exterior slab section is 176.5 ft-kips (see Table 2-26). Therefore,

$$v_u = \frac{64,400}{718.0} + \frac{0.38 \times 176.5 \times 12,000}{6962.3}$$

$$= 89.7 + 115.6 = 205.3 \text{ psi} < \phi v_c = 215.0 \text{ psi} \quad \text{O.K.}$$

- ***End Support—Design for Positive Moment***

The end span must be checked for positive moment. The controlling load combination will occur when gravity loads are at a minimum with seismic loads at a maximum:

$$M_D = M_{floorweight} + M_{superimposed}$$

$$= 0.26 w_D \ell_2 \ell_n^2 / 8$$

$$= 0.26 \times 0.145 \times 20 \times 18.17^2 / 8 = 31.1 \text{ ft-kips}$$

$$M_E = 94.7 \text{ ft-kips (see Fig. 2-26)}$$

$$M_u = 0.9 M_D - 1.43 M_E$$

$$= (0.9 \times 31.1) - (1.43 \times 94.7) = -107.4 \text{ ft-kips}$$

The required area of steel to carry this moment is 2.84 in.[2] (see Table 2-26). A total of 10-#5 bars are required to satisfy positive moment requirements at the exterior column. The 52-in. slab width in the column strip must resist 62% of the applied moment. This requires 6-#5 bars, with at least two of these three bars anchored into the column to satisfy the integrity steel requirements in UBC 1913.4.8.

As noted in Section 2.4.1, special reinforcement details are required for two-way slabs without beams in Seismic Zone 2A. Those provisions will be addressed once the reinforcement and shear stress requirements are satisfied at the other support locations along the span.

- *First Interior Support—Additional Flexural Reinforcement Required for Moment Transfer*

At this location:

1) Maximum factored moment due to gravity loads equals $0.70 \, M_o = 156.6$ ft-kips

2) Maximum unfactored moment due to Seismic Zone 2A loads is 93.4 ft-kips (see Fig. 2-26).

Therefore, the total moment in the slab is:

$$M_u = 0.75 \times [156.6 + (1.87 \times 93.4)] = 248.4 \text{ ft-kips} > 156.6 \text{ ft-kips}$$

The column strip moment is:

$$M_u = 0.75 \times 248.4 = 186.3 \text{ ft-kips}$$

The middle strip moment is:

$$M_u = 0.25 \times 248.4 = 62.1 \text{ ft-kips}$$

The fraction of the column strip moment to be resisted by the 52.0 in. wide strip is:

$$\gamma_f M_u = 0.6 \times 186.3 = 111.8 \text{ ft-kips}$$

where $\gamma_f = 0.6$ for $b_1 = b_2 = 30.63$ in.

Using Table 2-7, the required area of steel $A_s = 3.06$ in.^{2}

Total number of #5 bars = 3.06/0.31 = 9.9, use 10 bars.

Provide the required 10-#5 bars by concentrating 10 of the column strip bars (17-#5) within the 52.0 in. width over the column. For symmetry, add one column strip bar to the remaining 7 bars so that 4 bars will be on each side of the 52.0 in. strip. Check bar spacing:

        For 10-#5 within 52.0 in. width: 52.0/10 = 5.2 in. < 18.0 in.   O.K.

        For 8-#5 within 68.0 in. width: 68.0/8 = 8.5 in. < 18.0 in.   O.K.

No additional bars are required for moment transfer.

- *First Interior Support—Check for Shear Strength*

Combined shear stress at the face of the critical section must be checked. Properties of the critical section are as follows (see Case B in Fig. 2-11): $A_c = 1056.6$ in.^{2} and $J/c = 10{,}999.6$ in.^{3}

— Gravity:

$$V_u = w_u (\ell_1 \ell_2 - b_1 b_2) + (M_1 - M_2)/\ell_n$$

$$= 0.271 \, [20^2 - (30.63^2/144)] + (156.6 - 58.2)/18.17$$

$$= 106.6 + 5.4 = 112.0 \text{ kips}$$

The unbalanced moment is the difference between the negative moments on the faces of the column: $0.70 M_0 - 0.65 M_0 = 0.05 \times 223.7 = 11.2$ ft-kips.

Therefore, the combined shear stress is:

$$v_u = \frac{112,000}{1056.6} + \frac{0.4 \times 11.2 \times 12,000}{10,999.6}$$

$$= 106.0 + 4.9 = 110.9 \text{ psi} < 215.0 \text{ psi} \quad \text{O.K.}$$

— Gravity + seismic:

Using the provision in UBC 1921.8.3:

$$V_u = 0.75 \left[(1.4 V_D + 1.7 V_L) + (2 \times 1.87 V_E) \right]$$

where $1.4 V_D + 1.7 V_L = 112.0$ kips and $V_E = 10.3$ kips (see Fig. 2-26). Therefore,

$$V_u = 0.75 \times [112.0 + (2 \times 1.87 \times 10.3)] = 112.9 \text{ kips}$$

The unbalanced moment at the first interior support due to gravity loads is the difference between the moments acting on the two sides of the support, while the unbalanced moment due to seismic loads is the sum of the moments acting on the two sides of that same support:

$$M_u = 0.75 \times \left[(156.6 - 145.4) + 1.87(92.3 + 93.4) \right]$$

$$= 0.75 \times (11.2 + 347.3) = 268.8 \text{ ft-kips}$$

The combined shear stress is:

$$v_u = \frac{112,900}{1056.6} + \frac{0.4 \times 268.8 \times 12,000}{10,999.6}$$

$$= 106.9 + 117.3 = 224.2 \text{ psi} > 215.0 \text{ psi}$$

In this case, the strength of the concrete slab will be increased to $f'_c = 5000$ psi yielding an allowable shear stress of 240.0 psi, which is greater than the combined shear stress.

• *First Interior Support—Design for Positive Moment*

The first interior support must be designed for positive moment:

$$M_D = 0.70 \times 0.145 \times 20 \times 18.17^2/8 = 83.8 \text{ ft-kips}$$

$$M_E = 93.4 \text{ ft-kips (see Fig. 2-26)}$$

$$M_u = (0.9 \times 83.8) - (1.43 \times 93.4) = -58.2 \text{ ft-kips}$$

Table 2-26 shows the number of bars that must be fully developed at the support in the column and middle strips.

- *Second Interior Support—Additional Flexural Reinforcement Required for Moment Transfer*

The factored negative moment due to gravity on either side of the second interior support is $0.65 M_o = 145.4$ ft-kips. At the same support, the negative moment due to the seismic forces is 93.5 ft-kips (see Fig. 2-26). The total factored moment at the second support is then:

$$M_u = 0.75 \times [145.4 + (1.87 \times 93.5)] = 240.2 \text{ ft-kips} > 145.4 \text{ ft-kips}$$

This moment is less than 248.4 ft-kips which is the maximum factored negative moment due to combined gravity and seismic loads at the first interior support. Therefore, the first interior support reinforcement will be provided at the second and subsequent supports.

- *Midspan Region of Each Span*

Since the seismic loads cause negligible moments in the midspan regions, the reinforcement provided for gravity loads is sufficient. Reinforcement details for this region will be discussed below.

- *Reinforcement Details*

As noted in Section 2.4.1, special reinforcement details must be provided for two-way slabs without beams in Seismic Zone 2A. UBC 1921.8.6.3 requires that at least one-half of the reinforcement in the column strip at the supports be placed within the effective slab width b_e. At the edge column, 16 reinforcing bars are required in the column strip at the top of the slab. Ten of these 16 bars are located in b_e, which satisfies this requirement. Ten bars are required in the column strip at the bottom of the slab at this location. Six of these 10 bars are located in b_e which also satisfies the above requirement. At all interior columns, 18 reinforcing bars are required in the column strip at the top of the slab. Ten of these 18 bars are located in b_e, which satisfies this requirement. Note that the bars provided in the bottom of the slab at this location satisfy UBC 1921.8.6.3 as well.

UBC 1921.8.6.4 requires that at least one-fourth of the top reinforcement in the column strip at the supports be continuous throughout the span. Based on the number of bars at the first interior support, 5 bars will be made continuous to satisfy this requirement.

UBC 1921.8.6.5 requires that continuous bottom reinforcement in the column strip shall not be less than one-third of the top reinforcement in the column strip at the supports. Again, based on the top reinforcement at the first interior support, 6 bars will be made continuous in the column strip in the bottom of the slab.

UBC 1921.8.6.7 requires that all top and bottom reinforcement at discontinuous edges of the slab must be developed at the face of the exterior support. The development length for #5 bars with clear spacing $> 5\, d_b$ and with cover $= 1.2\, d_b$ is:

$$\ell_d = 0.8 \times 1.4 \times 0.04\, A_b f_y / \sqrt{f_c'} \qquad \textit{1912.2}$$
$$= 0.8 \times 1.4 \times 0.04 \times 0.31 \times 60{,}000/\sqrt{5000} = 11.8 \text{ in.}$$

but not less than

$$\ell_d = 0.03 d_b f_y / \sqrt{f_c'}$$
$$= 0.03 \times 0.625 \times 60{,}000/\sqrt{5000} = 15.9 \text{ in. (governs)}$$

Therefore, all reinforcement at the exterior edges must be extended at least 16.0 in. beyond the interior face of the exterior column.

UBC 1921.8.6.6 requires that not less than one-half of all bottom reinforcement at midspan shall be continuous and shall develop its yield strength at the face of the support. This requirement is automatically satisfied at the exterior edge of the slab by satisfying UBC 1921.8.6.7 as shown above. It is also satisfied by having at least one-half of all bottom reinforcement at midspan continue through the interior supports.

Reinforcement details for the slab designed in this section are shown in Fig. 2-30. Bar lengths were determined from UBC Fig. 19-1. A summary of the slab reinforcement is given in Table 2-27.

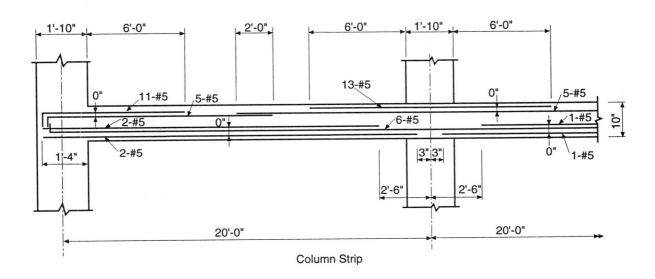

Column Strip

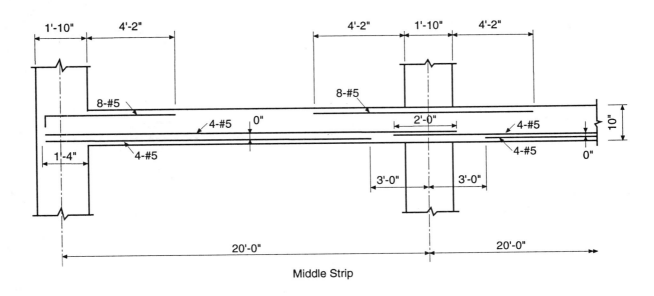

Middle Strip

Figure 2-30 Reinforcemnt Details for Interior Design Strip, Floor Level 4, Combined Gravity and Seismic Zone 2A Forces

Table 2-27 Summary of Slab Reinforcement for Interior Design Strip, Floor Level 4, Combined Gravity and Seismic Zone 2A Loads (10-in. Slab, 22-in. Square Columns)

Span Location		M_u (ft-kips)	No. of #5 bars
END SPAN			
Column Strip	Ext. Negative	176.5	16
	Positive	69.3	10
	Int. Negative	186.3	18
Middle Strip	Ext. Negative	0.0	8
	Positive	47.0	8
	Int. Negative	62.1	8
INTERIOR SPAN			
Column Strip	Positive	47.0	8
	Negative	180.1	18
Middle Strip	Positive	31.3	8
	Negative	60.0	8

2.4.3.2   Column Design

Table 2-28 summarizes the factored axial loads, bending moments, and shear forces for the first interior column and an exterior column located above the 2nd floor level in the first interior frame. In the case of the shear forces, the maximum shear was obtained for each load combination taking E to be twice that prescribed in UBC 1624 (UBC 1921.8.3). Figures 2-31 and 2-32 show the interaction diagrams for the interior and exterior column, respectively, and the strength demand corresponding to all of the applicable load combinations. It can be seen from the figures that 8-#11 bars ($\rho_g = 2.58\%$) and 4-#11 bars ($\rho_g = 1.29\%$) are adequate for the interior and the exterior column, respectively.

Table 2-28 Summary of Factored Axial Loads, Bending Moments, and Shear Forces for a Typical 22-in. Square Column above the 2nd Floor Level, Combined Gravity and Seismic Zone 2A Loads

(1)   U = 1.4D + 1.7L   Eq. (9-1), UBC 1909.2.1
(2)   U = 0.75 (1.4D + 1.7L ± 1.87E)   Eq. (9-2), UBC 1909.2.3*
(3)   U = 0.9D ± 1.43E   Eq. (9-3), UBC 1909.2.3*

Load Combination	INTERIOR COLUMN				EXTERIOR COLUMN			
	Axial Load (kips)	Bending Moment (ft-kips)		Shear Force (kips)	Axial Load (kips)	Bending Moment (ft-kips)		Shear Force (kips)
		Top	Bottom			Top	Bottom	
1	1297	7.9	-7.9	1.8	702	-29.1	29.1	-6.7
2: sidesway right	966	124.9	-154.6	57.7	405	37.6	-61.6	25.0
sidesway left	979	-113.0	142.7	-55.0	648	-81.3	105.3	-35.0
3: sidesway right	728	121.3	-151.6	62.9	271	46.6	-71.0	27.1
sidesway left	741	-121.3	151.6	-62.9	519	-74.7	99.1	-40.1

* For shear forces, E was taken as twice the value obtained from analysis (UBC 1921.8.3)

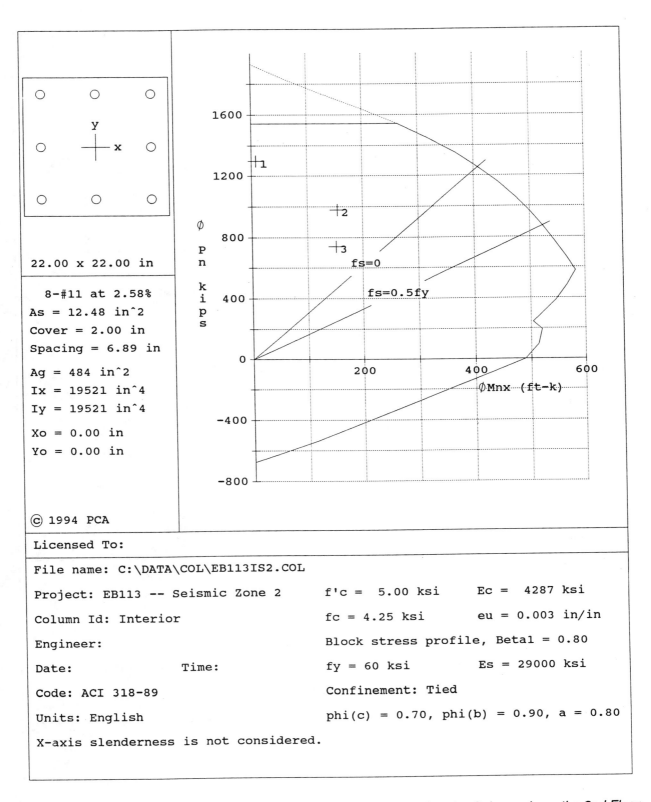

22.00 x 22.00 in

8-#11 at 2.58%
As = 12.48 in^2
Cover = 2.00 in
Spacing = 6.89 in

Ag = 484 in^2
Ix = 19521 in^4
Iy = 19521 in^4

Xo = 0.00 in
Yo = 0.00 in

© 1994 PCA

Licensed To:

File name: C:\DATA\COL\EB113IS2.COL

Project: EB113 -- Seismic Zone 2   f'c = 5.00 ksi   Ec = 4287 ksi

Column Id: Interior   fc = 4.25 ksi   eu = 0.003 in/in

Engineer: Block stress profile, Beta1 = 0.80

Date: Time: fy = 60 ksi   Es = 29000 ksi

Code: ACI 318-89   Confinement: Tied

Units: English   phi(c) = 0.70, phi(b) = 0.90, a = 0.80

X-axis slenderness is not considered.

Figure 2-31 Design Strength Interaction Diagram for a 22-in. Square Interior Column above the 2nd Floor Level, First Interior Frame

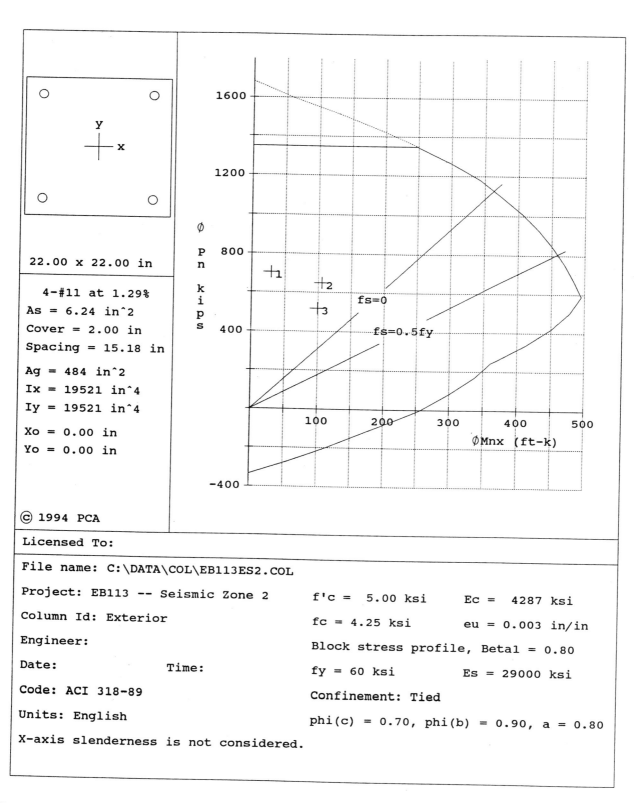

Figure 2-32 Design Strength Interaction Diagram for a 22-in. Square Exterior Column above the 2nd Floor Level, First Interior Frame

Transverse reinforcement and detailing of reinforcement will have to meet the additional requirements for Seismic Zone 2. UBC 1921.8.5.1 requires that for a length ℓ_o measured from the joint face, the tie spacing shall not exceed s_o which is the smallest of:

(1) 8 × diameter of smallest longitudinal bar = 8 × 1.41 = 11.3 in.

(2) 24 × diameter of tie bars = 24 × 0.5 = 12.0 in.

(3) 1/2 × smallest cross-sectional dimension of the column = 1/2 × 22.0 = 11.0 in. (governs)

(4) 12.0 in.

Therefore, the ties shall be spaced no farther than 11.0 in. over a distance ℓ_o which is the largest of:

(1) 1/6 the clear span of the member = 18.17 × 12/6 = 36.3 in. (governs)

(2) maximum cross-sectional dimension of the member = 22.0 in.

(3) 18.0 in.

UBC 1921.8.5.2 requires that the first tie be no more than $s_o/2$ (11/2 = 5.5 in.) from the joint face. Tie spacing for the remainder of the column shall not exceed twice s_o (2 × 11.0 = 22.0 in.; UBC 1921.8.5.4).

In UBC 1921.8, there is no specific requirement for column splices. For this example, the requirements for Seismic Zone 4 will be used. In Seismic Zone 4, UBC 1921.4.3.2 requires splices to be Class A within the center half of the member length.

The basic development length for a #11 bar in tension is:

$$\ell_{db} = 0.04 A_b f_y / \sqrt{f'_c}$$

$$= 0.04 \times 1.56 \times 60,000/\sqrt{5000} = 52.9 \text{ in.}$$

The cover to the column bars is not less than the minimum value specified in UBC 1907.7.1. Also, the transverse reinforcement satisfies UBC 1907.10.

The column bars have a clear spacing not less than three times the bar diameter. This gives a modification factor of 1.0 as specified in UBC 1912.2.3.1.

Therefore,

$$\text{Class A splice length } = 1.0 \times 1.0 \times \ell_{db}$$

$$= 1.0 \times 1.0 \times 52.9$$

$$= 52.9 \text{ in.} > 0.03 d_b f_y / \sqrt{f'_c} = 35.9 \text{ in.}$$

Use a 53.0 in. long splice as shown in Fig. 2-33 for an interior column in an interior frame. Note that the column shear strength is adequate to carry the maximum shear force of 62.9 kips (see Table 2-28).

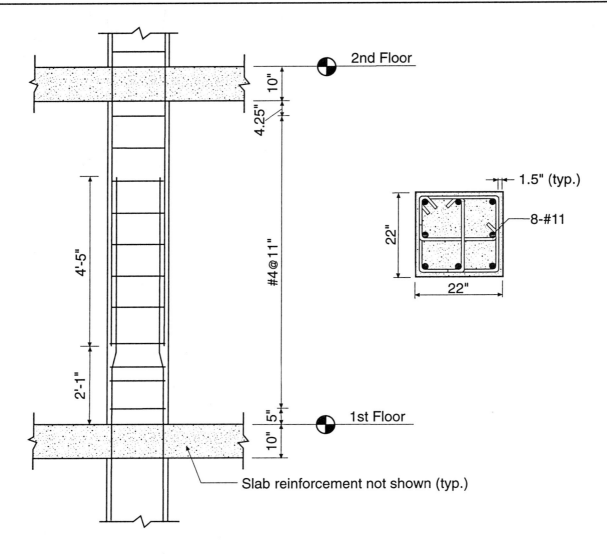

Figure 2-33 Reinforcement Details for a 22-in. Square Interior Column Above the 2nd Floor Level,
First Interior Frame

2.4.3.3   Story Drift Limitation and P-Δ Effects

As discussed in Section 2.3.4.3, the story drift must not exceed 0.46 in. Also, P-Δ effects can be ignored if the ratio of secondary moment to primary moment does not exceed 0.1. These requirements have been satisfied, as shown in Tables 2-29 and 2-30, respectively.

2.5   Design for Seismic Zone 2B

The design for Seismic Zone 2A required a 10-in. thick slab with an f'_c of 5000 psi. This is the practical limit of slab thickness and concrete strength for this type of a structural system. For this reason, this example building is not designed for Seismic Zone 2B loading. Of course, a flat plate-column frame is not allowed as part of the lateral force-resisting system in Seismic Zones 3 and 4.

Table 2-29 Lateral Displacements of Example Building under Seismic Zone 2A Forces

Floor Level	Total Lateral Displacement (in.)	Relative Lateral Displacement (in.)
R	2.16	0.05
14	2.11	0.07
13	2.04	0.10
12	1.94	0.12
11	1.82	0.14
10	1.68	0.15
9	1.53	0.17
8	1.36	0.18
7	1.18	0.20
6	0.98	0.20
5	0.78	0.21
4	0.57	0.21
3	0.36	0.21
2	0.17	0.17

Table 2-30 P-Δ Check under Seismic Zone 2A Forces

Floor Level	Gravity Loads * Σw_i (kips)	Story Drift Δ_i (in.)	$\Sigma w_i \Delta_i$ (in.-kips)	Story Shear V_i (kips)	Story Height h_{si} (in.)	$V_i h_{si}$ (in.-kips)	$\Sigma w_i \Delta_i / V_i h_{si}$
R	1831	0.05	91.6	115.1	114	13,121.4	0.007
14	3658	0.07	256.1	180.1	114	20,531.4	0.013
13	5485	0.10	548.5	240.2	114	27,382.8	0.020
12	7312	0.12	877.4	295.5	114	33,687.0	0.026
11	9139	0.14	1279.5	345.9	114	39,432.6	0.032
10	10,966	0.15	1644.9	391.4	114	44,619.6	0.037
9	12,793	0.17	2174.8	432.0	114	49,248.0	0.044
8	14,620	0.18	2631.6	467.7	114	53,317.8	0.049
7	16,447	0.20	3289.4	498.5	114	56,829.0	0.058
6	18,274	0.20	3654.8	524.5	114	59,793.0	0.061
5	20,101	0.21	4221.2	545.6	114	62,198.4	0.068
4	21,928	0.21	4604.9	561.8	114	64,045.2	0.072
3	23,755	0.21	4988.6	573.1	114	65,333.4	0.076
2	25,613	0.17	4354.2	579.6	150	86,940.0	0.050

* Includes floor weight, superimposed dead load, and reduced roof and floor live loads

References

2.1   *PCA-Frame — Three-Dimensional Static Analysis of Structures*, Portland Cement Association, Skokie, Illinois, 1992.

2.2   *PCACOL — Strength Design of Reinforced Concrete Column Sections*, Portland Cement Association, Skokie, Illinois, 1992.

Chapter 3

Beam-Column Frame

3.1   Introduction

3.1.1   General

The computation of the design loads for a 12-story reinforced concrete beam-column frame building under the requirements of the Uniform Building Code (UBC), 1994 edition, is illustrated. These computations are performed for UBC Seismic Zones 4, 2B, and 1, and also for a 70 mph wind, assuming Exposure B. Typical members are designed and detailed for each case.

3.1.2   Design Criteria

A typical floor plan and elevation of the building are shown in Figs. 3-1 and 3-2, respectively. The columns, beams, and slabs have constant cross-sections throughout the height of the building. Though the uniformity and symmetry used in this example have been adopted primarily for simplicity, these are generally considered to be sound engineering design concepts that should be utilized wherever practicable for seismic design. Although the member dimensions in this example are within the practical range, the structure itself is a hypothetical one, and has been chosen mainly for illustrative purposes. Other pertinent design data are as follows:

Service Loads:
    Live load: 50 psf
    Superimposed dead load: 42.5 psf

Material Properties:
    Concrete: f'_c = 4 ksi (6 ksi for columns in the bottom six stories for Seismic Zone 4)
                                    (6 ksi for columns in the bottom two stories for other seismic zones)
                    w_c = 150 pcf
    Reinforcement: f_y = 60 ksi

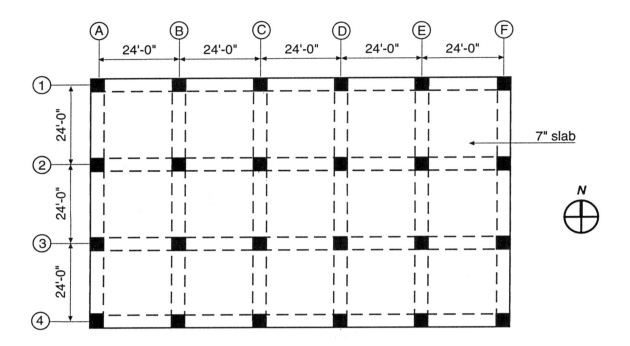

Figure 3-1 Typical Floor Plan of Example Building

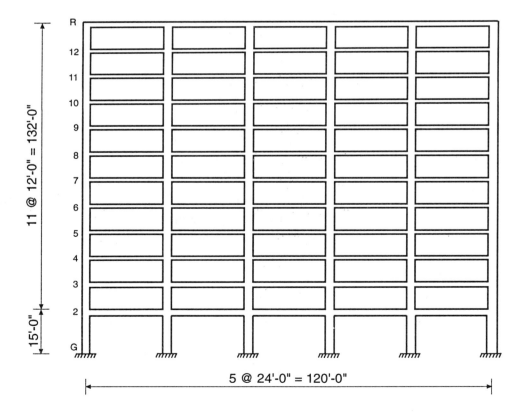

Figure 3-2 Elevation of Example Building

3.2   Design for Seismic Zone 4

This section will illustrate the design and detailing of members for Seismic Zone 4. Requirements for detailing in Zone 3 are identical to those for Zone 4. Also, the analysis for both zones is similar, except that Zone 4 design lateral forces are 33% higher than the corresponding Zone 3 forces. Therefore, the design of the structure in Seismic Zone 3 is not separately addressed here.

The member dimensions selected for this design are as follows: transverse beams: 24 × 26 in.; longitudinal beams: 24 × 20 in.; columns: 24 × 24 in.; slabs: 7 in.

3.2.1   Frame Analysis

On the basis of the given data and the dimensions of the building, the weights of the floors are listed in Table 3-1. It should be noted that the building has the same lateral load resisting system in both principal directions. Therefore, the seismic design forces will be the same in both the longitudinal and the transverse directions of the building. However, since the building is rectangular rather than square in plan, the lateral shears produced by torsion will most likely not be equal in the two directions. The weight of a typical floor includes all of the elements located between two imaginary parallel planes passing through the midheight of the columns above and below the floor considered. Calculations of the base shear, V, for the transverse and the longitudinal directions are shown at the bottom of Table 3-1. For this example, the importance factor, I, and the site soil coefficient, S, have been taken equal to 1.0 and 1.2, respectively. The value of $R_w = 12$ has been used in both directions. This is in accordance with UBC 1631.2.7 which requires that frames used as part of the lateral force resisting system in Seismic Zones 3 and 4 must be special moment-resisting frames.

The seismic design forces resulting from the distribution of the base shear in accordance with UBC Eqs. (28-6), (28-7), and (28-8) are also listed in Table 3-1. As an example, the seismic force F_x at the 11th floor level is given by:

$$F_{11} = \frac{(V - F_t) \times w_{11}h_{11}}{\sum\limits_{i=1}^{12} w_i h_i} = \frac{(852.5 - 75.7) \times 1669 \times 123}{1,609,956} = 99.0 \text{ kips}$$

UBC 1628.5 requires that the design story shear must be distributed to the various elements of the lateral force resisting system in proportion to their rigidities, considering the rigidity of the diaphragm. To account for uncertainties in the locations of the loads, the mass at each level must be assumed to be displaced from the calculated center of mass in each direction by a distance equal to five percent of the building dimension at that level, perpendicular to the direction of the force under consideration.

The results from the 3-D analysis, using *PCA-Frame* [2.1], for an exterior transverse frame and an interior longitudinal frame are given in Figs. 3-3 and 3-4, respectively. In each case, the gross areas of the columns and beams were used in the analysis, and rigid diaphragms were assigned at each floor level (see Section 3.2.2.2 for a discussion on diaphragm flexibility). The contribution of the effective slab width to the beam properties was ignored in the analysis. To comply with UBC 1628.5, the seismic forces were displaced a distance of 0.05 × 120.0 = 6.0 ft from the center of mass for analysis in the transverse direction, and 0.05 × 72.0 = 3.6 ft for analysis in the longitudinal direction.

Table 3-1 Seismic Forces for Zone 4

Floor Level	Weight, w_x (kips)	Height, h_x (ft)	$w_x h_x$ (ft-kips)	Lateral Force (kips)
R	1583	147	232,701	187.9
12	1669	135	225,315	108.6
11	1669	123	205,287	99.0
10	1669	111	185,259	89.3
9	1669	99	165,231	79.7
8	1669	87	145,203	70.7
7	1669	75	125,175	60.3
6	1669	63	105,147	50.7
5	1669	51	85,119	41.0
4	1669	39	65,091	31.4
3	1669	27	45,063	21.7
2	1691	15	25,365	12.2
Σ	19,964		1,609,956	852.5

Building height	h_n	= 147 ft
Building weight	W	= 19,964 kips
Fundamental period	T	= 0.030 $(h_n)^{3/4}$ = 1.27 sec.
Seismic zone factor	Z	= 0.4
Importance factor	I	= 1.0
Response coefficient	R_w	= 12
Soil factor	S	= 1.2
Coefficient	C	= 1.25 $S/T^{2/3}$ = 1.28
Base shear	V	= $ZICW/R_w$ = 852.5 kips
Top level force	F_t	= 0.07TV = 75.7 kips

For comparison purposes, a 2-D analysis was also performed. This structure was considered as four lumped frames in the longitudinal direction and six lumped frames in the transverse direction linked together by a rigid diaphragm at each floor level. The loads induced by torsion (due to seismic loads only) were calculated floor by floor as follows:

$$V_{Txi} = \frac{0.05 \times DIM_x \times V_x}{D_{xi}} \times \frac{R_{xi}D_{xi}^2}{\sum R_{xi}D_{xi}^2}$$

where

DIM_x = building dimension perpendicular to the direction of analysis, at floor level x.

R_{xi} = relative rigidity of frame i at floor level x.

V_x = story shear, in the direction of analysis, at floor level x.

D_{xi} = distance, perpendicular to the direction of analysis, from the center of rigidity to the centerline of frame i, at floor level x.

The values of DIM_x and D_{xi} are obtained from the building geometry. The relative rigidities are determined by finding the percentage of the story shear carried by each frame at a particular floor level. Since the frames in the example building are identical in a given direction, the seismic forces in that direction are carried equally by the frames. For instance, when the seismic forces are in the longitudinal direction, four frames share the load equally. Thus, the relative rigidity for a frame is 0.25, irrespective of the floor level, as shown in Table 3-2. With the building geometry and relative rigidities known, the story shear is adjusted for torsion (see Section 2.3.2). Results of this analysis are shown in Tables 3-2 and 3-3. The frames were reanalyzed under the revised seismic forces. The ratio of the 2-D to 3-D results are given in Figs. 3-5 and 3-6 for an exterior transverse frame and an interior longitudinal frame, respectively.

R	-31	-21	-49
12	-70	-60	-90
11	-106	-92	-118
10	-138	-121	-143
9	-166	-147	-166
8	-192	-171	-185
7	-213	-192	-205
6	-230	-209	-220
5	-246	-225	-230
4	-260	-237	-237
3	-269	-248	-243
2	-281	-252	-231

Bending Moments in Beams (ft-kips)

R	-2.4	-4.5
12	-5.9	-8.2
11	-9.0	-10.7
10	-11.8	-13.0
9	-14.3	-15.1
8	-16.5	-16.9
7	-18.4	-18.6
6	-20.0	-20.0
5	-21.4	-20.9
4	-22.6	-21.5
3	-23.5	-22.0
2	-24.2	-21.0

Shear Forces in Beams (kips)

R	41 / 24	84 / 66
12	69 / 51	116 / 104
11	89 / 74	151 / 140
10	108 / 95	182 / 172
9	125 / 113	209 / 202
8	140 / 132	232 / 224
7	148 / 141	258 / 251
6	162 / 155	272 / 267
5	170 / 165	286 / 283
4	177 / 176	295 / 292
3	178 / 171	305 / 311
2	204 / 345	286 / 386

Bending Moments in Columns (ft-kips)

R	5.4	12.5
12	10.1	18.3
11	13.6	24.3
10	16.9	29.5
9	19.8	34.3
8	22.7	38.0
7	24.1	42.4
6	26.4	45.0
5	27.9	47.4
4	29.4	48.9
3	29.0	51.3
2	36.6	44.8

Shear Forces in Columns (kips)

R	2.4	2.1
12	8.3	4.4
11	17.3	6.0
10	29.1	7.3
9	43.3	8.2
8	59.8	8.5
7	78.2	8.7
6	98.2	8.8
5	120.0	8.3
4	142.0	7.3
3	166.0	5.8
2	190.0	2.6

Axial Forces in Columns (kips)

Figure 3-3 Results of 3-D Analysis of an Exterior Transverse Frame under Seismic Zone 4 Forces

R					
R	-30	-26	-39	-38	-38
12	-60	-57	-72	-71	-71
11	-92	-86	-99	-99	-99
10	-119	-113	-124	-124	-124
9	-145	-137	-146	-146	-146
8	-168	-158	-165	-165	-165
7	-186	-177	-183	-183	-183
6	-202	-193	-198	-198	-198
5	-216	-206	-209	-209	-209
4	-227	-217	-217	-217	-217
3	-232	-222	-220	-220	-220
2	-225	-212	-204	-204	-205

Bending Moments in Beams (ft-kips)

R			
R	-2.6	-3.5	-3.5
12	-5.3	-6.5	-6.5
11	-8.1	-8.9	-8.9
10	-10.5	-11.2	-11.2
9	-12.8	-13.2	-13.2
8	-14.8	-15.0	-15.0
7	-16.5	-16.7	-16.7
6	-17.9	-18.0	-18.0
5	-19.2	-19.0	-19.0
4	-20.2	-19.7	-19.7
3	-20.6	-20.0	-20.0
2	-19.9	-18.5	-18.5

Shear Forces in Beams (kips)

R			
R	34	72	83
	6	42	50
12	61	101	107
	29	75	81
11	74	130	136
	46	106	112
10	88	156	161
	63	134	139
9	99	179	183
	77	160	164
8	111	198	201
	94	181	184
7	115	219	222
	99	202	205
6	127	231	233
	111	218	220
5	131	242	243
	120	233	233
4	135	248	247
	133	244	243
3	127	247	244
	135	267	263
2	118	195	192
	349	388	386

Bending Moments in Columns (ft-kips)

R			
R	3.3	9.5	11.1
12	7.5	14.6	15.7
11	10.0	19.7	20.7
10	12.5	24.2	25.0
9	14.7	28.3	28.9
8	17.1	31.6	32.1
7	17.8	35.1	35.6
6	19.8	37.4	37.7
5	20.9	39.6	39.7
4	22.3	41.0	40.9
3	21.8	42.9	42.3
2	31.1	38.8	38.5

Shear Forces in Columns (kips)

R			
R	2.6	1.0	0.1
12	7.9	2.1	0.1
11	15.9	3.0	0.1
10	26.5	3.7	0.1
9	39.3	4.1	0.1
8	54.1	4.3	0.1
7	70.6	4.4	0.1
6	88.5	4.5	0.1
5	108.0	4.3	0.1
4	128.0	3.8	0.1
3	149.0	3.2	0.1
2	168.0	1.8	0.1

Axial Forces in Columns (kips)

Figure 3-4 Results of 3-D Analysis of an Interior Longitudinal Frame under Seismic Zone 4 Forces

For comparison with the earthquake forces, the wind forces corresponding to a fastest mile wind speed of 70 mph and Exposure B, computed in accordance with UBC 1618, are shown in Tables 3-4 and 3-5 for the transverse and the longitudinal directions, respectively. The results from the analyses are given in Figs. 3-7 and 3-8.

The stiffnesses of the structure in this case are very similar in the two orthogonal principal directions (see Table 3-11). In this case, it can be assumed that the torsion caused by uncertainties in the location of the load is shared equally by the two principal directions of the structure. If the stiffnesses are not similar, the equal splitting of torsional effects between the principal directions of the structure is invalid. In this situation, the stronger direction of the structure would carry $\Delta_w \times 100 / (\Delta_s + \Delta_w)$ percent of the torsion and the weaker direction would carry $\Delta_s \times 100 / (\Delta_s + \Delta_w)$ percent of the torsion, where Δ_w is the lateral deflection of the structure in the weaker direction under a set of statically admissible lateral forces and Δ_s is the lateral deflection in the stronger direction of the structure under the same set of lateral forces. Therefore, when calculating torsional effects in the structure, the relative rigidities in the stronger direction of the structure can remain unaltered, while the relative rigidities in the weaker direction should be reduced by Δ_w/Δ_s to account for its larger flexibility.

It should be noted that for buildings located in Seismic Zones 2, 3, and 4 (i.e., moderate and high seismic risk areas), the detailing requirements for ductility provided in UBC 1921 are required even when the design of a member is governed by wind rather than by seismic loads.

Table 3-2 Distribution of Story Shear to Longitudinal Frames, Seismic Zone 4

Column Line Frame	% of Direct Seismic Story Shear	% of Torsional Story Shear	Revised % of Story Shear, Accounting for Torsion
1	25	+1.35	+26.35
2	25	+0.45	+25.45
3	25	-0.45	+24.55
4	25	-1.35	+23.65
A	0	+1.50	+1.50
B	0	+0.90	+0.90
C	0	+0.30	+0.30
D	0	-0.30	-0.30
E	0	-0.90	-0.90
F	0	-1.50	-1.50

Table 3-3 Distribution of Story Shear to Transverse Frames, Seismic Zone 4

Column Line Frame	% of Direct Seismic Story Shear	% of Torsional Story Shear	Revised % of Story Shear, Accounting for Torsion
A	16.67	+2.50	+19.17
B	16.67	+1.50	+18.17
C	16.67	+0.50	+17.17
D	16.67	-0.50	+16.17
E	16.67	-1.50	+15.17
F	16.67	-2.50	+14.17
1	0	+2.30	+2.30
2	0	+0.80	+0.80
3	0	-0.80	-0.80
4	0	-2.30	-2.30

Bending Moments in Beams

R	1.17	1.16	1.16
12	1.15	1.16	1.14
11	1.15	1.15	1.14
10	1.15	1.16	1.15
9	1.15	1.16	1.15
8	1.14	1.15	1.15
7	1.15	1.15	1.14
6	1.15	1.15	1.15
5	1.15	1.15	1.15
4	1.15	1.15	1.15
3	1.15	1.15	1.15
2	1.15	1.15	1.15

Shear Forces in Beams

R	1.15	1.15
12	1.15	1.15
11	1.16	1.15
10	1.14	1.15
9	1.15	1.15
8	1.15	1.15
7	1.15	1.15
6	1.15	1.15
5	1.15	1.15
4	1.15	1.15
3	1.15	1.15
2	1.15	1.15

Bending Moments in Columns

R	0.95 / 0.83	1.06 / 1.04
12	0.97 / 0.92	1.04 / 1.03
11	0.96 / 0.92	1.04 / 1.04
10	0.96 / 0.93	1.04 / 1.04
9	0.95 / 0.94	1.04 / 1.03
8	0.96 / 0.95	1.04 / 1.03
7	0.95 / 0.94	1.04 / 1.03
6	0.95 / 0.94	1.04 / 1.03
5	0.95 / 0.95	1.04 / 1.03
4	0.96 / 0.96	1.03 / 1.03
3	0.95 / 0.94	1.03 / 1.04
2	0.93 / 1.02	0.99 / 1.03

Shear Forces in Columns

R	0.91	1.05
12	0.94	1.04
11	0.94	1.04
10	0.95	1.03
9	0.95	1.03
8	0.95	1.03
7	0.95	1.04
6	0.95	1.03
5	0.95	1.03
4	0.95	1.03
3	0.95	1.04
2	0.99	1.02

Axial Forces in Columns

R	1.15	1.15
12	1.15	1.14
11	1.15	1.16
10	1.15	1.15
9	1.15	1.15
8	1.15	1.15
7	1.15	1.15
6	1.15	1.15
5	1.15	1.15
4	1.16	1.15
3	1.15	1.16
2	1.15	1.15

Figure 3-5 Ratio of 2-D to 3-D Results under Seismic Zone 4 Forces for an Exterior Transverse Frame

R				
R	1.03	1.02	1.03	1.02	1.00
12	1.02	1.01	1.01	1.02	1.02
11	1.01	1.02	1.01	1.01	1.01
10	1.03	1.02	1.02	1.02	1.02
9	1.01	1.02	1.02	1.02	1.02
8	1.02	1.02	1.02	1.02	1.02
7	1.02	1.02	1.02	1.02	1.02
6	1.02	1.02	1.02	1.02	1.02
5	1.02	1.02	1.02	1.02	1.02
4	1.02	1.02	1.02	1.02	1.02
3	1.02	1.02	1.02	1.02	1.02
2	1.02	1.02	1.01	1.01	1.02

Bending Moments in Beams

R			
R	1.00	1.02	1.00
12	1.02	1.02	1.02
11	1.02	1.02	1.02
10	1.02	1.02	1.02
9	1.02	1.02	1.02
8	1.02	1.02	1.02
7	1.02	1.02	1.02
6	1.02	1.02	1.02
5	1.02	1.02	1.02
4	1.02	1.02	1.02
3	1.02	1.02	1.02
2	1.02	1.02	1.02

Shear Forces in Beams

R			
R	0.99 / 0.90	1.01 / 0.99	1.01 / 1.01
12	1.00 / 0.97	1.00 / 1.00	1.01 / 1.00
11	0.99 / 0.98	1.01 / 1.00	1.01 / 1.00
10	0.99 / 0.98	1.00 / 1.01	1.01 / 1.01
9	1.00 / 0.98	1.01 / 1.01	1.01 / 1.01
8	0.99 / 0.99	1.01 / 1.00	1.01 / 1.00
7	0.99 / 0.99	1.01 / 1.01	1.01 / 1.00
6	0.99 / 0.99	1.00 / 1.00	1.00 / 1.00
5	0.99 / 0.98	1.00 / 1.00	1.00 / 1.00
4	0.99 / 0.99	1.00 / 1.00	1.00 / 1.00
3	0.98 / 0.99	1.00 / 1.01	1.00 / 1.00
2	0.98 / 1.01	1.00 / 1.01	0.99 / 1.01

Bending Moments in Columns

R			
R	0.98	1.00	1.01
12	0.99	1.00	1.00
11	0.99	1.01	1.01
10	0.99	1.00	1.00
9	0.99	1.00	1.00
8	0.99	1.00	1.00
7	0.99	1.00	1.00
6	0.99	1.00	1.00
5	0.99	1.01	1.00
4	0.99	1.00	1.00
3	0.99	1.00	1.00
2	1.00	1.00	1.00

Shear Forces in Columns

R			
R	1.00	1.00	1.00
12	1.01	1.03	1.00
11	1.02	1.02	1.00
10	1.02	1.01	1.00
9	1.02	1.02	1.00
8	1.02	1.02	1.00
7	1.02	1.03	1.00
6	1.02	1.02	1.00
5	1.02	1.01	1.00
4	1.02	1.02	1.00
3	1.01	1.01	1.00
2	1.02	1.04	1.00

Axial Forces in Columns

Figure 3-6 Ratio of 2-D to 3-D Results under Seismic Zone 4 Forces for an Interior Longitudinal Frame

Table 3-4 Wind Forces on Example Building in the Transverse Direction

Floor Level	Height (ft)	Tributary Height (ft)	Windward			Leeward			Total Force (kips)
			C_e	Pressure (psf)	Force (kips)	C_e	Pressure (psf)	Force (kips)	
R	147	6.0	1.27	12.8	9.3	1.27	8.0	5.8	15.1
12	135	12.0	1.24	12.5	18.0	1.27	8.0	11.5	29.5
11	123	12.0	1.21	12.2	17.5	1.27	8.0	11.5	29.0
10	111	12.0	1.17	11.8	17.0	1.27	8.0	11.5	28.5
9	99	12.0	1.13	11.4	16.4	1.27	8.0	11.5	27.9
8	87	12.0	1.07	10.8	15.6	1.27	8.0	11.5	27.1
7	75	12.0	1.02	10.3	14.8	1.27	8.0	11.5	26.3
6	63	12.0	0.96	9.7	14.0	1.27	8.0	11.5	25.5
5	51	12.0	0.90	9.1	13.1	1.27	8.0	11.5	24.6
4	39	12.0	0.84	8.5	12.2	1.27	8.0	11.5	23.7
3	27	12.0	0.74	7.5	10.8	1.27	8.0	11.5	22.3
2	15	13.5	0.62	6.3	10.2	1.27	8.0	13.0	23.2

Tributary width = 120.0 ft

Table 3-5 Wind Forces on Example Building in the Longitudinal Direction

Floor Level	Height (ft)	Tributary Height (ft)	Windward			Leeward			Total Force (kips)
			C_e	Pressure (psf)	Force (kips)	C_e	Pressure (psf)	Force (kips)	
R	147	6.0	1.27	12.8	5.6	1.27	8.0	3.5	9.1
12	135	12.0	1.24	12.5	10.8	1.27	8.0	6.9	17.7
11	123	12.0	1.21	12.2	10.5	1.27	8.0	6.9	17.4
10	111	12.0	1.17	11.8	10.2	1.27	8.0	6.9	17.1
9	99	12.0	1.13	11.4	9.8	1.27	8.0	6.9	16.7
8	87	12.0	1.07	10.8	9.3	1.27	8.0	6.9	16.2
7	75	12.0	1.02	10.3	8.9	1.27	8.0	6.9	15.8
6	63	12.0	0.96	9.7	8.4	1.27	8.0	6.9	15.3
5	51	12.0	0.90	9.1	7.8	1.27	8.0	6.9	14.7
4	39	12.0	0.84	8.5	7.3	1.27	8.0	6.9	14.2
3	27	12.0	0.74	7.5	6.5	1.27	8.0	6.9	13.4
2	15	13.5	0.62	6.3	6.1	1.27	8.0	7.8	13.9

Tributary width = 72.0 ft

R	-1.8	-0.2	-7.4
12	-7.9	-6.5	-14.0
11	-17.6	-15.1	-21.5
10	-27.2	-23.9	-29.5
9	-36.7	-32.5	-37.2
8	-46.0	-40.9	-44.5
7	-54.8	-49.4	-52.4
6	-62.9	-57.3	-59.7
5	-71.4	-65.2	-66.1
4	-79.8	-72.9	-71.9
3	-87.4	-80.5	-78.0
2	-96.5	-86.5	-78.4

*Bending Moments in
Beams (ft-kips)*

R	-0.1	-0.7
12	-0.7	-1.3
11	-1.5	-2.0
10	-2.3	-2.7
9	-3.1	-3.4
8	-4.0	-4.1
7	-4.7	-4.8
6	-5.5	-5.4
5	-6.2	-6.0
4	-6.9	-6.5
3	-7.6	-7.1
2	-8.3	-7.1

*Shear Forces in
Beams (kips)*

R	1.9 -0.8	8.4 5.5
12	9.4 4.6	17.0 13.7
11	17.0 13.7	14.5 9.8
10	19.7 15.1	26.4 23.0
9	24.7 20.2	44.2 41.1
8	29.8 26.1	52.4 49.0
7	33.4 29.4	62.3 58.5
6	39.0 34.5	69.5 66.2
5	43.2 39.0	77.3 74.2
4	47.7 44.6	84.1 80.9
3	50.4 44.8	92.3 92.1
2	60.0 108.0	88.3 122.0

*Bending Moments in
Columns (ft-kips)*

R	0.1	1.2
12	1.2	2.6
11	2.0	4.1
10	2.9	5.6
9	3.7	7.1
8	4.7	8.5
7	5.2	10.1
6	6.1	11.3
5	6.9	12.6
4	7.7	13.8
3	7.9	15.4
2	11.2	14.0

Shear Forces in Columns (kips)

R	0.1	0.6
12	0.8	1.2
11	2.2	1.7
10	4.6	2.0
9	7.7	2.3
8	11.7	2.4
7	16.4	2.4
6	21.9	2.4
5	28.1	2.2
4	35.0	1.8
3	42.6	1.2
2	51.0	0.0

*Axial Forces in
Columns (kips)*

Figure 3-7 Results of 3-D Analysis of an Exterior Transverse Frame under Wind Forces

R					
R	-1.2	-0.9	-3.0	-2.7	-2.7
12	-4.2	-3.9	-6.2	-6.1	-6.1
11	-9.1	-8.6	-10.5	-10.4	-10.4
10	-14.2	-13.3	-15.0	-14.9	-15.0
9	-19.2	-18.1	-19.5	-19.4	-19.5
8	-24.1	-22.8	-23.8	-23.8	-23.8
7	-28.9	-27.5	-28.3	-28.3	-28.3
6	-33.1	-31.7	-32.4	-32.3	-32.4
5	-37.6	-36.0	-36.2	-36.2	-36.2
4	-41.9	-40.0	-39.7	-39.7	-39.8
3	-45.1	-43.3	-42.6	-42.6	-42.7
2	-46.0	-43.4	-41.5	-41.6	-41.7

Bending Moments in Beams (ft-kips)

R			
R	-0.1	-0.3	-0.2
12	-0.4	-0.6	-0.6
11	-0.8	-1.0	-1.0
10	-1.3	-1.4	-1.4
9	-1.7	-1.8	-1.8
8	-2.1	-2.2	-2.2
7	-2.6	-2.6	-2.6
6	-3.0	-2.9	-2.9
5	-3.4	-3.3	-3.3
4	-3.7	-3.6	-3.6
3	-4.0	-3.9	-3.9
2	-4.1	-3.8	-3.8

Shear Forces in Beams (kips)

R			
R	1.3 / -1.6	4.2 / 1.2	5.9 / 2.6
12	6.2 / 1.1	9.8 / 5.7	10.8 / 6.6
11	8.9 / 3.8	15.1 / 10.8	16.1 / 11.7
10	11.7 / 6.6	20.2 / 15.9	21.0 / 16.6
9	14.3 / 9.2	25.2 / 21.0	25.8 / 21.6
8	17.0 / 12.5	29.9 / 25.6	30.4 / 26.0
7	18.9 / 14.0	35.3 / 30.5	35.7 / 30.9
6	22.1 / 16.8	39.5 / 35.0	39.7 / 35.1
5	24.2 / 19.3	43.8 / 39.7	43.8 / 39.6
4	26.3 / 23.0	47.4 / 44.0	47.2 / 43.8
3	26.1 / 24.6	49.9 / 51.4	49.3 / 50.4
2	25.4 / 72.6	41.3 / 80.5	40.5 / 80.1

Bending Moments in Columns (ft-kips)

R			
R	0.1	0.5	0.7
12	0.6	1.3	1.5
11	1.1	2.2	2.3
10	1.5	3.0	3.1
9	2.0	3.9	4.0
8	2.5	4.6	4.7
7	2.7	5.5	5.5
6	3.2	6.2	6.2
5	3.6	7.0	7.0
4	4.1	7.6	7.6
3	4.2	8.4	8.3
2	6.5	8.1	8.1

Shear Forces in Columns (kips)

R			
R	0.1	0.2	0.0
12	0.5	0.4	0.0
11	1.3	0.5	0.0
10	2.5	0.6	0.0
9	4.2	0.7	0.0
8	6.3	0.7	0.0
7	8.9	0.7	0.0
6	11.9	0.7	0.0
5	15.2	0.7	0.0
4	18.9	0.6	0.0
3	22.9	0.4	0.0
2	27.0	0.1	0.0

Axial Forces in Columns (kips)

Figure 3-8 Results of 3-D Analysis of an Interior Longitudinal Frame under Wind Forces

3.2.2   Design of Beams and Columns

The objective is to determine the required flexural and shear reinforcement for typical beams of the transverse and the longitudinal frames and for a typical column of the example structure subjected to dead loads, live loads, and wind or Seismic Zone 4 forces.

Two methods will be used to determine the moments due to the gravity loads acting on the beams. The first method utilizes the Direct Design Method of UBC 1913.6. Note that all of the limitations given in UBC 1913.6.1 are satisfied for the slab systems spanning in the transverse as well as in the longitudinal direction of the building. Since $\alpha_1 \ell_2 / \ell_1 > 1.0$ in both directions, the beam moments given in Table 3-6 are 85% of the column strip moments (UBC 1913.6.5.1).

The second method assumes that an equivalent uniformly distributed load acts on all spans of the beam. An equivalent uniformly distributed line load (plf) on a beam is determined by multiplying the area load (psf) with the corresponding triangular tributary area and then dividing by the clear span of the beam. The moments are computed using the coefficients in UBC 1908.3.3. Table 3-6 shows that the moments obtained from both methods are approximately equal at all locations along the span, except at the exterior support in the end span. At this location, the moment obtained from the equivalent uniform load method is about 3 times the moment obtained from the Direct Design Method. The moments from the Direct Design Method will be used in all subsequent calculations for the beam design.

Table 3-6 Comparison of Gravity Load Moments in the Beams

Method *	Factored Bending Moments (ft-kips)									
	Exterior Transverse Frame					Interior Longitudinal Frame				
	End Span			Interior Span		End Span			Interior Span	
	Ext. Neg.	Pos.	Int. Neg.	Pos.	Neg.	Ext. Neg.	Pos.	Int. Neg.	Pos.	Neg.
1	25.0	92.5	112.6	55.0	105.0	41.5	153.7	186.9	91.4	174.4
2	81.7	93.3	130.7	81.7	118.8	118.0	134.8	188.8	118.0	171.6

Method 1:   Direct Design Method

          Exterior Transverse Frame: $w_u = 0.318$ ksf, $M_o = w_u \ell_2 \ell_n^2 / 8 = 250.1$ ft-kips

          Interior Longitudinal Frame: $w_u = 0.286$ ksf, $M_o = w_u \ell_2 \ell_n^2 / 8 = 415.3$ ft-kips

Method 2:   Equivalent Uniform Load Method

          Exterior Transverse Frame: $w_u = 2.70$ kips/ft

          Interior Longitudinal Frame: $w_u = 3.90$ kips/ft

Table 3-7 gives the factored bending moments for a beam located on the second floor of an exterior frame in the transverse direction. The values given in parentheses are the factored moments when the equivalent uniform loads are used to compute the gravity moments. Note that for Seismic Zone 4, the load factors given in UBC 1921.2.7 are applicable (see Table 3-7). The moments due to the seismic loads and the wind loads are taken from Figs. 3-3 and 3-7, respectively. Similarly, Table 3-8 gives the factored moments for a beam located on the second floor of an interior longitudinal frame. In this case, seismic and wind moments were obtained from Figs. 3-4 and 3-8, respectively.

Table 3-9 gives the factored axial loads, bending moments, and shear forces for an exterior and an interior column, located between the 2nd and the 3rd floors in an exterior transverse frame. Live load reduction factors were computed in accordance with Eq. (6-1) of UBC 1606. The reactions due to the seismic and the wind loads were taken from Figs. 3-3 and 3-7, respectively. Table 3-10 lists the load combinations for the same exterior and interior columns located between the ground level and the second floor.

Table 3-7 Summary of Factored Bending Moments for a Beam on the 2nd Floor of an Exterior Transverse Frame (Seismic Zone 4) *

(1) U = 1.4D + 1.7L	Eq. (9-1), UBC 1909.2.1
(2) U = 1.4 (D + L ± E)	Eq. (9-2), UBC 1921.2.7
(3) U = 0.9D ± 1.4E	Eq. (9-3), UBC 1921.2.7
(4) U = 0.75 (1.4D + 1.7L ± 1.7W)	Eq. (9-2), UBC 1909.2.2
(5) U = 0.9D ± 1.3W	Eq. (9-3), UBC 1909.2.2

Load Combination	End Span Moments (ft-kips)						Interior Span Moments (ft-kips)			
	Exterior Negative		Positive		Interior Negative		Positive		Negative	
1	-25.0	(-81.7)	92.5	(93.3)	-112.6	(-130.7)	55.0	(81.7)	-105.0	(-118.8)
2: sidesway right	369.5	(315.5)	108.7	(109.4)	-460.3	(-477.5)	52.5	(77.9)	223.1	(210.1)
sidesway left	-417.3	(-471.3)	68.1	(68.8)	245.3	(228.1)	52.5	(77.9)	-423.7	(-436.7)
3: sidesway right	381.6	(353.6)	64.0	(65.7)	-406.0	(-416.4)	26.0	(39.8)	273.8	(265.6)
sidesway left	-405.2	(-433.2)	23.4	(25.1)	299.6	(289.2)	26.0	(39.8)	-373.0	(-381.2)
4: sidesway right	104.2	(61.7)	75.8	(76.6)	-194.8	(-208.6)	41.3	(61.4)	21.2	(10.7)
sidesway left	-141.8	(-184.3)	63.0	(63.8)	25.8	(12.0)	41.3	(61.4)	-178.8	(-189.3)
5: sidesway right	113.7	(85.9)	50.2	(51.7)	-165.7	(-175.9)	26.0	(39.8)	52.3	(44.3)
sidesway left	-137.3	(-165.1)	37.2	(38.7)	59.3	(49.1)	26.0	(39.8)	-151.5	(-159.5)

*Values in parentheses were determined using equivalent uniform loads

Table 3-8 Summary of Factored Bending Moments for a Beam on the 2nd Floor of an Interior Longitudinal Frame (Seismic Zone 4) *

(1) U = 1.4D + 1.7L	Eq. (9-1), UBC 1909.2.1
(2) U = 1.4 (D + L ± E)	Eq. (9-2), UBC 1921.2.7
(3) U = 0.9D ± 1.4E	Eq. (9-3), UBC 1921.2.7
(4) U = 0.75 (1.4D + 1.7L ± 1.7W)	Eq. (9-2), UBC 1909.2.2
(5) U = 0.9D ± 1.3W	Eq. (9-3), UBC 1909.2.2

Load Combination	End Span Moments (ft-kips)						Interior Span Moments (ft-kips)			
	Exterior Negative		Positive		Interior Negative		Positive		Negative	
1	-41.5	(-118.0)	153.7	(134.8)	-186.9	(-188.8)	91.4	(118.0)	-174.4	(-171.6)
2: sidesway right	275.7	(201.9)	154.0	(137.6)	-473.8	(-477.7)	86.5	(113.1)	120.4	(121.1)
sidesway left	-354.3	(-428.1)	137.2	(120.8)	119.8	(115.9)	86.5	(113.1)	-450.8	(-450.1)
3: sidesway right	296.2	(259.9)	77.8	(71.3)	-381.2	(-384.9)	41.3	(55.1)	206.8	(205.5)
sidesway left	-333.8	(-370.1)	61.0	(54.5)	212.4	(208.7)	41.3	(55.1)	-364.4	(-365.7)
4: sidesway right	27.6	(-30.5)	116.9	(103.6)	-195.6	(-198.2)	68.6	(89.2)	-77.9	(-76.9)
sidesway left	-89.8	(-147.9)	113.7	(100.4)	-84.8	(-87.4)	68.6	(89.2)	-183.7	(-182.7)
5: sidesway right	41.0	(4.7)	71.0	(64.5)	-140.8	(-144.5)	41.3	(55.1)	-24.8	(-26.1)
sidesway left	-78.6	(-114.9)	67.8	(61.3)	-28.0	(-31.7)	41.3	(55.1)	-132.8	(-134.1)

*Values in parentheses were determined using equivalent uniform loads

Table 3-9 Summary of Factored Axial Loads, Bending Moments, and Shear Forces for Columns between the 2nd and 3rd Floors of an Exterior Transverse Frame (Seismic Zone 4)

(1) $U = 1.4D + 1.7L$ Eq. (9-1), UBC 1909.2.1
(2) $U = 1.4(D + L \pm E)$ Eq. (9-2), UBC 1921.2.7
(3) $U = 0.9D \pm 1.4E$ Eq. (9-3), UBC 1921.2.7
(4) $U = 0.75(1.4D + 1.7L \pm 1.7W)$ Eq. (9-2), UBC 1909.2.2
(5) $U = 0.9D \pm 1.3W$ Eq. (9-3), UBC 1909.2.2

| Load Combination | Interior Column | | | | Exterior Column | | | |
| | Axial Load (kips) | Bending Moment (ft-kips) | | Shear Force (kips) | Axial Load (kips) | Bending Moment (ft-kips) | | Shear Force (kips) |
		Top	Bottom			Top	Bottom	
1	1061	-8.6	8.6	1.8	671	-15.0	15.0	3.1
2: sidesway right	1030	420.9	-429.3	86.5	423	235.2	-225.4	36.5
sidesway left	1046	-433.1	441.5	-89.0	888	-263.2	253.4	-44.7
3: sidesway right	588	421.3	-429.7	86.6	144	242.1	-232.3	48.3
sidesway left	604	-432.7	441.1	-88.9	608	-256.3	246.5	-51.1
4: sidesway right	794	111.2	-111.0	22.6	449	53.0	-45.9	10.1
sidesway left	797	-124.1	123.9	-25.2	558	-75.5	68.4	-14.6
5: sidesway right	594	114.3	-114.1	23.2	321	58.4	-51.1	11.1
sidesway left	597	-125.7	125.4	-25.5	431	-72.6	65.4	-14.0

Table 3-10 Summary of Factored Axial Loads, Bending Moments, and Shear Forces for Columns between Grade and the 2nd Floor of an Exterior Transverse Frame (Seismic Zone 4)

(1) $U = 1.4D + 1.7L$ Eq. (9-1), UBC 1909.2.1
(2) $U = 1.4(D + L \pm E)$ Eq. (9-2), UBC 1921.2.7
(3) $U = 0.9D \pm 1.4E$ Eq. (9-3), UBC 1921.2.7
(4) $U = 0.75(1.4D + 1.7L \pm 1.7W)$ Eq. (9-2), UBC 1909.2.2
(5) $U = 0.9D \pm 1.3W$ Eq. (9-3), UBC 1909.2.2

| Load Combination | Interior Column | | | | Exterior Column | | | |
| | Axial Load (kips) | Bending Moment (ft-kips) | | Shear Force (kips) | Axial Load (kips) | Bending Moment (ft-kips) | | Shear Force (kips) |
		Top	Bottom			Top	Bottom	
1	1159	-8.6	8.6	1.3	732	-15.0	15.0	2.3
2: sidesway right	1127	394.3	-534.3	72.4	447	271.6	-469.0	57.7
sidesway left	1135	-406.5	546.5	-74.3	979	-299.6	497.0	-62.1
3: sidesway right	646	394.7	-534.7	72.4	190	278.5	-475.9	58.8
sidesway left	653	-406.1	546.1	-74.2	722	-292.7	490.1	-61.0
4: sidesway right	869	106.1	-149.1	19.9	484	65.3	-126.5	12.6
sidesway left	869	-119.0	162.0	-21.9	614	-87.8	149.0	-18.5
5: sidesway right	650	109.1	-152.9	20.4	344	70.9	-133.3	15.9
sidesway left	650	-120.5	164.3	22.2	477	-85.1	147.5	-18.1

3.2.2.1   General Code Requirements for Seismic Zone 4

Concrete frames that are part of the lateral force system of structures in Seismic Zone 4 must be designed as Special Moment-Resisting Frames (SMRF). As such, design must be in accordance with all of the provisions of UBC 1921.2 through 1921.7, in addition to the requirements of UBC 1901 through 1917.

UBC 1631.1 requires that buildings located in Seismic Zones 2, 3, or 4 be designed for the effects of earthquake forces acting in a direction other than the principal axes if a column of a structure forms part of two or more intersecting lateral force-resisting systems. An exception to this is permitted if the axial load in the column due to seismic forces acting in either direction is less than 20% of the allowable column axial load.

The maximum column axial force from unfactored earthquake loads is 190.0 kips (see Figs. 3-3 and 3-4). The design axial load strength, $\phi P_{n(max)}$ of a 24 × 24 in. column with $f'_c = 6000$ psi and with 8-#10 bars ($\rho_g = 1.8\%$) symmetrically arranged is 1957 kips. The axial load strength of a 24 × 24 in. column with $f'_c = 4000$ psi and with 8-#10 bars symmetrically arranged is 1419 kips. Twenty percent of the strength of either of these columns exceeds the maximum axial force resulting from the earthquake loads; therefore, orthogonal effects due to seismic forces can be neglected for the columns being designed.

3.2.2.2   Diaphragm Flexibility

The distribution of horizontal forces to the vertical elements is a function of the in-plane rigidity of the floor system. A floor system acts like a diaphragm between the columns. If the diaphragm is relatively flexible, its deflection may be great enough so that the loads are distributed to the vertical elements according to their tributary areas. In this case, continuity of the diaphragm is lost and the vertical elements act separately to resist the lateral loads. On the other hand, if the diaphragm is rigid, the distribution of the load will be in proportion to the relative stiffnesses of the vertical elements, and all of these elements will work together to resist the lateral loads.

The rigidity or flexibility of a diaphragm can be ascertained by determining its flexibility factor, F. A slab is considered to be a rigid diaphragm if it has a flexibility factor of less than 1.0. A flexibility factor between 1 and 10 is considered to represent a semi-rigid diaphragm and a factor greater than 10 indicates a flexible diaphragm. A description of the effects of different flexibility factors on the distribution of the loads to the resisting elements is shown in Fig. 3-9. This figure is a summary of a detailed discussion on diaphragm flexibility from Chapter 5 of *The Tri-Services Manual* [1.9]. The flexibility factor for a concrete diaphragm is defined in that publication as:

$$F = \frac{10^6}{8.5 h w_c^{1.5} \sqrt{f'_c}}$$

where:

h = thickness of the slab, in.

w_c = weight of the concrete, pcf; minimum value of w_c allowed in buildings is 90 pcf

f'_c = compressive strength of the concrete at 28 days, psi

According to UBC 1909.5.3.3, the minimum allowable slab thickness for any structure is 3.5 in. Using a minimum value of 90 pcf for w_c, the above equation indicates a reinforced concrete slab to be rigid even for an f'_c of 1500 psi. Since these are extreme values, it is obvious that unless a concrete slab has large openings, it is unlikely that it will be anything but rigid.

Flexibility Category	Diaphragm Definition	Diaphragm Flexibilities Relative to Flexibilities of Walls
Rigid Diaphragm (F < 1)	A rigid diaphragm is assumed to distribute horizontal forces to the vertical resisting elements in proportion to their relative rigidities.	 a) Schematic Plan
Semi-Rigid Diaphragm (1 < F < 10)	Semi-rigid and semi-flexible diaphragms are those which have significant deflection under load but which also have sufficient stiffness to distribute a portion of their load to vertical elements in proportion to the rigidities of the vertical elements.	 b) Rigid Diaphragm
Semi-Flexible Diaphragm (10 < F < 70)	The action is analogous to a continuous concrete beam system of appreciable stiffness on yielding supports. The support reactions are dependent on the relative stiffnesses of both diaphragm and vertical elements.	 c) Semi-Rigid Diaphragm
Flexible Diaphragm (70 < F < 150)	A flexible diaphragm and a very flexible diaphragm are analogous to a shear deflecting continuous beam or series of beams spanning between supports. The supports are considered non-yielding, as the relative stiffness of the vertical resisting elements compared to that of the diaphragm is great. Thus a flexible diaphragm will be considered to distribute the lateral forces to the vertical resisting elements on a tributary load basis.	 d) Flexible Diaphragm
Very Flexible Diaphragm (F > 150)		

Figure 3-9 Diaphragm Flexibility

3.2.2.3 Story Drift Limitation and P-Δ Effects

UBC 1628.8.2 specifies that for buildings with a fundamental period of 0.7 seconds or greater, the calculated story drift (including torsional effects) shall not exceed $0.03/R_w$ nor 0.004 times the story height.

For h_s = 12 ft and R_w = 12:

$$\frac{0.03h_s}{R_w} = \frac{0.03 \times 12 \times 12}{12} = 0.36 \text{ in. (governs)}$$

$$0.004h_s = 0.004 \times 12 \times 12 = 0.58 \text{ in.}$$

Table 3-11 shows that for all stories, the lateral drifts obtained from the prescribed lateral forces in both directions are less than the limiting value.

Table 3-11 Lateral Displacements and Drifts of Example Building under Seismic Zone 4 Forces

Floor Level	Transverse Direction		Longitudinal Direction	
	Total Lateral Displacement (in.)	Interstory Drift (in.)	Total Lateral Displacement (in.)	Interstory Drift (in.)
R	2.25	0.08	3.18	0.09
12	2.17	0.11	3.09	0.14
11	2.06	0.13	2.95	0.19
10	1.93	0.17	2.76	0.23
9	1.76	0.18	2.53	0.26
8	1.58	0.21	2.27	0.29
7	1.37	0.20	1.98	0.31
6	1.17	0.22	1.67	0.32
5	0.95	0.22	1.35	0.34
4	0.73	0.24	1.01	0.35
3	0.49	0.23	0.66	0.35
2	0.26	0.26	0.31	0.31

According to UBC 1628.9, P-Δ effects need not be considered as long as the ratio of the secondary moment to the primary moment does not exceed 0.1 or, in Seismic Zones 3 and 4, where the story drift ratio is less than $0.02/R_w$ (0.02/12 = 0.0017). The secondary-to-primary moment ratios $\Sigma w_i \Delta_i / V_i h_{si}$ are calculated at all story levels for Seismic Zone 4 in Table 3-12. All values are less than 0.1, and P-Δ effects need not be included in the analysis.

Table 3-12 P-Δ Check under Seismic Zone 4 Forces

Floor Level	Gravity Loads* Σw_i (kips)	Transverse Story Drift Δ_i (in.)	Longitudinal Story Drift Δ_i (in.)	Story Shear V_i (kips)	Story Height h_{si} (in.)	Transverse $\Sigma w_i \Delta_i/V_i h_{si}$	Longitudinal $\Sigma w_i \Delta_i/V_i h_{si}$
R	1863	0.08	0.09	187.9	144	0.006	0.006
12	3723	0.11	0.14	296.5	144	0.010	0.012
11	5583	0.13	0.19	395.5	144	0.013	0.019
10	7443	0.17	0.23	484.8	144	0.018	0.025
9	9303	0.18	0.26	564.5	144	0.021	0.030
8	11,163	0.21	0.29	635.2	144	0.026	0.035
7	13,023	0.20	0.31	695.5	144	0.026	0.040
6	14,883	0.22	0.32	746.2	144	0.031	0.044
5	16,743	0.22	0.34	787.2	144	0.033	0.050
4	18,603	0.24	0.35	818.6	144	0.038	0.055
3	20,463	0.23	0.35	840.3	144	0.039	0.059
2	22,343	0.26	0.31	852.5	180	0.038	0.045

* Includes floor weight, superimposed dead load, and reduced roof and floor live loads

3.2.2.4 Proportioning and Detailing a Flexural Member of an Exterior Transverse Frame

Members are to be designed as flexural elements if they meet the following requirements (UBC 1921.3.1):

- Factored axial compressive force on the member shall not exceed $A_g f'_c/10$.

- Clear span of the member shall not be less than four times its effective depth.

- Width-to-depth ratio of the member shall not be less than 0.3.

- Width of the member shall not be (1) less than 10 in. and (2) more than the width of the supporting member (measured on a plane perpendicular to the longitudinal axis of the flexural member) plus distances on either side of the supporting member not exceeding three-quarters of the depth of the flexural member.

Since the beams under consideration have negligible axial forces, the first requirement is satisfied. A check of the other three requirements for the 24 × 26 in. beam is shown below.

$$\text{Clear span} = 22.0 \text{ ft} > 4d = \frac{4 \times 23.44}{12} = 7.81 \text{ ft} \quad \text{O.K.}$$

$$\frac{\text{Width}}{\text{Depth}} = \frac{24.0 \text{ in.}}{26.0 \text{ in.}} = 0.92 > 0.30 \quad \text{O.K.}$$

Beam width is 24.0 in. which is greater than 10.0 in. Beam width is less than width of supporting column + (1.5 × depth of beam) = 24.0 + (1.5 × 26.0) = 63.0 in. O.K.

3.2.2.4.1 Required Flexural Reinforcement

The maximum negative moment for a 2nd floor beam at an interior support of an exterior frame in the transverse direction is 460.3 ft-kips (see Table 3-7). The required reinforcement, ignoring the effects of any compression reinforcement, is $A_s = 4.71$ in.[2] Use 5-#9 bars at this location ($\phi M_n = 485.9$ ft-kips).

Check limitations on the reinforcement ratio:

$$\rho = A_s/bd = 5.0/(24.0 \times 23.44) = 0.0089$$

$$0.025 > 0.0089 > 200/f_y = 0.0033 \quad \text{O.K.} \qquad\qquad 1921.3.2.1$$

The maximum negative moment in the beam at an exterior support is 417.3 ft-kips. Use 5-#9 bars at this support also.

The positive moment strength at the joint face must be equal to at least 50% of the corresponding negative moment strength at that joint (UBC 1921.3.2.2). The negative moment strength at an interior or exterior joint face is $\phi M_n = 485.9$ ft-kips; the positive moment reinforcement must not give a moment strength less than $485.9/2 = 243.0$ ft-kips. The maximum positive moments in the beam at the interior and exterior supports are 299.6 and 381.6 ft-kips, respectively (see Table 3-7). Since both of these moments are larger than the 50% requirements, the actual moments will control the amount of required reinforcement. The required number of reinforcing bars is 4-#9 at the exterior support, giving a design moment strength of 395.4 ft-kips. The required number of reinforcing bars at the interior support is 3-#9 bars, giving a design moment strength of 301.5 ft-kips. Check the limitations on the reinforcement ratio:

$$0.025 > \rho = A_s/bd = 3.0/(24.0 \times 23.44) = 0.0053 > 0.0033 \quad \text{O.K.} \qquad\qquad 1921.3.2.1$$

The maximum positive moment at the midspan of either span is 108.7 ft-kips (see Table 3-7). According to UBC 1921.3.2.2, the positive moment strength at any section along the member length must not be less than 25% of the maximum moment strength provided at the face of either joint. Thus, the minimum design value of the positive moment for both spans is $485.9/4 = 121.5$ ft-kips which is larger than 108.7 ft-kips. This requires an area of steel equal to 1.17 in.[2]; therefore, use 2-#9 bars ($\phi M_n = 204.3$ ft-kips).

Similarly, the negative moment strength at any section along the member length shall not be less than 25% of the maximum moment strength provided at the face of either joint ($485.9/4 = 121.5$ ft-kips). Two #9 bars will be required, giving a moment strength of 204.3 ft-kips.

When reinforcing bars are extended through a joint, the column dimension parallel to the beam reinforcement shall not be less than 20 times the diameter of the longitudinal bar (UBC 1921.5.1.4). For this situation, the minimum required depth of the column in the direction of the continuous longitudinal bars is $20d_b = 20 \times 1.128 = 22.6$ in. which is less than the 24.0 in. column width provided.

3.2.2.4.2 Required Shear Reinforcement

The beams must be designed for shear forces according to UBC 1921.3.4. These shear forces are based on the probable moment strength at each joint face, using a strength reduction factor of 1.0 and a stress in the tensile flexural reinforcement equal to 1.25 f_y, combined with the shear caused by unfactored gravity loads. It should be noted that ACI 318-89 (Revised 1992) uses the same procedure, but with factored gravity loads.

Figure 3-10 shows the probable flexural strengths M_{pr} at the joint faces of the exterior span for sidesway to the right and to the left. As noted above, these flexural strengths are determined using a steel stress of 1.25f_y and a strength reduction factor of 1.0. Figure 3-10 also shows the unfactored gravity loads acting on the beam, as well as the shear forces at the faces of the columns. In general, the shear forces V_e are determined as follows:

$$V_e = \frac{M_{pre}^{\pm} + M_{pri}^{\mp}}{\ell_n} \pm \frac{W}{2}$$

where $M_{pre}^{\pm}$ = probable flexural strength at the exterior support

$M_{pri}^{\mp}$ = probable flexural strength at the interior support

ℓ_n = clear span length

W = resultant force of the unfactored gravity loads

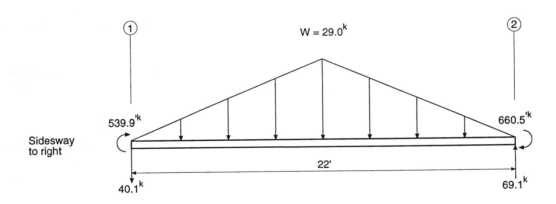

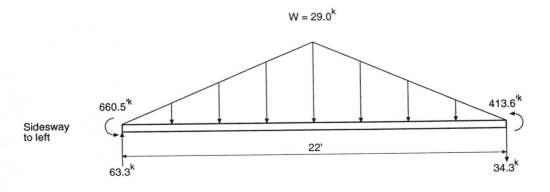

Figure 3-10 Design Shear Forces for Transverse Beams in an Exterior Frame (Seismic Zone 4)

The shear strength of the concrete is not considered in Seismic Zones 3 and 4 if the factored axial compressive force, including earthquake effects, is less than $A_g f'_c/20$ and if the earthquake-induced shear force represents one-half or more of the total design shear (UBC 1921.3.4.2). The beam being designed carries negligible axial forces, and the earthquake-induced shear force, which is $(660.5 + 539.9)/22 = 54.6$ kips, is greater than one-half the total design shear of 69.1 kips. Thus, V_c is taken as zero. The maximum shear force that shear reinforcement must be designed for is:

$$\phi V_s = V_u - \phi V_c \qquad\qquad\qquad\qquad\qquad\qquad\qquad \textit{1911.1.1}$$

$$V_s = \frac{69.1}{0.85} - 0 = 81.3 \text{ kips}$$

Shear strength contributed by shear reinforcement shall not be taken greater than $8\sqrt{f'_c} \times b_w d$:

$$V_{s(max)} = 8\sqrt{4000} \times 24 \times 23.44/1000 \qquad\qquad\qquad\qquad \textit{1911.5.6.8}$$

$$= 284.6 \text{ kips} > 81.3 \text{ kips} \quad \text{O.K.}$$

In Seismic Zones 3 and 4, the following detailing requirements must be satisfied in addition to the shear design requirements. Hoops shall be provided:

- Over a length equal to twice the member depth, starting from the face of the supporting member toward midspan, at both ends of the flexural member (UBC 1921.3.3.1).

- Over lengths equal to twice the member depth on both sides of a section where flexural yielding may occur in connection with inelastic lateral displacements of the frame (UBC 1921.3.3.1).

Where hoops are not required, stirrups with 135-degree seismic hooks shall be spaced at no more than d/2 throughout the length of the member (UBC 1921.3.3.4).

The maximum allowable hoop spacing within a distance 2h = 52.0 in. from the face of the supports is (UBC 1921.3.3.2):

$$s_{max} = \frac{d}{4} = \frac{23.44}{4} = 5.9 \text{ in. (governs)}$$

$$= 8 \times \text{(diameter of smallest longitudinal bar)} = 8 \times 1.128 = 9.0 \text{ in.}$$

$$= 24 \times \text{(diameter of hoop bars)} = 24 \times 0.5 = 12.0 \text{ in.}$$

$$= 12.0 \text{ in.}$$

The required spacing of #4 closed stirrups (hoops) for the factored shear force of 81.3 kips is:

$$s = \frac{A_v f_y d}{V_s} = \frac{0.40 \times 60 \times 23.44}{81.3} = 6.9 \text{ in.} \qquad\qquad \textit{1911.5.6.2}$$

Therefore, hoops shall be spaced at 5 in. on center (maximum allowable spacing) with the first one being located 2 in. from the face of the column. Provide 11 hoops at 5 in. spacing, the last one being at 52.0 in. from the support face.

The shear force at 52.0 in. from the face of the support is 66.9 kips. Therefore,

$$V_s = \frac{66.9}{0.85} = 78.7 \text{ kips}$$

The required spacing of #4 stirrups at this location is:

$$s = \frac{0.40 \times 60 \times 23.44}{78.7} = 7.2 \text{ in.} < \frac{23.44}{2} = 11.7 \text{ in.}$$

A 7 in. spacing, starting at 52 in. from the column face to the start of the longitudinal bar splices at midspan, will be sufficient.

Note that lap splices in longitudinal reinforcement are permitted only if hoop or spiral reinforcement is provided over the lap length (UBC 1921.3.2.3). The maximum spacing of the transverse reinforcement enclosing the lapped bars must not exceed d/4 or 4 in. Lap splices must not be used within a distance of twice the member depth from the joint face, nor at locations where analysis indicates flexural yielding caused by inelastic lateral displacements of the frame to be likely.

3.2.2.4.3 Reinforcing Bar Cutoff Points and Splices

The negative reinforcement at the supports consists of 5-#9 bars. The location where three of the five bars can be terminated away from an interior support will be determined. The loading used to find the cutoff points is 0.9 times the dead load in combination with the probable moment strengths, M_{pr}, at the member ends, which is conservative. The design bending moment strength, ϕM_n, provided by 2-#9 reinforcing bars in the beam is 204.3 ft-kips. Therefore, the three reinforcing bars can only be terminated after the required moment strength, M_u, has been reduced to 204.3 ft-kips.

With $\phi = 1.0$ and $f_s = 75$ ksi, $M^+_{pr} = 539.9$ ft-kips at the exterior end of the beam, and $M^-_{pr} = 660.5$ ft-kips at the interior end (see Fig. 3-10). The dead load is equal to $0.9 \times 2.04 = 1.84$ kips/ft at midspan. The distance from the face of the support to where the moment under the loading considered equals 204.3 ft-kips is readily obtained by summing moments about section a-a in Fig. 3-11:

$$\frac{x}{2}\left(\frac{1.84x}{11}\right)\left(\frac{x}{3}\right) + 660.5 - 64.7x = 204.3$$

Solving for x gives a distance of 7.2 ft. Therefore, the negative reinforcing bars can be terminated at a distance of $7.2 + d = 9.2$ ft from the face of the interior support (noting that d = 23.44 in. > 12 d_b = 12 × 1.128 = 13.5 in.; see UBC 1912.10.3).

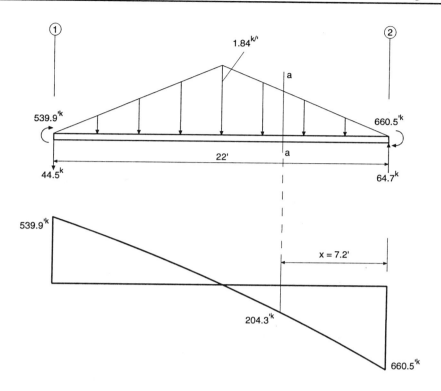

Fig. 3-11 Moment Diagram for Cutoff Location of Negative Bars at Interior Support of Transverse Beam (Seismic Zone 4)

In lieu of determining the cutoff point using the actual triangular load distribution on the beam, the bar cutoff point can be obtained by using an equivalent uniform load distribution. Figure 3-12 shows the moment diagram for the beam loaded with 90% of the equivalent uniform dead load. Using the same reinforcement determined previously, the following equation can be used to determine the location of the cutoff point:

$$\frac{1}{2}(1.31x^2) + 660.5 - 69.0x = 204.3$$

Solving for x gives 7.1 ft which is almost the same value obtained from the first method.

The #9 bars being terminated must be properly developed at the support. In Seismic Zone 4, the development length for reinforcement in tension, if the depth of concrete cast in one lift below the bar exceeds 12.0 in., is given by:

$$\ell_d = 3.5 f_y d_b / 65 \sqrt{f'_c} \hspace{3cm} 1921.5.4.2$$

$$= 3.5 \times 60,000 \times 1.128 / 65 \sqrt{4000}$$

$$= 57.6 \text{ in.} < (9.2 \times 12) = 110.4 \text{ in.} \quad \text{O.K.}$$

Using the same procedure outlined above, the cutoff point for two of the 4-#9 bars at the bottom near the exterior support will be determined. The same loading condition as used for the cutoff of the negative reinforcing bars gives the following (see Fig. 3-13):

$$-\frac{x}{2}\left(\frac{1.84x}{11}\right)\left(\frac{x}{3}\right) + 539.9 - 44.5x = 204.3$$

Solution of this equation gives x = 7.3 ft. The cutoff length is equal to 7.3 + d = 9.3 ft which is greater than the required development length.

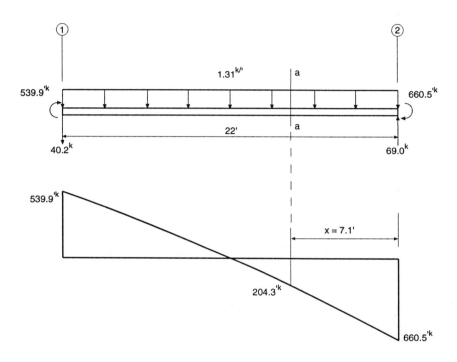

Figure 3-12 Moment Diagram for Cutoff Location of Negative Bars at Interior Support of Transverse Beam, Equivalent Uniform Load (Seismic Zone 4)

The calculations for cutoff points of the top bars at the exterior support and the bottom bars at the first interior support are performed in a similar manner. Three of the top bars at the exterior support are terminated at 10.0 ft from the face of the column. One of the bottom bars at the first interior support is terminated at 7.5 ft from the face of the column. These lengths are both greater than the required development lengths.

Lap splices of flexural reinforcement should not be placed within a distance of 2h from the faces of the supports, nor within regions of potential plastic hinging (UBC 1921.3.2.3). All lap splices have to be confined by hoops or spirals with a maximum spacing of d/4 or 4 in. over the length of the lap.

The required lap splice length is the product of the basic development length and applicable modification factors. The basic development length for #9 bars is:

$$\ell_d = 0.04\, A_b f_y / \sqrt{f'_c} \qquad\qquad 1912.2.2$$

$$= 0.04 \times 1.0 \times 60{,}000 / \sqrt{4000}$$

$$= 38.0 \text{ in.}$$

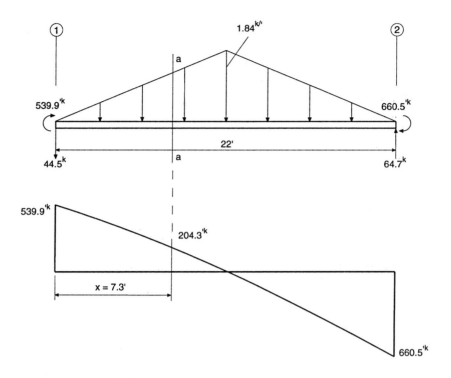

Figure 3-13 Moment Diagram for Cutoff Location of Positive Bars at the Exterior Support of Transverse Beam (Seismic Zone 4)

The development length modification factors are shown in Table 3-13. These factors account for the following:

- Factor I - Bar spacing, amount of cover, and enclosing transverse reinforcement (UBC 1912.2.3).

- Factor II - Top bars or other bars (UBC 1912.2.4.1).

- Factor III - Lightweight or normal weight concrete (UBC 1912.2.4.2).

- Factor IV - Epoxy-coated or uncoated bars (UBC 1912.2.4.3).

- Factor V - Amount of steel provided compared to amount required for flexure (UBC 1912.2.5). Note that this factor is not applicable in Seismic Zones 3 and 4.

The information required for determining the splice length is:

	Top Bars	Bottom Bars
Stress level	$< 0.5\,f_y$	$> 0.5\,f_y$
Splice type	A	B
Minimum top or bottom cover	2.0 in. ($1.8\,d_b$)	2.0 in. ($1.8\,d_b$)
Minimum side cover	large (slab)	2.0 in. ($1.8\,d_b$)

Table 3-13 Development Length Factors

Factor I

Criteria	Cover	Transverse Reinforcement Requirements	Clear Bar Spacing	Confinement provided UBC 1912.2.3.5	Applicable Factor
1	$\geq$ 1.5 in. clear interior or $\geq$ 2.0 in. clear exterior exposure	Meeting UBC 1907.10.5 or UBC 1911.5.4 and 1911.5.5.3	$\geq 3d_b$	YES	0.75
				NO	1.00
			$\geq 5d_b$ & side cover $\geq 2.5\,d_b$	YES	0.60
				NO	0.80
2	$\geq$ 1.5 in. clear interior or $\geq$ 2.0 in. clear exterior exposure	$A_{tr} \geq \dfrac{d_b s N}{40}$	—	YES	0.75
				NO	1.00
			$\geq 5d_b$ & side cover $\geq 2.5\,d_b$	YES	0.60
				NO	0.80
3	$\geq 2\,d_b$	—	$\geq 3d_b$	YES	0.75
				NO	1.00
			$\geq 5d_b$ & side cover $\geq 2.5\,d_b$	YES	0.60
				NO	0.80
4	Either $\leq d_b$	—	Or $< 2d_b$	YES	1.50
				NO	2.00
5	All other conditions		—	YES	1.05
				NO	1.40
			$\geq 5d_b$ & side cover $\geq 2.5\,d_b$	YES	0.84
				NO	1.12

Bar Size	#4	#5	#6	#7	#8	#9	#10	#11
Minimum Value of Factor I*	1.88	1.51	1.28	1.09	0.95	0.85	0.75	0.68

* Based on $0.03\, d_b\, f_y / \sqrt{f'_c}$

Factor II

Bar with more than 12 in. of fresh concrete cast in one lift below it	1.3
All others	1.0

Factor III

Bars Embedded in Lightweight Concrete	1.3 or $6.7\sqrt{f'_c}\,/f_{ct}$ but > 1.0
Bars Embedded in Normal Weight Concrete	1.0

Factor IV

Epoxy Coated Bars	Cover $< 3d_b$	1.5
	Clear Spacing $< 6d_b$	1.5
	All other	1.2
Uncoated Bars		1.0

Factor V

$$= \frac{A_s\ (\text{required})\,*}{A_s\ (\text{provided})}$$

* This factor cannot be used in Seismic Zones 3 & 4

ℓ_d = (Factor I) $\times$ (Factor II) $\times$ (Factor III) $\times$ (Factor IV) $\times$ (Factor V) $\times$ ℓ_{db}

Clear spacing between
bars being spliced   17.7 in. (15.7 d_b)   17.7 in. (15.7 d_b)

Transverse reinforcement meets requirements of UBC 1911.5.4 and 1912.2.3.5 for both top and
bottom bars.

This gives a value of 0.60 and 0.75 for Factor I for the top and bottom splices, respectively (see Table 3-13).
However, the limiting value of 0.85 for #9 bars will control in both cases. Factor II is 1.3 for top bars. All other
factors are 1.0; therefore,

Top splice length (Class A) = $1.0 \times (0.85 \times 1.3 \times 1.0 \times 1.0 \times 1.0 \times 38.0) = 42.0$ in.

Bottom splice length (Class B) = $1.3 \times (0.85 \times 1.0 \times 1.0 \times 1.0 \times 1.0 \times 38.0) = 42.0$ in.

Figure 3-14 shows the reinforcement details for the transverse beam.

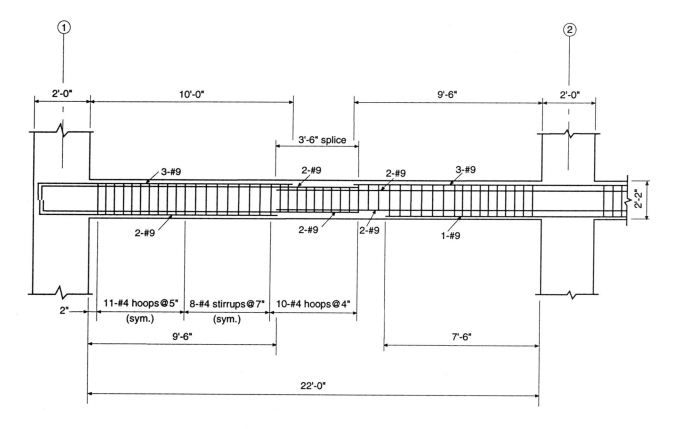

Figure 3-14 Reinforcement Details for Beam of Exterior Transverse Frame (Seismic Zone 4)

Termination of flexural reinforcement is permitted in a tension zone only if one of the three following conditions
is satisfied (UBC 1912.10.5):

• Shear at the cutoff point does not exceed two-thirds that permitted, including shear strength of shear reinforcement provided.

- Stirrup area in excess of that required for shear and torsion is provided along each terminated bar or wire over a distance from the termination point equal to three-fourths the effective depth of member. Excess stirrup area A_v shall not be less than $60b_w s/f_y$. Spacing s shall not exceed $d/8\beta_b$ where β_b is the ratio of area of reinforcement cut off to total area of tension reinforcement at the section.

- For #11 bars and smaller, continuing reinforcement provides double the area required for flexure at the cutoff point and shear does not exceed three-fourths that permitted.

The satisfaction of the above requirement, though it is technically applicable, is not checked in this design, since the termination points of the bars are determined conservatively using the probable moment strengths M_{pr} (based on $f_s = 1.25\ f_y$) at the member ends.

3.2.2.5 Proportioning and Detailing a Flexural Member of an Interior Longitudinal Frame

Check limitations on section dimensions:

$$\text{Clear span} = 22.0\ \text{ft} > 4d = \frac{4 \times 17.44}{12} = 5.8\ \text{ft} \quad \text{O.K.} \qquad\qquad 1921.3.1$$

$$\frac{\text{Width}}{\text{Depth}} = \frac{24.0\ \text{in.}}{20.0\ \text{in.}} = 1.20 > 0.30 \quad \text{O.K.}$$

Beam width is 24.0 in. which is greater than 10.0 in. Beam width is less than width of supporting column + (1.5 × depth of beam) = 24.0 + (1.5 × 20.0) = 54.0 in. O.K.

3.2.2.5.1 Required Flexural Reinforcement

The maximum negative moment at an interior support of a 2nd floor beam of an interior frame in the longitudinal direction is 473.8 ft-kips (see Table 3-8). The required reinforcement, ignoring the effects of any compressive reinforcement, is $A_s = 7.1$ in.[2] Use 8-#9 bars at this location ($\phi M_n = 521.6$ ft-kips).

Check limitations on the reinforcement ratio:

$$\rho = A_s/bd = 8/(24.0 \times 17.44) = 0.019$$

$$0.025 > 0.019 > 200/f_y = 0.0033 \quad \text{O.K.} \qquad\qquad 1921.3.2.1$$

The maximum negative moment in the beam at an exterior support is 354.3 ft-kips. Use 6-#9 bars at this location ($\phi M_n = 411.1$ ft-kips).

The positive moment strength at the joint face must be equal to at least 50% of the corresponding negative moment strength at that joint (UBC 1921.3.2.2). The negative moment strength at the interior joint face is 521.6 ft-kips; the positive moment strength must not be less than 521.6/2 = 260.8 ft-kips at this location. The maximum positive moment in the beam at the interior support is 212.4 ft-kips (see Table 3-8). Therefore, the requirements of UBC 1921.3.2.2 will control the design at the interior support. The required number of bars is 4-#9 ($\phi M_n = 287.4$ ft-kips).

Similarly, the negative moment strength at the exterior joint face is 411.1 ft-kips, so that the positive moment strength must not be less than 411.1/2 = 205.6 ft-kips. At this location, the maximum positive moment in the beam is 296.2 ft-kips (see Table 3-8) which is greater than 205.6 ft-kips. Therefore, use 5-#9 bars ($\phi M_n = 350.9$ ft-kips) at the exterior support. Note that at both the interior and exterior supports, the number of bars provided satisfies the limitations on the reinforcement ratio given in UBC 1921.3.2.1.

The maximum positive factored moment at the midspan of any of the spans is 154.0 ft-kips. According to UBC 1921.3.2.2, the positive moment strength at any section along the member length must not be less than 25% of the maximum moment strength provided at the face of either joint. Thus, the minimum design value of the positive moment for both spans is 521.6/4 = 130.4 ft-kips which is less than 154.0 ft-kips, so that the actual midspan moment governs. Therefore, use 3-#9 bars, giving a design moment strength of $\phi M_n = 220.5$ ft-kips.

Similarly, the negative moment strength at any section along the member length must not be less than 25% of the maximum moment strength provided at the face of either joint (521.6/4 = 130.4 ft-kips). Two #9 bars will be sufficient ($\phi M_n = 150.3$ ft-kips).

When reinforcing bars are extended through a joint, the column dimension parallel to the beam reinforcement shall not be less than 20 times the diameter of the longitudinal bar (UBC 1921.5.1.4). The minimum required depth of the column in the direction of the continuous longitudinal bars is $20d_b = 20 \times 1.128 = 22.6$ in. which is less than the 24.0 in. column width provided.

3.2.2.5.2 Required Shear Reinforcement

The beams must be designed for shear forces according to UBC 1921.3.4. These shear forces are based on the probable moment strength at each joint face, using a strength reduction factor of 1.0 and a stress in the tensile flexural reinforcement equal to $1.25f_y$, combined with the shear caused by unfactored gravity loads. Note that ACI 318-89 (Revised 1992) uses the same procedure, but with factored gravity loads.

Figure 3-15 shows the probable flexural strengths M_{pr} at the joint faces of the exterior span for sidesway to the right and to the left. This figure also shows the unfactored gravity loads, as well as the shear forces at the faces of the columns.

The shear strength of the concrete is not considered in Seismic Zones 3 and 4 if the factored axial compressive force, including earthquake effects, is less than $A_g f'_c/20$ and if the earthquake-induced shear force represents one-half or more of the total design shear (UBC 1921.3.4.2). The beam being designed carries negligible axial forces, and the earthquake-induced shear force, which is (473.0 + 687.6)/22 = 52.8 kips, is greater than one-half the total design shear of 78.3 kips. Thus, V_c is taken as zero. The maximum shear force that shear reinforcement must be designed for is:

$$\phi V_s = V_u - \phi V_c \qquad \qquad \textit{1911.1.1}$$

$$V_s = \frac{78.3}{0.85} - 0 = 92.1 \text{ kips}$$

Shear strength contributed by shear reinforcement shall not be taken greater than $8\sqrt{f'_c} \times b_w d$:

$$V_{s(max)} = 8\sqrt{4000} \times 24 \times 17.44 / 1000$$

1911.5.6.8

$$= 211.8 \text{ kips} > 92.1 \text{ kips} \quad \text{O.K.}$$

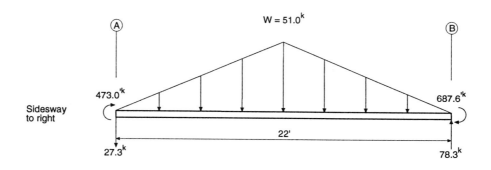

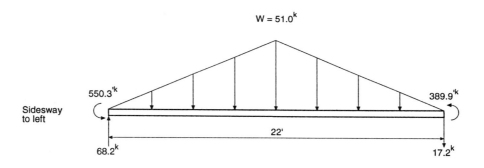

Figure 3-15 *Design Shear Forces for Longitudinal Beams in an Interior Frame (Seismic Zone 4)*

The maximum allowable hoop spacing within a distance 2h = 40.0 in. from the face of the supports is (UBC 1921.3.3.2):

$$s_{max} = \frac{d}{4} = \frac{17.44}{4} = 4.4 \text{ in. (governs)}$$

$$= 8 \times (\text{diameter of smallest longitudinal bar}) = 8 \times 1.125 = 9.0 \text{ in.}$$

$$= 24 \times (\text{diameter of hoop bars}) = 24 \times 0.5 = 12.0 \text{ in.}$$

$$= 12.0 \text{ in.}$$

The required spacing of #4 closed stirrups (hoops) for the factored shear force of 92.1 kips is:

$$s = \frac{A_v f_y d}{V_s} = \frac{0.40 \times 60 \times 17.44}{92.1} = 4.5 \text{ in.} \qquad\qquad 1911.5.6.2$$

Therefore, hoops should be spaced at 4 in. on center with the first one being located 2 in. from the face of the column. Provide 11 hoops at 4 in. spacing, the last one being at 42.0 in. from the support face. The shear force at 42.0 in. from the face of the support is 75.7 kips. Therefore,

$$V_s = \frac{75.7}{0.85} = 89.1 \text{ kips}$$

The required stirrup spacing of #4 stirrups at this location is:

$$s = \frac{0.40 \times 60 \times 17.44}{89.1} = 4.7 \text{ in.}$$

A 4 in. spacing of stirrups starting at 42.0 in. from the column face to the start of the longitudinal bar splices at midspan will be required. Note that if #5 stirrups are used, the required spacing would be 7 in. in this region. At this time, the designer may wish to increase the overall beam depth to increase the overall shear capacity. This, obviously, would require a complete reanalysis of the frame, which is not done here.

3.2.2.5.3 Reinforcing Bar Cutoff Points and Splices

The negative reinforcement at the interior support consists of 8-#9 bars. The location where six of the eight bars can be terminated away from an interior support will be determined. The loading used to find the cutoff points is 0.9 times the dead load in combination with the probable moment strengths, M_{pr}, at the member ends, which is conservative. The design moment strength, ϕM_n, provided by 2-#9 reinforcing bars in the beam is 150.3 ft-kips. Therefore, the six reinforcing bars can only be terminated after the required moment strength, M_u, has been reduced to 150.3 ft-kips.

With $\phi = 1.0$ and $f_s = 75$ ksi, $M^+_{pr} = 473.0$ ft-kips at the exterior end of the beam, and $M^-_{pr} = 687.6$ ft-kips at the interior end (see Fig. 3-15). The dead load is equal to $0.9 \times 3.44 = 3.10$ kips/ft at midspan. The distance from the face of the support to where the moment under the loading considered equals 150.3 ft-kips is readily obtained by summing moments about section a-a in Fig. 3-16:

$$\frac{x}{2}\left(\frac{3.10x}{11}\right)\left(\frac{x}{3}\right) + 687.6 - 69.8x = 150.3$$

Solving for x gives a distance of 8.1 ft. Therefore, the negative reinforcing bars can be terminated at a distance of 8.1 + d = 9.6 ft from the face of the interior support.

The #9 bars being terminated must be properly developed at the support. In Seismic Zone 4, the development length of tensile bars for joints of frames, if the depth of concrete cast in one lift below the bar exceeds 12 in., is given by:

$$\ell_d = 3.5 \times f_y \times d_b/(65\sqrt{f'_c})$$

$$= 3.5 \times 60{,}000 \times 1.128/(65\sqrt{4000})$$

$$= 57.6 \text{ in.} < (9.6 \times 12) = 115.2 \text{ in.} \quad \text{O.K.}$$

In a similar manner, the cutoff points for the other bars at the other locations can be determined. Figure 3-17 shows the reinforcement details for the longitudinal beam.

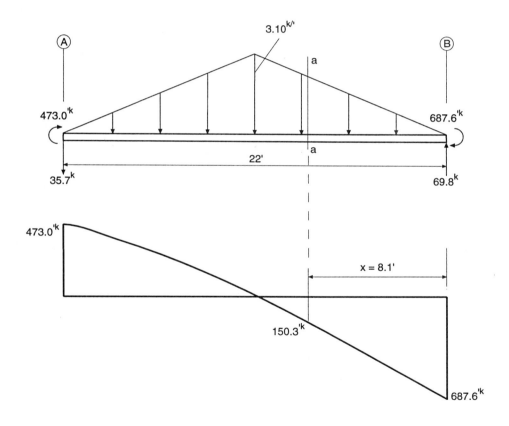

Figure 3-16 Moment Diagram for Cutoff Location of Negative Bars at Interior Support of Longitudinal Beam
(Seismic Zone 4)

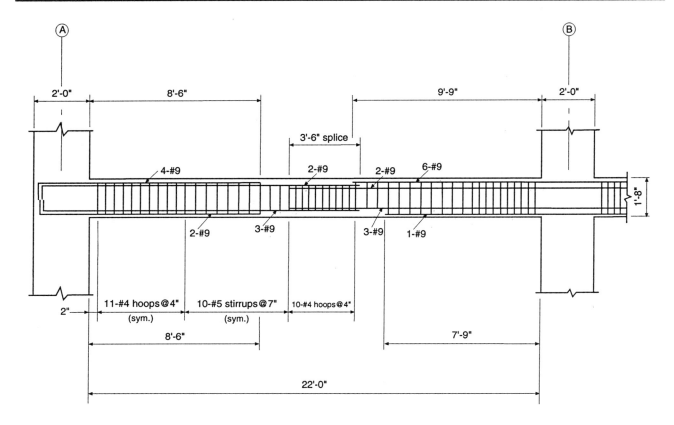

Figure 3-17 Reinforcement Details for Beam of Interior Longitudinal Frame (Seismic Zone 4)

3.2.2.5.4 Anchorage Length of Flexural Tension Reinforcement at Exterior Column

The reinforcing bars that terminate at exterior columns must be properly developed. This is accomplished by providing a hook at the end of the reinforcing bar that is embedded in the core of the concrete column. The length of the hook is a function of the yield strength of the reinforcing bars, the concrete strength and the reinforcing bar diameter. The designs for the beams in the transverse and the longitudinal directions utilize only #9 reinforcing bars, and the hook length is determined as the largest of the following:

$$\ell_{dh} = f_y\, d_b/65\sqrt{f_c'} = 60{,}000 \times 1.128/65\sqrt{6000} = 13.4 \text{ in. (governs)} \qquad \textit{1921.5.4.1}$$

$$= 8d_b = 8 \times 1.128 = 9.0 \text{ in.}$$

$$= 6.0 \text{ in.}$$

Therefore, the hook must be extended at least 13.5 in. into the column with a 90° bend plus 12 d_b extension beyond the bend (UBC 1907.1.2).

3.2.2.6   Proportioning and Detailing a Column

3.2.2.6.1 General Code Requirements

This section outlines the design of an interior column of an exterior transverse frame between levels 2 and 3. The maximum axial force on this column at this level is 1061 kips (see Table 3-9). The requirements in UBC 1921.4 apply to all members resisting earthquake forces and factored axial forces exceeding $A_g f'_c/10$. The factored axial force of 1061 kips exceeds $A_g f'_c/10 = (24 \times 24 \times 6)/10 = 346$ kips; therefore, the two following criteria need to be satisfied:

- Shortest column dimension shall not be less than 12.0 in. This criterion is met since the columns are 24.0 in. square.

- Ratio of the shortest cross-sectional dimension to the perpendicular dimension shall not be less than 0.4. This criterion is met since the columns are square.

Based on the load combinations in Table 3-9, a 24.0-in. square column with 8-#10 bars is adequate (see Fig. 3-18). The reinforcement ratio for columns must be greater than 0.01 but less than 0.06 (UBC 1921.4.3.1):

$$0.01 < \rho = (8 \times 1.27)/(24 \times 24) = 0.0176 < 0.06 \quad \text{O.K.}$$

3.2.2.6.2 Relative Flexural Strengths of Columns and Girders

UBC 1921.4.2.2 requires that the sum of the flexural strengths of columns at a joint must be greater than or equal to 6/5 times the sum of the flexural strengths of girders framing into that joint. Note that the column flexural strength is to be determined for the factored axial force resulting in the lowest flexural strength, consistent with the direction of the lateral forces being considered.

The interior column being designed is part of the exterior transverse frame. Only one beam frames into the column in the longitudinal direction, so that the relative flexural strength check need only be performed for the transverse direction.

The negative design flexural strength of the exterior transverse beam at the interior column is 485.9 ft-kips (see Section 3.2.2.4.1). The positive design flexural strength of the beam on the opposite face of the column is 301.5 ft-kips.

For the upper column framing into the top of the joint, the minimum flexural strength is 681.2 ft-kips, which corresponds to a factored axial load of 588 kips (see Table 3-9 and Fig. 3-18). Similarly, the minimum flexural strength is 686.8 ft-kips for the lower column framing into the bottom of the joint; this corresponds to a factored axial load of 1159 kips (see Table 3-10). Thus,

$$\Sigma M_e = 681.2 + 686.8 = 1368.0 \text{ ft-kips}$$

$$\Sigma M_g = 485.9 + 301.5 = 787.4 \text{ ft-kips}$$

$$1368.0 \text{ ft-kips} > \frac{6}{5} \times 787.4 = 944.9 \text{ ft-kips} \quad \text{O.K.}$$

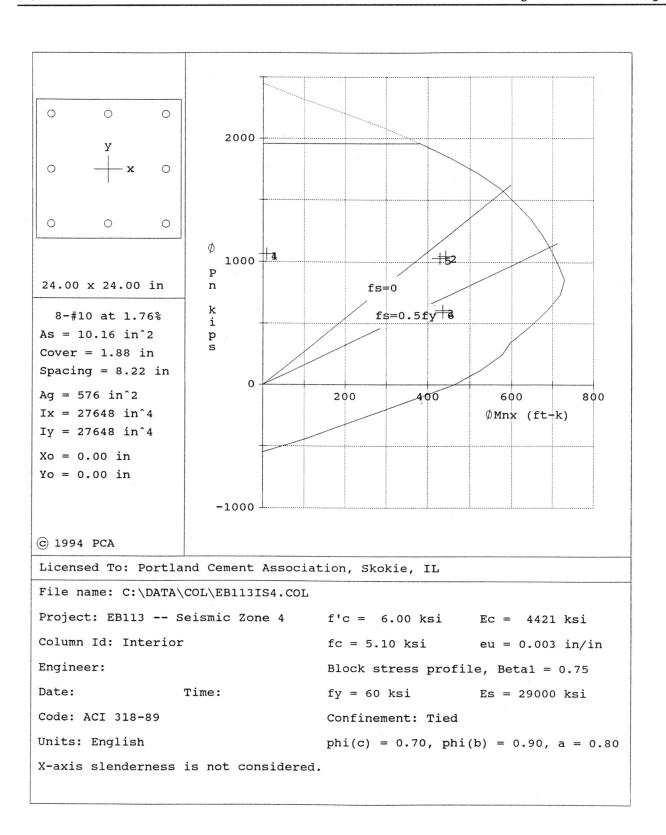

24.00 x 24.00 in

8-#10 at 1.76%
As = 10.16 in^2
Cover = 1.88 in
Spacing = 8.22 in

Ag = 576 in^2
Ix = 27648 in^4
Iy = 27648 in^4

Xo = 0.00 in
Yo = 0.00 in

© 1994 PCA

Licensed To: Portland Cement Association, Skokie, IL

File name: C:\DATA\COL\EB113IS4.COL

Project: EB113 -- Seismic Zone 4   f'c = 6.00 ksi   Ec = 4421 ksi

Column Id: Interior   fc = 5.10 ksi   eu = 0.003 in/in

Engineer: Block stress profile, Beta1 = 0.75

Date: Time: fy = 60 ksi   Es = 29000 ksi

Code: ACI 318-89   Confinement: Tied

Units: English   phi(c) = 0.70, phi(b) = 0.90, a = 0.80

X-axis slenderness is not considered.

Figure 3-18 Design Strength Interaction Diagram for a 24-in. Square Interior Column of an Exterior Transverse Frame above the 2nd Floor Level, Seismic Zone 4

3.2.2.6.3 Confinement Reinforcement

Special transverse reinforcement for confinement is required over a distance ℓ_o at the column ends where ℓ_o equals the maximum of:

Depth of member = 24.0 in. (governs)

$$\frac{1}{6}(\text{clear height}) = \frac{10.33 \times 12}{6} = 20.7 \text{ in.}$$

1921.4.4.4

18.0 in.

According to UBC 1921.4.4.2, the maximum allowable spacing of rectangular hoops within the 24.0 in. length is 4 in. The columns in the structure are subject to double curvature, and therefore the point of contraflexure is within the middle half of the clear height; therefore, special confinement will not be required for the full length of the column (UBC 1921.4.4.1).

The minimum required cross-sectional area of hoop reinforcement is the larger value obtained from the following two equations:

$$A_{sh} = 0.3 s h_c \left[\frac{A_g}{A_{ch}} - 1 \right] \frac{f'_c}{f_{yh}}$$

1921.4.4.1

$$A_{sh} = 0.09 s h_c \frac{f'_c}{f_{yh}}$$

where

s = spacing of transverse reinforcement (in.)

h_c = cross-sectional dimension of column core, measured center-to-center of confining reinforcement (in.)

A_{ch} = cross-sectional area of a structural member measured out-to-out of transverse reinforcement (in.^{2})

f_{yh} = specified yield strength of transverse reinforcement (psi)

Using a hoop spacing of 3 in., f_{yh} = 60,000 psi and tentatively assuming #4 bar hoops, the required cross-sectional area is:

$$A_{sh} = 0.3 \times 3 \times 20.5 \times 6000/60,000 \times [(576/441) - 1] = 0.57 \text{ in.}^2 \text{ (governs)}$$

$$A_{sh} = 0.09 \times 3 \times 20.5 \times 6000/60,000 = 0.55 \text{ in.}^2$$

Using #4 hoops with one #4 crosstie provides $A_{sh} = 3 \times 0.2 = 0.60$ in.^{2} Therefore, use 3 in. spacing for the hoops at the column ends.

3.2.2.6.4 Transverse Reinforcement for Shear

The design of columns for shear is similar to the shear design of beams. It is not based on the factored shear forces obtained from a lateral load analysis, but rather on the probable moment strengths at the column ends. UBC 1921.4.5 requires that the design shear force shall be determined from the consideration of the maximum forces that can be developed at the faces of the joints. These joint forces shall be determined using the probable moment strengths M_{pr}, calculated without strength reduction factors and assuming that the stress in the tensile reinforcement is equal to $1.25f_y$. The member shear need not exceed those determined from joint strengths based on the probable moment strength M_{pr} of the transverse members framing into the joint.

The largest probable moment strength that may develop in a column can conservatively be assumed to correspond to the balanced point of the column interaction diagram. The probable moment strength corresponding to the balanced point for the column under consideration is equal to $760.0/\phi = 760.0/0.70 = 1085.7$ ft-kips (see Fig. 3-19).

In Section 3.2.2.4.2, the probable moment strengths of the beams were determined. The probable positive moment strength at the face of the interior column is 413.6 ft-kips. The probable negative moment strength at the same location is 660.5 ft-kips. The largest moment that can develop from the beams is $413.6 + 660.5 = 1074.1$ ft-kips. This moment is less than the sum of the probable flexural strengths at the column ends ($2 \times 1085.7 = 2171.4$ ft-kips). Therefore, the columns need only be designed to resist the maximum shear than can be transferred through the beams. The design shear for the columns is:

$$V_u = \frac{1074.1/2 + 1074.1/2}{9.83} = 109.2 \text{ kips}$$

Note that this factored shear force is greater than all of the factored shear forces determined from analysis (see Table 3-9). Since the factored axial forces are greater than $A_g f'_c/20 = (24 \times 24 \times 6)/20 = 173$ kips, the shear strength of the concrete may be used (see Table 3-9 and UBC 1921.4.5.2):

$$V_c = 2\left[1 + \frac{N_u}{2000A_g}\right]\sqrt{f'_c}\, b_w d \qquad\qquad 1911.3.1.2$$

Conservatively use the minimum axial load from Table 3-9 (i.e., $N_u = 588$ kips); therefore,

$$V_c = 2\left[1 + \frac{588,000}{2000 \times 576}\right] \times \sqrt{6000} \times 24 \times 21.37/1000 = 120.0 \text{ kips}$$

$$\phi V_c = 0.85 \times 120.0 = 102.0 \text{ kips} \cong 109.2 \text{ kips}$$

Thus, use the 3 in. spacing over the distance $\ell_o = 24.0$ in. near the column ends that was determined by the requirements for confinement.

The remainder of the column length must contain hoop reinforcement with center-to-center spacing not to exceed either six times the diameter of the longitudinal column bars or 6 in. (UBC 1921.4.4.6). In this case the 6 in. controls.

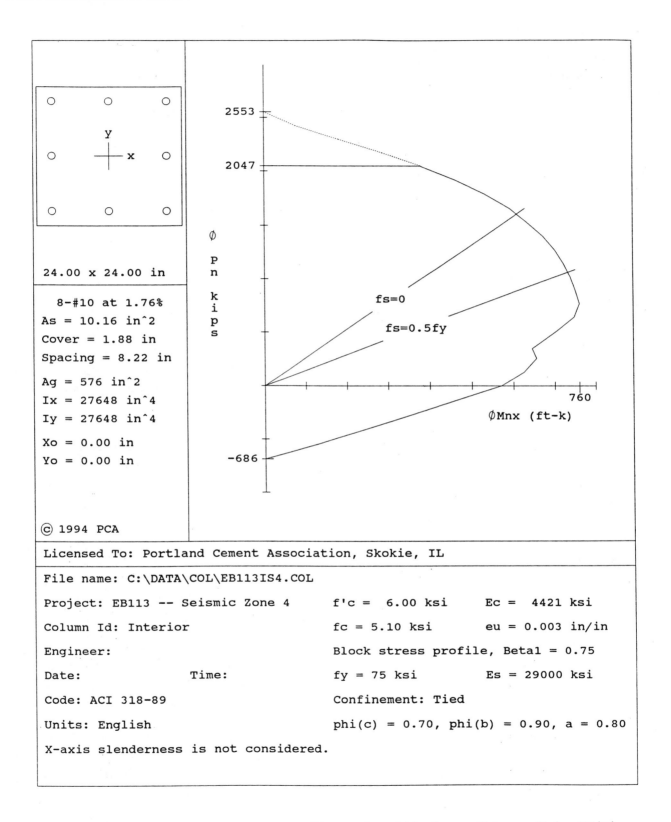

24.00 x 24.00 in

8-#10 at 1.76%
As = 10.16 in^2
Cover = 1.88 in
Spacing = 8.22 in

Ag = 576 in^2
Ix = 27648 in^4
Iy = 27648 in^4

Xo = 0.00 in
Yo = 0.00 in

© 1994 PCA

Licensed To: Portland Cement Association, Skokie, IL

File name: C:\DATA\COL\EB113IS4.COL

Project: EB113 -- Seismic Zone 4   f'c = 6.00 ksi   Ec = 4421 ksi

Column Id: Interior   fc = 5.10 ksi   eu = 0.003 in/in

Engineer: Block stress profile, Beta1 = 0.75

Date: Time: fy = 75 ksi   Es = 29000 ksi

Code: ACI 318-89   Confinement: Tied

Units: English   phi(c) = 0.70, phi(b) = 0.90, a = 0.80

X-axis slenderness is not considered.

Figure 3-19 Design Strength Interaction Diagram for a 24-in. Square Column with f_y = 75 ksi

3.2.2.6.5 Minimum Length of Lap Splices of Column Vertical Bars

Lap splices shall conform to UBC 1921.2.6.1 (see UBC 1921.4.3.2). It is clear from Fig. 3-18 that a Class B tension splice is required for this column. The basic development length for #10 bars is:

$$\ell_{db} = \frac{0.04A_b f_y}{\sqrt{f'_c}} = \frac{0.04 \times 1.27 \times 60,000}{\sqrt{6000}} = 39.3 \text{ in.}$$ *1912.2.2*

The information required for determining the development length is:

> Minimum cover to bars being spliced = 2.0 in. (1.6 d_b)
> Clear bar spacing = 8.1 in. (6.4 d_b)
> Transverse reinforcement meets requirements of UBC 1907.10.5 and 1912.2.3.5.

This would give a development length factor of 0.75 (see Table 3-13). The other factors are all equal to 1.0; therefore,

> Class B splice length = 1.3 ℓ_d = 1.3 × (0.75 × 1.0 × 1.0 × 1.0 × 1.0 × 39.3) = 38.3 in.

Use a 3 ft-4 in. splice. Reinforcement details for the column are shown in Fig. 3-20. Note that for simplicity, the 3 in. hoop spacing is continued to the location of the longitudinal bar lap splice (otherwise only 2 hoops would be required at a spacing of 6 in. at each end of the splice location.).

3.2.2.7   Proportioning and Detailing of an Interior Beam-Column Connection

This section will address the shear strength and detailing requirements for an interior beam-column connection of an exterior frame. The column considered is an interior column of the exterior transverse frame, and has beams framing in on three sides: two transverse beams and one longitudinal beam.

3.2.2.7.1 Transverse Reinforcement for Confinement

UBC 1921.5.2.1 requires that transverse hoop reinforcement as specified in UBC 1921.4.4 shall be provided within a joint, unless the joint is confined by structural members as specified in UBC 1921.5.2.2.

UBC 1921.5.2.2 specifies that within the depth of the shallowest framing member, transverse reinforcement equal to at least one-half the amount required by UBC 1921.4.4.1 shall be provided where members frame into all four sides of the joint and where each member width is at least three fourths the column width. At these locations, the maximum allowable spacing specified in UBC 1921.4.4.2 may be increased to 6 in.

The joint in question is obviously not confined by framing members according to the criteria of UBC 1921.5.2.2. Thus, assuming that the beam-column joint is located at the second floor level of the structure, the special column-end reinforcement shown for the 2nd story column in Fig. 3-20 (#4 closed hoops with one crosstie in each direction at 3 in. spacing) must continue through the beam-column joint being designed.

3.2.2.7.2 Shear Strength of Joint

Figure 3-21 shows the beam-column joint at the second floor level of the structure. The shear strength is checked in the transverse direction of the building. The shear at section x-x is obtained by subtracting the column horizontal shear from the sum of the tensile force in the top beam reinforcement and the compressive force at the top of the beam on the opposite face of the column.

The column horizontal shear, V_h, can be obtained by assuming that the beams in the adjoining floors are deformed so that plastic hinges form at their junctions with the column, with M^-_{pr} (beam) = 660.5 ft-kips and M^+_{pr} (beam) = 413.6 ft-kips. Therefore, the column moments above and below the joint will equal 413.6 + 660.5 = 1074.1 ft-kips.

It is assumed that the beam end moments are resisted equally by the columns above and below the joint; the horizontal shear at the column ends is then equal to:

$$V_h = \frac{2 \times \left(\dfrac{413.6 + 660.5}{2} \right)}{12} = 89.5 \text{ kips}$$

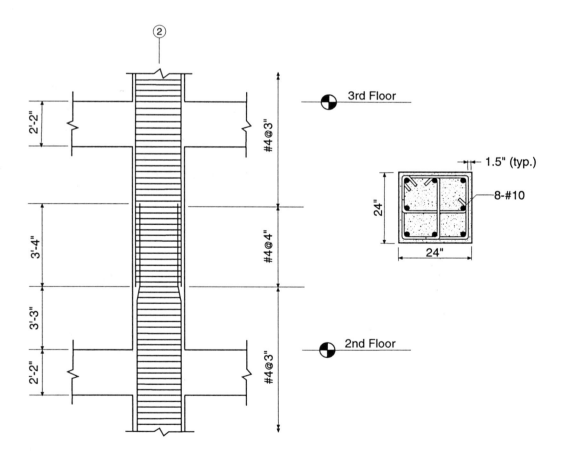

Figure 3-20 Reinforcement Details for a 24-in. Square Interior Column of an Exterior Transverse Frame (Seismic Zone 4)

According to UBC 1921.5.1.1, the tensile force in the top beam reinforcement (5-#9 bars) $= 5 \times 1.25 \times 60 = 375.0$ kips. The compressive force on the opposite side of the column is equal to the tensile force of 3-#9 bars which is $3 \times 1.25 \times 60 = 225.0$ kips.

The net shear at section x-x of the joint is $T_1 + C_2 - V_h = 375.0 + 225.0 - 89.5 = 510.5$ kips.

For a joint confined on three faces, the nominal shear strength is (UBC 1921.5.3.1):

$$\phi V_c = \phi 15 \sqrt{f_c'} \times A_j$$

$$= 0.85 \times 15 \sqrt{6000} \times 576.0/1000 = 568.9 \text{ kips} > 510.5 \text{ kips} \quad \text{O.K.}$$

A_j is equal to the effective cross-sectional area within a joint in a plane parallel to the plane of the reinforcement generating shear in the joint. The joint depth is the overall depth of the column. Note that since the transverse beams are as wide as the column, the effective joint width is equal to the column width. Therefore, $A_j = 24 \times 24 = 576.0$ in.^{2}

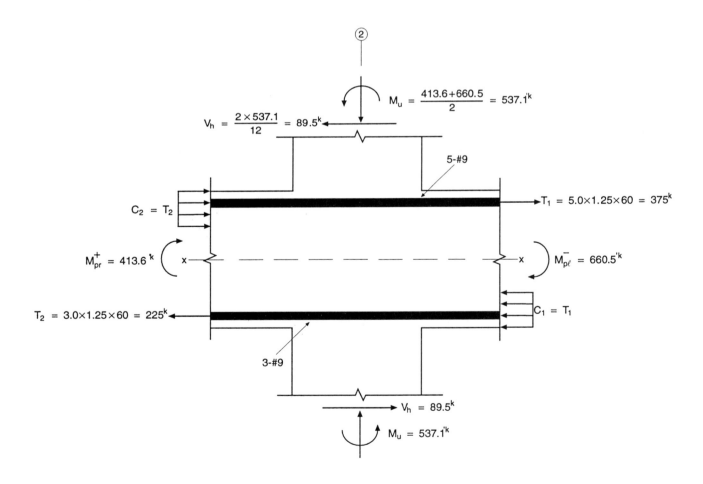

Figure 3-21 Shear Analysis of Interior Beam-Column Joint of an Exterior Frame (Seismic Zone 4)

3.2.2.8   Proportioning and Detailing an Exterior Beam-Column Connection

This section will address the shear strength and detailing requirements for an exterior beam-column connection of an exterior frame. The column considered is a corner column of the building and has beams framing in on two sides: one transverse and one longitudinal beam.

3.2.2.8.1 Transverse Reinforcement for Confinement

The joint is obviously not confined by framing members according to the criteria of UBC 1921.5.2.2. Thus, assuming that the beam-column joint is located at the second floor level of the structure, the special column-end reinforcement for the second story exterior column listed in Table 3-9 (this column was not designed) must continue through the beam-column joint in question.

3.2.2.8.2 Shear Strength of Joint

Figure 3-22 shows a section through a corner beam-column joint at the second floor level. The shear strength is checked in the transverse direction of the building. The shear across section x-x is obtained by subtracting the column horizontal shear from the tensile force in the top flexural reinforcement.

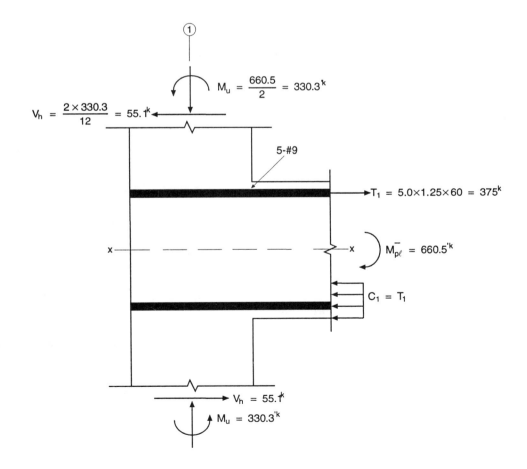

$$M_u = \frac{660.5}{2} = 330.3^k$$

$$V_h = \frac{2 \times 330.3}{12} = 55.1^k$$

5-#9

$$T_1 = 5.0 \times 1.25 \times 60 = 375^k$$

$$\overline{M_{p\ell}} = 660.5^{'k}$$

$$C_1 = T_1$$

$$V_h = 55.1^k$$

$$M_u = 330.3^{'k}$$

Figure 3-22 Shear Analysis of Exterior Beam-Column Joint of an Exterior Frame (Seismic Zone 4)

The column horizontal shear, V_h, can be obtained by assuming that the beam in the adjoining floor is deformed so that a plastic hinge forms at the junction with the column, with M^-_{pr} (beam) = 660.5 ft-kips.

It is assumed that the beam end moments are resisted equally by the columns above and below the joint; the horizontal shear at the column ends is then equal to:

$$V_h = \frac{2 \times \left(\dfrac{660.5}{2}\right)}{12} = 55.1 \text{ kips}$$

The tensile force in the top beam reinforcement (5-#9) = $5 \times 1.25 \times 60 = 375.0$ kips.

The net shear at section x-x of the joint is $T_1 - V_h = 375.0 - 55.1 = 319.9$ kips.

For a joint confined on two faces, the nominal shear strength is (UBC 1921.5.3.1):

$$\phi V_c = \phi 12 \sqrt{f'_c} \times A_j$$

$$= 0.85 \times 12 \sqrt{6000} \times 576.0/1000 = 455.1 \text{ kips} > 319.9 \text{ kips} \quad \text{O.K.}$$

3.3   Design for Seismic Zone 2B

For the Seismic Zone 2B design, the member dimensions were selected as follows: transverse beams: 22×20 in.; longitudinal beams: 22×18 in.; columns: 22×22 in.; slabs: 7 in.

3.3.1   Frame Analysis

On the basis of the given data and the dimensions of the building, the weights of the floors are listed in Table 3-14. This example uses the same values for the importance factor, I, and the site soil coefficient, S, as in the Seismic Zone 4 example (1.0 and 1.2, respectively.)

The lateral load resisting system of this building is an Intermediate Moment Resisting Frames (IMRF) of concrete. The value of R_w is 8 for this type of structural system. This lateral load resisting system was not used in Seismic Zone 4, since it is not allowed in zones of high seismicity.

The seismic design forces resulting from the distribution of the base shear computed in accordance with UBC Eqs. (28-6), (28-7), and (28-8) are also shown in Table 3-14.

Wind forces and story shears corresponding to a fastest mile wind speed of 70 mph and Exposure B, computed in accordance with UBC 1618 were shown previously in Section 3.2.1 in Tables 3-4 and 3-5 for the transverse and the longitudinal directions, respectively.

Three-dimensional analyses of the structure in both directions under the respective seismic and wind loads were carried out using *PCA-Frame*. Results from the seismic analysis are given in Figs. 3-23 and 3-24 for an exterior transverse frame and an interior longitudinal frame, respectively. The results from the wind analysis are given in Figs. 3-25 and 3-26. Two-dimensional analyses, including the effects of torsion, are not performed here since the ratio of the 2-D to 3-D results would be similar to the ones obtained previously in Section 3.2.1.

Table 3-14 Seismic Forces for Zone 2B

Floor Level	Weight, w_x (kips)	Height, h_x (ft)	$w_x h_x$ (ft-kips)	Lateral Force (kips)
R	1455	147	213,885	129.3
12	1527	135	206,145	74.6
11	1527	123	187,821	68.0
10	1527	111	169,497	61.3
9	1527	99	151,173	54.7
8	1527	87	132,849	48.1
7	1527	75	114,525	41.4
6	1527	63	96,201	34.8
5	1527	51	77,877	28.2
4	1527	39	59,553	21.6
3	1527	27	41,229	14.9
2	1546	15	23,190	8.4
Σ	18,271		1,473,945	585.0

Building height	h_n	$= 147\,ft$
Building weight	W	$= 18,271\,kips$
Fundamental period	T	$= 0.030\,(h_n)^{3/4} = 1.27\,sec.$
Seismic zone factor	Z	$= 0.2$
Importance factor	I	$= 1.0$
Response coefficient	R_w	$= 8$
Soil factor	S	$= 1.2$
Coefficient	C	$= 1.25\,S/T^{2/3} = 1.28$
Base shear	V	$= ZICW/R_w = 585.0\,kips$
Top level force	F_t	$= 0.07TV = 52.0\,kips$

R		
R	-25	-20	-32
12	-49	-45	-58
11	-73	-67	-78
10	-94	-87	-97
9	-113	-105	-113
8	-130	-121	-128
7	-145	-135	-140
6	-158	-148	-150
5	-169	-158	-158
4	-177	-166	-163
3	-180	-171	-167
2	-175	-164	-155

Bending Moments in Beams (ft-kips)

R		
R	-2.0	-2.9
12	-4.3	-5.3
11	-6.3	-7.1
10	-8.2	-8.7
9	-9.9	-10.2
8	-11.3	-11.5
7	-12.7	-12.6
6	-13.8	-13.5
5	-14.7	-14.2
4	-15.5	-14.7
3	-15.8	-15.0
2	-15.3	-13.9

Shear Forces in Beams (kips)

R		
R	31 / 14	62 / 42
12	52 / 32	83 / 68
11	64 / 47	107 / 93
10	77 / 61	128 / 116
9	87 / 74	147 / 137
8	97 / 86	164 / 154
7	105 / 96	177 / 170
6	111 / 104	189 / 183
5	116 / 111	198 / 195
4	121 / 123	202 / 199
3	114 / 120	208 / 219
2	117 / 260	173 / 288

Bending Moments in Columns (ft-kips)

R		
R	3.7	8.6
12	7.0	12.5
11	9.3	16.7
10	11.5	20.4
9	13.4	23.6
8	15.2	26.5
7	16.7	29.0
6	18.0	31.0
5	19.0	32.7
4	20.3	33.4
3	19.5	35.6
2	25.1	30.8

Shear Forces in Columns (kips)

R		
R	2.0	0.9
12	6.3	1.9
11	12.6	2.7
10	20.7	3.2
9	30.6	3.6
8	41.9	3.8
7	54.6	3.7
6	68.4	3.4
5	83.1	2.9
4	98.6	2.1
3	114.0	1.3
2	130.0	0.0

Axial Forces in Columns (kips)

Figure 3-23 Results of 3-D Analysis of an Exterior Transverse Frame under Seismic Zone 2B Forces

R	-22	-19	-27	-26	-26
12	-42	-40	-49	-49	-49
11	-64	-60	-68	-68	-68
10	-83	-79	-85	-85	-85
9	-100	-95	-101	-100	-100
8	-116	-110	-114	-114	-114
7	-129	-123	-126	-126	-126
6	-141	-134	-136	-136	-136
5	-151	-144	-144	-144	-144
4	-159	-151	-149	-149	-149
3	-161	-154	-151	-151	-151
2	-153	-145	-139	-140	-140

Bending Moments in Beams (ft-kips)

R	-1.9	-2.4	-2.4
12	-3.7	-4.4	-4.4
11	-5.6	-6.1	-6.1
10	-7.3	-7.7	-7.7
9	-8.8	-9.1	-9.1
8	-10.2	-10.3	-10.3
7	-11.4	-11.4	-11.4
6	-12.4	-12.2	-12.2
5	-13.3	-13.0	-13.0
4	-14.0	-13.4	-13.5
3	-14.2	-13.7	-13.7
2	-13.4	-12.6	-12.6

Shear Forces in Beams (kips)

R	24 / 4	51 / 28	57 / 33
12	43 / 19	70 / 51	74 / 54
11	52 / 31	90 / 72	94 / 76
10	61 / 43	108 / 92	111 / 95
9	69 / 53	123 / 109	126 / 112
8	76 / 62	137 / 125	139 / 127
7	82 / 70	149 / 139	150 / 140
6	87 / 78	158 / 150	159 / 150
5	90 / 83	166 / 161	165 / 160
4	94 / 96	169 / 166	168 / 165
3	83 / 94	169 / 185	167 / 182
2	77 / 244	129 / 270	127 / 269

Bending Moments in
Columns (ft-kips)

R	2.3	6.6	7.6
12	5.2	10.0	10.7
11	6.9	13.6	14.1
10	8.6	16.6	17.1
9	10.1	19.4	19.8
8	11.5	21.8	22.1
7	12.7	24.0	24.1
6	13.7	25.7	25.8
5	14.4	27.2	27.1
4	15.8	28.0	27.8
3	14.7	29.5	29.1
2	21.5	26.6	26.4

Shear Forces in Columns (kips)

R	1.9	0.6	0.1
12	5.6	1.3	0.1
11	11.2	1.8	0.1
10	18.4	2.2	0.1
9	27.3	2.4	0.1
8	37.4	2.5	0.1
7	48.8	2.5	0.1
6	61.2	2.3	0.1
5	74.5	2.0	0.1
4	88.5	1.4	0.1
3	103.0	0.9	0.1
2	116.0	0.0	0.0

Axial Forces in Columns (kips)

Figure 3-24 Results of 3-D Analysis of an Interior Longitudinal Frame under
Seismic Zone 2B Forces

R	-2.7	-1.9	-6.7
12	-8.7	-8.0	-12.8
11	-17.8	-16.4	-20.6
10	-27.2	-25.2	-28.8
9	-36.4	-33.9	-36.8
8	-45.5	-42.4	-44.6
7	-54.3	-50.7	-52.1
6	-62.9	-58.8	-59.3
5	-71.2	-66.7	-66.1
4	-79.3	-74.2	-72.1
3	-85.3	-80.8	-78.3
2	-87.2	-81.4	-76.4

Bending Moments in Beams (ft-kips)

R	-0.2	-0.6
12	-0.8	-1.2
11	-1.5	-1.9
10	-2.4	-2.6
9	-3.2	-3.3
8	-4.0	-4.0
7	-4.7	-4.7
6	-5.5	-5.4
5	-6.2	-6.0
4	-6.9	-6.5
3	-7.5	-7.1
2	-7.6	-6.9

Shear Forces in Beams (kips)

R	3 / -2	9 / 4
12	11 / 3	18 / 12
11	16 / 8	28 / 22
10	21 / 14	37 / 31
9	26 / 19	46 / 40
8	31 / 24	54 / 49
7	35 / 28	63 / 57
6	40 / 33	71 / 66
5	44 / 38	79 / 74
4	48 / 45	85 / 80
3	47 / 45	93 / 94
2	49 / 119	77 / 133

Bending Moments in Columns (ft-kips)

R	0.1	1.1
12	1.2	2.5
11	2.0	4.1
10	2.9	5.6
9	3.7	7.1
8	4.5	8.6
7	5.3	10.0
6	6.1	11.4
5	6.8	12.7
4	7.7	13.7
3	7.7	15.6
2	11.2	14.0

Shear Forces in Columns (kips)

R	0.2	0.4
12	1.0	0.8
11	2.5	1.1
10	4.9	1.3
9	8.0	1.5
8	12.0	1.5
7	16.7	1.5
6	22.2	1.4
5	28.5	1.1
4	35.4	0.7
3	42.9	0.3
2	50.5	0.4

Axial Forces in Columns (kips)

Figure 3-25 Results of 3-D Analysis of an Exterior Transverse Frame under Wind Forces

R	-1.4	-1.1	-3.0	-2.8	-2.7
12	-4.4	-4.1	-6.2	-6.1	-6.1
11	-9.3	-8.8	-10.5	-10.4	-10.4
10	-14.4	-13.6	-15.1	-15.0	-15.0
9	-19.4	-18.4	-19.6	-19.5	-19.5
8	-24.3	-23.1	-24.0	-23.9	-23.9
7	-29.1	-27.7	-28.2	-28.2	-28.2
6	-33.8	-32.1	-32.3	-32.3	-32.3
5	-38.3	-36.5	-36.2	-36.3	-36.3
4	-42.7	-40.6	-39.7	-39.8	-39.8
3	-45.6	-43.8	-42.8	-42.8	-42.8
2	-45.6	-43.2	-41.4	-41.5	-41.6

Bending Moments in Beams (ft-kips)

R	-0.1	-0.3	-0.2
12	-0.4	-0.6	-0.6
11	-0.8	-0.9	-0.9
10	-1.3	-1.4	-1.4
9	-1.7	-1.8	-1.8
8	-2.1	-2.2	-2.2
7	-2.6	-2.6	-2.6
6	-3.0	-2.9	-2.9
5	-3.4	-3.3	-3.3
4	-3.8	-3.6	-3.6
3	-4.0	-3.9	-3.9
2	-4.0	-3.7	-3.8

Shear Forces in Beams (kips)

R	2 / -2	4 / 1	6 / 2
12	6 / 1	10 / 6	6 / 10
11	9 / 4	15 / 11	16 / 11
10	12 / 6	20 / 16	21 / 16
9	15 / 9	25 / 21	26 / 21
8	17 / 12	30 / 26	31 / 26
7	20 / 15	35 / 31	35 / 31
6	22 / 17	39 / 35	39 / 35
5	24 / 20	44 / 40	43 / 40
4	27 / 24	47 / 44	47 / 44
3	25 / 25	50 / 52	49 / 51
2	24 / 74	40 / 82	39 / 81

Bending Moments in Columns (ft-kips)

R	0.1	0.5	0.7
12	0.6	1.3	1.4
11	1.1	2.2	2.3
10	1.5	3.0	3.1
9	2.0	3.9	3.9
8	2.4	4.7	4.7
7	2.8	5.5	5.5
6	3.3	6.2	6.2
5	3.6	7.0	6.9
4	4.2	7.6	7.5
3	4.2	8.5	8.3
2	6.6	8.1	8.0

Shear Forces in Columns (kips)

R	0.1	0.2	0.0
12	0.5	0.3	0.0
11	1.3	0.4	0.0
10	2.6	0.5	0.0
9	4.3	0.6	0.0
8	6.4	0.6	0.0
7	9.0	0.6	0.0
6	11.9	0.6	0.0
5	15.3	0.4	0.0
4	19.1	0.3	0.0
3	23.1	0.1	0.0
2	27.1	0.2	0.0

Axial Forces in Columns (kips)

Figure 3-26 Results of 3-D Analysis of an Interior Longitudinal Frame under Wind Forces

3.3.2   Design of Beams and Columns

The objective is to determine the required flexural and shear reinforcement for typical beams of both the transverse and the longitudinal frames, and for a typical column common to both frames. The structure will be designed for dead and live loads, in combination with either wind or Seismic Zone 2B forces.

The Direct Design Method in UBC 1913.6 is used to determine the moments due to gravity loads acting on the beams. It was shown in Section 3.2.2 that an equivalent uniform load method yielded approximately the same moments at all locations along the span except at the exterior support in the end span. Table 3-15 gives the factored beam moments in an exterior transverse frame and an interior longitudinal frame obtained from the Direct Design Method.

Table 3-15 Factored Gravity Load Bending Moments in the Beams

Location	M_u (ft-kips)
EXTERIOR TRANSVERSE FRAME	
End Span	
Exterior Negative	23.9
Positive	88.4
Interior Negative	107.6
Interior Span	
Positive	52.6
Negative	100.4
INTERIOR LONGITUDINAL FRAME	
End Span	
Exterior Negative	49.8
Positive	153.8
Interior Negative	186.9
Interior Span	
Positive	91.4
Negative	174.5

Table 3-16 gives the factored moments for the applicable combinations of gravity and lateral loads on a 2nd floor beam in an exterior frame in the transverse direction. Similarly, Table 3-17 gives the factored moments on a 2nd floor beam in an interior longitudinal frame.

Table 3-18 gives the factored axial loads, bending moments, and shear forces for an exterior and interior column, located between the 2nd and 3rd floors in an exterior transverse frame. Live load reduction factors were computed in accordance with Eq. (6-1) in UBC 1606. Table 3-19 lists the load combinations for exterior and interior columns located between the ground level and the second floor.

Table 3-16 Summary of Factored Bending Moments for a Beam on the 2nd Floor of an Exterior Transverse Frame (Seismic Zone 2B)

(1)	U = 1.4D + 1.7L	Eq. (9-1), UBC 1909.2.1
(2)	U = 0.75 (1.4D + 1.7L ± 1.87E)	Eq. (9-2), UBC 1909.2.3
(3)	U = 0.9D ± 1.43E	Eq. (9-3), UBC 1909.2.3
(4)	U = 0.75 (1.4D + 1.7L ± 1.7W)	Eq. (9-2), UBC 1909.2.2
(5)	U = 0.9D ± 1.3W	Eq. (9-3), UBC 1909.2.2

Load Combination	End Span Moments (ft-kips)			Interior Span Moments (ft-kips)	
	Exterior Negative	Positive	Interior Negative	Positive	Negative
1	-23.9	88.4	-107.6	52.6	-100.4
2: sidesway right	227.5	74.9	-310.7	39.5	142.1
sidesway left	-263.3	57.8	149.3	39.5	-292.7
3: sidesway right	239.3	49.4	-284.0	24.2	175.5
sidesway left	-261.2	32.0	185.0	24.2	-267.8
4: sidesway right	93.3	70.3	-184.5	39.5	22.1
sidesway left	-129.1	62.4	23.1	39.5	-172.7
5: sidesway right	102.4	44.7	-155.3	24.2	53.2
sidesway left	-124.3	36.7	56.3	24.2	-145.5

Table 3-17 Summary of Factored Bending Moments for a Beam on the 2nd Floor of an Interior Longitudinal Frame (Seismic Zone 2B)

(1)	U = 1.4D + 1.7L	Eq. (9-1), UBC 1909.2.1
(2)	U = 0.75 (1.4D + 1.7L ± 1.87E)	Eq. (9-2), UBC 1909.2.3
(3)	U = 0.9D ± 1.43E	Eq. (9-3), UBC 1909.2.3
(4)	U = 0.75 (1.4D + 1.7L ± 1.7W)	Eq. (9-2), UBC 1909.2.2
(5)	U = 0.9D ± 1.3W	Eq. (9-3), UBC 1909.2.2

Load Combination	End Span Moments (ft-kips)			Interior Span Moments (ft-kips)	
	Exterior Negative	Positive	Interior Negative	Positive	Negative
1	-49.8	153.8	-186.9	91.4	-174.5
2: sidesway right	177.2	121.1	-343.6	68.5	65.5
sidesway left	-251.9	109.6	63.2	68.5	-327.2
3: sidesway right	196.4	74.9	-291.2	41.0	121.9
sidesway left	-241.2	63.2	123.5	41.0	-278.5
4: sidesway right	20.8	116.9	-195.3	68.5	-77.8
sidesway left	-95.5	113.8	-85.1	68.5	-183.9
5: sidesway right	36.9	70.6	-140.0	41.0	-24.2
sidesway left	-81.7	67.5	-27.7	41.0	-132.4

Table 3-18 Summary of Factored Axial Loads, Bending Moments, and Shear Forces for Columns between the 2nd and 3rd Floors of an Exterior Transverse Frame (Seismic Zone 2B)

(1)	$U = 1.4D + 1.7L$	Eq. (9-1), UBC 1909.2.1
(2)	$U = 0.75 (1.4D + 1.7L \pm 1.87E)$	Eq. (9-2), UBC 1909.2.3
(3)	$U = 0.9D \pm 1.43E$	Eq. (9-3), UBC 1909.2.3
(4)	$U = 0.75 (1.4D + 1.7L \pm 1.7W)$	Eq. (9-2), UBC 1909.2.2
(5)	$U = 0.9D \pm 1.3W$	Eq. (9-3), UBC 1909.2.2

Load Combination	Interior Column				Exterior Column			
	Axial Load (kips)	Bending Moment (ft-kips)		Shear Force (kips)	Axial Load (kips)	Bending Moment (ft-kips)		Shear Force (kips)
		Top	Bottom			Top	Bottom	
1	998	-8.6	8.6	1.7	612	-14.1	14.1	2.7
2: sidesway right	747	285.3	-300.7	56.7	299	149.3	-157.7	33.8
sidesway left	750	-298.2	313.6	-59.2	619	-170.5	178.9	-39.2
3: sidesway right	553	294.2	-309.9	58.5	190	156.5	-165.1	31.1
sidesway left	557	-300.7	316.4	-59.7	516	-169.5	178.1	-33.7
4: sidesway right	748	112.1	-113.4	21.8	404	49.4	-46.8	9.3
sidesway left	749	-125.0	126.3	-24.3	514	-70.5	68.0	-13.4
5: sidesway right	555	117.7	-119.0	22.9	297	54.6	-52.0	10.3
sidesway left	556	-124.1	125.4	-24.2	409	-67.6	65.0	-12.8

Table 3-19 Summary of Factored Axial Loads, Bending Moments, and Shear Forces for Columns between Grade and the 2nd Floor of an Exterior Transverse Frame (Seismic Zone 2B)

(1)	$U = 1.4D + 1.7L$	Eq. (9-1), UBC 1909.2.1
(2)	$U = 0.75 (1.4D + 1.7L \pm 1.87E)$	Eq. (9-2), UBC 1909.2.3
(3)	$U = 0.9D \pm 1.43E$	Eq. (9-3), UBC 1909.2.3
(4)	$U = 0.75 (1.4D + 1.7L \pm 1.7W)$	Eq. (9-2), UBC 1909.2.2
(5)	$U = 0.9D \pm 1.3W$	Eq. (9-3), UBC 1909.2.2

Load Combination	Interior Column				Exterior Column			
	Axial Load (kips)	Bending Moment (ft-kips)		Shear Force (kips)	Axial Load (kips)	Bending Moment (ft-kips)		Shear Force (kips)
		Top	Bottom			Top	Bottom	
1	1090	-8.6	8.6	1.3	668	-14.1	14.1	2.1
2: sidesway right	818	236.2	-397.5	47.5	319	153.5	-354.1	44.2
sidesway left	818	-249.1	410.4	-49.5	683	-174.7	375.2	-49.6
3: sidesway right	606	244.2	-408.6	49.0	199	160.8	-365.3	39.5
sidesway left	606	-250.6	415.1	-49.9	571	-173.8	378.3	-41.4
4: sidesway right	817	91.7	-163.1	19.1	437	51.9	-141.2	14.5
sidesway left	818	-104.6	176.0	-21.1	565	-73.1	162.3	-17.7
5: sidesway right	606	96.9	-169.7	20.0	320	57.2	-148.2	15.4
sidesway left	607	-103.3	176.1	-21.0	451	-70.2	161.2	-17.4

3.3.2.1 General Code Requirements for Seismic Zone 2

According to UBC 1921.2.1.3, concrete frames that are part of the lateral force systems of structures in Seismic Zone 2 must be designed as Intermediate Moment-Resisting Frames (IMRF). As such, this structure must be designed in accordance with all the applicable provisions of UBC 1901 through 1918 and 1921.8.

Orthogonal effects must be checked as required in UBC 1631.1. This section requires that buildings located in Seismic Zone 2 shall be designed for the effects of earthquake forces acting in a direction other than along the principal axes if a column of a structure forms part of two or more intersecting lateral force resisting systems. An exception to this is permitted if the axial load in the column due to seismic forces acting in either principal direction is less than 20% of the column allowable axial load.

The maximum column axial force from unfactored earthquake forces is 130.0 kips (see Figs. 3-23 and 3-24). The design axial load strength, $\phi P_{n(max)}$ of a 22 × 22 in. column with f'_c = 6000 psi and with 8-#9 bars symmetrically arranged is 1628 kips. The design axial load strength of the same column with f'_c = 4000 psi is 1175 kips. Twenty percent of either of these strengths exceeds the maximum axial force resulting from the earthquake loads. Therefore, orthogonal effects due to seismic forces can be neglected.

3.3.2.2 Diaphragm Flexibility

UBC 1628.5 requires that the shears resulting from horizontal torsion must be included in design where diaphragms are not flexible. In Section 3.2.2.2 it was shown that the diaphragms of the example building are rigid and that the consideration of increased torsional effects is necessary.

3.3.2.3 Story Drift Limitation and P-Δ Effects

The member forces and the story drifts induced by P-Δ effects need to be considered in the evaluation of overall structural frame stability. According to UBC 1628.9, P-Δ effects can be neglected when the ratio of the secondary moment to the primary moment is less than or equal to 0.1. Based on the results shown in Table 3-20, the P-Δ effects may be neglected for this building.

For buildings having a fundamental period greater than 0.7 seconds, the calculated story drift (including torsional effects) must not exceed $0.03/R_w$ nor 0.004 times the story height (UBC 1628.8.2).

For h_s = 12 ft and R_w = 8:

$$\frac{0.03h_s}{R_w} = \frac{0.03 \times 12 \times 12}{8} = 0.54 \text{ in. (governs)}$$

$$0.004h_s = 0.004 \times 12 \times 12 = 0.58 \text{ in.}$$

It can be seen from Table 3-20 that for all stories, the lateral drifts obtained from the prescribed lateral forces in both directions are less than the limiting value.

Table 3-20 P-Δ Check under Seismic Zone 2B Forces

Floor Level	Gravity Loads * Σw_i (kips)	Transverse Story Drift Δ_i (in.)	Longitudinal Story Drift Δ_i (in.)	Story Shear V_i (kips)	Story Height h_{si} (in.)	Transverse $\Sigma w_i \Delta_i / V_i h_{si}$	Longitudinal $\Sigma w_i \Delta_i / V_i h_{si}$
R	1747	0.10	0.10	129.3	144	0.009	0.009
12	3477	0.13	0.15	203.6	144	0.015	0.018
11	5207	0.18	0.19	271.6	144	0.024	0.025
10	6937	0.22	0.23	332.9	144	0.032	0.033
9	8667	0.24	0.27	387.6	144	0.037	0.042
8	10,397	0.28	0.30	435.7	144	0.046	0.050
7	12,127	0.29	0.33	477.1	144	0.051	0.058
6	13,857	0.32	0.35	511.9	144	0.060	0.066
5	15,587	0.33	0.36	540.1	144	0.066	0.072
4	17,317	0.34	0.38	561.7	144	0.073	0.081
3	19,047	0.32	0.36	576.6	144	0.073	0.083
2	20,794	0.30	0.31	585.0	180	0.059	0.061

* Includes floor weight, superimposed dead load, and reduced roof and floor live loads

3.3.2.4   Proportioning and Detailing a Flexural Member of an Exterior Transverse Frame

Members designed for Seismic Zone 2 earthquake forces must meet the requirements in UBC 1921.8 for frames in regions of moderate seismic risk. According to these requirements, a member shall be designed as a beam if the factored axial load on the member is less than or equal to $A_g f'_c/10$. The beams in the example building carry negligible axial loads, and thus qualify as flexural members.

3.3.2.4.1 Required Flexural Reinforcement

The maximum negative moment for a beam on the 2nd floor of an exterior transverse frame at an interior support is 310.7 ft-kips (see Table 3-16). The required area of steel, ignoring the effects of any compression reinforcement, is $A_s = 4.37$ in.^{2} Use 5-#9 bars ($\phi M_n = 349.8$ ft-kips).

Check limitations on the reinforcement ratio:

$$\rho \quad = A_s/bd = 5.0/(22 \times 17.56) = 0.0129 > 200/f_y = 0.0033 \quad \text{O.K.} \qquad\qquad 1910.5.1$$

$$\rho_{max} = 0.75 \ \rho_{bal} \qquad\qquad 1910.3.3$$

$$= 0.75 \times (0.85 \times \beta_1 f'_c/f_y) \times \left(\frac{87,000}{87,000 + f_y} \right)$$

$$= 0.75 \times (0.85 \times 0.85 \times 4/60) \times \left(\frac{87,000}{87,000 + 60,000} \right)$$

$$= 0.0214 > 0.0129 \quad \text{O.K.}$$

The maximum negative moment in the beam at the exterior support is 263.3 ft-kips (see Table 3-16). Conservatively use 5-#9 bars at this location.

The positive moment strength at a joint face must be equal to at least 33% of the corresponding negative moment strength at that joint (UBC 1921.8.4.1). In this case, the positive moment strength must be at least 349.8/3 = 116.6 ft-kips at the faces of all columns. This is less than the 239.3 ft-kips and 185.0 ft-kips at the exterior and interior beam ends from the applied loads (see Table 3-16). The larger value will require 4-#9 bars, giving a positive design moment strength of 287.1 ft-kips at both ends of the beam. The reinforcement ratio = 4.0/(22 × 17.56) = 0.0104 which is greater than 0.0033 and less than 0.0214.

The maximum positive factored moment at the midspan of either span is 88.4 ft-kips. According to UBC 1921.8.4.1, the positive moment strength at any section along the member length must not be less than 20% of the maximum moment strength provided at the face of either joint. Thus, the minimum value of ϕM^+_n for both spans is: 349.8/5 = 70.0 ft-kips which is less than 88.4 ft-kips. Solving for the required positive moment reinforcement yields A_s = 1.15 in.^{2} The minimum required area of steel is $0.0033bd = 0.0033 \times 22 \times 17.56 = 1.28$ in.^{2} Use 2-#9 bars (ϕM_n = 150.8 ft-kips). Note that this same amount of reinforcement must be used for negative moment strength at any section along the member length as well (UBC 1921.8.4.1).

3.3.2.4.2 Required Shear Reinforcement

The shear strength of beams resisting moderate (Zone 2) earthquake effects must not be less than the sum of the shear associated with the development of nominal moment strengths of the member at each restrained end of the clear span and the unfactored shear from gravity loads (UBC 1921.8.3). Note that ACI 21.8.3 requires the use of the factored gravity loads. The shear forces in the end span of a typical transverse beam are shown in Fig. 3-27. The shear capacity of the concrete is:

$$V_c = 2\sqrt{f'_c} \times bd \qquad\qquad 1911.3.1.1$$

$$= \frac{2\sqrt{4000} \times 22 \times 17.56}{1000} = 48.9 \text{ kips}$$

The required strength to be provided by the shear reinforcement is:

$$\phi V_s = V_u - \phi V_c = 45.6 - (0.85 \times 48.9) = 4.0 \text{ kips}$$

$$V_s = 4.0/0.85 = 4.8 \text{ kips}$$

The required spacing of #3 closed stirrups is:

$$s = \frac{A_v f_y d}{V_s} = \frac{0.22 \times 60 \times 17.56}{4.8} = 48.3 \text{ in.} \qquad\qquad 1911.5.6.2$$

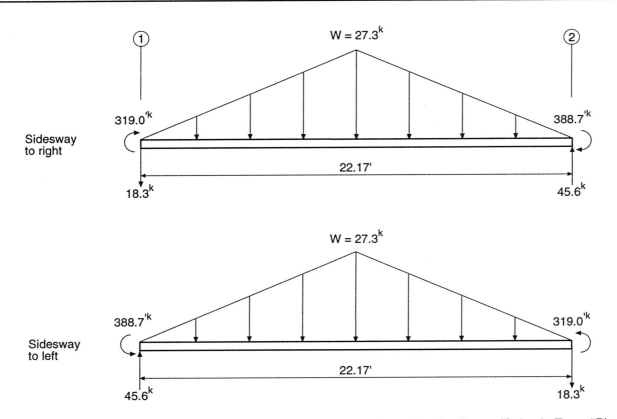

Figure 3-27 Design Shear Forces for Transverse Beams in an Exterior Frame (Seismic Zone 2B)

The maximum allowable stirrup spacing within a distance of 2h = 2 × 20 = 40 in. from the faces of supports is:

$$s_{max} = \frac{d}{4} = \frac{17.56}{4} = 4.4 \text{ in. (governs)}$$ *1921.8.4.2*

$$= 8 \times \text{(diameter of smallest longitudinal bar)} = 8 \times 1.128 = 9.0 \text{ in.}$$

$$= 24 \times \text{(diameter of stirrup bars)} = 24 \times 0.375 = 9.0 \text{ in.}$$

$$= 12.0 \text{ in.}$$

Thus, use #3 closed stirrups at a 4.0 in. spacing.

Beyond a distance of 2h from the supports, the maximum stirrup spacing is:

$$s_{max} = \frac{d}{2} = 8.8 \text{ in.}$$ *1921.8.4.3*

The minimum area of shear reinforcement, A_v, is:

$$A_v = \frac{50 \times b \times s}{f_y} = \frac{50 \times 22 \times 8.0}{60,000} = 0.15 \text{ in.}^2 < 0.22 \text{ in.}^2$$ *1911.5.5.3*

Thus, #3 stirrups at 8 in. will be adequate.

The shear strength of the member with an 8 in. spacing is:

$$\phi V_n = 0.85 \times (V_s + V_c)$$

$$= 0.85 \times \left(\frac{0.22 \times 60 \times 17.56}{8} + 48.9 \right)$$

$$= 0.85 \times (29.0 + 48.9) = 66.2 \text{ kips}$$

Two-thirds of this strength is:

$$\frac{2}{3}(\phi V_n) = \frac{2}{3} \times 66.2 = 44.1 \text{ kips}$$

The location where the applied shear is equal to 44.1 kips is 3.7 ft from the face of the support. Therefore, any reinforcing bars terminated in a tension zone beyond 3.7 ft from the column faces will meet the requirements of UBC 1912.10.5.1 since $\frac{2}{3}(\phi V_n) > V_u$.

3.3.2.4.3 Reinforcing Bar Cutoff Points and Splices

The negative reinforcement at the supports consists of 5-#9 bars. The location where three of the five bars can be terminated away from an interior support will be determined. The loading used to find the cutoff point is 0.9 times the dead load in combination with the nominal moment strength, M_n, at the member ends (using $f_s = f_y = 60$ ksi and $\phi = 1.0$). The design moment strength, ϕM_n, provided by 2-#9 reinforcing bars is 150.8 ft-kips.

With $\phi = 1.0$ and $f_y = 60$ ksi, $M^+_n = 287.1/0.9 = 319.0$ ft-kips at the exterior end of the beam, and $M^-_n = 349.8/0.9 = 388.7$ ft-kips at the interior end. The dead load is equal to $0.9 \times 1.86 = 1.67$ kips/ft at midspan. The distance from the face of the interior support to where the moment under the loading considered equals 150.8 ft-kips is readily obtained by summing moments about section a-a in Fig. 3-28:

$$\frac{x}{2} \left(\frac{1.67x}{11.085} \right) \left(\frac{x}{3} \right) + 388.7 - 41.2x = 150.8$$

Solution of this equation gives x = 5.9 ft. Three of the 5-#9 top bars near the interior support may be discontinued at (x + d) = (5.9 + 17.56/12) = 7.4 ft from the face of the support.

Due to the symmetry of the reinforcement, the cutoff point for the top #9 bars at the exterior support is the same as for the interior support.

The positive reinforcing bar cutoff location is determined in a similar manner. The positive moment strength of 2-#9 bars is 150.8 ft-kips. Using the moment diagram in Fig. 3-28 yields a distance of 7.0 ft from the exterior face where the factored moment equals the strength of 150.8 ft-kips. This gives a cutoff length equal to (x + d) = (7.0 + 17.56/12) = 8.5 ft. The cutoff locations for the positive reinforcing bars at the interior support is also 8.5 ft from the face of the support.

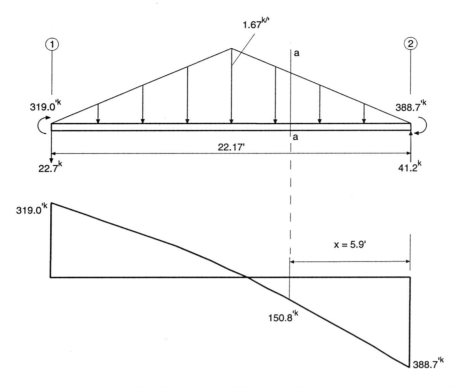

Figure 3-28 Moment Diagram for Cutoff Location of Negative Bars at Interior Support of Transverse Beam (Seismic Zone 2B)

Reinforcing bars must be properly developed at the supports. The development length is the product of the basic development length and the factors previously presented in Table 3-13. The basic development length for a #9 bar is:

$$\ell_{db} = 0.04 A_b f_y / \sqrt{f'_c} \qquad\qquad 1912.2.2$$

$$= 0.04 \times 1.0 \times 60,000 / \sqrt{4000}$$

$$= 38.0 \text{ in.}$$

The information required for determining the development length is:

	Top Bars	Bottom Bars
Minimum top or bottom cover to bars being developed	1.88 in. (1.7 d_b)	1.88 in. (1.7 d_b)
Minimum side cover	large (slab)	1.88 in. (1.7 d_b)
Clear spacing between bars being developed	3.15 in. (2.8 d_b)	4.58 in. (4.1 d_b)

Transverse reinforcing meets requirements of UBC 1911.5.4 and 1911.5.5.3 for both top and bottom bars.

Factor I will have a value of 1.4 for the top bars and 1.0 for the bottom bars (see Table 3-13). Factor II is 1.3 for the top bars. All other factors are 1.0; therefore, the development lengths are

Top bar $\ell_d = 1.4 \times 1.3 \times 1.0 \times 1.0 \times 1.0 \times 38.0/12 = 5.8$ ft < 7.4 ft   O.K.

Bottom bar $\ell_d = 1.0 \times 1.0 \times 1.0 \times 1.0 \times 1.0 \times 38.0/12 = 3.2$ ft < 8.5 ft   O.K.

The information required for determining the splice length of the #9 bars is:

	Top Bars	Bottom Bars
Stress level	$< 0.5 f_y$	$> 0.5 f_y$
Splice type	A	B
Minimum top or bottom cover to bars	1.88 in. (1.7 d_b)	1.88 in. (1.7 d_b)
Minimum side cover	large (slab)	1.88 in. (1.7 d_b)
Clear spacing between bars being spliced	16.0 in. (14.2 d_b)	16.0 in. (14.2 d_b)

Transverse reinforcement meets requirements of UBC 1911.5.4 and 1911.5.5.3 for both top and bottom bars.

This gives a value of 0.85 and 1.0 for Factor I for top and bottom splices, respectively. The splice lengths are:

Top splice length (Class A) = $1.0 \times (0.85 \times 1.3 \times 1.0 \times 1.0 \times 1.0 \times 38.0) = 42.0$ in.

Bottom splice length (Class B) = $1.3 \times (1.0 \times 1.0 \times 1.0 \times 1.0 \times 1.0 \times 38.0) = 49.4$ in.

Figure 3-29 shows the reinforcement details for the beam in the exterior transverse frame.

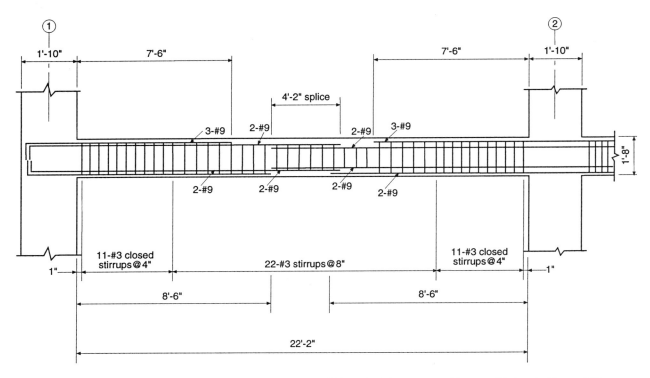

Figure 3-29 Reinforcement Details for Beam of Exterior Transverse Frame (Seismic Zone 2B)

3.3.2.5   Proportioning and Detailing a Flexural Member of an Interior Longitudinal Frame

3.3.2.5.1 Required Flexural Reinforcement

The maximum negative moment for a longitudinal beam of an interior frame at an interior support is 343.6 ft-kips (see Table 3-17). The required reinforcing steel, ignoring the effects of any compression reinforcement, is $A_s = 5.76$ in.^{2} Use 6-#9 bars ($\phi M_n = 355.0$ ft-kips).

Check limitations on the reinforcement ratio:

$$\rho = A_s/bd = 6.0/(22 \times 15.56) = 0.0175 \qquad\qquad\qquad \textit{1910.5.1}$$

$$0.0033 < 0.0175 < 0.0214 \quad \text{O.K.} \qquad\qquad\qquad \textit{1910.3.3}$$

The maximum negative moment in the beam at an exterior support is 251.9 ft-kips. Use 6-#9 bars at this support as well.

The positive moment strength at the joint face must be equal to at least 33% of the corresponding negative moment strength at that joint (UBC 1921.8.4.1). In this case, the positive moment strength must be at least $355.0/3 = 118.3$ ft-kips. This is less than the 196.4 ft-kips and 123.5 ft-kips at the exterior and interior supports, respectively, from the applied loads (see Table 3-17). This will require 3-#9 bars, giving a positive design moment strength of 193.8 ft-kips at both ends of the beam. The reinforcement ratio is $= 3.0/(22 \times 15.56) = 0.0088$ which is greater than 0.0033 but less than 0.0214.

The maximum positive factored moment in the midspan region of either span is 153.8 ft-kips. The positive moment strength at any section along the member length shall not be less than 20% of the maximum moment strength provided at the face of either joint. Thus, the minimum value of ϕM^+_n for both spans is: 355.0/5 = 71.0 ft-kips which is less than 153.8 ft-kips. Using 3-#9 bars gives a design moment capacity of 193.8 ft-kips. The minimum area of steel required is $A_s = 0.0033bd = 0.0033 \times 22 \times 15.56 = 1.13$ in.^{2} which is less than 3.0 in.^{2}

The negative moment strength at any section along the member length shall not be less than 20% of the maximum moment strength provided at the face of either joint. Thus, the minimum values of M^-_u for both spans is 355.0/5 = 71.0 ft-kips. Use 2-#9 bars ($\phi M_n = 132.8$ ft-kips).

3.3.2.5.2 Shear Reinforcement Requirements

The shear strength of beams resisting moderate (Zone 2) earthquake effects must not be less than the sum of the shear associated with the development of the nominal moment strengths of the member (M_n) at each restrained end and the unfactored shear calculated from gravity loads (UBC 1921.8.3). Note that ACI 21.8.3 requires the use of the factored gravity loads. The shear forces in the end span of a typical longitudinal beam are shown in Fig. 3-30. The shear capacity of the concrete is:

$$V_c = 2\sqrt{f'_c} \times bd$$

$$= \frac{2\sqrt{4000} \times 22 \times 15.56}{1000} = 43.3 \text{ kips}$$

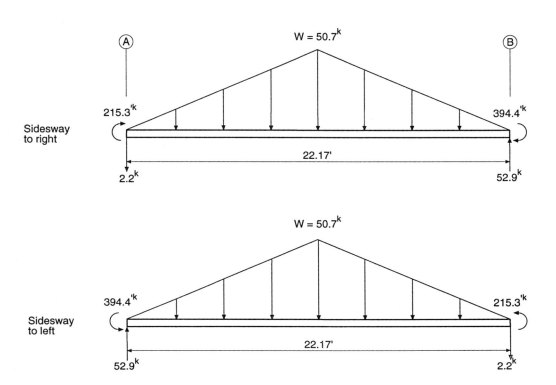

Figure 3-30 Design Shear Forces for Longitudinal Beams in an Interior Frame (Seismic Zone 2B)

The required shear strength to be provided by the shear reinforcement is:

$$\phi V_s = V_u - \phi V_c = 52.9 - (0.85 \times 43.3) = 16.1 \text{ kips}$$

$$V_s = 16.1/0.85 = 18.9 \text{ kips}$$

The required spacing of #3 closed stirrups is:

$$s = \frac{A_v f_y d}{V_s} = \frac{0.22 \times 60 \times 15.56}{18.9} = 10.9 \text{ in.} \qquad \textit{1911.5.6.2}$$

The maximum allowable stirrup spacing within a distance of 2h = 2 $\times$ 18 = 36 in. from the faces of supports is:

$$s_{max} = \frac{d}{4} = \frac{15.56}{4} = 3.9 \text{ in. (governs)} \qquad \textit{1921.8.4.2}$$

$$= 8 \times \text{(diameter of smallest longitudinal bar)} = 8 \times 1.128 = 9.0 \text{ in.}$$

$$= 24 \times \text{(diameter of stirrup bars)} = 24 \times 0.375 = 9.0 \text{ in.}$$

$$= 12.0 \text{ in.}$$

Thus, use #3 closed stirrups at a 4.0 in. spacing.

Beyond a distance of 2h from the supports, the maximum stirrup spacing is:

$$s_{max} = \frac{d}{2} = \frac{15.56}{2} = 7.8 \text{ in.} \qquad \textit{1921.8.4.3}$$

The minimum area of shear reinforcement, A_v, is:

$$A_v = \frac{50 \times b \times s}{f_y} = \frac{50 \times 22 \times 7}{60,000} = 0.13 \text{ in.}^2 < 0.22 \text{ in.}^2 \qquad \textit{1911.5.5.3}$$

Thus, #3 stirrups at 7 in. will be adequate.

The shear strength of the member with a 7 in. spacing is:

$$\phi V_n = 0.85 \times (V_s + V_c)$$

$$= 0.85 \times \left(\frac{0.22 \times 60 \times 15.56}{7} + 43.3 \right)$$

$$= 61.8 \text{ kips}$$

Two-thirds of this strength is:

$$\frac{2}{3}(\phi V_n) = \frac{2}{3} \times 61.8 = 41.2 \text{ kips}$$

The location where the applied shear is equal to 41.2 kips is 7.5 ft from the face of the support. Therefore, any reinforcing bars terminated in a tension zone beyond 7.5 ft from the column faces will meet the requirements of UBC 1912.10.5.1 since $\frac{2}{3}(\phi V_n) > V_u$.

3.3.2.5.3 Reinforcing Bar Cutoff Points and Splices

The negative reinforcement at the supports consists of 6-#9 bars. The location where four of the #9 bars can be terminated away from the interior support will be determined. The loading used to find the cutoff point is 0.9 times the dead load in combination with the nominal moment strength, M_n, at the member ends ($f_y = 60$ ksi and $\phi = 1.0$). The design moment strength, ϕM_n, provided by 2-#9 reinforcing bars in the beam is 132.8 ft-kips.

With $\phi = 1.0$ and $f_s = f_y = 60$ ksi, $M^+_n = 215.3$ ft-kips on the exterior end of the beam, and $M^-_n = 394.4$ ft-kips at the interior end. The dead load is equal to $0.9 \times 3.37 = 3.04$ kips/ft at midspan. The distance from the face of the right support to where the moment under the loading considered equals 132.8 ft-kips is readily obtained by summing moments about section a-a in Fig. 3-31:

$$\frac{x}{2}\left(\frac{3.04x}{11.085}\right)\left(\frac{x}{3}\right) + 394.4 - 44.4x = 132.8$$

Solution of this equation gives x = 6.1 ft. The 4-#9 top bars near the interior support may be discontinued at (x + d) = (6.1 + 15.56/12) = 7.4 ft from the face of the support. To satisfy the requirements of UBC 1912.10.5.1, use a 7 ft-9 in. bar length.

Due to the symmetry of the reinforcement, the cutoff point for the top #9 bars at the exterior support is the same as for the interior support.

Note that all of the positive bars must be made continuous due to the requirements given in UBC 1921.8.4.1 (see Section 3.3.2.5.1).

Reinforcing bars must be properly developed at the supports. The development length is the product of the basic development length and the factors previously presented in Table 3-13. The basic development length for a #9 bar is 38.0 in.

The information required for determining the top bar development length is:

Minimum top or bottom cover to bars being developed	1.88 in. (1.7 d_b)
Minimum side cover	large (slab)
Clear spacing between bars being developed	2.30 in. (2.0 d_b)

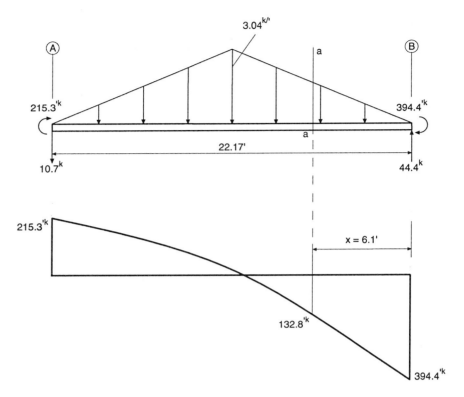

Figure 3-31 Moment Diagram for Cutoff Location of Negative Bars at Interior Support of Longitudinal Beam (Seismic Zone 2B)

Transverse reinforcement meets requirements of UBC 1911.5.4 and 1911.5.5.3 for top bars.

Factor I will have a value of 1.4 and Factor II will have a value of 1.3 for the top bars. Using the other applicable factors gives:

Top bar $\ell_d = 1.4 \times 1.3 \times 1.0 \times 1.0 \times 1.0 \times 38.0/12 = 5.8$ ft < 7.75 ft   O.K.

The information required for determining the splice length is:

	Top Bars	Bottom Bars
Stress level	$< 0.5 f_y$	$> 0.5 f_y$
Splice type	A	B
Minimum top or bottom cover to bars being spliced	1.88 in. $(1.7 d_b)$	1.88 in. $(1.7 d_b)$
Minimum side cover	large (slab)	1.88 in. $(1.7 d_b)$
Clear spacing between bars being spliced	16.0 in. $(14.2 d_b)$	7.43 in. $(6.6 d_b)$

Transverse reinforcement meets requirements of UBC 1911.5.4 and 1911.5.5.3 for both top and bottom bars.

This gives a value of 1.0 for Factor I for both top and bottom splices. The splice lengths are:

Top splice length (Class A) = 1.0 × (1.0 × 1.3 × 1.0 × 1.0 × 1.0 × 38.0) = 49.4 in.

Bottom splice length (Class B) = 1.3 × (1.0 × 1.0 × 1.0 × 1.0 × 1.0 × 38.0) = 49.4 in.

Figure 3-32 shows the reinforcement details for the beam in the interior longitudinal frame.

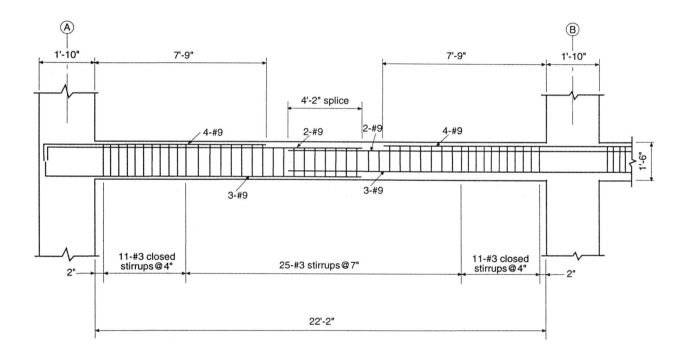

Figure 3-32 Reinforcement Details for Beam of Interior Longitudinal Frame (Seismic Zone 2B)

3.3.2.6   Proportioning and Detailing a Column

3.3.2.6.1 General Code Requirements

This section outlines the design of an interior column of an exterior transverse frame between levels 2 and 3. The maximum factored axial load for this column at the third floor level is 998 kips (see Table 3-18).

$$P_u \ = 998 \text{ kips} > \frac{A_g f'_c}{10} \ = \ \frac{22^2 \times 6}{10} = 290.4 \text{ kips}$$

Therefore, the reinforcement details must satisfy UBC 1921.8.5 for this member.

Based on the combinations in Table 3-18, a 22-in. square column with 8-#9 bars is adequate (see Fig. 3-33). Check the reinforcement ratio (UBC 1910.9.1):

$$0.08 > \rho = \frac{A_{st}}{A_g} = \frac{8 \times 1.0}{22^2} = 0.0165 > 0.01 \quad \text{O.K.}$$

3.3.2.6.2 Confinement Reinforcement

According to UBC 1921.8.5, transverse reinforcement for confinement is required over a distance ℓ_0 at column ends, where ℓ_0 equals the maximum of:

$$\frac{1}{6} \text{ (clear height)} = \frac{10.5 \times 12}{6} = 21.0 \text{ in.}$$

Depth of member = 22.0 in.   (governs)

18.0 in.

The maximum allowable spacing of #3 ties over the length ℓ_0 from the joint face is:

$$s_0 \quad = 8 \times \text{(diameter of smallest longitudinal bar)} = 8 \times 1.128 = 9.0 \text{ in.} \quad \text{(governs)}$$

$$= 24 \times \text{(diameter of tie bars)} = 24 \times 0.375 = 9.0 \text{ in.}$$

$$= \frac{h}{2} = \frac{22}{2} = 11.0 \text{ in.}$$

$$= 12.0 \text{ in.}$$

The maximum spacing of ties outside of the distance ℓ_0 is $2s_0 = 2 \times 9.0 = 18.0$ in. (UBC 1921.8.5.4). These ties must be arranged such that every corner and alternate longitudinal bar shall have support provided by the corner of a tie or crosstie.

3.3.2.6.3 Transverse Reinforcement for Shear

The design shear in columns resisting earthquake effects is not based on the factored shears obtained from a lateral load analysis, but rather on the nominal moment strengths that can be developed at the faces of the joints (UBC 1921.8.3). The largest probable moment strength that may develop can conservatively be taken as the moment at the balanced point; from Fig. 3-33, this moment is 551/0.7 = 787 ft-kips. The design moment strengths of the beams on either side of the joint are 349.8 ft-kips and 287.1 ft-kips. Since the sum of the nominal moments at the ends of the column is greater than the largest moment that can develop from the beams, the beam moment strengths, M_n, will control the design. The factored shear is:

$$V_u \quad = 2 \times (349.8/2 + 287.1/2)/(0.9 \times 10.33) = 68.5 \text{ kips}$$

Note that this shear force is greater than all of the factored shear forces determined from analysis (see Table 3-18).

Since the member is subject to axial compression, the shear strength provided by the concrete, V_c, is calculated as:

$$V_c = 2\left(1 + \frac{N_u}{2000\,A_g}\right)\sqrt{f_c'}\,b_w d$$

1911.3.1.2

Conservatively take $N_u = 553$ kips which is the minimum value of axial compression for the column (see Table 3-18).

$$V_c = 2 \times \left(1 + \frac{553}{2 \times 484}\right) \times \sqrt{6000} \times 22 \times 19.56/1000 = 104.8 \text{ kips}$$

$$\phi V_c = 0.85 \times 104.8 = 89.1 \text{ kips} > V_u = 68.5 \text{ kips}$$

Determine the minimum spacing of #3 ties based on shear:

$$s = A_v f_y/50 b_w = (3 \times 0.11 \times 60,000)/(50 \times 22) = 18 \text{ in.}$$

1911.5.5.3

Thus, the transverse reinforcement spacing over the distance $\ell_o = 22$ in. at each column end is governed by the requirements for confinement rather than shear. Use #3 ties at 9 in. within a distance of 22 in. from the column ends and #3 ties spaced at a maximum of 18 in. over the remainder of the column height.

In UBC 1921.8, no specific requirements for column splices are provided. For this design, the requirements for Seismic Zone 4 will be used rather than the alternative of using those for regions of low or no seismicity. In Seismic Zone 4, UBC 1921.4.3.2 requires tension splices to be within the center of the member length.

The basic development length for #9 bars is 31.0 in.

The information required for determining the splice length is:

Minimum clear cover to bars being spliced = 1.88 in. ($1.7\,d_b$)

Clear spacing between bars being spliced = 6.3 in. ($5.6\,d_b$)

Transverse reinforcement meets requirements of UBC 1907.10.5

This gives a value of 1.0 for Factor I (see Table 3-13). The other factors are all equal to 1.0, giving:

Class B splice length = $1.3 \times (1.0 \times 1.0 \times 1.0 \times 1.0 \times 1.0 \times 31.0) = 40.3$ in.

Reinforcement details for the column are shown in Fig. 3-34.

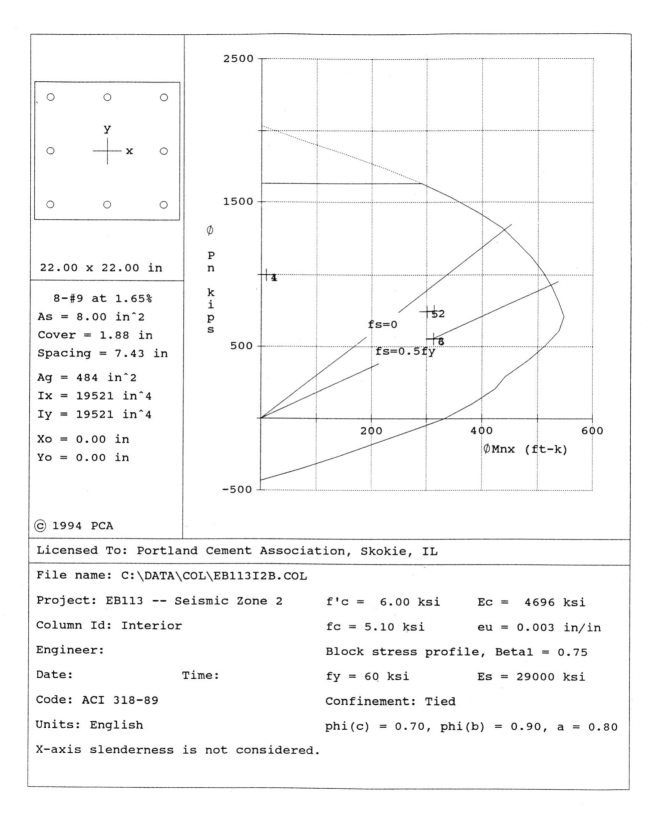

22.00 x 22.00 in

8-#9 at 1.65%
As = 8.00 in^2
Cover = 1.88 in
Spacing = 7.43 in

Ag = 484 in^2
Ix = 19521 in^4
Iy = 19521 in^4

Xo = 0.00 in
Yo = 0.00 in

© 1994 PCA

Licensed To: Portland Cement Association, Skokie, IL

File name: C:\DATA\COL\EB113I2B.COL

Project: EB113 -- Seismic Zone 2   f'c = 6.00 ksi   Ec = 4696 ksi

Column Id: Interior   fc = 5.10 ksi   eu = 0.003 in/in

Engineer: Block stress profile, Beta1 = 0.75

Date: Time: fy = 60 ksi   Es = 29000 ksi

Code: ACI 318-89   Confinement: Tied

Units: English   phi(c) = 0.70, phi(b) = 0.90, a = 0.80

X-axis slenderness is not considered.

Figure 3-33 Design Strength Interaction Diagram for a 22-in. Square Interior Column of an Exterior Transverse Frame above the 2nd Floor Level (Seismic Zone 2B)

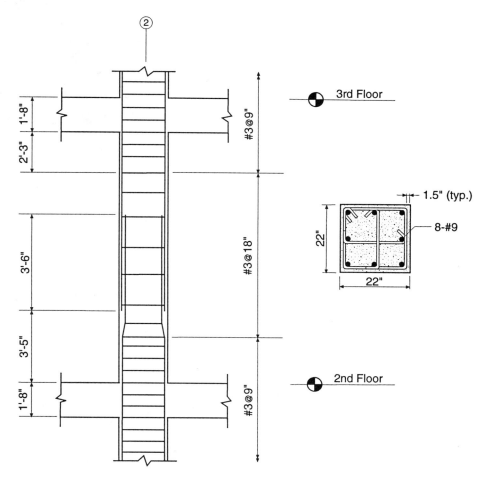

Figure 3-34 Reinforcement Details for a 22-in. Square Interior Column of an Exterior Transverse Frame
(Seismic Zone 2B)

3.3.2.7   Proportioning and Detailing of an Interior Beam-Column Connection

Proportioning and detailing of beam-column connections is required for Seismic Zones 3 and 4 only. Nevertheless, beam-column connections were checked for this example. The requirements of UBC 1921.5 were utilized with the exception of using a steel stress equal to f_y rather than $1.25f_y$ as required for Seismic Zones 3 and 4.

This section will check the shear strength and detailing requirements for an interior beam-column joint that is part of an interior transverse and an interior longitudinal frame. The joint has beams framing into all four sides: two transverse and two longitudinal beams. All the beams are as wide as the column.

3.3.2.7.1 Transverse Reinforcement for Confinement

The columns above and below this joint were not designed. Assuming that the transverse reinforcement in the critical end regions is the same as for the column designed in Section 3.3.2.6, #3 ties at 9 in. will be continued through the joint. Although the joint in question is confined by beams on all four faces according to the criteria of UBC 1921.5.2.2, any relaxation in the moderate amount of transverse reinforcement provided is felt to be unjustified.

3.3.2.7.2 Shear Strength of Joint

Figure 3-35 shows the beam-column joint at the second floor level of the structure. The shear strength is checked in the longitudinal direction of the building. The shear across section x-x is obtained by subtracting the column horizontal shear from the sum of the tensile force in the top flexural reinforcement of the beam on one side and the concrete compressive force near the top of the beam on the opposite face of the column.

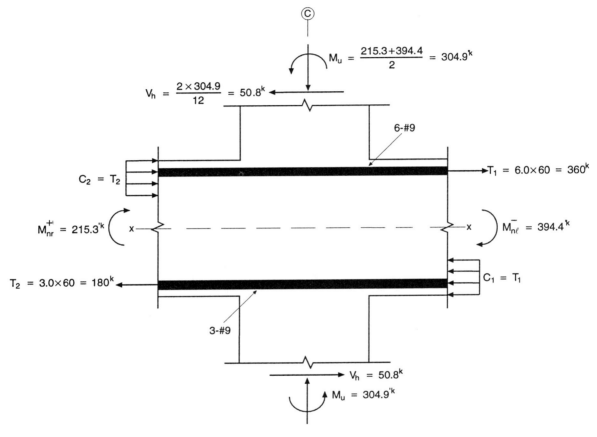

Figure 3-35 Shear Analysis of an Interior Beam-Column Joint of an Interior Longitudinal Frame
(Seismic Zone 2B)

The column horizontal shear, V_h, can be obtained by assuming that the beams in the adjoining floors are deformed so that plastic hinges form at their junctions with the column, with M^-_n (beam) = 394.4 ft-kips and M^+_n (beam) = 215.3 ft-kips. Therefore, the sum of the column moments above and below the joint will equal 394.4 + 215.3 = 609.7 ft-kips.

It is assumed that beam end moments are resisted equally by the columns above and below the joint. The horizontal shear at the column ends is then equal to:

$$V_h \qquad = 2 \times (394.4/2 + 215.3/2)/12 = 50.8 \text{ kips}$$

The tensile force in the top beam reinforcement (6-#9 bars) = $6.0 \times 60 = 360.0$ kips. The compression on the opposite side of the column is equal to the tensile force in 3-#9 bars which is $3.0 \times 60 = 180.0$ kips.

The net shear at section x-x of the joint is $T_1 + C_2 - V_h = 360.0 + 180.0 - 50.8 = 489.2$ kips.

For a joint confined on all four faces, the nominal shear strength is (UBC 1921.5.3.1):

$$\phi V_c = \phi\, 20 \sqrt{f'_c} \times A_j$$

$$= 0.85 \times 20 \sqrt{6000} \times 484.0/1000 = 637.3 \text{ kips} > 489.2 \text{ kips} \quad \text{O.K.}$$

A_j is equal to the effective cross-sectional area within a joint in a plane parallel to the plane of the reinforcement generating shear in the joint. The joint depth is the overall depth of the column. The effective joint width is equal to the width of the column. Therefore, $A_j = 22 \times 22 = 484.0$ in.^{2}

3.3.3.8   Proportioning and Detailing of an Exterior Beam-Column Connection

This section will address the shear strength and detailing requirements for an exterior beam-column joint that is part of an exterior transverse and in interior longitudinal frame. The joint has beams framing into three sides: two transverse beams and one longitudinal beam. The beams are all as wide as the column.

3.3.3.8.1 Transverse Reinforcement for Confinement

Assume that the column above this joint has the same reinforcement as the column designed in Section 3.3.2.6. The transverse reinforcement in the critical end regions, #3 ties at 9 in., will be continued through the depth of the joint. It may be noted as a matter of interest that the joint in question is unconfined according to the criteria of UBC 1921.5.2.2.

3.3.3.8.2 Shear Strength of Joints

Figure 3-36 shows the beam-column joint being designed at the second floor level of the structure. The shear strength is checked in the longitudinal direction of the building. The shear across section x-x is obtained by subtracting the column horizontal shear from the tensile force in the top flexural reinforcement.

The column horizontal shear, V_h, can be obtained by assuming that the beam in the adjoining floor is deformed so that a plastic hinge forms at the junctions with the column, with M^-_n (beam) = 394.4 ft-kips.

It is assumed that the beam end moments are resisted equally by the columns above and below the joint. The horizontal shear at the column ends is then equal to:

$$V_h = 2 \times (394.4/2)/12 = 32.9 \text{ kips}$$

The tensile force in the top beam reinforcement (6-#9 bars) = 6.0 × 60 = 360.0 kips. The net shear at section x-x of the joint is $T_1 - V_h = 360.0 - 32.9 = 327.1$ kips.

For a joint confined on three faces, the nominal shear strength is (UBC 1921.5.3.1):

$$\phi V_c = \phi\, 15 \sqrt{f'_c} \times A_j$$

$$= 0.85 \times 15 \sqrt{6000} \times 484.0/1000 = 478.0 \text{ kips} > 327.1 \text{ kips} \quad \text{O.K.}$$

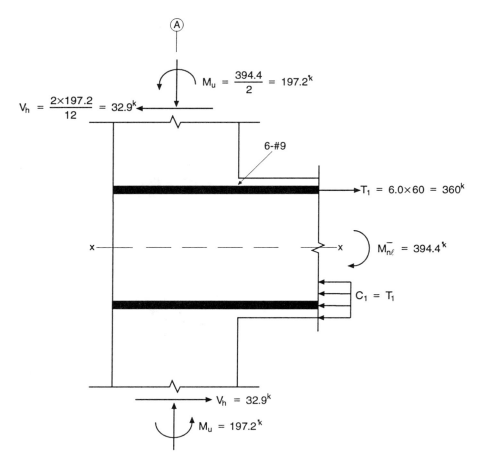

Figure 3-36 Shear Analysis of an Exterior Beam-Column Joint of an Interior Longitudinal Frame (Seismic Zone 2B)

3.4   Design for Seismic Zone 1

For Seismic Zone 1, the transverse as well as the longitudinal beam dimensions are reduced to 18×18 in., and the column dimensions are also reduced to 18×18 in. The slab remains 7 in. thick.

3.4.1   Frame Analysis

On the basis of the given data and the dimensions of the building, the weights of the floors are listed in Table 3-21. This example uses an importance factor I of 1.0 and a site soil coefficient S of 1.2 (same as before).

The lateral load resisting system of this building consists of Ordinary Moment-Resisting Frames (OMRF). This type of structural system has an R_w value of 5 and no special detailing requirements.

The seismic design forces resulting from the distribution of the base shear computed in accordance with UBC Eqs. (28-6), (28-7), and (28-8) are also shown in Table 3-21.

Table 3-21 Seismic Forces for Zone 1

Floor Level	Weight, w_x (kips)	Height, h_x (ft)	$w_x h_x$ (ft-kips)	Lateral Force (kips)
R	1381	147	203,077	73.2
12	1430	135	193,050	41.9
11	1430	123	175,890	38.1
10	1430	111	158,730	34.4
9	1430	99	141,570	30.7
8	1430	87	124,410	27.0
7	1430	75	107,250	23.3
6	1430	63	90,090	19.5
5	1430	51	72,930	15.8
4	1430	39	55,770	12.1
3	1430	27	38,610	8.4
2	1442	15	21,630	4.7
Σ	17,123		1,382,937	329.1

Building height	h_n	=	*147 ft*
Building weight	W	=	*17,123 kips*
Fundamental period	T	=	$0.030 (h_n)^{3/4}$ *= 1.27 sec.*
Seismic zone factor	Z	=	*0.075*
Importance factor	I	=	*1.0*
Response coefficient	R_W	=	*5*
Soil factor	S	=	*1.2*
Coefficient	C	=	$1.25 \, S/T^{2/3}$ *= 1.28*
Base shear	V	=	$ZICW/R_W$ *= 329.1 kips*
Top level force	F_t	=	$0.07TV$ *= 29.2 kips*

Wind forces and story shears corresponding to a fastest mile wind speed of 70 mph and Exposure B are computed in accordance with UBC 1618.

Three-dimensional analyses of the structure in both directions under the respective seismic and wind loads were carried out using *PCA-Frame*. Results from the seismic analysis are given in Figs. 3-37 and 3-38 for an exterior transverse frame and an interior longitudinal frame, respectively. The results from the wind analysis are given in Figs. 3-39 and 3-40.

For buildings having a fundamental period greater than 0.7 seconds, the calculated story drift (including torsional effects) must not exceed $0.03/R_w$ nor 0.004 times the story height (UBC 1628.8.2):

$$\frac{0.03h_s}{R_w} = \frac{0.03 \times 12 \times 12}{5} = 0.86 \text{ in.}$$

$$0.004h_s = 0.004 \times 12 \times 12 = 0.58 \text{ in. (governs)}$$

It can be seen from Table 3-22 that for all stories, the lateral drifts obtained from the prescribed lateral forces in both directions are less than the limiting value.

R	-14	-11	-17
12	-29	-26	-33
11	-43	-39	-44
10	-55	-50	-54
9	-66	-60	-63
8	-76	-70	-71
7	-84	-78	-78
6	-92	-85	-84
5	-98	-90	-88
4	-103	-95	-91
3	-105	-98	-94
2	-105	-97	-89

Bending Moments in Beams (ft-kips)

R	-1.1	-1.5
12	-2.4	-2.9
11	-3.6	-3.9
10	-4.7	-4.8
9	-5.6	-5.6
8	-6.5	-6.4
7	-7.2	-7.0
6	-7.8	-7.5
5	-8.4	-7.9
4	-8.8	-8.1
3	-9.1	-8.4
2	-9.0	-8.0

Shear Forces in Beams (kips)

R	18 / 9	33 / 24
12	28 / 19	45 / 39
11	36 / 28	59 / 53
10	43 / 36	71 / 66
9	49 / 43	82 / 77
8	55 / 50	91 / 87
7	59 / 55	99 / 95
6	63 / 60	105 / 102
5	66 / 64	110 / 109
4	69 / 70	113 / 111
3	66 / 66	118 / 122
2	73 / 140	104 / 155

Bending Moments in Columns (kips)

R	2.2	4.8
12	4.0	7.0
11	5.3	9.3
10	6.6	11.4
9	7.7	13.2
8	8.7	14.8
7	9.5	16.2
6	10.3	17.3
5	10.8	18.3
4	11.5	18.7
3	11.0	20.0
2	14.1	17.3

Shear Forces in Columns (ft-kips)

R	1.1	0.4
12	3.6	0.9
11	7.2	1.2
10	11.8	1.3
9	17.4	1.4
8	23.9	1.3
7	31.1	1.0
6	38.9	0.6
5	47.3	0.1
4	56.1	0.6
3	65.1	1.2
2	74.1	2.3

Axial Forces in Columns (kips)

Figure 3-37 Results of 3-D Analysis of an Exterior Transverse Frame under Seismic Zone 1 Forces

Bending Moments in Beams (ft-kips)

R					
R	-12	-9	-15	-14	-14
12	-25	-23	-28	-28	-28
11	-38	-34	-39	-39	-39
10	-49	-45	-49	-48	-48
9	-59	-55	-57	-57	-57
8	-69	-63	-65	-65	-65
7	-77	-71	-71	-71	-71
6	-84	-77	-77	-77	-77
5	-90	-83	-81	-81	-81
4	-95	-87	-84	-84	-84
3	-97	-90	-87	-87	-87
2	-99	-91	-84	-85	-85

Shear Forces in Beams (kips)

R			
R	-0.9	-1.3	-1.2
12	-2.1	-2.5	-2.5
11	-3.2	-3.4	-3.4
10	-4.2	-4.3	-4.3
9	-5.1	-5.1	-5.0
8	-5.9	-5.7	-5.7
7	-6.5	-6.3	-6.3
6	-7.2	-6.8	-6.8
5	-7.7	-7.2	-7.2
4	-8.1	-7.5	-7.5
3	-8.3	-7.7	-7.8
2	-8.4	-7.5	-7.6

Bending Moments in Columns (ft-kips)

R			
R	13 / 5	26 / 18	29 / 21
12	22 / 14	37 / 31	39 / 33
11	28 / 21	48 / 43	50 / 44
10	33 / 27	58 / 54	60 / 55
9	38 / 33	67 / 63	68 / 64
8	43 / 38	75 / 72	76 / 72
7	46 / 42	82 / 79	82 / 79
6	49 / 46	88 / 85	87 / 84
5	52 / 49	92 / 91	91 / 89
4	55 / 55	95 / 93	93 / 92
3	51 / 49	99 / 101	97 / 98
2	59 / 122	89 / 137	87 / 136

Shear Forces in Columns (kips)

R			
R	1.5	3.7	4.2
12	3.0	5.7	6.0
11	4.0	7.6	7.9
10	5.0	9.3	9.5
9	5.9	10.9	11.0
8	6.7	12.2	12.3
7	7.4	13.4	13.4
6	8.0	14.4	14.3
5	8.4	15.2	15.0
4	9.1	15.7	15.4
3	8.4	16.7	16.2
2	12.0	15.0	14.8

Axial Forces in Columns (kips)

R			
R	0.9	0.3	0.1
12	3.0	0.7	0.1
11	6.2	1.0	0.2
10	10.4	1.1	0.2
9	15.5	1.1	0.2
8	21.3	1.0	0.2
7	27.9	0.7	0.2
6	35.0	0.4	0.2
5	42.7	0.1	0.2
4	50.8	0.7	0.2
3	59.1	1.3	0.1
2	67.5	2.2	0.1

Figure 3-38 Results of 3-D Analysis of an Interior Longitudinal Frame under Seismic Zone 1 Forces

R	-2.7	-1.7	-6.2
12	-9.0	-8.0	-12.5
11	-18.5	-16.7	-20.4
10	-28.1	-25.7	-28.6
9	-37.7	-34.5	-36.6
8	-47.0	-43.1	-44.4
7	-56.1	-51.6	-51.9
6	-65.0	-59.8	-59.0
5	-73.6	-67.9	-65.9
4	-82.0	-75.5	-71.9
3	-88.6	-82.8	-78.9
2	-93.3	-85.6	-78.8

Bending Moments in Beams (ft-kips)

R	-0.2	-0.6
12	-0.8	-1.1
11	-1.6	-1.8
10	-2.4	-2.5
9	-3.2	-3.3
8	-4.0	-3.9
7	-4.8	-4.6
6	-5.6	-5.3
5	-6.3	-5.9
4	-7.0	-6.4
3	-7.6	-7.0
2	-8.0	-7.0

Shear Forces in Beams (kips)

R	2.8 / -0.8	8.4 / 4.6
12	10.3 / 4.2	17.3 / 12.7
11	15.4 / 9.4	26.8 / 22.1
10	20.6 / 14.6	35.9 / 31.2
9	25.5 / 19.7	44.7 / 40.2
8	30.3 / 24.7	53.3 / 48.9
7	35.0 / 29.6	61.6 / 57.4
6	39.5 / 34.5	69.5 / 65.5
5	43.8 / 38.8	77.3 / 73.8
4	48.5 / 45.6	83.7 / 79.6
3	48.7 / 44.7	93.1 / 93.1
2	54.5 / 114.0	82.5 / 128.0

Bending Moments in Columns (kips)

R	0.2	1.1
12	1.2	2.5
11	2.1	4.1
10	2.9	5.6
9	3.8	7.1
8	4.6	8.5
7	5.4	9.9
6	6.2	11.3
5	6.9	12.6
4	7.8	13.6
3	7.8	15.5
2	11.2	14.0

Shear Forces in Columns (ft-kips)

R	0.2	0.4
12	1.0	0.7
11	2.5	1.0
10	4.9	1.1
9	8.1	1.2
8	12.1	1.1
7	16.9	0.9
6	22.5	0.6
5	28.7	0.2
4	35.7	0.4
3	43.4	1.0
2	51.3	2.0

Axial Forces in Columns (kips)

Figure 3-39 Results of 3-D Analysis of an Exterior Transverse Frame under Wind Forces

R	-1.0	-0.6	-2.8	-2.5	-2.3
12	-4.3	-3.8	-6.2	-6.1	-6.0
11	-9.7	-8.7	-10.6	-10.5	-10.4
10	-15.0	-13.7	-15.2	-15.1	-15.0
9	-20.4	-18.6	-19.7	-19.6	-19.6
8	-25.5	-23.5	-24.1	-24.0	-24.0
7	-30.6	-28.2	-28.3	-28.3	-28.3
6	-35.6	-32.8	-32.3	-32.4	-32.4
5	-40.4	-37.3	-36.2	-36.3	-36.4
4	-45.3	-41.7	-39.8	-39.9	-40.1
3	-48.9	-45.7	-43.7	-43.8	-43.9
2	-52.6	-48.2	-44.6	-45.0	-45.3

Bending Moments in Beams (ft-kips)

R	-0.1	-0.2	-0.2
12	-0.4	-0.6	-0.5
11	-0.8	-0.9	-0.9
10	-1.3	-1.4	-1.3
9	-1.7	-1.8	-1.7
8	-2.2	-2.1	-2.1
7	-2.6	-2.5	-2.5
6	-3.0	-2.9	-2.9
5	-3.5	-3.2	-3.2
4	-3.9	-3.5	-3.6
3	-4.2	-3.9	-3.9
2	-4.5	-4.0	-4.0

Shear Forces in Beams (kips)

R	1.1 / -0.9	3.6 / 1.7	5.1 / 2.9
12	5.5 / 2.0	9.0 / 6.4	9.9 / 7.3
11	8.3 / 4.9	14.3 / 11.6	15.0 / 12.3
10	11.1 / 7.8	19.3 / 16.7	19.9 / 17.2
9	13.9 / 10.6	24.2 / 21.7	24.6 / 22.0
8	16.5 / 13.4	29.0 / 26.6	29.2 / 26.7
7	19.2 / 16.2	33.7 / 31.4	33.6 / 31.2
6	21.7 / 18.9	38.2 / 35.9	37.9 / 35.6
5	24.1 / 21.2	42.6 / 40.6	42.0 / 39.8
4	26.9 / 25.3	46.4 / 44.1	45.6 / 43.4
3	26.7 / 23.5	51.4 / 50.8	50.1 / 49.1
2	32.4 / 65.8	48.4 / 73.8	47.2 / 73.2

Bending Moments in Columns (ft-kips)

R	0.1	0.5	0.7
12	0.6	1.3	1.4
11	1.1	2.2	2.3
10	1.6	3.0	3.1
9	2.0	3.8	3.9
8	2.5	4.6	4.7
7	2.9	5.4	5.4
6	3.4	6.2	6.1
5	3.8	6.9	6.8
4	4.4	7.5	7.4
3	4.2	8.5	8.3
2	6.5	8.1	8.0

Shear Forces in Columns (kips)

R	0.1	0.2	0.0
12	0.4	0.4	0.1
11	1.3	0.5	0.1
10	2.5	0.5	0.1
9	4.3	0.6	0.1
8	6.4	0.5	0.1
7	9.1	0.5	0.1
6	12.1	0.3	0.1
5	15.5	0.0	0.1
4	19.4	0.3	0.1
3	23.6	0.6	0.1
2	28.1	1.1	0.0

Axial Forces in Columns (kips)

Figure 3-40 Results of 3-D Analysis of an Interior Longitudinal Frame under Wind Forces

Table 3-22 Lateral Displacements and Drifts due to Seismic Zone 1 Forces

Level	Transverse Direction		Longitudinal Direction	
	Displacement (in.)	Interstory Drift (in.)	Displacement (in.)	Interstory Drift (in.)
R	3.25	0.09	2.89	0.08
12	3.16	0.15	2.81	0.12
11	3.01	0.19	2.69	0.17
10	2.82	0.22	2.52	0.20
9	2.60	0.26	2.32	0.23
8	2.34	0.29	2.09	0.25
7	2.05	0.32	1.84	0.28
6	1.73	0.33	1.56	0.30
5	1.40	0.35	1.26	0.31
4	1.05	0.36	0.95	0.33
3	0.69	0.34	0.62	0.30
2	0.35	0.35	0.32	0.32

Table 3-23 P-Δ Check under Seismic Zone 1 Forces

Floor Level	Gravity Loads * Σw_i (kips)	Transverse Story Drift Δ_i (in.)	Longitudinal Story Drift Δ_i (in.)	Story Shear V_i (kips)	Story Height h_{si} (in.)	Transverse $\Sigma w_i \Delta_i / V_i h_{si}$	Longitudinal $\Sigma w_i \Delta_i / V_i h_{si}$
R	1654	0.09	0.08	73.2	144	0.014	0.013
12	3268	0.15	0.12	115.1	144	0.030	0.024
11	4882	0.19	0.17	153.2	144	0.042	0.038
10	6496	0.22	0.20	187.6	144	0.053	0.048
9	8110	0.26	0.23	218.3	144	0.067	0.059
8	9724	0.29	0.25	245.3	144	0.080	0.069
7	11,338	0.32	0.28	268.6	144	0.094	0.082
6	12,952	0.33	0.30	288.1	144	0.103	0.094
5	14,566	0.35	0.31	303.9	144	0.117	0.103
4	16,180	0.36	0.33	316.0	144	0.128	0.117
3	17,794	0.34	0.30	324.4	144	0.130	0.114
2	19,419	0.35	0.32	329.1	180	0.115	0.105

* Includes floor weight, superimposed dead load, and reduced roof and floor live loads

In Seismic Zone 1, P-Δ effects are potentially much more significant than in higher seismic zones because the stiffness of the lateral load resisting systems in the higher zones is required to be greater than that in the lower zones. Therefore, it is prudent to check the P-Δ effects prior to any member design for Seismic Zone 1. UBC 1628.9 allows P-Δ effects to be neglected when the ratio of the secondary moment to the primary moment is less than or equal to 0.1. The results of the P-Δ check are shown in Table 3-23. Based on these results, the P-Δ effects for this frame cannot be ignored.

The SEAOC-88 Blue Book Commentary Section 1E.9 presents a method to modify the lateral forces to account for P-Δ effects [1.1]. Using this method, the story shears are increased by an amplification factor, F_d. The amplification factor accounts for the secondary effect caused by the initial story drift, which causes an additional increment of drift, which in turn leads to more drift, and so on. The initial drift is then actually increased by the factor $(1 + \theta + \theta^2 + \theta^3...)$ which gives an amplification factor equal to $1/(1 - \theta)$. It should be noted that when calculating θ, the lateral deflection is multiplied by the factor $3R_w/8$ to approximate the actual drift of the structure including inelastic action of the structural elements. The revised lateral forces for the frames in both directions with P-Δ effects included are shown in Table 3-24.

The structure was reanalyzed using the revised seismic forces given in Table 3-24. Figures 3-41 and 3-42 give the ratio of P-Δ results to first order results for the transverse and longitudinal frames, respectively. Note that similar results can be obtained using the second order analysis option in *PCA-Frame*. The design of the members in the subsequent sections will be based on the results obtained from the P-Δ analysis as required in UBC 1628.9.

Table 3-24 Revised Lateral Forces Accounting for P-Δ Effects (Seismic Zone 1)

Level	Transverse				Longitudinal			
	$\Sigma w_i \, \Delta_i / V_i h_{si}$ *	$\theta = \dfrac{3}{8} R_w \dfrac{\Sigma w_i \Delta_i}{V_i h_{si}}$	$F_d = \dfrac{1}{1-\theta}$	Revised Lateral Force (kips)	$\Sigma w_i \, \Delta_i / V_i h_{si}$ *	$\theta = \dfrac{3}{8} R_w \dfrac{\Sigma w_i \Delta_i}{V_i h_{si}}$	$F_d = \dfrac{1}{1-\theta}$	Revised Lateral Force (kips)
R	0.014	0.026	1.03	75.4	0.013	0.024	1.03	75.4
12	0.030	0.056	1.06	44.4	0.024	0.045	1.05	44.0
11	0.042	0.079	1.09	41.5	0.038	0.071	1.08	41.2
10	0.053	0.099	1.11	38.2	0.048	0.090	1.10	37.8
9	0.067	0.126	1.14	35.0	0.059	0.111	1.12	34.4
8	0.080	0.150	1.18	31.9	0.069	0.129	1.15	31.1
7	0.094	0.176	1.21	28.2	0.082	0.154	1.18	27.5
6	0.103	0.193	1.24	24.2	0.094	0.176	1.21	23.6
5	0.117	0.219	1.28	20.2	0.103	0.193	1.24	19.6
4	0.128	0.240	1.32	16.0	0.117	0.219	1.28	15.5
3	0.130	0.244	1.32	11.1	0.114	0.214	1.27	10.6
2	0.115	0.216	1.28	6.0	0.105	0.197	1.25	5.9

*see Table 3-23

3.4.2   Design of Beams and Columns

The objective is to determine the required flexural and shear reinforcement for typical beams of both the transverse and the longitudinal frames, and for a typical column common to both frames. The structure will be designed for dead and live loads, in combination with either wind or Seismic Zone 1 forces.

As was done in the previous sections, the Direct Design Method of UBC 1913.6 is used to determine the moments due to gravity loads acting on the beams. Table 3-25 gives the factored bending moments in an exterior transverse frame and an interior longitudinal frame obtained from this method.

Table 3-26 gives the factored bending moments for the applicable combinations of gravity and lateral loads on a 2nd floor beam in an exterior frame in the transverse direction. In this case, the seismic moments were taken from Figs. 3-37 and 3-41 (including P-Δ effects), and the wind moments were taken from Fig. 3-39. Similarly, Table 3-27 gives the factored moments on a 2nd floor beam in an interior longitudinal frame. Here, seismic moments were taken from Figs. 3-38 and 3-42, and the wind moments were taken from Fig. 3-40.

R	1.02	1.02	1.05
12	1.04	1.03	1.05
11	1.05	1.05	1.05
10	1.06	1.06	1.06
9	1.07	1.07	1.07
8	1.08	1.08	1.08
7	1.09	1.09	1.09
6	1.10	1.10	1.10
5	1.11	1.11	1.11
4	1.13	1.12	1.12
3	1.13	1.13	1.12
2	1.13	1.13	1.13

Bending Moments in Beams

R	1.03	1.05
12	1.04	1.05
11	1.05	1.05
10	1.06	1.06
9	1.07	1.07
8	1.08	1.08
7	1.09	1.09
6	1.10	1.10
5	1.11	1.11
4	1.12	1.12
3	1.13	1.13
2	1.13	1.13

Shear Forces in Beams

R	1.03 / 1.01	1.04 / 1.03
12	1.05 / 1.03	1.04 / 1.04
11	1.06 / 1.05	1.06 / 1.05
10	1.07 / 1.06	1.07 / 1.06
9	1.08 / 1.07	1.08 / 1.07
8	1.09 / 1.08	1.09 / 1.08
7	1.10 / 1.09	1.10 / 1.09
6	1.11 / 1.10	1.11 / 1.11
5	1.12 / 1.11	1.12 / 1.11
4	1.13 / 1.12	1.13 / 1.12
3	1.13 / 1.13	1.13 / 1.12
2	1.13 / 1.13	1.14 / 1.14

Bending Moments in Columns

R	1.02	1.04
12	1.04	1.04
11	1.05	1.05
10	1.06	1.06
9	1.07	1.08
8	1.09	1.09
7	1.09	1.09
6	1.11	1.10
5	1.12	1.12
4	1.13	1.12
3	1.14	1.13
2	1.14	1.13

Shear Forces in Columns

R	1.03	1.11
12	1.03	1.11
11	1.04	1.11
10	1.05	1.12
9	1.06	1.12
8	1.06	1.13
7	1.07	1.14
6	1.08	1.15
5	1.08	1.22
4	1.09	1.12
3	1.09	1.13
2	1.10	1.13

Axial Forces in Columns

Figure 3-41 Ratio of P- Δ Results to First Order Results for an Exterior Transverse Frame under Seismic Zone 1 Forces

Bending Moments in Beams

R/Floor					
R	1.02	1.02	1.03	1.04	1.03
12	1.03	1.03	1.04	1.04	1.04
11	1.04	1.04	1.05	1.04	1.04
10	1.05	1.05	1.05	1.05	1.06
9	1.06	1.06	1.07	1.06	1.06
8	1.07	1.07	1.07	1.07	1.07
7	1.08	1.08	1.08	1.08	1.08
6	1.09	1.09	1.09	1.09	1.09
5	1.10	1.10	1.10	1.10	1.10
4	1.11	1.10	1.10	1.10	1.10
3	1.11	1.11	1.11	1.11	1.11
2	1.12	1.12	1.11	1.11	1.11

Shear Forces in Beams

R/Floor			
R	1.02	1.04	1.03
12	1.03	1.04	1.04
11	1.04	1.05	1.05
10	1.05	1.05	1.05
9	1.06	1.06	1.06
8	1.07	1.07	1.07
7	1.08	1.08	1.08
6	1.09	1.09	1.09
5	1.10	1.10	1.10
4	1.10	1.10	1.10
3	1.11	1.11	1.11
2	1.11	1.11	1.11

Bending Moments in Columns

R/Floor			
R	1.02	1.03	1.03
	1.00	1.03	1.03
12	1.03	1.03	1.04
	1.05	1.04	1.04
11	1.05	1.05	1.05
	1.04	1.04	1.05
10	1.06	1.06	1.06
	1.05	1.05	1.06
9	1.07	1.07	1.07
	1.06	1.06	1.06
8	1.08	1.08	1.08
	1.07	1.07	1.07
7	1.09	1.09	1.09
	1.08	1.08	1.08
6	1.10	1.10	1.10
	1.09	1.09	1.09
5	1.10	1.11	1.10
	1.10	1.10	1.10
4	1.11	1.11	1.11
	1.11	1.11	1.11
3	1.11	1.11	1.12
	1.11	1.11	1.11
2	1.12	1.11	1.12
	1.12	1.11	1.11

Shear Forces in Columns

R/Floor			
R	1.01	1.03	1.04
12	1.03	1.04	1.04
11	1.05	1.05	1.05
10	1.06	1.06	1.06
9	1.07	1.06	1.06
8	1.08	1.08	1.07
7	1.08	1.08	1.08
6	1.09	1.09	1.09
5	1.10	1.11	1.10
4	1.11	1.10	1.11
3	1.11	1.11	1.12
2	1.12	1.11	1.12

Axial Forces in Columns

R/Floor			
R	1.02	1.10	1.08
12	1.03	1.10	1.08
11	1.03	1.11	1.08
10	1.04	1.10	1.08
9	1.05	1.11	1.09
8	1.06	1.12	1.09
7	1.06	1.13	1.08
6	1.07	1.15	1.09
5	1.07	0.90	1.08
4	1.08	1.09	1.06
3	1.08	1.11	1.06
2	1.09	1.11	0.98

Figure 3-42 Ratio of P-Δ Results to First Order Results for an Interior Longitudinal Frame under Seismic Zone 1 Forces

Table 3-28 gives the factored axial loads, bending moments, and shear forces (including P-Δ effects) for an exterior and interior column between the 2nd and 3rd floors in an exterior transverse frame. Live load reduction factors were computed in accordance with Eq. (6-1) in UBC 1606. Table 3-29 lists the load combinations for exterior and interior columns located between the ground level and the second floor.

Table 3-25 Factored Gravity Load Bending Moments in the Beams

Location	M_u (ft-kips)
EXTERIOR TRANSVERSE FRAME	
End Span	
Exterior Negative	28.5
Positive	88.0
Interior Negative	107.0
Interior Span	
Positive	52.3
Negative	99.9
INTERIOR LONGITUDINAL FRAME	
End Span	
Exterior Negative	55.1
Positive	156.8
Interior Negative	190.7
Interior Span	
Positive	93.2
Negative	178.0

Table 3-26 Summary of Factored Bending Moments for a Beam on the 2nd Floor of an Exterior Transverse Frame (Seismic Zone 1)

(1)	U = 1.4D + 1.7L	Eq. (9-1), UBC 1909.2.1
(2)	U = 0.75 (1.4D + 1.7L ± 1.87E)	Eq. (9-2), UBC 1909.2.3
(3)	U = 0.9D ± 1.43E	Eq. (9-3), UBC 1909.2.3
(4)	U = 0.75 (1.4D + 1.7L ± 1.7W)	Eq. (9-2), UBC 1909.2.2
(5)	U = 0.9D ± 1.3W	Eq. (9-3), UBC 1909.2.2

Load Combination	End Span Moments (ft-kips)			Interior Span Moments (ft-kips)	
	Exterior Negative	Positive	Interior Negative	Positive	Negative
1	-28.5	88.0	-107.0	52.3	-99.9
2: sidesway right	145.1	76.0	-234.0	39.2	66.2
sidesway left	-187.8	56.0	73.5	39.2	-216.0
3: sidesway right	156.7	50.2	-205.3	23.8	98.4
sidesway left	-182.7	29.8	108.1	23.8	-189.2
4: sidesway right	97.6	73.1	-189.4	39.2	25.6
sidesway left	-140.3	58.9	28.9	39.2	-175.4
5: sidesway right	108.3	47.3	-159.9	23.8	33.4
sidesway left	-134.3	32.7	62.7	23.8	-124.2

Table 3-27 Summary of Factored Bending Moments for a Beam on the 2nd Floor of an Interior Longitudinal Frame (Seismic Zone 1)

(1)	U = 1.4D + 1.7L	Eq. (9-1), UBC 1909.2.1
(2)	U = 0.75 (1.4D + 1.7L ± 1.87E)	Eq. (9-2), UBC 1909.2.3
(3)	U = 0.9D ± 1.43E	Eq. (9-3), UBC 1909.2.3
(4)	U = 0.75 (1.4D + 1.7L ± 1.7W)	Eq. (9-2), UBC 1909.2.2
(5)	U = 0.9D ± 1.3W	Eq. (9-3), UBC 1909.2.2

Load Combination	End Span Moments (ft-kips)			Interior Span Moments (ft-kips)	
	Exterior Negative	Positive	Interior Negative	Positive	Negative
1	-55.1	156.8	-190.7	93.2	-178.0
2: sidesway right	114.2	126.9	-286.0	69.9	-1.2
sidesway left	-196.9	108.3	-0.1	69.9	-265.8
3: sidesway right	133.9	79.8	-231.3	41.8	55.1
sidesway left	-183.3	60.8	60.3	41.8	-214.7
4: sidesway right	25.7	121.7	-204.5	69.9	-75.8
sidesway left	-108.4	113.6	-81.6	69.9	-191.3
5: sidesway right	43.7	74.5	-148.2	41.8	-20.9
sidesway left	-93.1	66.1	-22.8	41.8	-138.7

Table 3-28 Summary of Factored Axial Loads, Bending Moments, and Shear Forces for Columns between the 2nd and 3rd Floors of an Exterior Transverse Frame (Seismic Zone 1)

(1)	U = 1.4D + 1.7L	Eq. (9-1), UBC 1909.2.1
(2)	U = 0.75 (1.4D + 1.7L ± 1.87E)	Eq. (9-2), UBC 1909.2.3
(3)	U = 0.9D ± 1.43E	Eq. (9-3), UBC 1909.2.3
(4)	U = 0.75 (1.4D + 1.7L ± 1.7W)	Eq. (9-2), UBC 1909.2.2
(5)	U = 0.9D ± 1.3W	Eq. (9-3), UBC 1909.2.2

Load Combination	Interior Column				Exterior Column			
	Axial Load (kips)	Bending Moment (ft-kips)		Shear Force (kips)	Axial Load (kips)	Bending Moment (ft-kips)		Shear Force (kips)
		Top	Bottom			Top	Bottom	
1	929	-9.0	9.0	1.7	558	-15.5	15.5	3.0
2: sidesway right	695	180.3	-184.9	34.8	319	93.0	-93.0	20.5
sidesway left	699	-193.8	198.4	-37.4	518	-116.3	116.3	-26.5
3: sidesway right	509	187.5	-192.2	36.2	202	99.7	-99.7	19.0
sidesway left	513	-193.8	198.6	-37.4	406	-113.7	113.7	-21.7
4: sidesway right	695	112.0	-112.0	21.3	363	50.5	-45.4	9.1
sidesway left	698	-125.5	125.5	-23.9	474	-73.7	68.6	-13.6
5: sidesway right	510	118.2	-118.2	22.5	248	56.3	-51.1	10.2
sidesway left	512	-123.9	123.9	-23.6	361	-70.3	65.1	-12.9

Table 3-29 Summary of Factored Axial Loads, Bending Moments, and Shear Forces for Columns between Grade and the 2nd Floor of an Exterior Transverse Frame (Seismic Zone 1)

(1) U = 1.4D + 1.7L	Eq. (9-1), UBC 1909.2.1
(2) U = 0.75 (1.4D + 1.7L ± 1.87E)	Eq. (9-2), UBC 1921.2.3
(3) U = 0.9D ± 1.43E	Eq. (9-3), UBC 1921.2.3
(4) U = 0.75 (1.4D + 1.7L ± 1.7W)	Eq. (9-2), UBC 1909.2.2
(5) U = 0.9D ± 1.3W	Eq. (9-3), UBC 1909.2.2

Load Combination	Interior Column				Exterior Column			
	Axial Load (kips)	Bending Moment (ft-kips)		Shear Force (kips)	Axial Load (kips)	Bending Moment (ft-kips)		Shear Force (kips)
		Top	Bottom			Top	Bottom	
1	1013	-9.0	9.0	1.3	608	-15.5	15.5	2.2
2: sidesway right	1008	159.5	-241.1	28.1	342	104.1	-210.2	27.9
sidesway left	1018	-173.0	254.6	-30.0	570	-127.4	233.5	-32.3
3: sidesway right	615	166.3	-249.5	29.2	251	111.0	-219.2	23.2
sidesway left	623	-172.7	255.9	-30.1	485	-125.0	233.2	-25.1
4: sidesway right	1009	98.4	-156.5	17.9	391	57.9	-133.7	13.5
sidesway left	1017	-111.9	170.0	-19.8	521	-81.1	157.0	-16.7
5: sidesway right	554	104.4	-163.5	18.8	264	63.9	-141.2	14.4
sidesway left	560	-110.1	169.3	-19.6	398	-77.9	155.2	-16.4

3.4.2.1   General Code Requirements for Seismic Zone 1

Concrete frames that are part of the lateral force resisting system of structures in Seismic Zone 1 may be designed as Ordinary Moment-Resisting Frames. As such, this structure must be designed in accordance with all the applicable provisions of UBC 1624 and UBC 1901 through 1918, as illustrated in the following sections. Note that the ductility related detailing requirements of UBC 1921 are not applicable in Seismic Zone 1. The UBC 1631.1 requirements concerning orthogonal effects also do not apply in Seismic Zone 1.

3.4.2.2   Diaphragm Flexibility

UBC 1628.5 requires that the shears resulting from horizontal torsion must be included in design where diaphragms are not flexible. As shown in calculations for Seismic Zone 4, the floors of the building example are rigid diaphragms.

3.4.2.3   Proportioning and Detailing a Flexural Member of an Exterior Transverse Frame

3.4.2.3.1 Required Flexural Reinforcement

The maximum negative moment for a beam of an exterior transverse frame at an interior support is 234.0 ft-kips (see Table 3-26). The required area of steel, ignoring the effects of any compression reinforcement, is $A_s = 3.80$ in.^{2} Use 4-#9 bars ($\phi M_n = 244.7$ ft-kips).

Check limitations on the reinforcement ratio:

$$\rho = A_s/bd = 4.0/(18 \times 15.56) = 0.0143 \qquad\qquad 1910.5.1$$

$$0.0033 < 0.0143 < 0.0214 \quad \text{O.K.}$$

The maximum negative moment in the beam at the exterior support is 187.8 ft-kips (see Table 3-26). Use 4-#9 bars at this support also.

The maximum positive moment at the exterior support is 156.7 ft-kips. Ignoring the effects of any compression reinforcement, the required area of steel is $A_s = 2.42$ in.^{2} Use 3-#9 bars ($\phi M_n = 190.1$ ft-kips).

The maximum positive moment at the first interior support is 108.1 ft-kips. The maximum positive moment at midspan is 88.0 ft-kips. Therefore, the 3-#9 bottom bars may be used for the full length of the beam.

3.4.2.3.2 Required Shear Reinforcement

The loading to be used in determining shear reinforcement requirements in Seismic Zone 1 differs from that used in higher seismic zones. In Seismic Zones 3 and 4, the probable moment strengths of the members are used for the end moments. In Seismic Zone 2, the nominal moment strengths of the members are used. In Seismic Zone 1, however, the moments produced by the actual factored loads are used to determine the shear forces.

Since the beams have $(\alpha_1 \ell_2 / \ell_1) = 1.5 > 1.0$, they must carry the entire shear caused by the factored loads on tributary areas bounded by 45-degree lines drawn from the corners of the panels and the center lines of the adjacent panels parallel to the long sides (UBC 1913.6.8.1). The shear forces caused by the various types of loads at a distance d from the face of an interior support are as follows:

$$V_D = 9.8 \text{ kips}$$

$$V_L = 3.4 \text{ kips}$$

$$V_E = 1.13 \times 9.0 = 10.2 \text{ kips} \quad \text{(see Figs. 3-37 and 3-41)}$$

$$V_W = 8.0 \text{ kips} \quad \text{(see Fig. 3-39)}$$

Substituting these forces into the five load combinations yields:

$$V_u = 1.4V_D + 1.7V_L = 19.5 \text{ kips}$$

$$V_u = 0.75 \times (1.4V_D + 1.7V_L + 1.87V_E) = 28.9 \text{ kips} \quad \text{(governs)}$$

$$V_u = 0.9V_D + 1.43V_E = 23.4 \text{ kips}$$

$$V_u = 0.75 \times (1.4V_D + 1.7V_L + 1.7V_W) = 24.8 \text{ kips}$$

$$V_u = 0.9V_D + 1.3V_W = 19.2 \text{ kips}$$

The shear strength provided by the concrete is:

$$V_c = 2\sqrt{f'_c} \times b \times d \qquad\qquad 1911.3.1.1$$

$$= 2\sqrt{4000} \times 18 \times 15.56/1000 = 35.4 \text{ kips}$$

The shear strength that must be provided by stirrups is:

$$V_s = V_u/\phi - V_c$$

$$= 28.9/0.85 - 35.4 < 0$$

Therefore, use #3 closed stirrups at a $d/2 = 7$ in. spacing at the ends of the beams.

Stirrups can be discontinued where $\phi V_c/2 \geq V_u$. Thus, stirrups can be discontinued when V_u is less than (0.85 $\times$ 35.4/2) = 15.1 kips. The shear force from the seismic load will be constant over the length of the beam. The factored shear force from seismic loads alone = 1.4 $\times$ 10.2 = 14.3 kips. This is approximately equal to 15.1 kips; thus, stirrups will be provided over the entire length of the beam.

The shear strength of the members with 7 in. spacing of #3 stirrups is:

$$\phi V_n = 0.85 \times (V_s + V_c)$$

$$= 0.85 \times \left(\frac{0.22 \times 60 \times 15.56}{7} + 35.4 \right)$$

$$= 0.85 \times (29.3 + 35.4) = 55.0 \text{ kips}$$

Two-thirds of this capacity is:

$$\frac{2}{3}(\phi V_n) = \frac{2}{3} \times 55.0 = 36.7 \text{ kips}$$

This value is larger than the maximum shear, and the reinforcing bars may be terminated anywhere within the tension zone (UBC 1912.10.5).

3.4.2.3.3 Reinforcing Bar Cutoff Points and Splices

The negative reinforcement at the supports is 4-#9 bars. The location where two of the four bars can be terminated away from an interior support will be determined.

The negative design moment strength of a section with 2-#9 top bars is 131.2 ft-kips. The distance from the face of the right support to where the factored moment is reduced to 131.2 ft-kips must be determined. Based on the various loading conditions from Table 3-26, the location where the bars can be cut off is obtained by summing moments about section a-a in Fig. 3-43. Load combination 2 governs, yielding $x = 3.3$ ft. Two of the 4-#9 top bars near the interior support can be discontinued at $(x + d) = (3.3 + 15.56/12) = 4.6$ ft from the face of the interior support. Similarly, 2-#9 bars at the exterior support can be discontinued at a distance of 3.7 ft away from the face of the support.

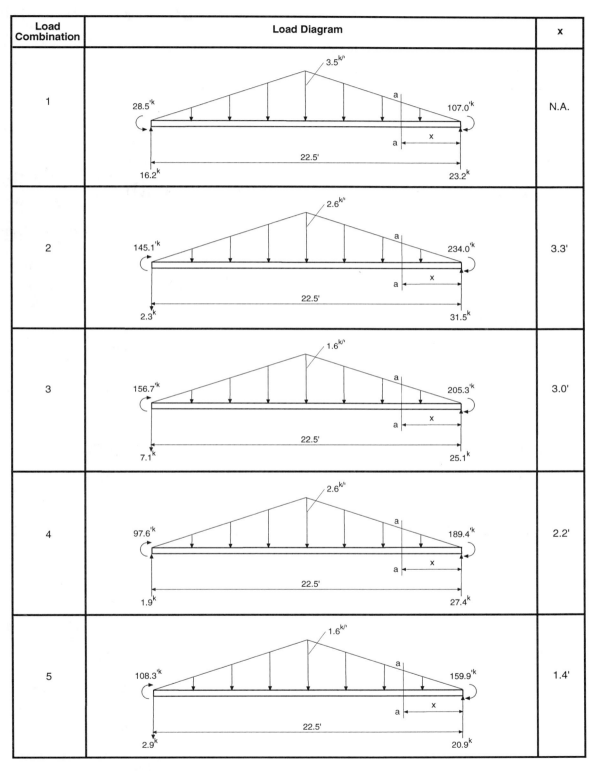

Load Combination	Load Diagram	x
1		N.A.
2		3.3'
3		3.0'
4		2.2'
5		1.4'

Figure 3-43 Loading Diagrams for Cutoff Locations of Negative Bars at Interior Support of Transverse Beam (Seismic Zone 1)

Reinforcing bars must be properly developed at the supports. The development length is the product of the basic development length and the factors given in Table 3-13. The basic development length for a #9 bar is 38.0 in.

The information required for determining the development length of the top bars is:

Minimum top cover to bars being developed	$= 1.88$ in. $(1.67\ d_b)$
Minimum side cover	large (slab)
Clear spacing between bars being developed	$= 3.25$ in. $(2.88\ d_b)$

Transverse reinforcement meets requirements of UBC 1911.5.4 and 1911.5.5.3.

Factor I equals 1.4. Factor II equals 1.3 for the top bars. Using the other applicable factors, the development length for the top bars is:

Top bar $\ell_d = 1.4 \times 1.3 \times 1.0 \times 1.0 \times 1.0 \times 38.0/12 = 5.8$ ft > 4.6 ft

The information required for determining the splice length is:

	Top Bars	Bottom Bars
Stress level	$< 0.5\ f_y$	$> 0.5\ f_y$
Splice type	A	B
Minimum top or bottom cover	1.88 in. $(1.67\ d_b)$	1.88 in. $(1.67\ d_b)$
Minimum side cover	large (slab)	1.88 in. $(1.67\ d_b)$
Clear spacing between bars being spliced	12.0 in. $(10.64\ d_b)$	5.4 in. $(4.82\ d_b)$

Transverse reinforcement meets requirements of UBC 1911.5.4 and 1911.5.5.3 for both top and bottom bars.

This gives a value of 0.85 and 1.0 for Factor I for the top and bottom splices, respectively. The splice lengths are:

Top splice length (Class A) $= 1.0 \times (0.85 \times 1.3 \times 1.0 \times 1.0 \times 1.0 \times 38.0) = 42.0$ in.

Bottom splice length (Class B) $= 1.3 \times (1.0 \times 1.0 \times 1.0 \times 1.0 \times 1.0 \times 38.0) = 49.4$ in.

Figure 3-44 shows the reinforcement details for the beam in the exterior transverse frame.

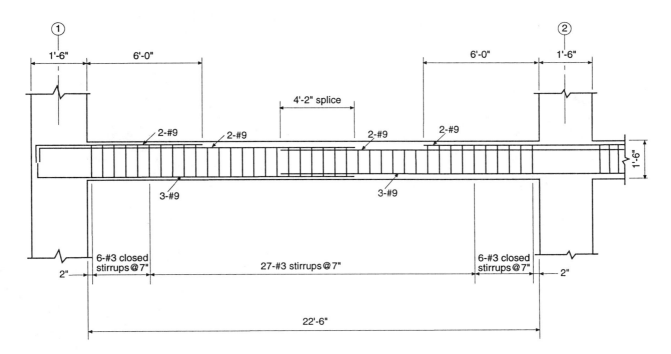

Figure 3-44 Reinforcement Details for Beam of Exterior Transverse Frame (Seismic Zone 1)

3.4.2.4 Proportioning and Detailing a Flexural Member of an Interior Longitudinal Frame

3.4.2.4.1 Required Flexural Reinforcement

The maximum negative moment for a longitudinal beam in an interior frame at an interior support is 286.0 ft-kips (see Table 3-27). The required area of steel is $A_s = 4.81$ in.^{2} Use 5-#9 bars ($\phi M_n = 294.8$ ft-kips).

The maximum negative moment in the beam at the exterior support is 196.9 ft-kips. Use 5-#9 bars at this support as well.

The maximum positive moment at the exterior support is 133.9 ft-kips. This requires 3-#8 bars ($\phi M_n = 153.5$ ft-kips).

The maximum positive moment at the first interior support is 60.3 ft-kips. The maximum positive moment at midspan is 156.8 ft-kips. Therefore, the 3-#8 bars can be used for the full length of the beam.

3.4.2.4.2 Required Shear Reinforcement

The shear forces caused by the various types of loads at a distance d from the face of an interior support are as follows:

$$V_D = 18.5 \text{ kips}$$

$$V_L = 6.7 \text{ kips}$$

$$V_E = 1.11 \times 8.4 = 9.3 \text{ kips (see Figs. 3-38 and 3-42)}$$

$$V_W = 4.5 \text{ kips (see Fig. 3-40)}$$

Substituting these forces into the five load combinations yields:

$$V_u = 1.4V_D + 1.7V_L = 37.3 \text{ kips}$$

$$V_u = 0.75 \times (1.4V_D + 1.7V_L + 1.87V_E) = 41.0 \text{ kips} \text{ (governs)}$$

$$V_u = 0.9V_D + 1.43V_E = 30.0 \text{ kips}$$

$$V_u = 0.75 \times (1.4V_D + 1.7V_L + 1.7V_W) = 33.7 \text{ kips}$$

$$V_u = 0.9V_D + 1.3V_W = 22.5 \text{ kips}$$

The shear strength provided by the concrete is:

$$V_c = 2\sqrt{f'_c} \times b \times d$$ *1911.3.1.1*

$$= 2\sqrt{4000} \times 18 \times 15.56/1000 = 35.4 \text{ kips}$$

The shear strength that must be provided by stirrups is:

$$V_s = V_u/\phi - V_c$$

$$= 41.0/0.85 - 35.4 = 12.8 \text{ kips}$$

The shear strength provided by a #3 stirrup using a 7 in. spacing is:

$$V_s = A_v \times f_y \times d/s$$

$$= 0.22 \times 60,000 \times 15.56/(7.0 \times 1000) = 29.3 \text{ kips} > 12.8 \text{ kips}$$

Therefore, use #3 closed stirrups at a 7 in. spacing at the ends of the beams.

Stirrups can be discontinued where $\phi V_c/2 \geq V_u$. Thus, stirrups can be discontinued when V_u is less than (0.85 $\times$ 35.4/2) = 15.1 kips. The shear force from the seismic load will be constant over the length of the beam. The factored shear force from seismic loads alone = 1.4 $\times$ 9.3 = 13.0 kips. This is approximately equal to 15.1 kips; thus, stirrups will be provided over the entire length of the beam.

The shear strength of the members with a 7 in. spacing of #3 stirrups is:

$$\phi V_n = 0.85 \times (V_s + V_c)$$

$$= 0.85 \times \left(\frac{0.22 \times 60 \times 15.56}{7} + 35.4 \right)$$

$$= 0.85 \times (29.3 + 35.4) = 55.0 \text{ kips}$$

Two-thirds of this capacity is:

$$\frac{2}{3}(\phi V_n) = \frac{2}{3} \times 55.0 = 36.7 \text{ kips} < 41.0 \text{ kips}$$

Using the second load combination, the location where the applied shear is equal to 36.7 kips is 4.6 ft from the face of the support. Therefore, any reinforcing bars terminated in a tension zone beyond 4.6 ft from the column face will meet the requirements of UBC 1912.10.5 since $\frac{2}{3}(\phi V_n) > V_u$.

3.4.2.4.3   Reinforcing Bar Cutoff Points and Splices

The negative reinforcement at the supports consists of 5-#9 bars. The location where three of the five bars can be terminated away from an interior support will be determined.

The negative design moment strength of a section with 2-#9 top bars is 131.2 ft-kips. The distance from the face of the right support to where the factored moment is reduced to 131.2 ft-kips must be determined. Based on the various loading conditions from Table 3-27, the location where the bars can be cut off is obtained by summing moments about section a-a in Fig. 3-45. Load combination 2 governs, yielding x = 3.5 ft. Three of the 5-#9 top bars near the interior support can be discontinued at (x + d) = (3.5 + 15.56/12) = 4.8 ft from the face of the interior support. Similarly, 3-#9 bars at the exterior support can be discontinued at a distance of 3.2 ft away from the face of the support.

Reinforcing bars must be properly developed at the supports. The development length is the product of the basic development length and the factors given in Table 3-13. The basic development length for a #9 bar is 38.0 in.

The information required for determining the development length of the top bars is:

Minimum top cover to bars being developed	= 1.88 in. (1.67 d_b)
Minimum side cover	large (slab)
Clear spacing between bars being developed	= 2.15 in. (1.91 d_b)

Transverse reinforcement meets requirements of UBC 1911.5.4 and 1911.5.5.3.

Factor I equals 2.0. Factor II equals 1.3 for the top bars. Using the other applicable factors gives:

Top bar ℓ_d = 2.0 × 1.3 × 1.0× 1.0 × 1.0 × 38.0/12 = 8.2 ft > 4.8 ft and 3.2 ft

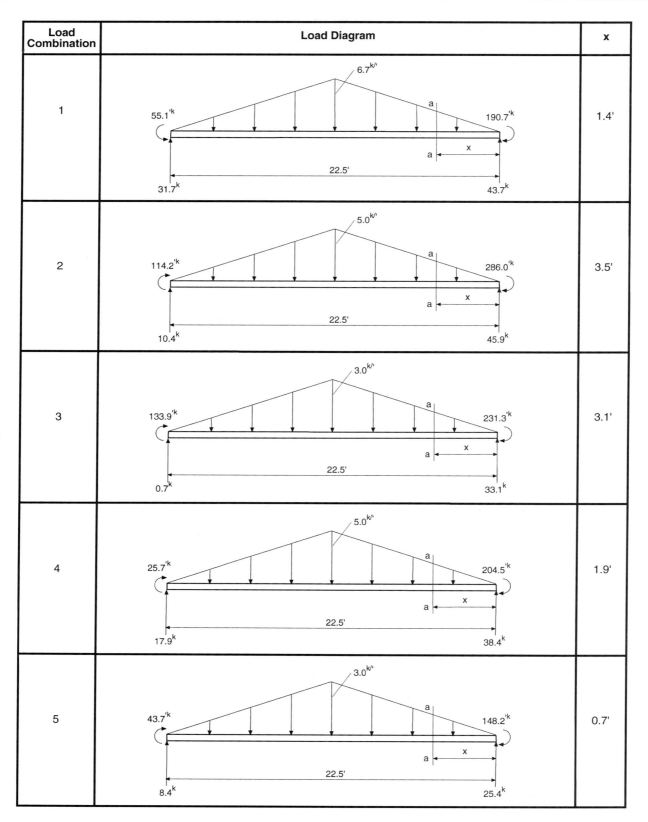

Figure 3-45 Loading Diagrams for Cutoff Locations of Negative Bars at Interior Support of Longitudinal Beam (Seismic Zone 1)

The information required for determining the splice length is:

	Top Bars	Bottom Bars
Stress level	$< 0.5 \, f_y$	$> 0.5 \, f_y$
Splice type	A	B
Minimum top or bottom cover to bars being spliced	1.88 in. (1.67 d_b)	1.88 in. (1.88 d_b)
Minimum side cover	large (slab)	1.88 in. (1.88 d_b)
Clear spacing between bars being spliced	12.0 in. (10.64 d_b)	5.63 in. (5.63 d_b)

Transverse reinforcement meets requirements of UBC 1911.5.4 and 1911.5.5.3 for both top and bottom bars.

This gives a value of 0.85 and 1.0 for Factor I for the top and bottom splices, respectively. The splice lengths are:

Top splice length (Class A) = $1.0 \times (0.85 \times 1.3 \times 1.0 \times 1.0 \times 1.0 \times 38.0) = 42.0$ in.

Bottom splice length (Class B) = $1.3 \times (1.0 \times 1.0 \times 1.0 \times 1.0 \times 1.0 \times 30.0) = 39.0$ in.

Figure 3-46 shows the reinforcement details for the beam in the interior longitudinal frame.

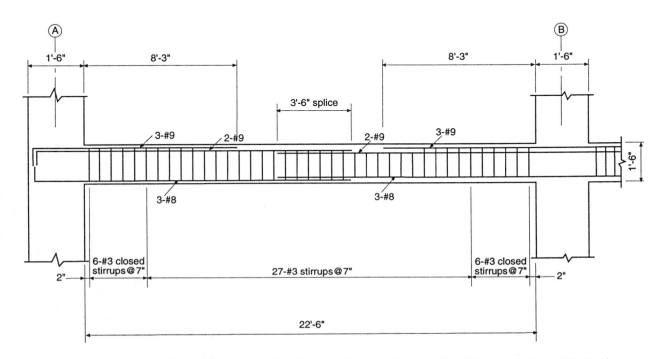

Figure 3-46 Reinforcement Details for Beam of Interior Longitudinal Frame (Seismic Zone 1)

3.4.2.5   Proportioning and Detailing a Column

3.4.2.5.1 General Code Requirements

This section will present the design of an interior column of the exterior transverse frame supporting the third floor of the structure. The maximum factored axial load on this column at this level is 929 kips (see Table 3-28).

An 18-in. square tied column with 8-#9 bars will be adequate for the load combinations shown in Table 3-28. An interaction diagram for this column is shown in Fig. 3-47. Note that slenderness effects need not be considered since P-Δ effects were included in the analysis. The reinforcement ratio for this column is 2.5% which is between 1% and 8%.

3.4.2.5.2 Transverse Reinforcement Requirements

Tie reinforcement for compression members must conform to the following maximum spacing requirements, for #9 longitudinal bars and #3 ties (UBC 1907.10.5):

$16 \times$ longitudinal bar diameter = $16 \times 1.128 = 18.1$ in.

$48 \times$ tie bar diameter = $48 \times 0.375 = 18.0$ in. (governs)

minimum column dimension = 18.0 in.

The maximum factored shear force on this column is 37.4 kips (see Table 3-28). For members subject to axial compression, the shear strength provided by the concrete is (UBC 1911.3.1.2):

$$V_c = 2\left(1 + \frac{N_u}{2000 A_g}\right)\sqrt{f_c'}\, b_w d$$

With $N_u = 510$ kips, which is the minimum value of the axial load for this column:

$$V_c = 2 \times \left(1 + \frac{513,000}{2000 \times 324}\right)\sqrt{6000} \times 18.0 \times 15.56/1000 = 77.7 \text{ kips}$$

$$\phi V_c = 0.85 \times 77.7 = 66.1 \text{ kips}$$

Since $\phi V_c = 66.1$ kips is less than $2V_u = 74.8$ kips, minimum shear reinforcement must be provided. Assuming #4 ties, the maximum allowable spacing is:

$$s = A_v f_y/50 b_w = 0.40 \times 60,000/(50 \times 18) = 26.7 \text{ in.}$$

This spacing is larger than the maximum spacing requirements given in UBC 1907.10.5. Therefore, use #3 ties spaced at 18.0 in.

From the interaction diagram given in Fig. 3-47, it is clear that a Class A splice will be required for the column bars. However, since more than one-half of the bars will be spliced at one location, provide a Class B splice (UBC 1912.15.2).

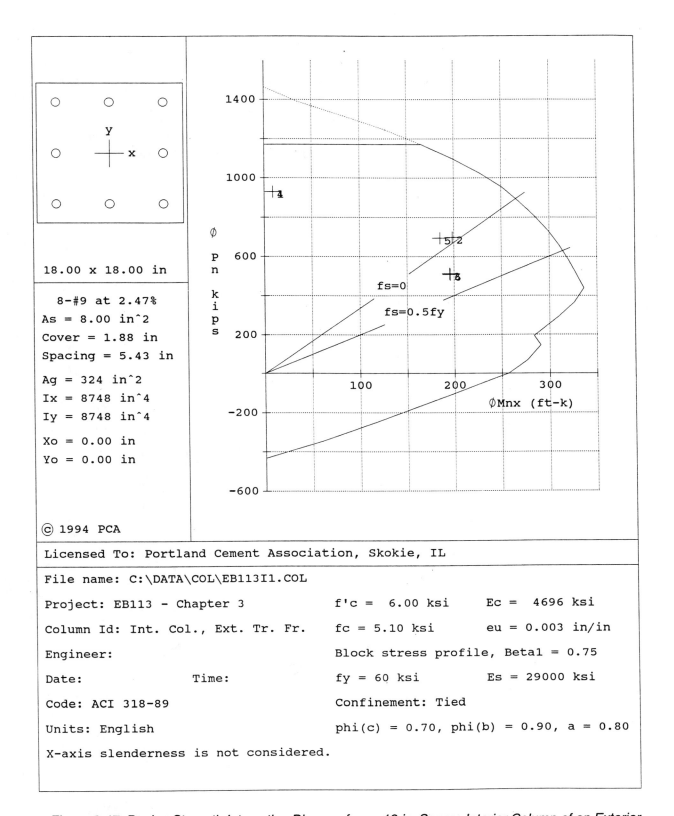

Figure 3-47 Design Strength Interaction Diagram for an 18-in. Square Interior Column of an Exterior Transverse Frame above the 2nd Floor Level, Seismic Zone 1

The information required for determining the splice length is:

Minimum cover to bars being spliced   = 2.0 in. (1.77 d_b)

Clear spacing between bars being spliced   = 4.9 in. (4.3 d_b)

Transverse reinforcement meets requirements of UBC 1907.10.5.

This gives a value of 1.0 for Factor I (see Table 3-13). The other factors are all equal to 1.0.

Class B splice length = 1.3 × (1.0 × 1.0 × 1.0× 1.0 × 1.0 × 31.0) = 40.3 in.

Figure 3-48 shows the reinforcement details for this column.

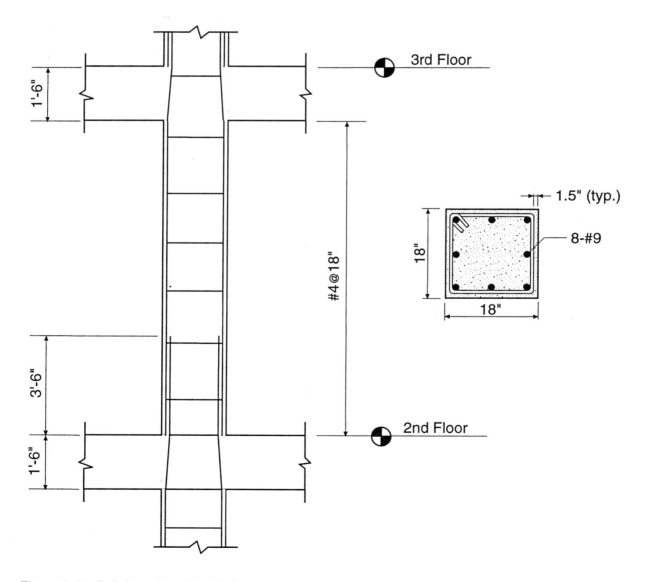

Figure 3-48 Reinforcement Details for an 18-in. Square Interior Column of an Exterior Transverse Frame
(Seismic Zone 1)

3.5 Design for Seismic Zone 0 (Wind Loads)

3.5.1 Frame Analysis

This section will present the design of the structure for wind forces corresponding to a fastest mile wind speed of 70 mph and Exposure B computed in accordance with UBC 1618. The calculations of the lateral forces are shown in Tables 3-4 and 3-5 in Section 3.2.1. The same member sizes and material strengths used in Section 3.4 (Seismic Zone 1) will be used here.

Story drift limitations and the check for P-Δ effects are not applicable to structures in Seismic Zone 0. Also, accidental torsion from eccentricities of lateral forces need not be considered in Seismic Zone 0.

3.5.2 Design of Beams and Columns

As noted above, the member sizes and material strengths are the same as those used in Section 3.4. Therefore, the loading combinations determined for wind and gravity loads shown in Tables 3-26 through 3-29 (load combinations 1, 4, and 5) are also applicable to this design.

3.5.2.1 Proportioning and Detailing a Flexural Member of an Exterior Transverse Frame

3.5.2.1.1 Required Flexural Reinforcement

The maximum negative moment for a beam in an exterior transverse frame at an interior support is 189.4 ft-kips (see Table 3-26). The required area of steel is $A_s = 2.99$ in.^{2} Use 4-#8 bars ($\phi M_n = 199.2$ ft-kips).

The maximum negative moment in the beam at the exterior support is 140.3 ft-kips. Conservatively use 4-#8 bars at this support as well.

The maximum positive moment at the exterior support is 108.3 ft-kips. This requires 2-#8 bars.

The maximum positive moment at the first interior support is 62.7 ft-kips. The maximum positive moment at midspan is 88.0 ft-kips. Therefore, the 2-#8 bottom bars may be used for the full length of the beam.

3.5.2.1.2 Required Shear Reinforcement

In Seismic Zone 0, the moments produced by the actual factored loads are used to determine the design shear forces.

The shear forces caused by the various types of loads at a distance d from the face of an interior support are as follows:

$$V_D = 9.8 \text{ kips}$$

$$V_L = 3.4 \text{ kips}$$

$$V_W = 8.0 \text{ kips (see Fig. 3-39)}$$

Substituting these forces into the three load combinations yields:

$$V_u = 1.4V_D + 1.7V_L = 19.5 \text{ kips}$$

$$V_u = 0.75 \times (1.4V_D + 1.7V_L + 1.7V_W) = 24.8 \text{ kips} \quad \text{(governs)}$$

$$V_u = 0.9V_D + 1.3V_W = 19.2 \text{ kips}$$

The shear strength provided by the concrete is:

$$V_c = 2\sqrt{f'_c} \times b \times d \qquad\qquad\qquad 1911.3.1.1$$

$$= 2\sqrt{4000} \times 18 \times 15.56/1000 = 35.4 \text{ kips}$$

The shear strength that must be provided by stirrups is:

$$V_s = V_u/\phi - V_c$$

$$= 24.8/0.85 - 35.4 < 0$$

The maximum spacing of #3 stirrups is:

$$s_{max} = (A_v \times f_y)/(50 \times b)$$

$$= (0.22 \times 60,000)/(50 \times 18) = 14.7 \text{ in.}$$

but not greater than $\dfrac{d}{2} = \dfrac{15.56}{2} = 7.8$ in.

Therefore, use #3 stirrups at a 7 in. spacing at the ends of the beams.

Stirrups can be discontinued where $\phi V_c/2 \geq V_u$. Thus, stirrups can be discontinued when V_u is less than $(0.85 \times 35.4/2) = 15.1$ kips. The shear force from the wind load will be constant over the length of the beam. The factored shear force from wind forces alone $= 1.3 \times 8.0 = 10.4$ kips which is not much smaller than 15.1 kips. Therefore, stirrups will be used for the entire length of the beam.

The shear strength of the members with a 7 in. spacing is:

$$\phi V_n = 0.85 \times (V_s + V_c)$$

$$= 0.85 \times \left(\frac{0.22 \times 60 \times 15.56}{7} + 35.4 \right)$$

$$= 0.85 \times (29.3 + 35.4) = 55.0 \text{ kips}$$

Two-thirds of this strength is:

$$\frac{2}{3}(\phi V_n) = \frac{2}{3} \times 55.0 = 36.7 \text{ kips}$$

This value is larger than the maximum shear, and the reinforcing bars may be terminated anywhere within the tension zone (UBC 1912.10.5.1).

3.5.2.1.3   Reinforcing Bar Cutoff Points and Splices

The negative reinforcement at the supports consists of 4-#8 bars. The location where two of the four bars can be terminated away from an interior support will be determined. The negative design moment strength of a section with 2-#8 top bars is 105.1 ft-kips. The distance from the face of the right support to where the factored moment is reduced to 105.1 ft-kips must be determined. Based on the critical loading combinations, the location where the bars can be cut off is obtained by summing moments about section a-a in Fig. 3-49. Two of the 4-#8 top bars near the interior support can be discontinued at $(x + d) = (3.2 + 15.56/12) = 4.5$ ft from the face of the interior support. Similarly, 2-#8 bars at the exterior support can be discontinued at a distance 3.0 ft away from the support.

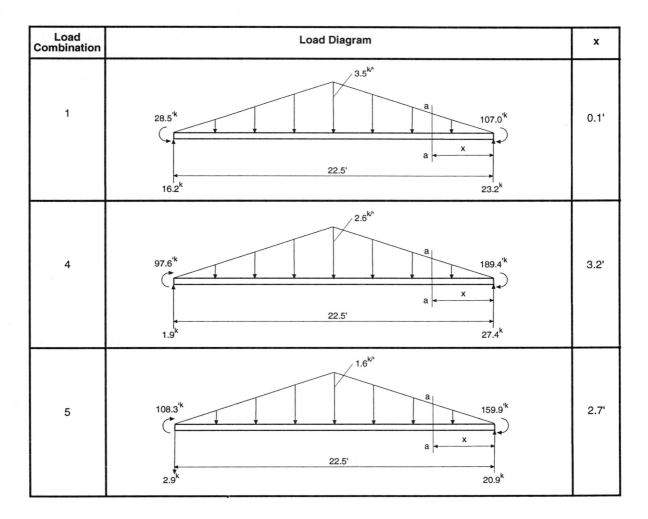

Figure 3-49 Loading Diagrams for Cutoff Locations of Negative Bars at Interior Support of Transverse Beam (Seismic Zone 0)

Reinforcing bars must be properly developed at the supports. The development length is the product of the basic development length and the factors given in Table 3-13. The basic development length for a #8 bar is 30 in.

The information required for determining the development length for the top bars is:

Minimum top cover to bars being developed   = 1.88 in. (1.88 d_b)

Minimum side cover   large (slab)

Clear spacing between bars being developed   = 3.42 in. (3.42 d_b)

Transverse reinforcement meets requirements of UBC 1911.5.4 and 1911.5.5.3.

Factor I equals 1.0. Factor II equals 1.3 for the top bars. Using the other applicable factors gives:

Top bar ℓ_d = 1.0 × 1.3 × 1.0 × 1.0 × 1.0 × 30.0/12 = 3.3 ft

The information required for determining the splice length is:

	Top Bars	Bottom Bars
Stress level	< 0.5 f_y	> 0.5 f_y
Splice type	A	B
Minimum top or bottom cover	1.88 in. (1.88 d_b)	1.88 in. (1.88 d_b)
Minimum side cover	large (slab)	1.88 in. (1.88 d_b)
Clear spacing between bars being spliced	12.25 in. (12.25 d_b)	12.25 in. (12.25 d_b)

Transverse reinforcement meets requirements of UBC 1911.5.4 and 1911.5.5.3 for both top and bottom bars.

This gives a value of 0.95 and 1.0 for Factor I for the top and bottom splices, respectively. The splice lengths are:

Top splice length (Class A) = 1.0 × (0.95 × 1.3 × 1.0 × 1.0 × 1.0 × 30.0) = 37.1 in.

Bottom splice length (Class B) = 1.3 × (1.0 × 1.0 × 1.0 × 1.0 × 1.0 × 30.0) = 39.0 in.

Figure 3-50 shows the reinforcement details for the beam in the exterior transverse frame.

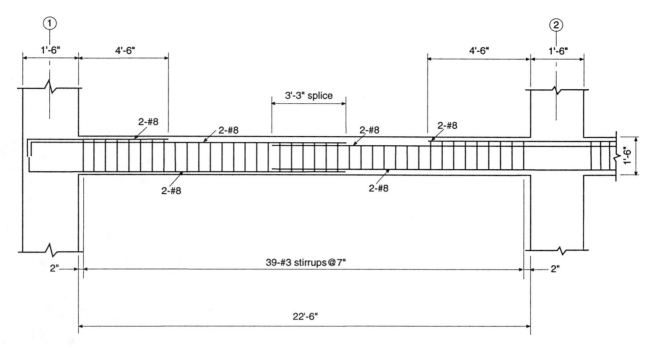

Figure 3-50 Reinforcement Details for Beam of Exterior Transverse Frame (Seismic Zone 0)

3.5.2.2   Proportioning and Detailing a Flexural Member of an Interior Longitudinal Frame

3.5.2.2.1 Required Flexural Reinforcement

The maximum negative moment for a longitudinal beam of an interior frame at an interior support is 204.5 ft-kips (see Table 3-27). The required area of steel is $A_s = 3.25$ in.^{2} Use 4-#9 bars ($\phi M_n = 244.7$ ft-kips).

The maximum negative moment in the beam at the exterior support is 108.4 ft-kips. Use 3-#9 bars.

There is no positive moment at the first interior support. The maximum positive moment at midspan is 156.8 ft-kips which is greater than 25.7 ft-kips which is the maximum positive moment at the exterior support. Three #9 bars with a moment strength of 190.1 ft-kips will be adequate.

3.5.2.2.2 Required Shear Reinforcement

The shear forces caused by the various types of loads at a distance d from the face of an interior support are as follows:

$$V_D = 18.5 \text{ kips}$$

$$V_L = 6.7 \text{ kips}$$

$$V_W = 4.5 \text{ kips (see Fig. 3-40)}$$

Substituting these forces into three load combinations yields:

$$V_u = 1.4V_D + 1.7V_L = 37.3 \text{ kips (governs)}$$

$$V_u = 0.75 \times (1.4V_D + 1.7V_L + 1.7V_W) = 33.7 \text{ kips}$$

$$V_u = 0.9V_D + 1.3V_W = 22.5 \text{ kips}$$

The shear strength provided by the concrete is:

$$V_c = 2\sqrt{f'_c} \times b \times d$$

$$= 2\sqrt{4000} \times 18 \times 15.56/1000 = 35.4 \text{ kips}$$

The shear strength that must be provided by the stirrups is:

$$V_s = V_u/\phi - V_c$$

$$= 37.3/0.85 - 35.4 = 8.5 \text{ kips}$$

The maximum spacing of #3 stirrups with 2 legs is governed by:

$$s_{max} = (A_v \times f_y)/(50 \times b)$$

$$= (0.22 \times 60,000)/(50 \times 18) = 14.7 \text{ in.}$$

but not greater than $\dfrac{d}{2} = \dfrac{15.56}{2} = 7.8 \text{ in.}$

The shear strength provided by #3 stirrups using a 7 in. spacing is:

$$V_s = A_v \times f_y \times d/s$$

$$= 0.22 \times 60,000 \times 15.56/(7.0 \times 1000) = 29.3 \text{ kips} > 8.5 \text{ kips}$$

Therefore, use #3 stirrups at a 7 in. spacing at the ends of the beams.

Stirrups can be discontinued where $\phi V_c/2 \geq V_u$. Thus, stirrups can be discontinued when V_u is less than (0.85 $\times$ 35.4/2) = 15.1 kips. The shear force from the wind load will be constant over the length of the beam. Use stirrups throughout the entire length of the beam.

The shear strength of the members with a 7 in. spacing is:

$$\phi V_n = 0.85 \times (V_s + V_c)$$

$$= 0.85 \times \left(\frac{0.22 \times 60 \times 15.56}{7} + 35.4 \right)$$

$$= 0.85 \times (29.3 + 35.4) = 55.0 \text{ kips}$$

Two-thirds of this strength is:

$$\frac{2}{3}(\phi V_n) = \frac{2}{3} \times 55.0 = 36.7 \text{ kips}$$

This value is approximately equal to V_u, and the reinforcing bars may be terminated anywhere within the tension zone (UBC 1912.10.5.1).

3.5.2.2.3 Reinforcing Bar Cutoff Points and Splices

The negative reinforcement at the interior support consists of 4-#9 bars. The location where two of the four bars can be terminated away from an interior support will be determined. The negative moment strength of a section with 2-#9 top bars is 131.2 ft-kips. The distance from the face of the right support to where the factored moment is reduced to 131.2 ft-kips must be determined. Based on the critical loading conditions, the location where the bars can be cut off is obtained by summing moments about section a-a in Fig. 3-51. Two of the 4-#9 top bars near the interior support can be discontinued at $(x + d) = (1.9 + 15.56/12) = 3.2$ ft from the face of the interior support. Similarly, 1-#9 bar at the exterior support can be discontinued at 3.1 ft away from that support.

Reinforcing bars must be properly developed at the supports. The development length is the product of the basic development length and the factors given in Table 3-13. The basic development length for a #9 bar is 38.0 in.

The information required for determining the development length of the top bars is:

Minimum top cover to bars being developed	= 1.88 in. (1.67 d_b)
Minimum side cover	large (slab)
Clear spacing between bars being developed	= 3.25 in. (2.88 d_b)

Transverse reinforcement meets requirements of UBC 1911.5.4 and 1911.5.5.3.

Factor I equals 1.4. Factor II equals 1.3 for the top bars. Using the other applicable factors gives:

Top bar $\ell_d = 1.4 \times 1.3 \times 1.0 \times 1.0 \times 1.0 \times 38.0/12 = 5.8$ ft

The information required for determining the splice length is:

	Top Bars	Bottom Bars
Stress level	< 0.5 f_y	> 0.5 f_y
Splice type	A	B
Minimum top or bottom cover to bars being spliced	1.88 in. (1.67 d_b)	1.88 in. (1.67 d_b)
Minimum side cover	large (slab)	1.88 in. (1.67 d_b)

Clear spacing between
bars being spliced   12.0 in. (10.64 d_b)   5.43 in. (4.8 d_b)

Transverse reinforcement meets requirements of UBC 1911.5.4 and 1911.5.5.3 for both top and bottom bars.

This gives a value of 1.0 for Factor I for the top and bottom splices. Thus,

Top splice length (Class A) = $1.0 \times (1.0 \times 1.3 \times 1.0 \times 1.0 \times 1.0 \times 38.0) = 49.4$ in.

Bottom splice length (Class B) = $1.3 \times (1.0 \times 1.0 \times 1.0 \times 1.0 \times 1.0 \times 38.0) = 49.4$ in.

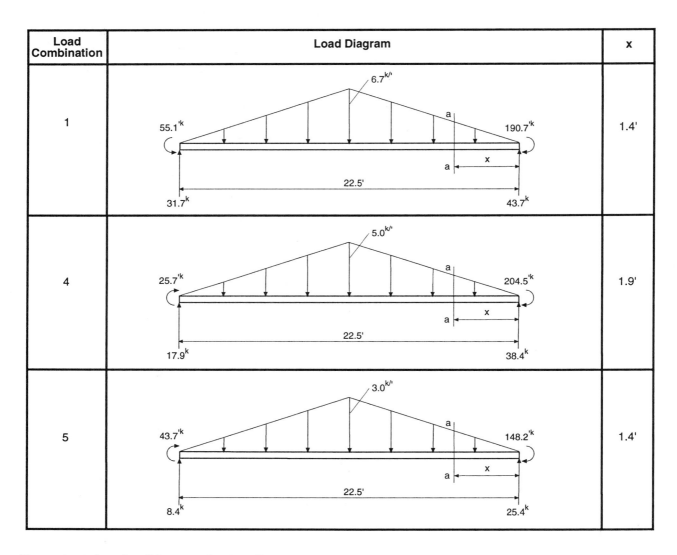

*Figure 3-51 Loading Diagrams for Cutoff Locations of Negative Bars at Interior Support of Longitudinal Beam
(Seismic Zone 0)*

Figure 3-52 shows the reinforcement details for the interior longitudinal beam.

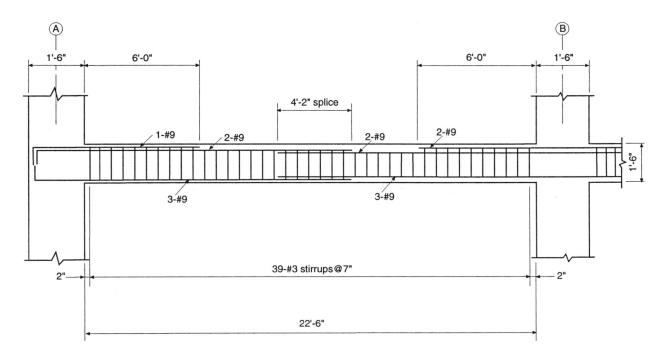

Figure 3-52 Reinforcement Details for Beam of Interior Longitudinal Frame (Seismic Zone 0)

3.5.2.3   Proportioning and Detailing a Column

3.5.2.3.1 General Requirements

This section addresses the design of an interior column of an exterior transverse frame, supporting the third floor of the structure. Tables 3-28 and 3-29 show the design load combinations for this column for both Seismic Zone 1 forces and wind forces. Examination of these results shows the axial loads and bending moments for the wind force combinations are not much different than those for Seismic Zone 1. Therefore, the details shown in Fig. 3-48 can be used here as well.

Chapter 4

Building Frame System

4.1 Introduction

4.1.1 General

The computation of the design loads for a 10-story reinforced concrete shearwall-frame building under the requirements of the Uniform Building Code (UBC), 1994 edition, is illustrated. These computations are performed for UBC Seismic Zones 4 and 2B. Typical members are designed and detailed for each of the seismic zones.

Buildings in high and moderate seismic zones are designed to carry only a portion of the elastic lateral forces that might result from the maximum ground motion, since designing for the full elastic forces would be uneconomical and is generally unnecessary. For this reason, special detailing is required for the structure to prevent collapse in the event that the design forces are exceeded. This special detailing will increase construction costs and time. The Building Frame System provides an alternative to special detailing for the entire structure. This is a structural system with an essentially complete frame providing support for gravity loads, with resistance to lateral loads being provided by shearwalls. Since beam-column frames are not part of the lateral load resisting system, they do not require special detailing. However, they must conform to the requirements in UBC 1921.7.

The Building Frame System and the Dual System (discussed in Chapter 5) are both shearwall-frame systems. However, in addition to the requirement that there be an essentially complete space frame providing support for gravity loads, the Dual System is required to have the following features (UBC 1627.6.5):

1) Resistance to lateral loads is provided by shearwalls and moment-resisting frames (Special Moment Resisting Frame or Intermediate Moment Resisting Frame). The moment-resisting frames must be designed to independently resist at least 25% of the design base shear.

2) The two systems must be designed to resist the total design base shear in proportion to their relative rigidities considering the interaction of the dual system at all levels.

In Seismic Zones 0 and 1, where concrete shearwall-frame systems would typically include ordinary moment frames of concrete, the concept of designing the moment-resisting frames to independently resist at least 25% of the design base shear loses its validity, in view of the limited inelastic deformability of the ordinary moment frame of concrete. Normal design practice in the lower seismic zones is to design the shearwalls and the frames to carry their share of gravity loads based on tributary areas, and to design these subsystems to also resist their share of the design lateral forces based on relative lateral rigidities. These shearwall-frame systems in lower seismic zones are thus neither Building Frame Systems, nor Dual Systems. This type of a structure, simply called the shearwall-frame system, is designed and detailed in Chapter 6.

In a Building Frame System, even though the shearwalls are designed to carry the total loads, both the shearwalls and beam-column frames form one structural system and will laterally deflect by virtually the same amount. The deflection of the beam-column system must be compatible with that of the shearwalls and any forces resulting from the need to ensure this compatibility must be addressed in the design of the components so that the beam-column frames do not collapse under large lateral seismic deflections.

4.1.2   Design Criteria

A typical plan and elevation of the structure considered are shown in Figs. 4-1 and 4-2. The member sizes are held constant throughout the building height. Other pertinent design data are as follows:

Service Loads:
    Live load: 50 psf
    Superimposed dead load: 32 psf

Material Properties:
    Concrete: f'_c = 4 ksi (Beams and Columns)

                           = 6 ksi (Shearwalls)

                w_c = 150 pcf

    Reinforcement: f_y = 60 ksi

4.2   Design for Seismic Zone 4

This section will present the design and detailing of members for Seismic Zone 4. Requirements for detailing in Seismic Zone 3 are identical to those for Zone 4. Also, the analysis for both zones is similar, except that Seismic Zone 4 design lateral forces are 33% higher than the corresponding Seismic Zone 3 forces. Therefore, the design of the structure in Seismic Zone 3 is not separately addressed here.

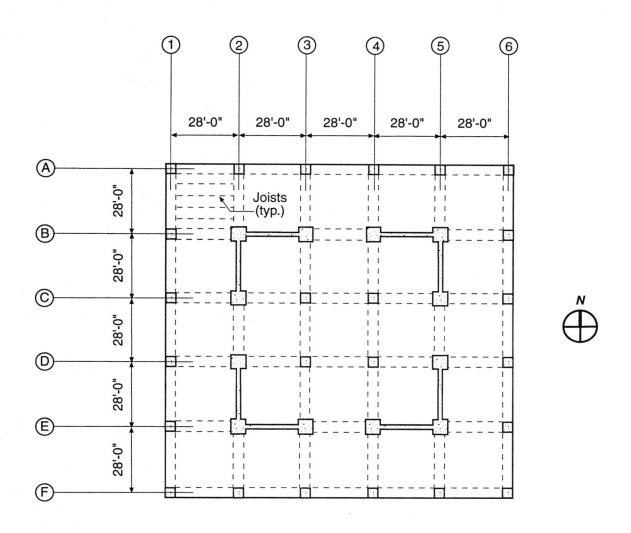

Beams: 20 x 22 in. (Seismic Zone 4)
20 x 20 in. (Seismic Zone 2B)

Columns: 20 x 20 in.

Shearwalls: boundary elements: 38 x 38 in. (Seismic Zone 4)
30 x 30 in. (Seismic Zone 2B)

wall thickness: 12 in. (Seismic Zone 4)
12 in. (Seismic Zone 2B)

Figure 4-1 Typical Floor Plan of Example Building

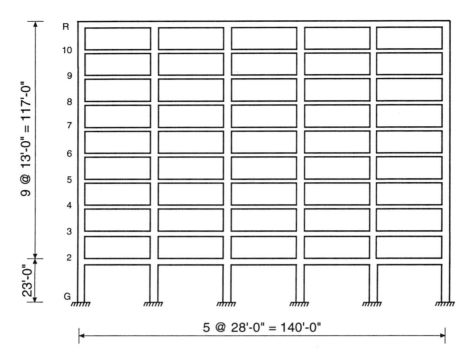

R
10
9
8
7
6
5
4
3
2
G

9 @ 13'-0" = 117'-0"

23'-0"

5 @ 28'-0" = 140'-0"

Figure 4-2 Elevation of Example Building

4.2.1   Frame Analysis

On the basis of the given data and the dimensions of the building, the weights of the floors are listed in Table 4-1. The building has the same lateral load resisting system in both principal directions. Therefore, the seismic forces will be the same in both the longitudinal and the transverse directions of the building.

The weight of a typical floor includes all elements located between two imaginary planes passing through the mid-height of the columns above and below the floor considered. The calculations for the base shear, V, for the transverse and longitudinal directions are shown at the bottom of Table 4-1. For this example, the importance factor, I, and the site soil factor, S, have been taken equal to 1.0 and 1.2, respectively. The value of R_w is equal to 8 in accordance with UBC Table 16-N. The seismic design forces resulting from the distribution of the base shear in accordance with UBC Eqs. (28-6), (28-7), and (28-8) are also listed in Table 4-1.

The member forces resulting from the seismic loads are determined in the following manner. A lateral load analysis is performed on the portion of the structure that is the lateral load resisting system. For this structure, the lateral load resisting system is composed of the four shearwall units and the four coupling beams framing into the walls. The shearwalls must also meet story drift limitations and be designed for P-Δ effects if applicable (UBC 1628.8). Table 4-2 shows the results from the structural analysis for lateral loads on the lateral load resisting system. The 3-D structural analysis was performed using *SAP90* [4.1]. The lateral loads were applied at a distance of 5% of the building dimension from the center of gravity to allow for accidental torsion; also, rigid diaphragms were specified at each floor level.

For structures having a fundamental period greater than 0.7 seconds, the story drift must not exceed 0.03/R_w nor 0.004 times the story height (UBC 1628.8.2).

Table 4-1 Seismic Forces for Zone 4

Floor Level	Weight, w_x (kips)	Height, h_x (ft)	$w_x h_x$ (ft-kips)	Lateral Force (kips)
R	3326	140	465,640	649.9
10	3691	127	468,757	472.6
9	3691	114	420,774	424.2
8	3691	101	372,791	375.8
7	3691	88	324,808	327.4
6	3691	75	276,825	279.1
5	3691	62	228,842	230.7
4	3691	49	180,859	182.3
3	3691	36	132,876	134.0
2	3972	23	91,356	92.1
Σ	36,826		2,963,528	3168.1

Building height $\quad h_n = 140$ ft
Building weight $\quad W = 36,826$ kips
Fundamental period $\quad T = 0.020 (h_n)^{3/4} = 0.814$ sec.
Seismic zone factor $\quad Z = 0.4$
Importance factor $\quad I = 1.0$
Response coefficient $\quad R_w = 8$
Soil factor $\quad S = 1.2$
Coefficient $\quad C = 1.25\, S/T^{2/3} = 1.72$
Base shear $\quad V = ZICW/R_w = 3168.1$ kips
Top level force $\quad F_t = 0.07TV = 181$ kips

For typical floors:

$$\frac{0.03h_s}{R_w} = \frac{0.03 \times 13 \times 12}{8} = 0.59 \text{ in. (governs)}$$

$$0.004h_s = 0.004 \times 13 \times 12 = 0.62 \text{ in.}$$

For the lower level:

$$\frac{0.03h_s}{R_w} = \frac{0.03 \times 23 \times 12}{8} = 1.04 \text{ in. (governs)}$$

$$0.004h_s = 0.004 \times 23 \times 12 = 1.10 \text{ in.}$$

As shown in Table 4-3, the story drift limits are not exceeded at any floor level.

In Seismic Zone 4, P-Δ effects need not be considered where the story drift does not exceed $0.02h_s/R_w$ (UBC 1628.9):

$$\frac{0.02h_s}{R_w} = \frac{0.02 \times 13 \times 12}{8} = 0.39 \text{ in.}$$

*Table 4-2 Results of Analysis of Lateral Load Resisting System Due to
Seismic Zone 4 Forces*

Shearwall D2-E2				
Level	Axial Load (kips)	Bending Moment (ft-kips)		Shear Force (kips)
		Top	Bottom	
R	12.5	456	1867	178.7
10	25.0	-1399	5367	305.2
9	37.6	-4909	10,341	417.9
8	49.9	-9887	16,636	519.2
7	61.8	-16,196	24,109	608.5
6	73.2	-23,691	32,602	685.3
5	83.3	-32,222	41,961	749.1
4	92.4	-41,624	52,018	799.7
3	99.8	-51,747	62,600	834.8
2	105.1	-62,404	82,024	853.1

Coupling Beam C2-D2	
Level	Bending Moment (ft-kips)
R	-184
10	-189
9	-186
8	-184
7	-177
6	-168
5	-154
4	-135
3	-109
2	-81

This value is greater than any of the interstory drifts in Table 4-3, so that P-Δ effects need not be considered.

*Table 4-3 Lateral Displacements and Drifts of the Lateral
Load Resisting System Due to Seismic Zone 4 Forces*

Floor Level	Total Lateral Displacement (in.)	Interstory Drift (in.)
R	1.11	0.13
10	0.98	0.13
9	0.85	0.13
8	0.72	0.13
7	0.59	0.12
6	0.47	0.12
5	0.35	0.11
4	0.24	0.09
3	0.15	0.09
2	0.07	0.07

All framing elements not required by design to be part of the lateral force resisting system must be investigated and shown to be adequate for vertical load carrying capacity when displaced $3R_w/8$ times the displacement resulting from the required lateral forces acting on the lateral force resisting system (UBC 1631.2.4). This will provide an added degree of safety for non-seismically detailed members of the framing system to assure that they can also withstand the possible deformations without losing vertical load-carrying capacity. P-Δ effects on these elements must also be taken into account.

The deformation compatibility analysis requires that the amplified lateral displacements $\Delta_i' = 3R_w\Delta_i/8$ be imposed at every floor level i on the non-lateral-load-resisting system, where Δ_i are the displacements of the lateral load resisting system (see Table 4-3). The analysis for these prescribed displacements was performed using *SAP90*. To account for cracking due to the large deformations, values of $I = 0.4\,I_g$ were used for the beams and $I = 0.8\,I_g$ were used for the columns. P-Δ effects were also included in the analysis; typically, the magnitude of the reactions increased only slightly (at most by approximately 1.5%). Table 4-4 shows the results for a column and beam accounting for deformation compatibility and P-Δ effects.

Table 4-4 Results of Analysis of Non-Lateral-Load-Resisting System Accounting for Deformation Compatibility and P-Δ Effects (Seismic Zone 4)

Column A3				
Level	Axial Load (kips)	Bending Moment (ft-kips)		Shear Force (kips)
		Top	Bottom	
R	4.2	61.7	-45.6	8.2
10	9.8	33.7	-37.2	5.4
9	15.0	38.7	-38.3	5.9
8	20.4	38.2	-40.8	6.0
7	25.6	33.6	-37.1	5.4
6	30.6	34.5	-39.4	5.6
5	35.3	28.2	-36.6	4.9
4	39.5	23.9	-35.1	4.5
3	43.1	16.0	-32.2	3.7
2	45.6	4.4	-28.5	1.4

Beam A3-B3 Bending Moment (ft-kips)		
Level	Exterior	Interior
R	-61.7	55.9
10	-79.3	76.0
9	-75.9	72.3
8	-76.4	72.9
7	-74.4	70.9
6	-71.6	68.2
5	-67.6	64.4
4	-60.5	57.7
3	-51.2	48.9
2	-36.6	34.3

4.2.2   Design of Structural Members

The objective is to determine the required shear and flexural reinforcement for a beam (A3-B3) and a column (A3) that are not part of the lateral load resisting system, and a coupling beam (C2-D2) and shearwall (D2-E2) that are part of the lateral load resisting system (note that segment E2-E3 of the shearwall resists a very small portion of the seismic loads acting in the N-S direction and thus will be neglected in this direction).

The gravity loads acting on beam A3-B3 are as follows:

Tributary area $= 784 \text{ ft}^2$

Dead loads:

Beams and slab	= 115.5 psf
Superimposed	= 32 psf
Total	= 147.5 psf
Load per foot	= 147.5 × 784/28 = 4130 plf

Live load:

Reduction factor $= r(A - 150) = 0.08 \times (784 - 150) = 50.7\% > 40\%$ *1606*
Load per foot $= (1.0 - 0.4) \times 784 \times 50/28 = 840$ plf

The factored load due to gravity on this beam is:

$$w_u \quad = (1.4 \times 4.13) + (1.7 \times 0.84) = 7.21 \text{ kips/ft}$$

The interior end of this beam frames into a shearwall. This end of the beam is considered fixed, with a bending moment equal to $w_u \ell_n^2/12$. The bending moment at the exterior end of the beam is taken as $w_u \ell_n^2/16$; the midspan moment then equals $w_u \ell_n^2/19$.

The factored moments caused by gravity loads are:

Exterior negative moment $= w_u \ell_n^2/16 = 7.21 \times (28 - 20/24 - 38/24)^2/16 = 294.9$ ft-kips

Positive moment $= w_u \ell_n^2/19 = 7.21 \times (28 - 20/24 - 38/24)^2/19 = 248.4$ ft-kips

Interior negative moment $= w_u \ell_n^2/12 = 7.21 \times (28 - 20/24 - 38/24)^2/12 = 393.3$ ft-kips

Table 4-5 shows the bending moments for different combinations of gravity and compatibility effects for beam A3-B3 on the fifth floor of the frame. The gravity moments are obtained from the calculations above and the compatibility moments are taken from Table 4-4. Note that the compatibility moments represent an extreme probability design condition so that load factors are not required in this case.

Beam C2-D2 is part of the lateral load resisting system. Since the beam frames into shearwalls, both ends of the beam are considered fixed, with a moment of $w_u \ell_n^2/12$ at each beam end. Note that the gravity loads for this beam are the same as for Beam A3-B3.

Table 4-5 Summary of Factored Bending Moments for Beam A3-B3 on
Level 5 (Seismic Zone 4)

(1) U = 1.4D + 1.7L   Eq. (9-1), UBC 1909.2.1
(2) U = (D + L ± E)   Compatibility (No load factors)

| Load | Bending Moment (ft-kips) | | |
Combination	Exterior Negative	Positive	Interior Negative
1	-294.9	248.4	-393.3
2: sidesway right	-135.8	172.8	-335.5
sidesway left	-271.0	169.6	-206.7

The factored moments caused by gravity loads are:

Exterior negative moment $= w_u \ell_n^2/12 = 7.21 \times (28 - 38/12)^2 /12 = 370.5$ ft-kips

Positive moment $= w_u \ell_n^2/24 = 7.21 \times (28 - 38/12)^2 /24 = 185.3$ ft-kips

Interior negative moment $= w_u \ell_n^2/12 = 7.21 \times (28 - 38/12)^2 /12 = 370.5$ ft-kips

Table 4-6 shows the bending moments for different combinations of gravity and seismic loads for this beam. The gravity moments are obtained from the calculations above and the seismic moments are taken from Table 4-2.

Table 4-6 Summary of Factored Bending Moments for Beam C2-D2 on
Level 5 (Seismic Zone 4)

(1) U = 1.4D + 1.7L   Eq. (9-1), UBC 1909.2.1
(2) U = 1.4(D + L ± E)   Eq. (9-2), UBC 1921.2.7
(3) U = 0.9D ± 1.4E   Eq. (9-3), UBC 1921.2.7

| Load | Bending Moment (ft-kips) | | |
Combination	Exterior Negative	Positive	Interior Negative
1	-370.5	185.3	-370.5
2: sidesway right	-142.1	178.8	-573.3
sidesway left	-573.3	178.8	-142.1
3: sidesway right	24.5	95.5	-406.7
sidesway left	-406.7	95.5	24.5

Column A3 is not part of the lateral load resisting system. The gravity loads on this column are as follows:

Tributary area $= 415$ ft^2

Dead load $= 147.5$ psf (plus the weight of the column)

Live load $= 50$ psf

The different combinations of axial forces and bending moments corresponding to the appropriate load combinations are listed in Table 4-7. The values for gravity load effects are based on the information above (including live load reduction); the values for compatibility effects are taken from Table 4-4.

Table 4-7 Summary of Factored Axial Loads and Bending Moments for Column A3 between Levels 4 and 5 (Seismic Zone 4)

(1) U = 1.4D + 1.7L   Eq. (9-1), UBC 1909.2.1
(2) U = (D + L ± E)   Compatibility (No load factors)

Load Combination	Axial Load (kips)	Bending Moment (ft-kips)	
		Top	Bottom
1	767.2	-147.5	147.5
2: sidesway right	498.3	-73.5	65.2
sidesway left	568.9	-129.9	138.3

Shearwall D2-E2 is part of the lateral load resisting system. This shearwall is 28 ft long from centerline to centerline of the boundary elements. The boundary elements are 38 in. square and the wall thickness is 12 in. The gravity loads for this shearwall are as follows:

Tributary area = 1568 ft^2

Dead load   = 147.5 psf (plus the weight of the wall)

Live load   = 50 psf

The different combinations of axial loads and bending moments corresponding to the appropriate load combinations are listed in Table 4-8. The reactions due to the gravity loads are determined from the information provided above (including live load reduction), and the reactions due to the seismic loads are taken from Table 4-2.

4.2.2.1   Proportioning and Detailing Beam A3-B3

4.2.2.1.1 Required Flexural Reinforcement

The maximum negative moment for this beam at the interior support is 393.3 ft-kips (see Table 4-5). The required reinforcement, ignoring the effects of any compression reinforcement, is $A_s = 5.04$ in.^{2} Use 7-#8 bars at this location ($\phi M_n = 425.9$ ft-kips).

For frame members that are not part of the lateral load resisting system and whose compatibility moments do not exceed the design moment strength of the member, the longitudinal reinforcement must conform to the provisions in UBC 1921.3.2.1 (UBC 1921.7).

Table 4-8 Summary of Factored Axial Loads and Bending Moments for Shearwall D2-E2 between Ground and Level 2 (Seismic Zone 4)

(1) U = 1.4D + 1.7L Eq. (9-1), UBC 1909.2.1
(2) U = 1.4(D + L ± E) Eq. (9-2), UBC 1921.2.7
(3) U = 0.9D ± 1.4E Eq. (9-3), UBC 1921.2.7

| Load Combination | Shearwall D2-E2 | | | Boundary Element |
| | Axial Load (kips) | Bending Moment (ft-kips) | | Axial Load (kips) |
		Top	Bottom	
1	5063	—	—	2532
2: sidesway right	4817	-87,366	114,834	-1693
sidesway left	5112	87,366	-114,834	6657
3: sidesway right	2748	-87,366	114,834	-2727
sidesway left	3042	87,366	-114,834	5622

Check limitations on the reinforcement ratio:

$$\rho = A_s/bd = 5.53/(20 \times 19.56) = 0.0141$$

$$\rho_{max} = 0.0250 > 0.0141 > 200/f_y = 0.0033 \quad \text{O.K.}$$ *1921.3.2.1*

The maximum negative moment at the exterior support is 294.9 ft-kips. Use 5-#8 bars at this location (ϕM_n = 316.6 ft-kips).

The maximum positive moment is 248.4 ft-kips. Use 4-#8 bars, providing a moment strength of 258.3 ft-kips.

It can be seen from Table 4-5 that no positive moments occur at either the interior or exterior support. However, at least two bars shall be provided continuously at both the top and bottom of the section (UBC 1921.3.2.1).

4.2.2.1.2 Required Shear Reinforcement

The shear forces caused by gravity loads and compatibility requirements at a distance d from the face of the first interior support are calculated below:

V_D @ support $= 1.15 \, w_D \, \ell_n/2$

$= 1.15 \times 4.13 \times (28 - 20/24 - 38/24)/2 = 60.8$ kips

V_D @ d $= 60.8 - (4.13 \times 19.56/12) = 54.1$ kips

V_L @ d $= 54.1 \times 0.84/4.13 = 11.0$ kips

$V_{compatibility}$ (see Table 4-4) $= (67.6 + 64.4)/(28 - 20/24 - 38/24) = 5.2$ kips

Load combinations:

(1) U $= 1.4D + 1.7L$

$$V_u = (1.4 \times 54.1) + (1.7 \times 11.0) = 94.4 \text{ kips (governs)}$$

(2) $U = D + L + \text{compatibility}$

$$V_u = 54.1 + 11.0 + 5.2 = 70.3 \text{ kips}$$

The shear force carried by the concrete is:

$$V_c = 2\sqrt{f'_c} \times b \times d \hfill 1911.3.1.1$$

$$= 2\sqrt{4000} \times 20 \times 19.56/1000 = 49.5 \text{ kips}$$

The shear force carried by the stirrups at the support is:

$$V_s = V_u / \phi - V_c$$

$$= (94.4/0.85) - 49.5 = 61.6 \text{ kips} \hfill 1911.1.1$$

$$< 4\sqrt{f'_c} \times bd = 99.0 \text{ kips} \hfill 1911.5.4.3$$

Using #4 stirrups, the required spacing near the interior support is:

$$s = \frac{A_v f_y d}{V_s} = \frac{0.40 \times 60 \times 19.56}{61.6} = 7.6 \text{ in.} \hfill 1911.5.6.2$$

$$= 0.40 \times 60 \times 19.56/61.6 = 7.6 \text{ in.}$$

The maximum allowable stirrup spacing s_{max} is:

$$s_{max} = A_s \times f_y/(50 \times b_w) = 0.40 \times 60,000/(50 \times 20) = 24.0 \text{ in.} \hfill 1911.5.5.3$$

$$= \frac{d}{2} = \frac{19.56}{2} = 9.8 \text{ in. (governs)} \hfill 1911.5.4.1$$

$$= 24 \text{ in.}$$

Therefore, use #4 stirrups at 7 in. Stirrups are no longer required when $V_u \le \phi V_c/2 = 0.85 \times 49.5/2 = 21.0$ kips. Consequently, the length over which stirrups are required is (94.4 - 21.0)/7.21 = 10.2 ft. However, stirrups will be provided over the full length of the beam.

4.2.2.1.3 Reinforcing Bar Cutoff Points and Splices

The negative reinforcement at the interior support is 7-#8 bars. The location where five of the seven bars can be terminated will be determined. The loading used to find the cutoff point is 0.9 times the dead load in combination with the end moments from the compatibility load combination (see Table 4-5).

The design moment strength of the section with 2-#8 top bars is 134.1 ft-kips. The distance from the face of the interior support to where the moment is reduced to 134.1 ft-kips must be determined. Based on a moment of 335.5 ft-kips at the interior support and 135.8 ft-kips at the exterior end and a factored dead load of $0.9 \times 4.13 = 3.72$ kips/ft, the location where the bars can be cut off is obtained by summing moments about section a-a in Fig. 4-3:

$$\frac{3.72x^2}{2} + 335.5 - 55.4x = 134.1$$

Solving this equation gives x = 4.2 ft. Five of the 7-#8 top bars at the interior support can be discontinued at (x + d) = (4.2 + 19.56/12) = 5.83 ft (UBC 1912.10.3). By similar calculations, three of the 5-#8 bars at the exterior support can be discontinued at 5.0 ft away from that support.

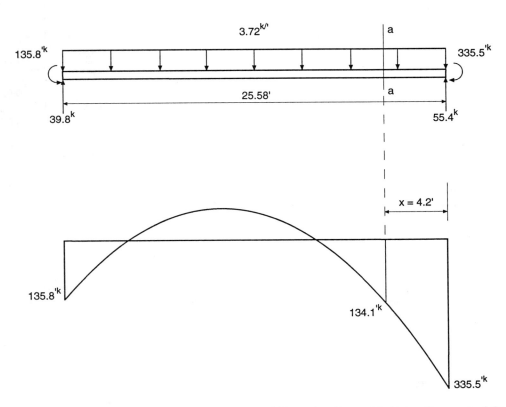

Figure 4-3 Moment Diagram for Cutoff Location of Negative Bars at Interior Support of Beam A3-B3
(Seismic Zone 4)

Reinforcing bars must be properly developed at the supports. The development length is the product of the basic development length and the factors presented in Table 4-9. The basic development length for a #8 bar is (UBC 1912.2.2):

$$\ell_{db} = 0.04 \times A_b \times f_y / \sqrt{f'_c}$$

$$= 0.04 \times 0.79 \times 60{,}000 / \sqrt{4000} = 30.0 \text{ in.}$$

Table 4-9 Development Length Factors

Factor I

Criteria	Cover	Transverse Reinforcement Requirements	Clear Bar Spacing	Confinement provided UBC 1912.2.3.5	Applicable Factor
1	$\geq$ 1.5 in. clear interior or $\geq$ 2.0 in. clear exterior exposure	Meeting UBC 1907.10.5 or UBC 1911.5.4 and 1911.5.5.3	$\geq$ 3d_b	YES	0.75
				NO	1.00
			$\geq$ 5d_b & side cover $\geq$ 2.5 d_b	YES	0.60
				NO	0.80
2	$\geq$ 1.5 in. clear interior or $\geq$ 2.0 in. clear exterior exposure	$A_{tr} \geq \dfrac{d_b sN}{40}$	—	YES	0.75
				NO	1.00
			$\geq$ 5d_b & side cover $\geq$ 2.5 d_b	YES	0.60
				NO	0.80
3	$\geq 2\,d_b$	—	$\geq$ 3d_b	YES	0.75
				NO	1.00
			$\geq$ 5d_b & side cover $\geq$ 2.5 d_b	YES	0.60
				NO	0.80
4	Either $\leq d_b$	—	Or $< 2d_b$	YES	1.50
				NO	2.00
5	All other conditions		—	YES	1.05
				NO	1.40
			$\geq$ 5d_b & side cover $\geq$ 2.5 d_b	YES	0.84
				NO	1.12

Bar Size	#4	#5	#6	#7	#8	#9	#10	#11
Minimum Value of Factor I*	1.88	1.51	1.28	1.09	0.95	0.85	0.75	0.68

* Based on 0.03 $d_b\, f_y/\sqrt{f'_c}$

Factor II

Bar with more than 12 in. of fresh concrete cast in one lift below it	1.3
All others	1.0

Factor III

Bars Embedded in Lightweight Concrete	1.3 or 6.7 $\sqrt{f'_c}/f_{ct}$ but > 1.0
Bars Embedded in Normal Weight Concrete	1.0

Factor IV

Epoxy Coated Bars	Cover < 3d_b	1.5
	Clear Spacing < 6d_b	1.5
	All other	1.2
Uncoated Bars		1.0

Factor V

$$= \frac{A_s \text{ (required) *}}{A_s \text{ (provided)}}$$

* This factor cannot be used in Seismic Zones 3 & 4

ℓ_d = (Factor I) $\times$ (Factor II) $\times$ (Factor III) $\times$ (Factor IV) $\times$ (Factor V) $\times$ ℓ_{db}

The information required for determining the development length factor is:

Minimum clear cover   = 2.0 in. ($2.0d_b$)

Side cover   large (slab)

Clear bar spacing   = 1.5 in. ($1.5 d_b$)

Transverse reinforcement meets requirements of UBC 1911.5.4 and 1911.5.5.3.

Factor I has a value of 2.0. Factor II has a value of 1.3 for the top bars. All other factors are 1.0; thus,

Top bar $\ell_d = 1.3 \times 2.0 \times 1.0 \times 1.0 \times 1.0 \times 30.0/12 = 6.5$ ft > 5.0 ft

In continuous members, at least one-fourth of the positive reinforcement must extend into the support (UBC 1912.11.1). Since four bottom reinforcing bars are required at midspan and two bars are provided at both supports, this criterion is satisfied.

Cutoff points for two of the four positive reinforcing bars at midspan will be determined. The load combination used in this case is 1.4D + 1.7L. The moment at the exterior support is 294.9 ft-kips, and at the interior support it is 393.3 ft-kips; the distributed load is 7.21 kips/ft. The capacity of 2-#8 bars with the beam section being designed is 134.1 ft-kips. The location where the bars can be cut off is obtained by summing moments about section a-a in Fig. 4-4:

$$\frac{7.21x^2}{2} + 393.3 - 96.1x = -134.1$$

Solution of this equation gives x = 7.7 ft. Two of the 4-#8 bars can be terminated at [(25.58/2) - 7.7 + (19.56/12)] = 6.7 ft > 6.5 ft. To avoid cutting the bars off in a tension zone, extend them to the location of the inflection point which is about 5 ft from the face of the support (see Fig. 4-4).

The information required for determining the splice lengths is:

	Top Bars	Bottom Bars
Stress level	$< 0.5 f_y$ (midspan)	$< 0.5 f_y$ (support)
Splice type	A	A
Minimum clear cover	2.0 in. ($2.0 d_b$)	2.0 in. ($2.0 d_b$)
Side Cover	large (slab)	2.0 in. ($2.0 d_b$)
Clear bar spacing	14.0 in. ($14.0 d_b$)	14.0 in. ($14.0 d_b$)

Transverse reinforcement meets requirements of UBC 1911.5.4 and 1911.5.5.3 for both top and bottom bars.

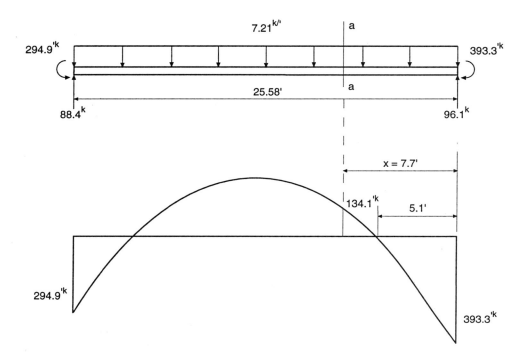

Figure 4-4 Moment Diagram for Cutoff Location of Positive Bars at Interior Support of Beam A3-B3 (Seismic Zone 4)

This gives a value of 1.0 for the bottom bars and 0.95 for the top bars. The splice lengths are:

Top splice length (Class A) = 1.0 × (0.95 × 1.3 × 1.0 × 1.0 × 1.0 × 30.0) = 37.0 in.

Bottom splice length (Class A) = 1.0 × (1.0 × 1.0 × 1.0 × 1.0 × 30.0) = 30.0 in.

Figure 4-5 shows the reinforcement details for beam A3-B3.

4.2.2.2   Proportioning and Detailing Beam C2-D2

Members are to be designed as flexural elements if they meet the following requirements (UBC 1921.3.1):

- Factored axial compressive force on the member does not exceed $A_g f'_c/10$.

- Clear span of the member is not less than four times its effective depth.

- Width-to-depth ratio of the member is not less than 0.3.

- Width of the member is not (1) less than 10 in. and (2) more than the width of the supporting member (measured on a plane perpendicular to the longitudinal axis of the flexural member) plus distances on either side of the supporting member not exceeding three-fourths of the depth of the flexural member.

Since the beams under consideration have no axial forces, the first requirement is satisfied. A check of the other three requirements is shown below.

$$\text{Clear span} = 24.83 \text{ ft} > 4d = \frac{4 \times 19.56}{12} = 6.5 \text{ ft} \quad \text{O.K.}$$

$$\frac{\text{Width}}{\text{Depth}} = \frac{20 \text{ in.}}{22 \text{ in.}} = 0.91 > 0.30 \quad \text{O.K.}$$

The beam width is 20 in. which is greater than 10 in. Also, the beam width is less than the width of the supporting column + (1.5 × depth of beam) = 38 + (1.5 × 22) = 71 in.   O.K.

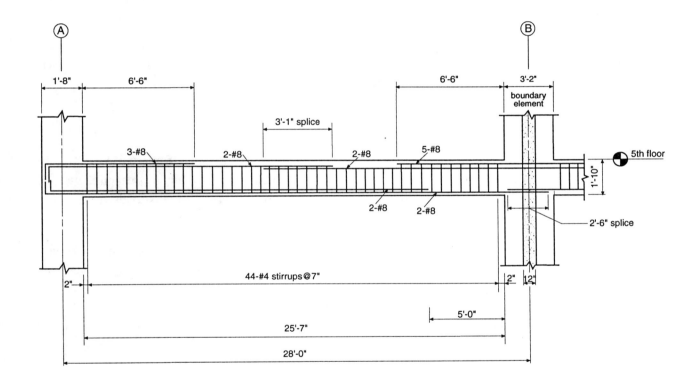

Figure 4-5 Reinforcement Details for Beam A3-B3 (Seismic Zone 4)

4.2.2.2.1 Required Flexural Reinforcement

The maximum moment at both ends of the beam is 573.3 ft-kips (see Table 4-6). The required amount of reinforcement, ignoring the effects of any compressive reinforcement, is $A_s = 7.93$ in.[2] Use 8-#9 bars at these locations ($\phi M_n = 576.7$ ft-kips).

Check limitations on the reinforcement ratio:

$$\rho \quad = A_s/bd = 8.0/(20 \times 19.56) = 0.0204$$

$$\rho_{max} = 0.0250 > 0.0204 > 200/f_y = 0.0030 \quad \text{O.K.} \hspace{2cm} 1921.3.2.1$$

The positive moment strength at a joint face must be equal to at least 50% of the corresponding negative moment strength at that joint (UBC 1921.3.2.2). The positive moment reinforcement must not give a moment strength less than 576.7/2 = 288.4 ft-kips. The maximum positive moment in the beam at the supports is 24.5 ft-kips. This moment is obviously less than the 50% requirement. Therefore, the required number of reinforcing bars is 4-#9, giving a design moment strength of 320.2 ft-kips.

The maximum positive moment at midspan is 185.3 ft-kips. The positive moment strength at any section along the member length must not be less than 25% of the maximum moment strength provided at the face of either joint. Thus, the minimum design value of the positive moment is 576.7/4 = 144.2 ft-kips which is less than 185.3 ft-kips. This later moment requires an area of steel equal to 2.22 in.^{2}; therefore, use 3-#9 bars, giving a design moment strength of ϕM_n = 246.1 ft-kips.

The negative moment strength at any section along the member length must not be less than 25% of the maximum moment strength provided at the face of either joint (576.7/4 = 144.2 ft-kips). This minimum requirement will control, and 2-#9 bars will be used (ϕM_n = 168.1 ft-kips).

4.2.2.2.2 Required Shear Reinforcement

The beam is to be designed for shear forces according to UBC 1921.3.4. These shear forces are based on the probable moment strength of each joint face using a strength reduction factor of 1.0 and a stress in the tensile flexural reinforcement equal to $1.25f_y$, combined with the shear caused by unfactored gravity loads. It should be noted that ACI 318-89 (Revised 1992) uses the same procedure, but with factored gravity loads.

Figure 4-6 shows the probable flexural strengths M_{pr} at the joint faces for sidesway to the right and to the left. The unfactored gravity loads and the shear forces at the faces of the boundary elements are shown as well.

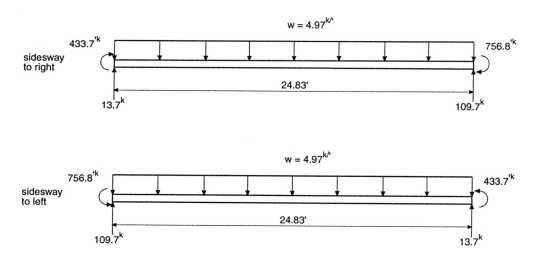

Figure 4-6 Design Shear Forces for Beam C2-D2 (Seismic Zone 4)

The shear strength of the concrete is not considered in Seismic Zones 3 and 4 if the factored axial compressive force, including earthquake effects, is less than $A_g f'_c/20$ and if the earthquake-induced shear force represents one-half or more of the total design shear (UBC 1921.3.4.2).

The shear due to seismic forces is:

$$V_E = (756.8 + 433.7)/24.83 = 48.0 \text{ kips}$$

$$2 \times 48.0 = 96.0 \text{ kips} < 109.7 \text{ kips}$$

Therefore, at the faces of the supports, the earthquake-induced shear force is less than one-half of the total shear. However, away from the support faces, the earthquake forces soon become more than one-half of the total design shear, so that V_c is taken equal to zero in this design.

The maximum shear force that must be designed for is:

$$\phi V_s = V_u - \phi V_c$$

$$V_s = (109.7/0.85) - 0 = 129.1 \text{ kips}$$

According to UBC 1911.5.6.8, the shear strength contributed by shear reinforcement cannot be taken greater than $8\sqrt{f_c'} \times b_w d$:

$$V_{s,max} = 8\sqrt{4000} \times (20 \times 19.56)/1000$$

$$= 197.9 \text{ kips} > 129.1 \text{ kips} \quad \text{O.K.}$$

In Seismic Zones 3 and 4, the following detailing requirements must be satisfied in addition to the shear design requirements. Hoops shall be provided:

- Over a length equal to twice the member depth, starting from the face of the supporting member toward midspan, at both ends of the flexural member (UBC 1921.3.3.1).

- Over lengths equal to twice the member depth where flexural yielding may occur in connection with inelastic lateral displacements of the frame (UBC 1921.3.3.1).

Where hoops are not required, stirrups with 135-degree seismic hooks shall be spaced at no more than d/2 throughout the length of the member (UBC 1921.3.3.4).

The maximum allowable hoop spacing s_{max} within a distance 2h = 44 in. from the face of the supports is (UBC 1921.3.3.2):

$$s_{max} = \frac{d}{4} = \frac{19.56}{4} = 4.9 \text{ in. (governs)}$$

$$= 8 \times (\text{diameter of smallest longitudinal bar}) = 8 \times 1.128 = 9.0 \text{ in.}$$

$$= 24 \times (\text{diameter of hoop bars}) = 24 \times 0.5 = 12.0 \text{ in.}$$

$$= 12.0 \text{ in.}$$

The required spacing of #4 closed stirrups (hoops) for the factored shear of 129.1 kips is:

$$s = \frac{A_v f_y d}{V_s} = \frac{0.40 \times 60 \times 19.56}{129.1} = 3.6 \text{ in.}$$

Therefore, hoops should be spaced at 3.5 in. on center with the first being located 2 in. from the face of the support. Thirteen stirrups are to be placed at this spacing. At a distance of 44 in. from the face of the support, the maximum factored shear is:

$$V_u = 109.7 - (4.97 \times 44/12) = 91.5 \text{ kips}$$

$$V_s = 91.5/0.85 = 107.6 \text{ kips} > 4 \times \sqrt{4000} \times (20 \times 19.56)/1000 = 99.0 \text{ kips}$$

The required spacing of #4 stirrups at this location is:

$$s = \frac{0.40 \times 60 \times 19.56}{107.6} = 4.4 \text{ in.} < \frac{d}{4} = \frac{19.56}{4} = 4.9 \text{ in.} \qquad \textit{1911.5.4.3}$$

A 4 in. spacing, starting at 44 in. from the faces of each support to the midspan region of the beam, will be sufficient. At this time, the designer may consider increasing the beam depth to increase the overall shear capacity. This, of course, would require reanalysis of the structure, which is not done here.

4.2.2.2.3 Reinforcing Bar Cutoff Points and Splices

The negative reinforcement at the supports is 8-#9 bars. The location where six of the eight can be terminated will be determined. The loading used to find the cutoff points is 0.9 times the dead load in combination with the probable moment strengths, M_{pr}, at the member ends (using $f_s = 1.25\, f_y$). The design bending moment strength, ϕM_n, provided by 2-#9 reinforcing bars in the beam is 168.1 ft-kips. Therefore, the six reinforcing bars can only be terminated after the required moment strength, M_u, has been reduced to 168.1 ft-kips.

With $\phi = 1.0$ and $f_s = 1.25\, f_y = 75$ ksi, $M^+_{pr} = 433.7$ ft-kips, and $M^-_{pr} = 756.8$ ft-kips. The factored dead load is equal to $0.9 \times 4.13 = 3.72$ kips/ft. The distance from the face of the support to where the moment under the loading considered equals 168.1 ft-kips is readily obtained by summing moments about section a-a in Fig. 4-7:

$$\frac{3.72x^2}{2} + 756.8 - 94.1x = 168.1$$

Solving for x gives a distance of 7.3 ft. Therefore, the negative reinforcing bars can be terminated at a distance of $7.3 + d = 8.9$ ft from the face of either support (noting that $d = 19.56$ in. $> 12\, d_b = 12 \times 1.128 = 13.5$ in.).

The #9 bars being terminated must be properly developed at the support. In Seismic Zone 4, the development length of tensile bars for joints of frames, if the depth of concrete cast in one lift below the bars exceeds 12 in., is given by:

$$\ell_d = 3.5 f_y d_b / 65 \sqrt{f'_c} \qquad\qquad\qquad\qquad \textit{1921.5.4.2}$$

$$= 3.5 \times 60,000 \times 1.128/(65 \sqrt{4000})$$

$$= 57.6 \text{ in.} < (8.9 \times 12) = 106.8 \text{ in.} \quad \text{O.K.}$$

Given the length of the beam and its position between the boundary elements of adjacent shearwalls, splicing of the flexural reinforcement is not required. Figure 4-8 shows the reinforcement details for beam C2-D2.

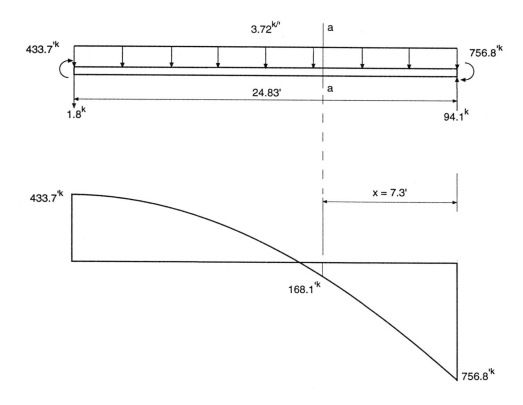

Figure 4-7 *Moment Diagram for Cutoff Location of Negative Bars at Support of Beam C2-D2 (Seismic Zone 4)*

4.2.2.3   Proportioning and Detailing Column A3

4.2.2.3.1 General Requirements

This section will detail the design of exterior column A3 supporting the fifth floor of the structure. The maximum factored axial force on this column at this level is 767.2 kips (see Table 4-7). A 20 in. square tied column with 8-#7 bars will be adequate for the load combinations shown in Table 4-7. An interaction diagram for this column is shown in Fig. 4-9. The reinforcement ratio for this column is 1.2%, which is between 1% and 8%. Note that slenderness effects need not be considered since P-Δ effects were included in the analysis.

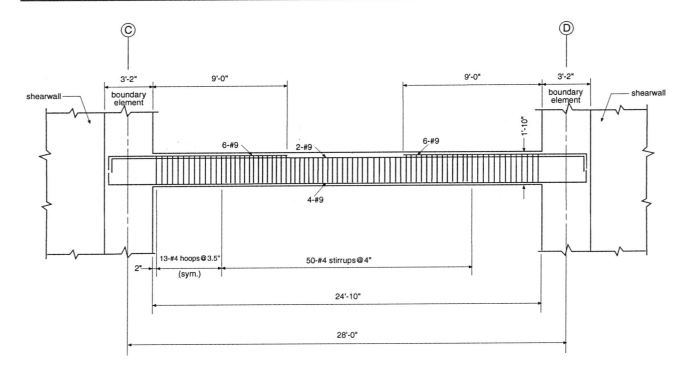

Figure 4-8 Reinforcement Details for Beam C2-D2 (Seismic Zone 4)

4.2.2.3.2 Confinement Reinforcement

Special transverse reinforcement for confinement is required over a distance ℓ_o at the column ends where ℓ_o equals the maximum of:

$$\frac{1}{6}\text{ (clear height)} = \frac{1}{6}(11.17 \times 12) = 22.3 \text{ in. (governs)} \hspace{2cm} \textit{1921.7.2.2}$$

Depth of member = 20.0 in.

18.0 in.

The maximum allowable spacing of #3 ties is:

$$s_o \quad = 8 \times \text{(diameter of smallest longitudinal bar)} = 8 \times 0.875 = 7.0 \text{ in. (governs)} \qquad \textit{1921.7.2.2}$$

$$= 24 \times \text{(diameter of tie bars)} = 24 \times 0.375 = 9.0 \text{ in.}$$

$$= \frac{h}{2} = \frac{20}{2} = 10.0 \text{ in.}$$

The maximum spacing of ties outside of the distance ℓ_o is $2s_o = 14.0$ in. (UBC 1921.7.2.4).

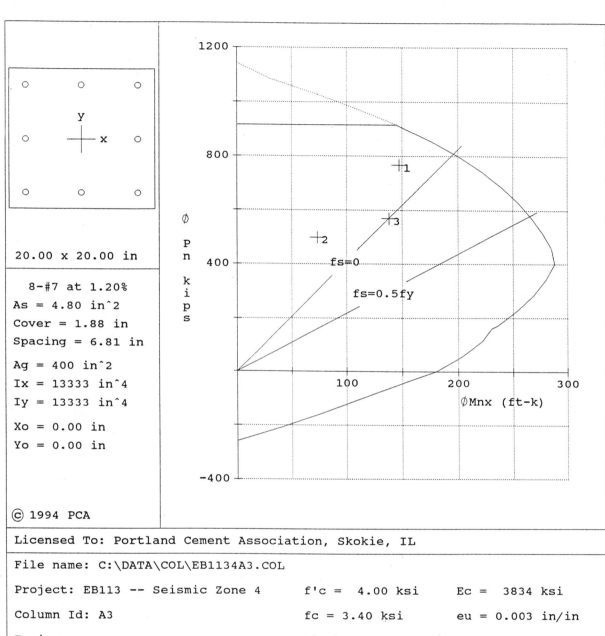

20.00 x 20.00 in

8-#7 at 1.20%
As = 4.80 in^2
Cover = 1.88 in
Spacing = 6.81 in

Ag = 400 in^2
Ix = 13333 in^4
Iy = 13333 in^4

Xo = 0.00 in
Yo = 0.00 in

© 1994 PCA

File name: C:\DATA\COL\EB1134A3.COL

Project: EB113 -- Seismic Zone 4   f'c = 4.00 ksi   Ec = 3834 ksi

Column Id: A3   fc = 3.40 ksi   eu = 0.003 in/in

Engineer: Block stress profile, Beta1 = 0.85

Date: Time: fy = 60 ksi   Es = 29000 ksi

Code: ACI 318-89   Confinement: Tied

Units: English   phi(c) = 0.70, phi(b) = 0.90, a = 0.80

X-axis slenderness is not considered.

Figure 4-9 Design Strength Interaction Diagram for Column A3, 5th Floor Level (Seismic Zone 4)

4.2.2.3.3 Transverse Reinforcement for Shear

The column shear caused by gravity loads is determined by taking the difference in beam moments on each side of the column and dividing by the clear column height. An exterior column will have a moment on one side only. The beam moment at the column face is equal to $w_u \ell_n^2/16$ (see Section 4.2.2). Therefore, the unfactored column shear caused by the dead load is:

$$V_D = \frac{w_D \ell_n^2}{16\ell_c}$$

$$= \frac{4.13 \times (28 - 20/24 - 38/24)^2}{16 \times (13 - 22/12)} = 15.1 \text{ kips}$$

The unfactored column shear caused by live loads is:

$$V_L = \frac{w_L \ell_n^2}{16\ell_c}$$

$$= \frac{0.84 \times (28 - 20/24 - 38/24)^2}{16 \times (13 - 22/12)} = 3.1 \text{ kips}$$

From Table 4-4, the compatibility shear is:

$$V_{\text{compatibility}} = 4.9 \text{ kips}$$

The load combinations for the column are:

$$V_u = 1.4V_D + 1.7V_L = 26.4 \text{ kips (governs)}$$

$$V_u = V_D + V_L + V_{\text{compatibility}} = 23.1 \text{ kips}$$

For members subject to axial compression, the shear strength provided by the concrete is:

$$V_c = 2\left(1 + \frac{N_u}{2000A_g}\right)\sqrt{f'_c}\, b_w d \qquad\qquad 1911.3.1.2$$

From Table 4-7, the minimum factored axial load on the column is $N_u = 498.3$ kips. Therefore,

$$\phi V_c = 0.85 \times 2\left(1 + \frac{498,300}{2000 \times 400}\right)\sqrt{4000} \times 20 \times 17.69/1000$$

$$= 61.7 \text{ kips} > 2V_u = 2 \times 26.4 = 52.8 \text{ kips}$$

Since $\phi V_c = 61.7$ kips is greater than $2V_u = 52.8$ kips, shear reinforcement is not required.

Therefore, a 14.0 in. spacing is sufficient in the region of the column outside of ℓ_o.

Since column A3 is not part of the lateral load resisting system, the splicing requirements in UBC 1912.17 govern. It is clear from Fig. 4-9 that none of the bars has a tensile stress greater than 0.5 f_y for any of the given load combinations. However, a Class B splice is required since all of the bars will be spliced just above the floor level (UBC 1912.17.2.2). The basic development length is:

$$\ell_{db} = \frac{0.04 A_b f_y}{\sqrt{f'_c}} = \frac{0.04 \times 0.6 \times 60,000}{\sqrt{4000}} = 22.8 \text{ in.} \qquad \textit{1912.2.2}$$

The information required for determining the development length is:

Minimum clear cover   = 1.88 in. (2.15 d_b)

Side cover   = 1.88 in. (2.15 d_b)

Clear bar spacing   = 6.81 in. (7.79 d_b)

Transverse reinforcement meets requirements of UBC 1907.10.5.

This gives a value of 1.09 for Factor I (see Table 4-9). The other factors are all equal to 1.0, giving:

Class B splice length = $1.3 \times (1.09 \times 1.0 \times 1.0 \times 1.0 \times 1.0 \times 22.8) = 32.3$ in.

Figure 4-10 shows the reinforcement details for column A3.

4.2.2.4 Proportioning and Detailing a Shearwall

A significant change in the design of shearwalls occurred with the adoption of the 1994 edition of the Uniform Building Code. In UBC-91 Sect. 2625(f) 3, the boundary elements of the shearwall, whenever required, must be designed to carry all of the factored gravity loads tributary to the wall plus the vertical force required to resist the overturning moment at the base of the wall due to the factored seismic loads. In other words, the boundary elements are to be designed completely ignoring the contribution of the web. As a result of this provision, large boundary elements with congested reinforcement are usually obtained.

To alleviate this problem, the provisions in UBC-94 Sect. 1921.6.5 give a design procedure which utilizes the entire cross-section of the wall in resisting the factored axial loads and moments. Since the current edition of the ACI Code uses the provisions in UBC-91, and for comparison purposes, both design methods will be presented in the following two sections. In Sect. 4.2.2.4.1, the section numbers given are from UBC-91 and ACI 318-89 (Revised 1992).

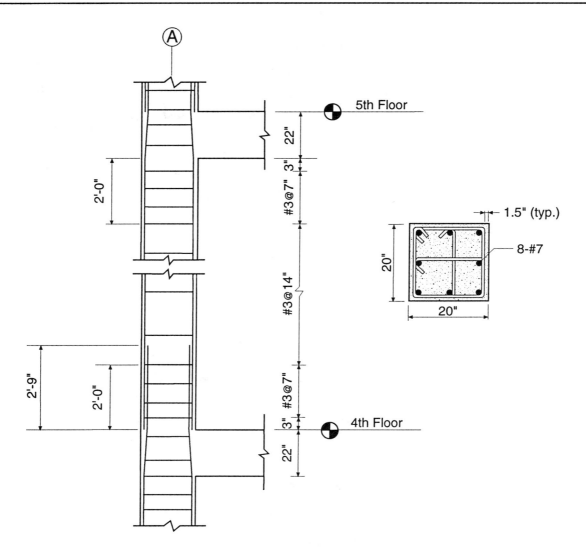

Figure 4-10 Reinforcement Details for Column A3 (Seismic Zone 4)

4.2.2.4.1 Design of Shearwall D2-E2 Based on UBC-91 Provisions

The shearwalls are part of the lateral load resisting system and must meet the applicable special detailing requirements of UBC 2625.

According to UBC 2625(f) 3A, a shearwall is required to have boundary elements if the maximum compressive stress in the shearwall due to the factored forces exceeds $0.2 f'_c$. A boundary element is a column-like member provided at the end of a shearwall; it is contiguous with the wall section. If the compressive stress at the edges of a shearwall exceeds $0.2 f'_c$ and boundary elements are not provided, the entire shearwall must have transverse reinforcement meeting the requirements specified for members subjected to bending and axial load (UBC 2625(e) 4A through C). It is usually more economical to provide boundary elements than to meet the special transverse reinforcement requirement.

The maximum stress in the shearwall occurs at the ends of the shearwall and is given by:

$$f_c = \frac{P_u}{A_g} \pm \frac{M_u \, (\ell_w / 2)}{I_g}$$

The stress is computed using factored loads and gross section properties.

Assuming that the length of the wall is 28.0 ft + 38.0 in. = 31.17 ft, the section properties of the wall are:

$$A_g = 31.17 \times \frac{12}{12} = 31.17 \text{ ft}^2$$

$$I_g = \frac{1}{12} \times \frac{12}{12} \times 31.17^3 = 2522.8 \text{ ft}^4$$

Using the factored loads from Table 4-8, the maximum fiber stress is:

$$f_c = \frac{5112}{31.17} + \frac{114,834 \, (31.17 / 2)}{2522.8}$$

$$= 164.0 + 709.4 = 873.4 \text{ ksf} = 6.1 \text{ ksi}$$

$$6.1 \text{ ksi} > 0.2 \, f_c' = 0.2 \times 6 = 1.2 \text{ ksi}$$

Therefore, provide boundary elements per UBC 2625 (f) 3.

•   ***Design of Boundary Elements***

Boundary elements must be designed to carry all factored gravity loads on the wall plus the vertical force required to resist the overturning moment calculated from factored forces related to earthquake effects (UBC 2625(f) 3C; ACI 21.6.5.3).

The maximum factored axial compressive force on a boundary element at the base of the wall is 6657 kips (see Table 4-8). This value was determined as follows:

$$P_u = 1.4 \, (P_D + P_L + P_E)$$

$$P_{D \text{ (boundary element)}} = P_{D \text{ (shearwall)}}/2$$

$$= \frac{1}{2} \, [\, (0.1475 \times 1568 \times 10) + (9 \times 84.0) + 148.0] = 1608.4 \text{ kips}$$

$$P_{L \text{ (boundary element)}} = P_{L \text{ (shearwall)}}/2$$

$$= \frac{1}{2} \times 0.050 \, [0.6 + (9 \times 0.4)] \times 1568 = 164.6 \text{ kips}$$

$$P_{E \text{ (boundary element)}} = \frac{M_{E \text{ (shearwall)}}}{\ell} + \frac{P_E}{2} = \frac{82,024}{28} + \frac{105.1}{2} = 2982.0 \text{ kips}$$

$$P_{u \text{ (boundary element)}} = 1.4 \times (1608.4 + 164.6 + 2982.0) = 6657.0 \text{ kips}$$

The axial load strength of the boundary element acting as a short column is:

$$P_u = \phi P_n = 0.8\phi \left[0.85 f'_c (A_g - A_{st}) + f_y A_{st} \right] \qquad \begin{array}{r} \textit{UBC 2610(d) 5B} \\ \textit{ACI 10.3.5.2} \end{array}$$

Solving for the required area of steel:

$$6657 = 0.8 \times 0.7 \times \left[0.85 \times 6.0 \times (38^2 - A_{st}) + 60 A_{st} \right]$$

$$A_{st} = 82.4 \text{ in.}^2$$

Use 54-#11 bars ($A_{st} = 84.2 \text{ in.}^2$).

The reinforcement ratio, ρ_g, must be greater than 1% and less than 6%:

$$\rho_g = \frac{A_{st}}{A_g} = \frac{84.2}{38^2} = 0.058 < 0.060 \quad \text{O. K.} \qquad \begin{array}{r} \textit{UBC 2625(e) 3a} \\ \textit{ACI 21.4.3.1} \end{array}$$

Although 54-#11 bars gives a reinforcement ratio less than the maximum allowed, the 54 bars will not adequately fit within the 38 × 38 in. section. The clear spacing between bars on the face of the section that has 15 bars is [38 - 2(1.5 + 0.625) - (15 × 1.41)]/14 = 0.9 in. which is less than the minimum allowed value of 1.5 d_b = 1.5 × 1.41 = 2.12 in. (UBC 2607(g) and ACI 7.6.3). Therefore, increase the size of the boundary element to 42 × 42 in. Note that at this time, the structure should be reanalyzed and the axial forces on the boundary elements should be recomputed based on the increase in the size of the boundary elements. However, it is assumed for our purposes here that the reactions on the wall do not change significantly. Consequently, the required A_{st} is:

$$6657 = 0.8 \times 0.7 \times \left[0.85 \times 6.0 \times (42^2 - A_{st}) + 60 A_{st} \right]$$

$$A_{st} = 52.7 \text{ in.}^2$$

Use 34-#11 bars ($\rho_g = 3.0\%$; clear space between bars = 2.63 in. > 2.12 in.).

The maximum factored tensile force on a boundary element is 2727 kips (see Table 4-8). The amount of steel required to carry the tensile force is:

$$A_s = \frac{P_u}{\phi f_y} = \frac{2727}{0.9 \times 60} = 50.5 \text{ in.}^2 < 52.7 \text{ in.}^2 \quad \text{O. K.}$$

The required area of confinement reinforcement in the form of rectangular hoops is:

$$A_{sh} \geq \begin{cases} 0.09 s h_c \dfrac{f'_c}{f_{yh}} \\[2ex] 0.3 s h_c \left[\dfrac{A_g}{A_{ch}} - 1 \right] \dfrac{f'_c}{f_y} \end{cases}$$

UBC 2625(e) 4A
ACI 21.4.4.1

where

s = spacing of transverse reinforcement, in.

h_c = cross-sectional dimension of column core measured center-to-center of transverse reinforcement, in.

f_{yh} = specified yield strength of transverse reinforcement, psi

A_g = gross area of section, in.^{2}

A_{ch} = cross-sectional area of a column measured out-to-out of transverse reinforcement, in.^{2}

Using the maximum allowed hoop spacing of 4 in. and assuming #5 bar hoops with 1.5 in. clear cover all around, the required cross-sectional area is:

$$A_{sh} \geq \begin{cases} 0.09 \times 4 \times 38.4 \times 6000 / 60{,}000 = 1.38 \text{ in.}^2 \text{ (governs)} \\[2ex] 0.3 \times 4 \times 38.4 \times [1764 / 1521 - 1] \times 6000 / 60{,}000 = 0.74 \text{ in.}^2 \end{cases}$$

#5 hoops with crossties around every other longitudinal bar provides $A_{sh} = 6 \times 0.31 = 1.86$ in.$^2 > 1.38$ in.^{2}

- ***Design of Wall***

At least two curtains of reinforcement must be used in the wall between the boundary elements if the in-plane factored shear force V_u for the wall exceeds $2A_{cv}\sqrt{f'_c}$ where A_{cv} is the net area of concrete section bounded by the wall thickness and length in the direction of shear force (UBC 2625(f) 2B; ACI 21.6.2.2). From Table 4-2, the shear force at the base of the wall is 853.1 kips. Using Eq. (9-2), the factored shear V_u is (UBC 2625(c)4):

$$V_u = 1.4 \times 853.1 = 1194.3 \text{ kips}$$

$$1194.3 \text{ kips} > 2A_{cv}\sqrt{f'_c} = 2 \times 12 \times (31.5 \times 12) \sqrt{6000}/1000 = 702.7 \text{ kips}$$

Therefore, use two curtains of reinforcement.

The reinforcement ratio along the longitudinal and the transverse direction must not be less than 0.0025; also, the spacing along these directions must not exceed 18 in. (UBC 2625(f) 2A; ACI 21.6.2).

The upper limit on the shear strength of a shearwall is given by:

$$\phi V_n = \phi 8 A_{cv} \sqrt{f'_c}$$

UBC 2625(h) 3D
ACI 21.6.4.6

$$= 0.6 \times 8 \times 12 \times (31.5 \times 12) \sqrt{6000}/1000 = 1686.5 \text{ kips} > 1194.3 \text{ kips} \quad \text{O.K.}$$

Typically, the shear strength reduction factor is 0.85. However, this factor must be taken as 0.6 for any structural member if its nominal shear strength is less than the shear corresponding to the development of the nominal flexural strength of the member (UBC 2609(d)4; ACI 9.3.4.1). In this case a conservative value of 0.6 is used in the design of the wall.

The nominal shear strength is given by:

$$V_n = A_{cv} \times \left[\alpha_c \sqrt{f'_c} + \rho_n f_y \right]$$

UBC 2625(h) 3
ACI 21.6.4

where α_c varies linearly from 3.0 for $h_w / \ell_w = 1.5$ to 2.0 for $h_w / \ell_w = 2.0$

or $\alpha_c = 2.0$ if $h_w / \ell_w \geq 2.0$

The α_c factor adjusts for a higher allowable shear strength for walls with low ratios of height to horizontal length. In this case, $\alpha_c = 2.0$ since the ratio $h_w / \ell_w = 140/31.5 = 4.4$.

Using 2-#6 horizontal bars at 14 in. gives the shear strength of the wall as follows:

$$\rho_n = \frac{2 \times 0.44}{12 \times 14} = 0.0052$$

$$\phi V_n = 0.6 \times (31.5 \times 12 \times 12) \times [2\sqrt{6000} + (0.0052 \times 60,000)]/1000$$

$$= 1270.8 \text{ kips} > V_u = 1194.3 \text{ kips} \quad \text{O.K.}$$

The vertical reinforcement ratio ρ_v shall not be less than the horizontal reinforcement ratio, ρ_n, when the ratio h_w / ℓ_w is less than 2.0 (UBC 2625(h) 3C; ACI 21.6.4.5). Since $h_w / \ell_w = 4.4 > 2.0$, the minimum reinforcement ratio $\rho_v = 0.0025$ will be used. The required area of vertical steel is:

$$A_{sv} = 0.0025 \times 12 \times 12 = 0.36 \text{ in.}^2/\text{ft}$$

Using 2-#5 bars:

$$s_{req} = \frac{12}{0.36} \times 0.62 = 20.7 \text{ in.} > 18 \text{ in.}$$

Use #5 vertical bars on each face at 18 in. o.c.

The adequacy of the shearwall section at the base under combined factored axial loads and bending moments in the plane of the wall must be checked. An interaction diagram for the shearwall with the reinforcement previously determined is shown in Fig. 4-11. It can be seen from the figure that the wall is adequate for the load combinations given in Table 4-8.

- ### *Splice Lengths for Reinforcement*

The splice length must be determined for the vertical bars in the web portion and the boundary element of the shearwall. For the boundary element, use a Class B splice, which has a length of 1.3 ℓ_d. The basic development length for #11 bars is:

$$\ell_{db} = 0.04\, A_b f_y / \sqrt{f'_c} \qquad\qquad \textit{UBC 2612(c) 2}$$
$$\textit{ACI 12.2.2}$$

$$= 0.04 \times 1.56 \times 60{,}000/\sqrt{6000} = 48.3 \text{ in.}$$

The information required to find the applicable factors is:

Minimum cover to bars being spliced = 2.13 in. (1.51 d_b)

Clear spacing between bars being spliced = 2.63 in. (1.87 d_b)

Transverse reinforcement meets requirements of UBC 2611(f) 4 & 5 and UBC 2612(c) 3E.

Minimum clear cover meets requirement of UBC 2607(h).

This gives a value of 1.5 for Factor I (see Table 4-9). The other factors are all equal to 1.0, giving:

Class B splice length = 1.3 × (1.5 × 1.0 × 1.0 × 1.0 × 1.0 × 48.3) = 94.2 in.

The lap splices for the vertical reinforcement in the web will be a Class B splice. The basic development length for #5 bars is:

$$\ell_{db} = 0.04\, A_b f_y / \sqrt{f'_c}$$

$$= 0.04 \times 0.31 \times 60{,}000/\sqrt{6000} = 9.6 \text{ in.}$$

The information required to find the applicable factors is:

Minimum clear cover = 2.25 in. (3.6 d_b)

Side cover   large

Clear bar spacing   = 6.25 in. (10.0d_b)

Reinforcement is an inner layer of wall reinforcement.

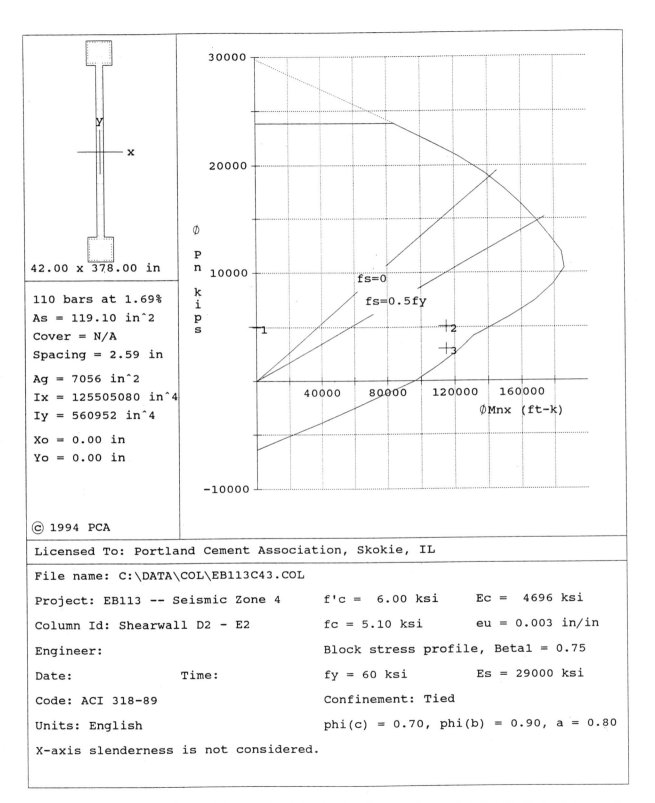

42.00 x 378.00 in

110 bars at 1.69%
As = 119.10 in^2
Cover = N/A
Spacing = 2.59 in

Ag = 7056 in^2
Ix = 125505080 in^4
Iy = 560952 in^4

Xo = 0.00 in
Yo = 0.00 in

© 1994 PCA

Licensed To: Portland Cement Association, Skokie, IL

File name: C:\DATA\COL\EB113C43.COL

Project: EB113 -- Seismic Zone 4   f'c = 6.00 ksi   Ec = 4696 ksi

Column Id: Shearwall D2 - E2   fc = 5.10 ksi   eu = 0.003 in/in

Engineer: Block stress profile, Beta1 = 0.75

Date: Time: fy = 60 ksi   Es = 29000 ksi

Code: ACI 318-89   Confinement: Tied

Units: English   phi(c) = 0.70, phi(b) = 0.90, a = 0.80

X-axis slenderness is not considered.

Figure 4-11 Design Strength Interaction Diagram for Shearwall D2-E2, UBC-91 (Seismic Zone 4)

A maximum value of 1.51 will control for Factor I. The other factors are equal to 1.0, giving:

$$\text{Class B splice length} = 1.3 \times (1.51 \times 1.0 \times 1.0 \times 1.0 \times 1.0 \times 9.6) = 18.9 \text{ in.}$$

The development length for the #6 horizontal bars will be calculated assuming that no hooks are used in the boundary element. According to UBC 2625(g) 4B and ACI 21.5.4.2, the basic development length is:

$$\ell_{dh} = f_y d_b / 65\sqrt{f'_c}$$

$$= (60,000 \times 0.75)/(65\sqrt{6000}) = 8.9 \text{ in.} > 8d_b = 6.0 \text{ in.}$$

For a straight bar, the required development length $\ell_d = 2.5 \times 8.9 = 22.3$ in.

This length can be accommodated within the confined core of the boundary element. Figure 4-12 shows the reinforcement details for this shearwall.

4.2.2.4.2 Design of Shearwall D2-E2 Based on UBC-94 Provisions

The design procedure for shearwalls based on the provisions given in UBC-94 can be summarized in three steps:

1) Design the shearwall for flexure and axial load by considering the entire cross-section of the wall to act as a short column; this will yield the required flexural reinforcement (UBC 1921.6.5.1). Check the shear resistance of the wall provided by the web (UBC 1921.6.4).

2) Provide the boundary zone detail requirements given in UBC 1921.6.5.6 at each end of the wall unless $P_u \leq 0.10\,A_g\,f'_c$ (for geometrically symmetrical walls) and either $M_u/V_u\,\ell_w \leq 1.0$ or $V_u \leq 3\,\ell_w h\sqrt{f'_c}$ and $M_u/V_u\,\ell_w \leq 3.0$ (UBC 1921.6.5.4). In any case, shearwalls with $P_u > 0.35\,P_o$ are not permitted to resist earthquake-induced forces (UBC 1921.6.5.3).

3) For shearwalls where boundary elements with specially detailed reinforcement have to be provided, determine the length of the boundary zone by using one of two methods. In the conservative method, the boundary zones are provided over a maximum distance of $0.25\,\ell_w$ at each end of the wall (UBC 1921.6.5.4). Alternatively, the boundary zone length may be based on the compressive strain levels at the edges of the wall when the wall is subjected to displacements resulting from the ground motions specified in UBC 1629.2, using cracked section properties of the wall (UBC 1921.6.5.5). Confinement is to be provided wherever the compressive strain exceeds 0.003.

• *Design of Boundary Zone*

Assuming a 1 ft-0 in. × 31 ft-2 in. wall, determine if boundary zones are required. From Table 4-8, the maximum factored axial load is 5112 kips. Check the criterion in UBC 1921.6.5.4:

$$0.10\,A_g\,f'_c = 0.10\,(12 \times 31.167 \times 12) \times 6.0 = 2693 \text{ kips} < P_u = 5112 \text{ kips}$$

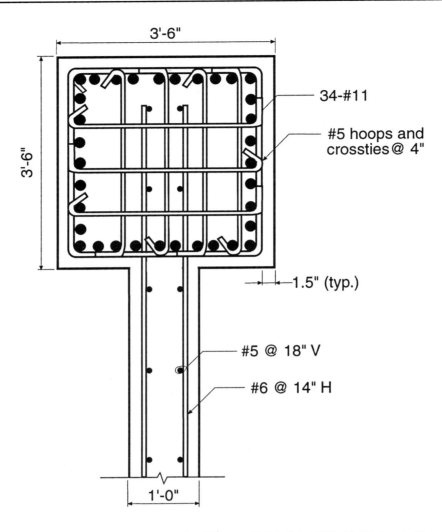

3'-6"

3'-6"

34-#11

#5 hoops and
crossties @ 4"

1.5" (typ.)

#5 @ 18" V

#6 @ 14" H

1'-0"

Figure 4-12 Reinforcement Details for Shearwall D2-E2, UBC-91 (Seismic Zone 4)

Therefore, boundary zones are required. Assume 38 × 38 in. boundary elements reinforced with 36-#11 bars. Since the provisions in UBC 1921.6.4 for the nominal shear strength of shearwalls is the same as in UBC-91 Sect. 2625(f) 2B, use 2-#5 @ 18 in. vertical bars in the web (see Sect. 4.2.2.4.1). Based on this reinforcement, the nominal axial load capacity of the wall at zero eccentricity is:

$$P_o = 0.85 \, f'_c \, (A_g - A_{st}) + f_y A_{st}$$

$$= 0.85 \times 6 \, (6464 - 125.34) + (60 \times 125.34) = 39{,}848 \text{ kips}$$

Since $P_u = 5112$ kips $< 0.15 \, P_o = 5977$ kips, conservatively provide boundary zones at each end at a distance of $0.15 \, \ell_w = 0.15 \times 31.167 = 4.7$ ft (UBC 1921.6.5.4). The special reinforcement details given in UBC 1921.6.5.6 are to be provided in this area at each end of the wall.

Alternatively, determine the length of the boundary zone using the provisions in UBC 1921.6.5.5. Before proceeding with the strain compatibility analysis, determine if the wall can be considered to contribute to the strength of the structure for resisting earthquake-induced forces. Since $P_u = 5112$ kips $< 0.35\,P_o = 13{,}974$ kips, the wall can be used to resist its portion of the seismic loads.

When performing the strain compatibility analysis for the wall, it is important to note that the compressive strain in the concrete is not allowed to be larger than ε_{max} where

$$\varepsilon_{max} = \frac{3R_w}{8}\,(0.004) \le 0.015$$

In this case,

$$\varepsilon_{max} = \frac{3 \times 8}{8}\,(0.004) = 0.012$$

For shearwalls in which the flexural limit state is governed by yielding at the base of the wall, the total curvature demand ϕ_t is given in UBC Eq. (21-9):

$$\phi_t = \frac{\Delta_i}{(h_w - \ell_p / 2)\ell_p} + \phi_y$$

where Δ_i = inelastic deflection at the top of the wall

$$= \Delta_t - \Delta_y$$

Δ_t = total deflection at the top of the wall equal to $3R_w/8$ times the elastic displacement using cracked section properties, or may be taken as $2(3R_w/8)\,\Delta_E$.

Δ_y = displacement at top of wall corresponding to the extreme fiber compressive strain of 0.003 at the critical section, or may be taken as $\left(M'_n / M_E\right)\Delta_E$ where M_E is the moment at the critical section when the top of the wall is displaced Δ_E and M'_n is the nominal flexural strength of the critical section at an axial load $P'_u = 1.2D + 0.5L + E$.

Δ_E = elastic displacement at the top of the wall using gross section properties and code-specified seismic forces

ℓ_p = height of the plastic hinge above the critical section, which may be taken as $0.5\,\ell_w$.

ϕ_y = $0.003/c'_u$

c'_u = neutral axis depth at P'_u and M'_n

The first step in the analysis will be determining the location of the neutral axis c'_u. Using the design loads listed in Sect. 4.2.2 (including live load reduction) and the seismic load given in Table 4-2, the load P'_u acting at the base of the wall is:

$$P'_u = (1.2 \times 3216.6) + (0.5 \times 329.2) + 105.1 = 4129.6 \text{ kips}$$

In lieu of hand computations, the PCA computer program *PCACOL* [2.2] was used to obtain the neutral axis depth c_u'. For 38 × 38 in. boundary elements reinforced with 36-#11 bars and a 12 in. web with 2-#5 @ 18 in., $c_u' = 42.1$ in. The nominal moment strength M_n' corresponding to $P_u' = 4129.6$ kips is 165,979 ft-kips (see Fig. 4-13). Figure 4-14 shows the interaction diagram with the governing load combinations given in Table 4-8. As can be seen from this figure, the wall is adequate to carry the factored loads.

From Table 4-3, $\Delta_E = 1.11$ in. Therefore,

$$\Delta_t = 2 \times \left(\frac{3 \times 8}{8}\right) \times 1.11 = 6.66 \text{ in.}$$

Also, using $M_E = 82,024$ ft-kips from Table 4-2:

$$\Delta_y = \left(\frac{M_n'}{M_E}\right) \Delta_E$$

$$= \left(\frac{165,979}{82,024}\right) \times 1.11 = 2.25 \text{ in.}$$

The inelastic deflection at the top of the wall is:

$$\Delta_i = \Delta_t - \Delta_y = 6.66 - 2.25 = 4.41 \text{ in.}$$

Assuming that $\ell_p = 0.5\,\ell_w = 0.5 \times 31.167 \times 12 = 187$ in., the total curvature demand is:

$$\phi_t = \frac{4.41}{[(140 \times 12) - (187/2)](187)} + \frac{0.003}{42.1}$$

$$= 1.49 \times 10^{-5} + 7.13 \times 10^{-5} = 8.62 \times 10^{-5}$$

Therefore, assuming the compressive strains to vary linearly over the depth c_u', the maximum compressive strain ε_u is:

$$\varepsilon_u = c_u'\,\phi_t$$

$$= 42.1 \times 8.62 \times 10^{-5} = 0.0036 > 0.0030 < \varepsilon_{max} = 0.0120 \quad \text{O.K.}$$

Confinement is required over

$$\left(42.1 - \frac{0.0030}{0.0036} \times 42.1\right) = 7.0 \text{ in.} < 38.0 \text{ in.}$$

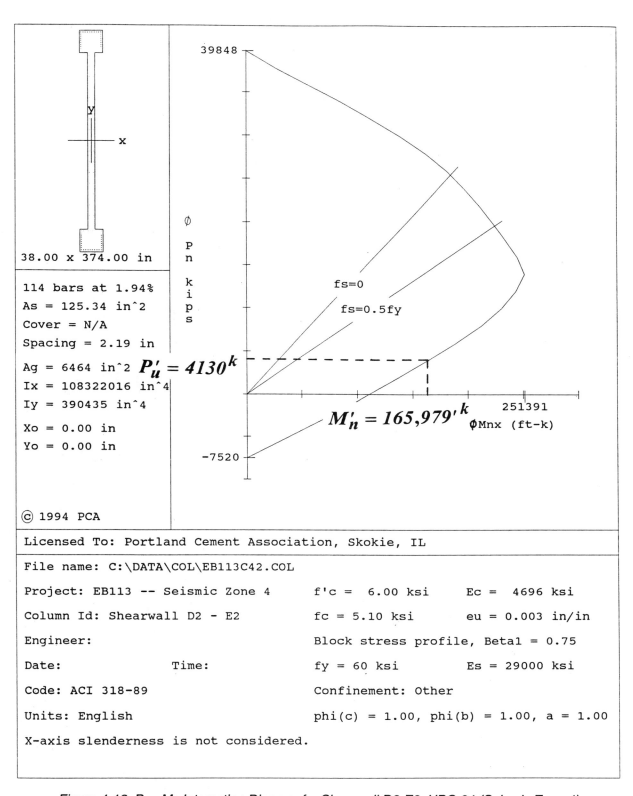

Figure 4-13 P_n - M_n Interaction Diagram for Shearwall D2-E2, UBC-94 (Seismic Zone 4)

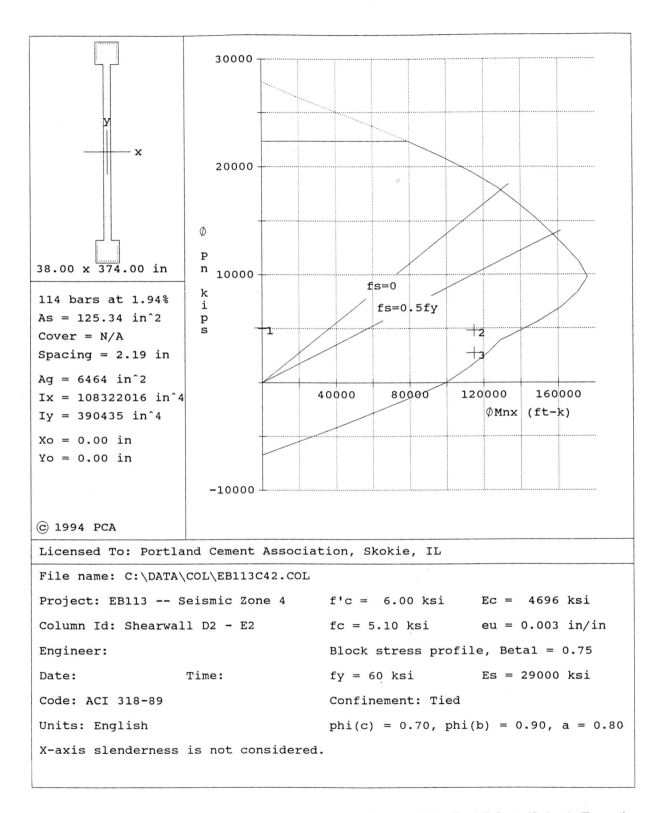

38.00 x 374.00 in

114 bars at 1.94%
As = 125.34 in^2
Cover = N/A
Spacing = 2.19 in

Ag = 6464 in^2
Ix = 108322016 in^4
Iy = 390435 in^4

Xo = 0.00 in
Yo = 0.00 in

© 1994 PCA

Licensed To: Portland Cement Association, Skokie, IL

File name: C:\DATA\COL\EB113C42.COL

Project: EB113 -- Seismic Zone 4   f'c = 6.00 ksi   Ec = 4696 ksi

Column Id: Shearwall D2 - E2   fc = 5.10 ksi   eu = 0.003 in/in

Engineer: Block stress profile, Beta1 = 0.75

Date: Time: fy = 60 ksi   Es = 29000 ksi

Code: ACI 318-89   Confinement: Tied

Units: English   phi(c) = 0.70, phi(b) = 0.90, a = 0.80

X-axis slenderness is not considered.

Figure 4-14 Design Strength Interaction Diagram for Shearwall D2-E2, UBC-94 (Seismic Zone 4)

For practical purposes, the entire 38×38 in. boundary element will be confined according to the provisions in UBC 1921.6.5.6. Check minimum dimensions of the boundary zone:

$$\text{Minimum thickness} = \frac{\ell_u}{10} = \frac{(23 \times 12) - 22}{10} = 25.4 \text{ in.} < 38.0 \text{ in.} \qquad \text{O.K.}$$

$$\text{Minimum length} \quad = \frac{\ell_w}{10} = \frac{31.167 \times 12}{10} = 37.4 \text{ in.} < 38.0 \text{ in.} \qquad \text{O.K.}$$

The minimum area of confinement reinforcement A_{sh} is given in UBC Eq. (21-10):

$$A_{sh} \quad = 0.09 sh_c f'_c / f_{yh}$$

Using the maximum allowable spacing of 6 in. and #5 hoops:

$$A_{sh} \quad = 0.09 \times 6 \times 34.375 \times \frac{6}{60} = 1.86 \text{ in.}^2$$

#5 hoops with crossties around every other longitudinal bar provides $A_{sh} = 6 \times 0.31 = 1.86$ in.^{2} Note that over the lap splice length of the vertical bars in the boundary zone, the hoops must be spaced at 4 in. Also, the minimum area of vertical reinforcement in the boundary zone is $0.005 \times 38^2 = 7.22$ in.^{2} which is less than the area provided by 36-#11 bars.

- ***Design of Wall***

As noted above, the nominal shear strength provisions for shearwalls in UBC 1921.6.4 are the same as those given in UBC-91. Therefore, calculations similar to those shown in Sect. 4.2.2.4.1 yield 2-#6 horizontal bars at 14 in. and 2-#5 vertical bars at 18 in.

- ***Splice Lengths for Reinforcement***

For the boundary elements, use a Class B splice. The information required to find the applicable factors is:

      Minimum cover to bars being spliced   = 2.13 in. (1.51 d_b)

      Clear space between bars being spliced   = 2.18 in. (1.55 d_b)

      Transverse reinforcement meets requirements of UBC 1911.5.4 and 1911.5.5.3 and UBC 1907.10.5.

      Minimum clear cover meets requirements of UBC 1907.7.1.

This gives a value of 1.5 for Factor I (see Table 4-9). The other factors are all equal to 1.0, giving:

      Class B splice length = $1.3 \times (1.5 \times 1.0 \times 1.0 \times 1.0 \times 1.0 \times 48.3) = 94.2$ in.

The splice lengths for the web reinforcement is the same given in Sect. 4.2.2.4.1. Figure 4-15 shows the reinforcement details for this shearwall. UBC 1921.6.5.6 gives other detailing requirements for the shearwall.

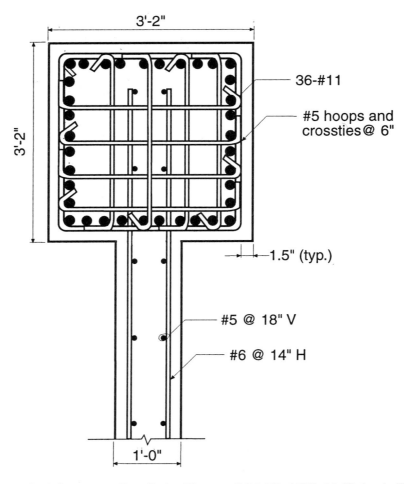

Figure 4-15 Reinforcement Details for Shearwall D2-E2, UBC-94 (Seismic Zone 4)

4.2.2.4.3 Comparison of Results for Shearwall D2-E2 using UBC-91 and UBC-94

In Sect. 4.2.2.4.1, it was determined using UBC-91 provisions that for 38×38 in. boundary elements, the required reinforcement is 54-#11 bars ($A_{st} = 84.2$ in.[2]). Even though these bars cannot fit within the section, it is interesting to compare this required A_{st} with that obtained from UBC-94 provisions. From Sect. 4.2.2.4.2, the required A_{st} is 56.2 in.[2] for the same boundary element size. Thus, in this design example, using the UBC-91 design provisions would require approximately 1.5 times more reinforcement in the boundary elements. The web reinforcement would remain the same.

It is important to note that for shearwalls subjected to an axial load to bending moment ratio larger than the one given here, the ratio of required steel using UBC-91 to UBC-94 provisions would be larger than the one shown above.

4.3   Design for Seismic Zone 2B

4.3.1   Frame Analysis

Given the data in Fig. 4-1, the weights of the floors are listed in Table 4-10. As before, the building has the same lateral load resisting system in both principal directions. Therefore, the seismic forces are the same in both the longitudinal and the transverse directions of the building.

The calculations for the base shear, V, for the transverse and longitudinal directions are shown at the bottom of Table 4-10. For this example, the importance factor, I, and the site soil factor, S, have been taken equal to 1.0 and 1.2, respectively. The value of R_w is equal to 8 in accordance with UBC Table 16-N. The seismic forces resulting from the distribution of the base shear in accordance with UBC Eqs. (28-6), (28-7), and (28-8) are also listed in Table 4-10.

Table 4-10 Seismic Forces for Zone 2B

Floor Level	Weight, w_x (kips)	Height, h_x (ft)	$w_x h_x$ (ft-kips)	Lateral Force (kips)
R	3257	140	455,980	317.6
10	3594	127	456,438	229.9
9	3594	114	409,716	206.4
8	3594	101	362,994	182.8
7	3594	88	316,272	159.3
6	3594	75	269,550	135.8
5	3594	62	222,828	112.2
4	3594	49	176,106	88.7
3	3594	36	129,384	65.2
2	3853	23	88,619	44.6
Σ	35,862		2,887,887	1542.6

Building height	h_n	$= 140\ ft$
Building weight	W	$= 35,862\ kips$
Fundamental period	T	$= 0.020\ (h_n)^{3/4} = 0.81\ sec.$
Seismic zone factor	Z	$= 0.2$
Importance factor	I	$= 1.0$
Response coefficient	R_w	$= 8$
Soil factor	S	$= 1.2$
Coefficient	C	$= 1.25\ S/T^{2/3} = 1.72$
Base shear	V	$= ZICW/R_w = 1542.6\ kips$
Top level force	F_t	$= 0.07TV = 88\ kips$

The member forces resulting from the seismic loads are determined in the same manner as for Zone 4 (see Section 4.2.1). Table 4-11 shows the results from the structural analysis for the lateral load resisting system subjected to the seismic forces.

For structures having a fundamental period greater than 0.7 seconds, the story drift must not exceed $0.03/R_w$, nor 0.004 times the story height (UBC 1628.8.2).

*Table 4-11 Results of Analysis of Lateral Load Resisting System Due to
Seismic Zone 2B Forces*

Shearwall D2-E2				
Level	Axial Load (kips)	Bending Moment (ft-kips)		Shear Force (kips)
		Top	Bottom	
R	5.0	184.3	952.7	87.2
10	10.1	-761.9	2699.0	148.9
9	15.0	-2513.6	5161.4	203.8
8	20.0	-4979.3	8269.2	253.0
7	24.8	-8092.0	11,948.2	296.5
6	29.3	-11,780.2	16,117.2	333.7
5	33.4	-15,968.3	20,709.0	365.1
4	37.0	-20,576.2	25,638.6	389.4
3	39.8	-25,530.0	30,814.7	406.6
2	42.1	-30,735.1	40,287.5	415.3

Coupling Beam C2-D2	
Level	Bending Moment (ft-kips)
R	-73.9
10	-75.6
9	-74.2
8	-73.7
7	-70.8
6	-67.1
5	-61.1
4	-53.8
3	-43.1
2	-31.8

For typical floors:

$$\frac{0.03h_s}{R_w} = \frac{0.03 \times 13 \times 12}{8} = 0.59 \text{ in. (governs)}$$

$$0.004h_s = 0.004 \times 13 \times 12 = 0.62 \text{ in.}$$

For the lower level:

$$\frac{0.03h_s}{R_w} = \frac{0.03 \times 23 \times 12}{8} = 1.04 \text{ in. (governs)}$$

$$0.004h_s = 0.004 \times 23 \times 12 = 1.10 \text{ in.}$$

As shown in Table 4-12, the story drift limits are not exceeded at any floor level.

Table 4-12 Lateral Displacements and Drifts of the Lateral
Load Resisting System Due to Seismic Zone 2B Forces

Floor Level	Total Lateral Displacement (in.)	Interstory Drift (in.)
R	0.58	0.07
10	0.51	0.06
9	0.45	0.07
8	0.38	0.07
7	0.31	0.07
6	0.24	0.06
5	0.18	0.06
4	0.12	0.04
3	0.08	0.04
2	0.04	0.04

The member forces and the story drifts induced by the P-Δ effects must be considered in the evaluation of overall structural frame stability. According to UBC 1628.9, P-Δ effects can be neglected when the ratio of the secondary moment to the primary moment is less than or equal to 0.1. Based on the results shown in Table 4-13, the P-Δ effects can be ignored.

Table 4-13 P-Δ Check for the Lateral Load Resisting System under Seismic Zone 2B Forces

Floor Level	Gravity Loads * Σw_i (kips)	Story Drift Δ_i (in.)	$\Sigma w_i \Delta_i$ (in.-kips)	Story Shear V_i (kips)	Story Height h_{si} (in.)	$V_i h_{si}$ (in.-kips)	$\Sigma w_i \Delta_i / V_i h_{si}$
R	3859	0.07	270.1	317.6	156	49,545.6	0.006
10	7855	0.06	471.3	547.5	156	85,410.0	0.006
9	11,851	0.07	829.6	753.9	156	117,608.4	0.007
8	15,847	0.07	1109.3	936.7	156	146,125.2	0.008
7	19,843	0.07	1389.0	1096.0	156	170,976.0	0.008
6	23,839	0.06	1430.3	1231.8	156	192,160.8	0.008
5	27,835	0.06	1670.1	1344.0	156	209,664.0	0.008
4	31,831	0.04	1273.2	1432.7	156	223,501.2	0.006
3	35,827	0.04	1433.1	1497.9	156	233,672.4	0.006
2	40,082	0.04	1603.3	1542.6	276	425,757.6	0.004

* Includes floor weight, superimposed dead load, and reduced roof and floor live loads

All framing elements not required by design to be part of the lateral force resisting system must be investigated and shown to be adequate for vertical load carrying capacity when displaced $3R_w/8$ times the displacement resulting from the required lateral forces acting on the lateral force resisting system (UBC 1631.2.4). This will provide an added degree of safety for non-seismically detailed members of the framing system to assure that they can withstand the possible deformation without losing vertical load-carrying capacity. P-Δ effects on these elements must also be taken into account.

The deformation compatibility analysis requires that the amplified lateral displacements $\Delta_i' = 3R_w \Delta_i / 8$ be imposed on every floor level i of the non-lateral-load-resisting system, where the Δ_i are given in Table 4-12. As was done for Seismic Zone 4, *SAP90* was used to perform the analysis. Table 4-14 shows the results for a column and a beam accounting for deformation compatibility and P-Δ effects (which in this case are negligible).

Table 4-14 Results of Analysis of Non-Lateral-Load-Resisting System Accounting for Deformation Compatibility and P-Δ Effects (Seismic Zone 2B)

Column A3				
Level	Axial Load (kips)	Bending Moment (ft-kips)		Shear Force (kips)
		Top	Bottom	
R	3.6	53.6	-48.0	7.8
10	8.0	16.5	-18.5	2.7
9	12.3	43.2	-37.3	6.2
8	17.1	32.1	-32.3	5.0
7	21.9	37.2	-42.2	6.1
6	26.2	21.5	-24.5	3.5
5	30.4	35.7	-44.6	6.2
4	33.8	4.8	-13.7	1.4
3	36.5	25.2	-28.6	4.1
2	38.8	5.6	-21.6	1.2

Beam A3-B3 Bending Moment (ft-kips)		
Level	Exterior	Interior
R	-53.6	46.4
10	-64.5	60.3
9	-61.7	57.3
8	-69.4	64.4
7	-69.5	64.6
6	-63.7	59.3
5	-60.2	55.9
4	-49.4	45.9
3	-38.9	36.4
2	-34.2	31.0

4.3.2   Design of Structural Members

The design objective is to determine the required shear and flexural reinforcement for a beam (A3-B3) and a column (A3) that are not part of the lateral load resisting system, and a coupling beam (C2-D2) and shearwall (D2-E2) that are part of the lateral load resisting system.

In Seismic Zone 2, reinforced concrete frames resisting forces induced by earthquake motions must be intermediate moment-resisting frames proportioned to satisfy only UBC 1921.8 in addition to the requirements of UBC 1901 through 1918. According to UBC 1921.8, no special requirements for shearwalls are needed in Seismic Zone 2.

The gravity loads acting on beam A3-B3 are as follows:

Tributary area $= 784 \text{ ft}^2$

Dead loads:

    Beams and slab $= 113.5$ psf
    Superimposed $= 32$ psf
    Total $= 145.5$ psf
    Load per foot $= 145.5 \times 784/28 = 4074$ plf

Live Load:

    Reduction factor $= r(A - 150) = 0.08 \times (784 - 150) = 50.7\% > 40\%$ *1606*
    Load per foot $= (1.0 - 0.4) \times 784 \times 50/28 = 840$ plf

The factored load due to gravity on this beam is:

$$w_u = (1.4 \times 4.07) + (1.7 \times 0.84) = 7.13 \text{ kips/ft}$$

The interior end of this beam frames into a shearwall. This end of the beam is considered fixed, with a bending moment equal to $w_u \ell_n^2/12$. The bending moment at the exterior end of the beam is taken as $w_u \ell_n^2/16$; the midspan moment then equals $w_u \ell_n^2/19$.

The factored moments caused by gravity loads are:

    Exterior negative moment $= w_u \ell_n^2/16 = 7.13 \times (28 - 20/24 - 30/24)^2 /16 = 299.3$ ft-kips

    Positive moment $= w_u \ell_n^2/19 = 7.13 \times (28 - 20/24 - 30/24)^2 /19 = 252.1$ ft-kips

    Interior negative moment $= w_u \ell_n^2/12 = 7.13 \times (28 - 20/24 - 30/24)^2 /12 = 399.1$ ft-kips

Table 4-15 shows the bending moments for different combinations of gravity and compatibility loads for beam A3-B3 on the fifth floor of the frame. The gravity load moments are obtained from the calculations above and the compatibility moments are taken from Table 4-14. Note that the compatibility moments represent an extreme probability design condition so that load factors are not required in this case.

Table 4-15 Summary of Factored Bending Moments for Beam A3-B3 on Level 5 (Seismic Zone 2B)

    (1) U $= 1.4D + 1.7L$ Eq. (9-1), UBC 1909.2.1
    (2) U $= (D + L \pm E)$ Compatibility (No load factors)

Load Combination	Bending Moment (ft-kips)		
	Exterior Negative	Positive	Interior Negative
1	-299.3	252.1	-399.1
2: sidesway right	-146.0	175.8	-330.7
sidesway left	-266.4	171.4	-218.9

Beam C2-D2 is part of the lateral load resisting system. Since the beam frames into shearwalls, both ends of the beam are considered fixed, with a moment of $w_u \ell_n^2 / 12$ at each beam end. Note that the gravity loads for this beam are the same as for Beam A3-B3.

The factored moments caused by gravity loads are:

Exterior negative moment $= w_u \ell_n^2 / 12 = 7.13 \times (28 - 30/12)^2 / 12 = 386.4$ ft-kips

Positive moment $= w_u \ell_n^2 / 24 = 7.13 \times (28 - 30/12)^2 / 24 = 193.2$ ft-kips

Interior negative moment $= w_u \ell_n^2 / 12 = 7.13 \times (28 - 30/12)^2 / 12 = 386.4$ ft-kips

Table 4-16 shows the bending moments for different combinations of gravity and seismic loads for this beam. The gravity moments are obtained from the calculations above and the seismic moments are taken from Table 4-11.

Table 4-16 Summary of Factored Bending Moments for Beam C2-D2 on Level 5 (Seismic Zone 2B)

(1) U = 1.4D + 1.7L   Eq. (9-1), UBC 1909.2.1
(2) U = 0.75(1.4D + 1.7L ± 1.87E)   Eq. (9-2), UBC 1909.2.3
(3) U = 0.9D ± 1.43E   Eq. (9-3), UBC 1909.2.3

Load Combination	Bending Moment (ft-kips)		
	Exterior Negative	Positive	Interior Negative
1	-386.4	193.2	-386.4
2: sidesway right	-203.9	144.9	-375.2
sidesway left	-375.2	144.9	-203.9
3: sidesway right	-111.1	99.3	-285.8
sidesway left	-285.8	99.3	-111.1

Column A3 is not part of the lateral load resisting system. The gravity loads for this column are as follows:

Tributary area $= 415$ ft^2

Dead load $= 145.5$ psf (plus the weight of the column)

Live load $= 50$ psf

The different combinations of axial forces and bending moments corresponding to the appropriate load combinations are listed in Table 4-17. The values for gravity load effects are based on the information above (including live load reduction); the values for compatibility effects are taken from Table 4-14.

Shearwall D2-E2 is part of the lateral load resisting system. Boundary elements are not required for shearwalls in Seismic Zone 2B; however, they are provided in this building, partly to maintain a regular column layout and partly to provide a larger bending moment capacity for the shearwalls. The ends are 30 × 30 in. and the wall is 12 in. thick. The gravity loads for this shearwall are as follows:

Tributary area = 1568 ft^2

Dead load   = 145.5 psf (plus the weight of the wall)

Live load   = 50 psf

Table 4-17 Summary of Factored Axial Loads and Bending Moments for
Column A3 between Levels 4 and 5 (Seismic Zone 2B)

(1) U = 1.4D + 1.7L   Eq. (9-1), UBC 1909.2.1
(2) U = (D + L ± E)   Compatibility (No load factors)

Load Combination	Axial Load (kips)	Bending Moment (ft-kips)	
		Top	Bottom
1	759.1	-149.5	149.5
2: sidesway right	497.4	-67.3	58.4
sidesway left	558.2	-138.7	147.6

The different combinations of axial loads and bending moments corresponding to the appropriate load combinations are listed in Table 4-18. The reactions due to the gravity loads are based on the information provided above (including live load reduction), and the reactions due to the seismic loads are taken from Table 4-11.

Table 4-18 Summary of Factored Axial Loads and Bending Moments for Shearwall
D2-E2 between Grade and Level 2 (Seismic Zone 2B)

(1) U = 1.4D + 1.7L   Eq. (9-1), UBC 1909.2.1
(2) U = 0.75(1.4D + 1.7L ± 1.87E)   Eq. (9-2), UBC 1909.2.3
(3) U = 0.9D ± 1.43E   Eq. (9-3), UBC 1909.2.3

Load Combination	Shearwall D2-E2		
	Axial Load (kips)	Bending Moment (ft-kips)	
		Top	Bottom
1	4871	—	—
2: sidesway right	3594	-43,106	56,503
sidesway left	3712	43,106	-56,503
3: sidesway right	2711	-43,951	57,611
sidesway left	2832	43,951	-57,611

4.3.2.1   Proportioning and Detailing Beam A3-B3

4.3.2.1.1 Required Flexural Reinforcement

The maximum negative moment for this beam at the interior support is 399.1 ft-kips (see Table 4-15). The required reinforcement, ignoring the effects of any compression reinforcement, is $A_s = 5.94$ in.^{2} Use 6-#9 bars at this location ($\phi M_n = 402.4$ ft-kips).

Check limitations on the reinforcement ratio:

$$\rho \quad = A_s/bd = 6.0/(20 \times 17.56) = 0.0171$$

$$\rho_{max} \quad = 0.75\ \rho_{bal} = 0.0213 > 0.0171 > 200/f_y = 0.0033 \quad \text{O.K.} \qquad\qquad 1921.3.2.1$$

The maximum negative moment for this beam at the exterior support is 299.3 ft-kips. Use 5-#9 bars at this location ($\phi M_n = 345.3$ ft-kips).

The maximum positive moment for the beam is 252.1 ft-kips Use 4-#9 bars, providing a moment strength of 284.2 ft-kips.

4.3.2.1.2 Required Shear Reinforcement

The shear forces caused by gravity loads and compatibility requirements at a distance d from the face of the first interior support are calculated below:

$$V_D \text{ @ support} \quad = 1.15 w_D\ \ell_n/2$$

$$= 1.15 \times 4.07 \times (28 - 20/24 - 30/24)/2 = 60.7 \text{ kips}$$

$$V_D \text{ @ d} \quad = 60.7 - (4.07 \times 17.56/12) = 54.7 \text{ kips}$$

$$V_L \text{ @ d} \quad = 54.7 \times 0.84/4.07 = 11.3 \text{ kips}$$

$$V_{compatibility} \quad = (60.2 + 55.9)/(28 - 20/24 - 30/24) = 4.5 \text{ kips (see Table 4-14)}$$

Load combinations:

(1) U $= 1.4D + 1.7L$

V_u $= (1.4 \times 54.7) + (1.7 \times 11.3) = 95.8$ kips   (governs)

(2) U $= D + L + \text{compatibility}$

V_u $= 54.7 + 11.3 + 4.5 = 70.5$ kips

The shear force carried by the concrete is:

$$V_c = 2\sqrt{f'_c} \times b \times d \qquad \text{1911.3.1.1}$$

$$= 2\sqrt{4000} \times 20 \times 17.56/1000 = 44.4 \text{ kips}$$

The shear force carried by the stirrups at the support is:

$$V_s = V_u/\phi - V_c$$

$$= (95.8/0.85) - 44.4 = 68.3 \text{ kips} \qquad \text{1911.1.1}$$

$$< 4\sqrt{f'_c} \times bd = 88.8 \text{ kips} \qquad \text{1911.5.4.3}$$

Using #4 stirrups, the required spacing near the interior support is:

$$s = A_v \times f_y \times d/V_s \qquad \text{1911.5.6.2}$$

$$= 0.40 \times 60 \times 17.56/68.3 = 6.2 \text{ in.}$$

The maximum allowable stirrup spacing s_{max} is:

$$s_{max} = A_s \times f_y/(50 \times b_w) = 0.4 \times 60,000/(50 \times 20) = 24.0 \text{ in.} \qquad \text{1911.5.5.3}$$

$$= \frac{d}{2} = \frac{17.56}{2} = 8.8 \text{ in. (governs)} \qquad \text{1911.5.4.1}$$

$$= 24 \text{ in.}$$

Therefore, use #4 stirrups at 6 in. Stirrups are no longer required when $V_u < \phi V_c/2 = 0.85 \times 44.4/2 = 18.9$ kips. Consequently, the length over which stirrups are required is $(95.8 - 18.9)/7.13 = 10.8$ ft. However, stirrups will be provided over the entire length of the beam.

4.3.2.1.3 Reinforcing Bar Cutoff Points and Splices

The negative reinforcement at the interior support consists of 6-#9 bars. The location where four of the six bars can be terminated will be determined. The loading used to find the cutoff points is 0.9 times the dead load in combination with the end moments from the compatibility load combination (see Table 4-15).

The design moment strength of the section with 2-#9 top bars is 150.1 ft-kips. The distance from the face of the interior support to where the moment is reduced to 150.1 ft-kips must be determined. Based on a moment of 330.7 ft-kips at the interior support and 146.0 ft-kips at the exterior end and a factored dead load of $0.9 \times 4.07 = 3.66$ kips/ft, the location where the bars can be cut off is obtained by summing moments about section a-a in Fig. 4-16:

$$\frac{3.66x^2}{2} + 330.7 - 54.6x = 150.1$$

Solving this equation gives x = 3.8 ft. Four of the 6-#9 top bars at the interior support can be discontinued at (x + d) = (3.8 + 17.56/12) = 5.3 ft (UBC 1912.10.3). By similar calculations, three of the five #9 bars at the exterior support can be discontinued at 4.1 ft away from that support.

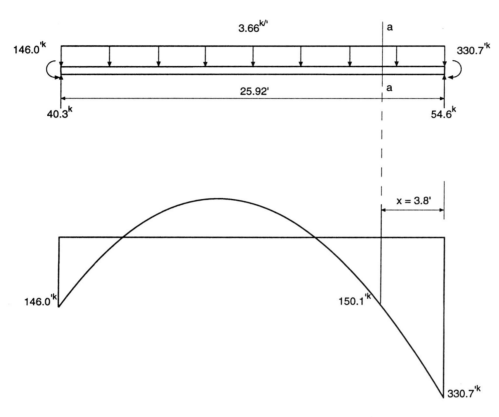

Figure 4-16 Moment Diagram for Cutoff Location of Negative Bars at Interior Support of Beam A3-B3
(Seismic Zone 2B)

Reinforcing bars must be properly developed at the support. The development length is the product of the basic development length and the factors presented in Table 4-9. The basic development length for a #9 bar is:

$$\ell_{db} = 0.04 \times A_b \times f_y / \sqrt{f'_c}$$

$$= 0.04 \times 1.0 \times 60,000 / \sqrt{4000} = 38.0 \text{ in.}$$

The information required for determining the development length factor is:

Minimum clear cover   = 2.0 in. ($1.77d_b$)

Side cover   large (slab)

Clear bar spacing   = 1.85 in. ($1.64 \, d_b$)

Transverse reinforcement meets requirements of UBC 1911.5.4 and 1911.5.5.3.

Factor I has a value of 2.0. Factor II has a value of 1.3 for the top bars. All other factors are 1.0; therefore,

Top bar $\ell_d = 1.3 \times 2.0 \times 1.0 \times 1.0 \times 1.0 \times 38.0/12 = 8.2 \text{ ft} > 5.3 \text{ ft}$

In continuous members at least one-fourth of the positive reinforcement must extend into the support (UBC 1912.11.1). Since four bottom reinforcing bars are required at midspan and at least two bars are provided at both supports, this criterion is satisfied.

Cutoff points for two of the four positive reinforcing bars at midspan will be determined. The load combination used in this case is 1.4D + 1.7L. The moment at the exterior column is 299.3 ft-kips, at the interior column it is 399.1 ft-kips (see Table 4-15), and the distributed load is 7.13 kips/ft. The capacity of 2-#9 bars with the beam section being designed is 150.1 ft-kips. The location where the bars can be cut off is obtained by summing moments about section a-a in Fig. 4-17:

$$\frac{7.13x^2}{2} + 399.1 - 96.2x = -150.1$$

Solution of this equation gives x = 8.2 ft. Two of the 4-#9 bars can be terminated at $[(25.92/2) - 8.2 + (17.56/12)]$ = 6.2 ft from the midspan section, which is less than the required ℓ_d of 8.2 ft.

The information required for determining the splice lengths is:

	Top Bars	Bottom Bars
Stress level	< 0.5 f_y (midspan)	< 0.5 f_y (support)
Splice type	A	A
Minimum clear cover	2.0 in. (1.77 d_b)	2.0 in. (1.77 d_b)
Side Cover	large (slab)	2.0 in. (1.77 d_b)
Clear bar spacing	13.74 in. (12.2 d_b)	13.74 in. (12.2 d_b)

Transverse reinforcement meets requirements of UBC 1911.5.4 and 1911.5.5.3 for both top and bottom bars.

This gives a value of 1.0 for Factor I for the bottom bars and 0.85 for the top bars. The splice lengths are:

Top splice length (Class A) = $1.0 \times (0.85 \times 1.3 \times 1.0 \times 1.0 \times 1.0 \times 38.0) = 42.0$ in.

Bottom splice length (Class A) = $1.0 \times (1.0 \times 1.0 \times 1.0 \times 1.0 \times 38.0) = 38.0$ in.

Figure 4-18 shows the reinforcement details for beam A3-B3.

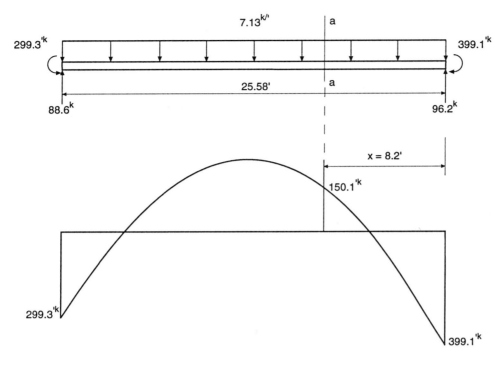

Figure 4-17 Moment Diagram for Cutoff Location of Positive Bars at Interior Support of Beam A3-B3 (Seismic Zone 2B)

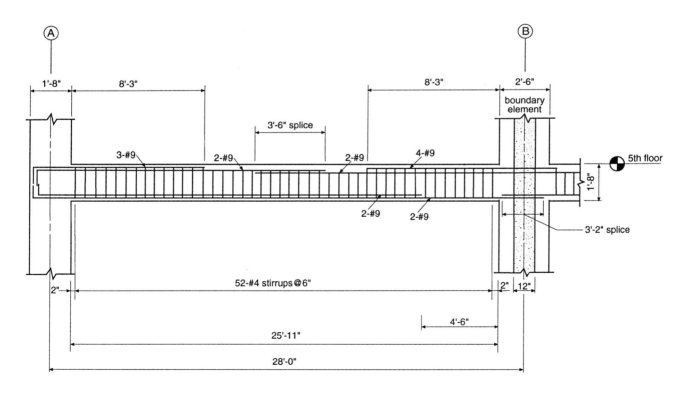

Figure 4-18 Reinforcement Details for Beam A3-B3 (Seismic Zone 2B)

4.3.2.2 Proportioning and Detailing Beam C2-D2

4.3.2.2.1 Required Flexural Reinforcement

The maximum moment at both ends of the beam is 386.4 ft-kips (see Table 4-16). The required amount of reinforcement, ignoring the effects of any compressive reinforcement, is $A_s = 5.71$ in.^{2} Use 6-#9 bars at these locations ($\phi M_n = 402.4$ ft-kips).

Check limitations on the reinforcement ratio:

$$\rho \quad = A_s/bd = 6.0/(20 \times 17.56) = 0.0171$$

$$\rho_{max} \quad = 0.0213 > 0.0171 > 200/f_y = 0.0030 \quad \text{O.K.}$$

1910.3.3
1910.5.1

According to UBC 1921.8.4.1, the positive moment strength at a joint face shall not be less than one-third of the corresponding negative moment strength at that joint. Thus, the positive moment reinforcement must not give a design moment strength less than 402.4/3 = 134.1 ft-kips. From Table 4-16 it can be seen that none of the load combinations produce a positive moment at the supports. Therefore, the required reinforcement is 2-#9, giving a design moment strength of 150.1 ft-kips.

The maximum positive moment is 193.2 ft-kips. The positive moment strength at any section along the member length must not be less than 20% of the maximum moment strength provided at the face of either joint (UBC 1921.8.4.1); thus, 402.4/5 = 80.5 ft-kips which is less than 193.2 ft-kips. Use 3-#9 bars, giving a design moment strength of $\phi M_n = 219.1$ ft-kips.

Similarly, the negative moment strength at any section along the member length must not be less than 20% of the maximum moment strength provided at the face of either joint. Thus, the minimum value is 80.5 ft-kips. Considering minimum reinforcement requirements, use 2-#9 bars, which gives a design moment strength of $\phi M_n = 150.1$ ft-kips.

4.3.2.2.2 Required Shear Reinforcement

The design shear strength of beams resisting Seismic Zone 2 earthquake effects must not be less than the sum of the shear associated with the development of the nominal moment strengths at the member ends plus the gravity shear forces (UBC 1921.8.3). Figure 4-19 shows the nominal moments at the joint faces for sidesway to the right and to the left. The unfactored gravity loads and the resulting shear forces at the joint faces are shown as well.

The shear strength provided by the concrete is:

$$V_c \quad = 2\sqrt{f'_c} \times b_w d \qquad \qquad \text{1911.3.1.1}$$

$$= 2\sqrt{4000} \times (20 \times 17.56)/1000 = 44.4 \text{ kips}$$

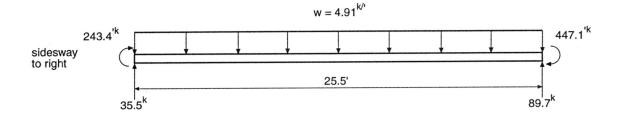

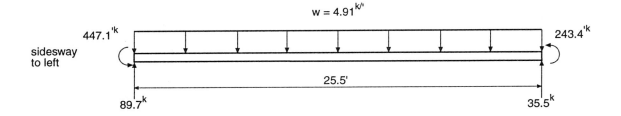

Figure 4-19 Design Shear Forces for Beam C2-D2 (Seismic Zone 2B)

The required strength to be provided by shear reinforcement becomes:

$$\phi V_s = V_u - \phi V_c = 89.7 - (0.85 \times 44.4) = 52.0 \text{ kips}$$

or $V_s = 52.0/0.85 = 61.2$ kips.

The required spacing of #4 stirrups is:

$$s = \frac{A_v f_y d}{V_s} = \frac{0.40 \times 60 \times 17.56}{61.2} = 6.9 \text{ in.}$$

The maximum allowable stirrup spacing within a distance of 2h = 2 × 20 = 40 in. from the faces of supports is:

$$s_{max} = \frac{d}{4} = \frac{17.56}{4} = 4.4 \text{ in.} \quad \text{(governs)} \qquad\qquad 1921.8.4.2$$

$$= 8 \times \text{(diameter of smallest longitudinal bar)} = 8 \times 1.128 = 9.0 \text{ in.}$$

$$= 24 \times \text{(diameter of stirrup bars)} = 24 \times 0.5 = 12.0 \text{ in.}$$

$$= 12.0 \text{ in.}$$

Use #4 stirrups at a 4.0 in. spacing within 40 in. from the face of each support.

Beyond a distance 2h from the supports, the maximum stirrup spacing is d/2 = 17.56/2 = 8.8 in. (UBC 1921.8.4.3).

The shear at a distance of 2h = 40 in. = 3.3 ft from the face of the support is:

$$V_u \quad = 89.7 - (4.91 \times 3.3) = 73.5 \text{ kips}$$

$$\phi V_s \quad = 73.5 - (0.85 \times 44.4) = 35.8 \text{ kips}$$

$$V_s \quad = 35.8/0.85 = 42.1 \text{ kips}$$

$$s \quad = \frac{0.40 \times 60 \times 17.56}{42.1} = 10.0 \text{ in.} > 8.8 \text{ in.}$$

Use #4 stirrups at an 8.0 in. spacing throughout the remaining portion of the beam.

4.3.2.2.3 Reinforcing Bar Cutoff Points and Splices

The negative reinforcement at the supports is 6-#9 bars. The location where three of the #9 bars can be terminated will be determined. The loading used to find the cutoff points is 0.9 times the dead load in combination with the nominal moment strength, M_n, at the member ends (using $f_y = f_s = 60$ ksi and $\phi = 1.0$). The design moment strength, ϕM_n, provided by 3-#9 reinforcing bars is 219.1 ft-kips.

With $\phi = 1.0$ and $f_s = f_y = 60$ ksi, $M^+_n = 243.4$ ft-kips and $M^-_n = 447.1$ ft-kips. The dead load is equal to $0.9 \times 4.07 = 3.66$ kips/ft. The distance from the face of the right support to where the moment under the loading considered equals 219.1 ft-kips is readily obtained by summing moments about section a-a in Fig. 4-20:

$$\frac{3.66x^2}{2} + 447.1 - 73.7x = 219.1$$

Solution of this equation gives x = 3.4 ft. Therefore, the negative reinforcing bars can be terminated at (x + d) = (3.4 + 17.56/12) = 4.9 ft from the face of support.

The #9 bars must be properly developed at the support. The development length is the product of the basic development length and the factors previously presented in Table 4-9. The basic development length for a #9 bar is:

$$\ell_{db} = 0.04 A_b f_y / \sqrt{f'_c} = 0.04 \times 1.0 \times 60,000/\sqrt{4000} = 38.0 \text{ in.}$$

The information required for determining the development length is:

Minimum clear cover   = 2.0 in. ($1.77d_b$)

Side cover   large (slab)

Clear bar spacing   = 1.85 in. ($1.64 \ d_b$)

Transverse reinforcement meets requirements of UBC 1911.5.4, 1911.5.5.3, and 1912.2.3.5.

Factor I has a value of 1.5 and Factor II had a value of 1.3 for the top bars. All the other factors are 1.0; thus,

Top bar $\ell_d = 1.3 \times 1.5 \times 1.0 \times 1.0 \times 1.0 \times 38.0/12 = 6.2$ ft > 4.9 ft

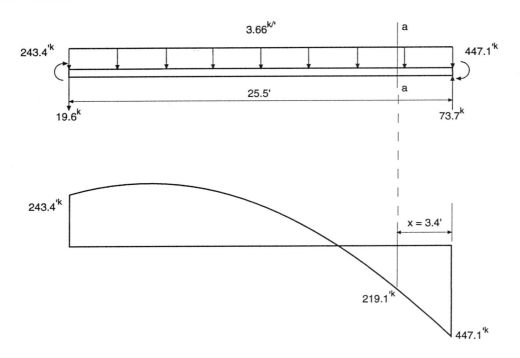

Figure 4-20 Moment Diagram for Cutoff Location of Negative Bars at Support of Beam C2-D2
(Seismic Zone 2B)

Reinforcing bars can be terminated in tension zones if one of the criteria given in UBC 1912.10.5 is satisfied. In this case, the governing criterion is UBC 1912.10.5.1, which allows termination in a tension zone if the shear at that location is less than two-thirds the shear capacity ϕV_n. Using #4 stirrups at 8 in., two-thirds of ϕV_n is:

$$\frac{2}{3}\phi V_n = \frac{2}{3} \times 0.85 \left(44.4 + \frac{0.4 \times 60 \times 17.56}{8.0} \right) = 55.0 \text{ kips}$$

The location from the face of the support where $V_u = 55.0$ kips is:

$$55.0 = 89.7 - (4.91x)$$

$$x = 7.1 \text{ ft}$$

Therefore, 3 of the 6-#9 top bars can be terminated at 7 ft-3 in. from the faces of the supports.

Given the length of the beam and its position between the shearwalls, splicing of the flexural reinforcement is not required. Figure 4-21 shows the reinforcement details for beam C2-D2.

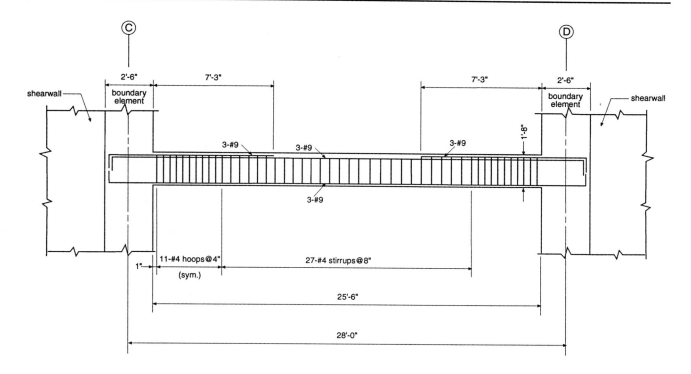

Figure 4-21 Reinforcement Details for Beam C2-D2 (Seismic Zone 2B)

4.3.2.3   Proportioning and Detailing Column A3

4.3.2.3.1 General Requirements

This section will detail the design of exterior column A3 supporting the fifth floor of the structure. The maximum factored axial force on this column at this level is 759.1 kips (see Table 4-17). A 20 in. square tied column with 8-#7 bars (ρ_g = 1.2%) will be adequate for the load combinations shown in Table 4-17. An interaction diagram for this column is shown in Fig. 4-22. Note that slenderness effects need not be considered since P-Δ effects were included in the analysis.

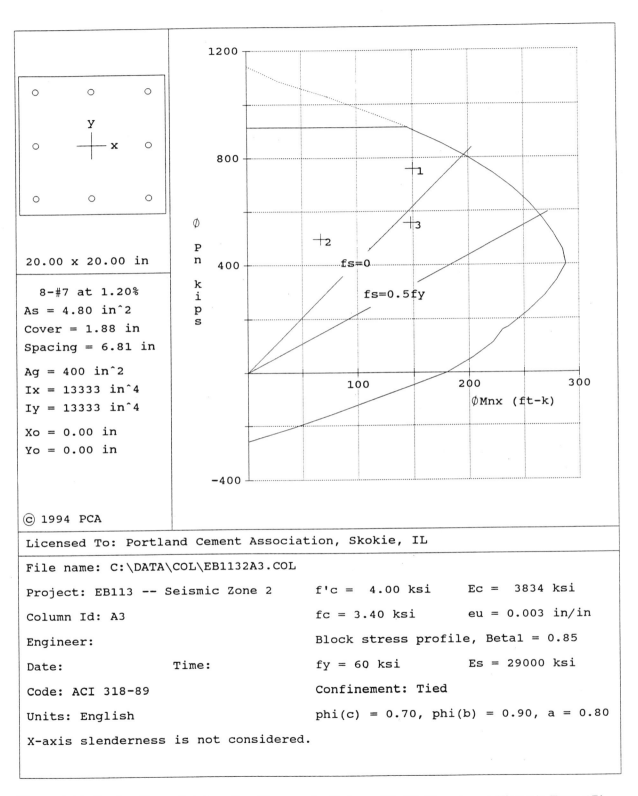

Licensed To: Portland Cement Association, Skokie, IL

File name: C:\DATA\COL\EB1132A3.COL

Project: EB113 -- Seismic Zone 2   f'c = 4.00 ksi   Ec = 3834 ksi

Column Id: A3   fc = 3.40 ksi   eu = 0.003 in/in

Engineer: Block stress profile, Beta1 = 0.85

Date: Time: fy = 60 ksi   Es = 29000 ksi

Code: ACI 318-89   Confinement: Tied

Units: English   phi(c) = 0.70, phi(b) = 0.90, a = 0.80

X-axis slenderness is not considered.

Figure 4-22 Design Strength Interaction Diagram for Column A3, 5th Floor Level (Seismic Zone 2B)

4.3.2.3.2 Confinement Reinforcement

Special transverse reinforcement for confinement is required over a distance ℓ_o at column ends where ℓ_o equals the maximum of:

$$\frac{1}{6} \text{ (clear height of column)} = \frac{(11.33 \times 12)}{6} = 22.7 \text{ in. (governs)} \qquad \textit{1921.8.5.1}$$

Depth of member = 20.0 in.

18.0 in.

The maximum allowable spacing of #3 ties over the length ℓ_o is:

$$s_o \quad = 8 \times \text{(diameter of smallest longitudinal bar)} = 8 \times 0.875 = 7.0 \text{ in. (governs)}$$

$$= 24 \times \text{(diameter of tie bars)} = 24 \times 0.375 = 9.0 \text{ in.}$$

$$= \frac{h}{2} = \frac{20}{2} = 10.0 \text{ in.}$$

The maximum spacing of ties outside of the distance ℓ_o is $2s_o = 14.0$ in. (UBC 1921.8.5.4).

4.3.2.3.3 Transverse Reinforcement for Shear

The column shear caused by gravity loads is determined by taking the difference in beam moments on each side of the column and dividing by the clear column height. An exterior column will have a moment on one side only. The beam moment at the column face is equal to $w_u \ell_n^2/16$. Therefore, the unfactored column shear caused by the dead load is:

$$V_D \quad = \frac{w_D \ell_n^2}{16\ell_c}$$

$$= \frac{4.07 \times (28 - 20/24 - 30/24)^2}{16 \times (13 - 20/12)} = 15.1 \text{ kips}$$

The unfactored column shear caused by live loads is:

$$V_L \quad = \frac{w_L \ell_n^2}{16\ell_c}$$

$$= \frac{0.84 \times (28 - 20/24 - 30/24)^2}{16 \times (13 - 20/12)} = 3.1 \text{ kips}$$

From Table 4-14, the compatibility shear is:

$$V_{compatibility} = 6.2 \text{ kips}$$

The load combinations for the column are:

$$V_u = 1.4V_D + 1.7V_L = 26.4 \text{ kips (governs)}$$

$$V_u = V_D + V_L + V_{compatibility} = 24.4 \text{ kips}$$

For members subject to axial compression, the shear strength provided by the concrete is:

$$V_c = 2\left(1 + \frac{N_u}{2000A_g}\right)\sqrt{f'_c}\, b_w d \qquad\qquad\qquad 1911.3.1.2$$

From Table 4-17, the minimum factored axial load on the column is 497.4 kips. Therefore,

$$\phi V_c = 0.85 \times 2\left(1 + \frac{497,400}{2000 \times 400}\right)\sqrt{4000} \times 20 \times 17.69 \,/\, 1000$$

$$= 61.7 \text{ kips} > 2V_u = 2 \times 24.4 = 48.8 \text{ kips}$$

Since $\phi V_c = 61.7$ kips is greater than $2V_u = 48.8$ kips, shear reinforcement is not required. Therefore, a 14.0 in. spacing is sufficient in the region of the column outside of ℓ_o.

Class B splices will be used, assuming that more than one-half of the bars are spliced at any one section (UBC 1912.17.2.2). The basic development length is:

$$\ell_{db} = \frac{0.04A_b f_y}{\sqrt{f'_c}} = \frac{0.04 \times 0.6 \times 60,000}{\sqrt{4000}} = 22.8 \text{ in.} \qquad\qquad 1912.2.2$$

The information required for determining the development length is:

          Minimum clear cover   = 1.88 in. ($2.15\, d_b$)

          Side cover   = 1.88 in. ($2.15\, d_b$)

          Clear bar spacing   = 6.81 in. ($7.79\, d_b$)

          Transverse reinforcement meets requirements of UBC 1907.10.5.

This gives a value of 1.09 for Factor I (see Table 4-9). The other factors are all equal to 1.0, giving:

          Class B splice length = $1.3 \times (1.09 \times 1.0 \times 1.0 \times 1.0 \times 1.0 \times 22.8) = 32.3$ in.

Figure 4-23 shows the reinforcement details for column A3.

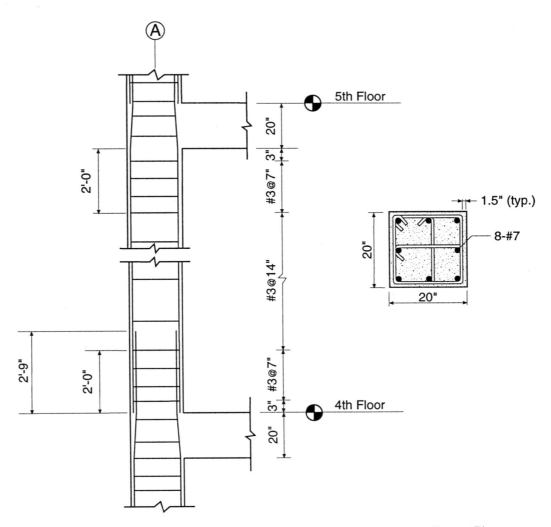

Figure 4-23 Reinforcement Details for Column A3 (Seismic Zone 2B)

4.3.2.4 Proportioning and Detailing a Shearwall

4.3.2.4.1 General

The shearwalls are part of the lateral load resisting system. Since this design is for Seismic Zone 2B, the only special detailing requirements are those in UBC 1921.8. Review of this section shows that none of these detailing requirements applies to shearwalls. This points out the advantage of the Building Frame System in Seismic Zone 2.

4.3.2.4.2 Design of Shearwall D2-E2

Boundary elements are not required in Seismic Zone 2B. The ends of the shearwall in this example are thickened to maintain a regular column layout and to provide increased moment strength of the shearwall. It is assumed that there are 24-#10 bars in each of the 30 × 30 in. segments of the wall.

The design shear strength of a shearwall is limited to:

$$\phi V_n = \phi 10\sqrt{f'_c} \times hd \qquad\qquad\qquad 1911.10.3$$

where d equals 80% of the length of the wall (UBC 1911.10.4). In this case, ϕV_n is:

$$\phi V_n = 0.85 \times 10\sqrt{6000} \times 12 \times (0.8 \times 366)/1000 = 2313.4 \text{ kips}$$

Using Table 4-11, the maximum factored shear at the base of the wall is:

$$V_u = (0.75 \times 1.87) \times 415.3 = 581.4 \text{ kips} < 2313.4 \text{ kips} \quad \text{O.K.}$$

The shear strength provided by the concrete is:

$$\phi V_c = \phi 2\sqrt{f'_c} \times hd \qquad\qquad\qquad 1911.10.5$$

$$= 0.85 \times 2\sqrt{6000} \times (0.8 \times 366) \times 12/1000$$

$$= 462.7 \text{ kips} < V_u = 581.4 \text{ kips}$$

Therefore, shear reinforcement is required. The maximum spacing for #5 horizontal bars is:

$$s_{max} \le \ell_w/5 = 366/5 = 73.2 \text{ in.} \qquad\qquad 1911.10.9.3$$

$$\le 3h = 3 \times 12 = 36.0 \text{ in.}$$

$$\le 18.0 \text{ in. (governs)}$$

The design shear strength of the wall with #5 horizontal bars spaced at 18 in. on both faces of the wall is:

$$\phi V_n = \phi(V_s + V_c)$$

$$\phi V_n = \phi A_{vh}f_yd/s + \phi V_c \qquad\qquad\qquad 1911.10.9.1$$

$$= 0.85 \times 0.62 \times 60 \times (0.8 \times 366)/18 + 462.7$$

$$= 977.1 \text{ kips} > V_u = 581.4 \text{ kips} \quad \text{O.K.}$$

The ratio of horizontal shear reinforcement area to gross concrete area of the vertical cross section must not be less than 0.0025 (UBC 1911.10.9.2). Therefore, the minimum area of horizontal steel per foot length of wall is:

$$A_{vh} = 0.0025 \times 12 \times 12 = 0.36 \text{ in.}^2/\text{ft}$$

For 2-#5 @ 18 in., $A_{vh} = 0.62 \times 12/18 = 0.41 \text{ in.}^2/\text{ft}$ which is greater than the minimum allowed. Therefore, use two layers of #5 horizontal bars spaced at 18 in.

The ratio of vertical shear reinforcement area to gross area of horizontal section must not be less than:

$$\rho_n = 0.0025 + 0.5 \, (2.5 - h_w/\ell_w) \, (\rho_h - 0.0025) \text{ or } 0.0025 \qquad \textit{1911.10.9.4}$$

With $\rho_h = 0.41/(12 \times 12) = 0.0028$, the minimum ρ_n is:

$$\rho_n = 0.0025 + 0.5 \, [2.5 - 140/(28 + 30/12)] \times (0.0028 - 0.0025)$$

$$= 0.0022 < 0.0025$$

Therefore, use $\rho_n = 0.0025$. The required vertical steel area is:

$$A_n = 0.0025 \times 12 \times 12 = 0.36 \text{ in.}^2/\text{ft}$$

Assuming two layers of #5 bars, the required spacing is $(0.62 \times 12)/0.36 = 20.7$ in.

The maximum spacing for the vertical #5 bars is:

$$s_{max} \le \ell_w/3 = 366/3 = 122.0 \text{ in.} \qquad \textit{1911.10.9.5}$$

$$\le 3h = 3 \times 12 = 36.0 \text{ in.}$$

$$\le 18.0 \text{ in. (governs)}$$

Therefore, use two layers of #5 vertical bars spaced at 18 in.

Based on the wall dimensions and vertical reinforcement given in this section, the interaction diagram shown in Fig. 4-24 can be obtained from *PCACOL*. Also given in the figure are the points representing the load combinations obtained from Table 4-18. It is clear from the figure that the wall is adequate to carry the loads. Reinforcement details for shearwall D2-E2 are given in Fig. 4-25.

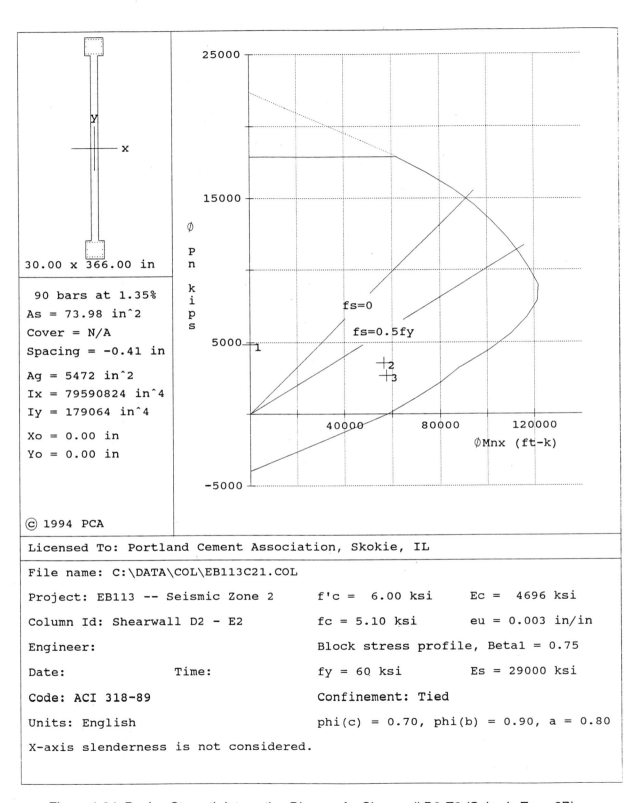

30.00 x 366.00 in

90 bars at 1.35%
As = 73.98 in^2
Cover = N/A
Spacing = -0.41 in

Ag = 5472 in^2
Ix = 79590824 in^4
Iy = 179064 in^4

Xo = 0.00 in
Yo = 0.00 in

ⓒ 1994 PCA

Licensed To: Portland Cement Association, Skokie, IL

File name: C:\DATA\COL\EB113C21.COL

Project: EB113 -- Seismic Zone 2   f'c = 6.00 ksi   Ec = 4696 ksi

Column Id: Shearwall D2 - E2   fc = 5.10 ksi   eu = 0.003 in/in

Engineer: Block stress profile, Beta1 = 0.75

Date: Time: fy = 60 ksi   Es = 29000 ksi

Code: ACI 318-89   Confinement: Tied

Units: English   phi(c) = 0.70, phi(b) = 0.90, a = 0.80

X-axis slenderness is not considered.

Figure 4-24 *Design Strength Interaction Diagram for Shearwall D2-E2 (Seismic Zone 2B)*

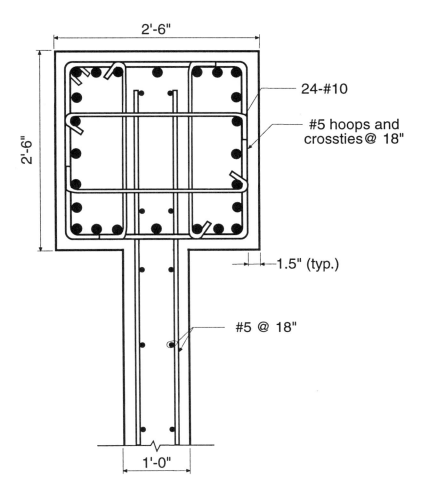

Figure 4-25 Reinforcement Details for Shearwall D2-E2 (Seismic Zone 2B)

Reference

4.1   *SAP90 — A Series of Computer Programs for the Finite Element Analysis of Structures,* Computers and
Structures, Inc., Berkeley, CA, 1992.

Chapter 5

Dual System

5.1   Introduction

5.1.1   General

The computation of the design loads for a 20-story reinforced concrete structure under the requirements of the Uniform Building Code (UBC), 1994 edition, is illustrated. These computations are performed for UBC Seismic Zones 4 and 2B. Typical members are designed and detailed for both seismic zones. The building employs a dual shearwall-frame interactive lateral load resisting system.

In a dual system, the specially detailed moment-resisting frames and the shearwalls must be designed to resist the total design base shear in proportion to their relative rigidities considering interaction between the two subsystems at all levels. An additional safeguard against total collapse is provided by requiring the moment-resisting frames to be capable of resisting at least 25% of the base shear without the benefit of the shearwalls (UBC 1627.6.5). The R_w value for the dual system is 12 or 9 when the moment-resisting frames are detailed as Special Moment-Resisting Frames or Intermediate Moment-Resisting Frames, respectively (UBC Table 16-N).

It is worthwhile to compare the properties of the dual system and the building frame system as shown in Table 5-1. A major difference in these systems is the trade-off between additional required special detailing with the significant reduction in the base shear when using the dual system.

5.1.2   Design Criteria

A typical plan and elevation of the structure are shown in Figs. 5-1 and 5-2, respectively. The member sizes for the structure are as follows:

	Seismic Zone 4	Seismic Zone 2B
Spandrel beams	27 × 36 in.	32 × 20 in.
Interior beams	24 × 24 in.	32 × 20 in.
Columns	36 × 36 in.	32 × 32 in.
Joists and topping	86 psf	86 psf
Shearwalls		
Grade to 9th floor	16 in. thick	12 in. thick
10th floor to 16th floor	14 in. thick	12 in. thick
17th floor to roof	12 in. thick	12 in. thick
Shearwall boundary elements	40 × 40 in.	—

Table 5-1 Comparison of Dual System to Building Frame System

	Dual System	Building Frame System
Lateral Load Resisting System	Shearwalls and beam-column frames carry lateral loads in proportion to their relative rigidities.	Shearwalls carry entire lateral loads.
Additional Requirements	Beam-column frames alone must be able to resist 25% of the base shear.	Beam-column frames must be able to withstand $(3R_w/8)$ times the deflection of the shearwalls.
Special Detailing Requirements	All members must be specially detailed.	Beam-column frames do not require special detailing.
R_w Value	$R_w = 12$ when beam-column frames are Special Moment-Resisting Frames. $R_w = 9$ when beam-column or slab-column frames are Intermediate Moment-Resisting Frames. ** Lateral loads are less than those for the same structure employing a Building Frame System	$R_w = 8$
Height Limitations*	$R_w = 12$: No limit $R_w = 9$: 160 ft	240 ft

*Height limits applicable to Seismic Zones 3 and 4 only. See UBC 1627.7.
** Prohibited in Seismic Zones 3 and 4.

Other pertinent design data are as follows:

Service Loads:
    Superimposed dead load: 20 psf
    Live load: 50 psf
    Roofing: 10 psf

Material Properties:

Concrete: f'_c = 6000 psi (vertical members up to 6th floor)
= 4000 psi (all others)

All horizontal members are constructed of lightweight concrete (w_c = 115 pcf).

Reinforcement: f_y = 60,000 psi

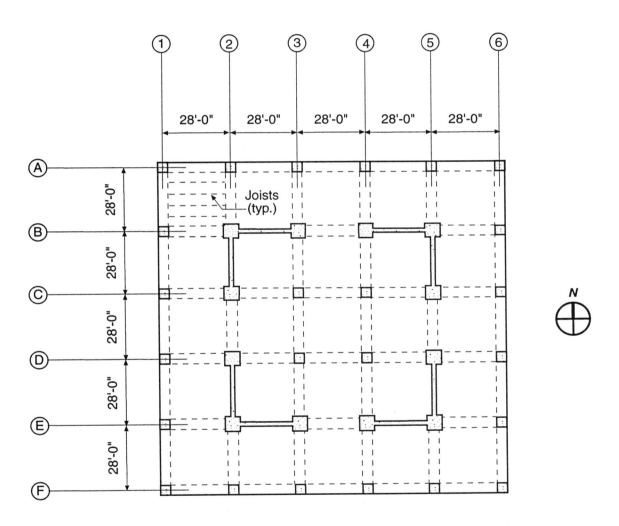

Figure 5-1 Typical Floor Plan of Example Building

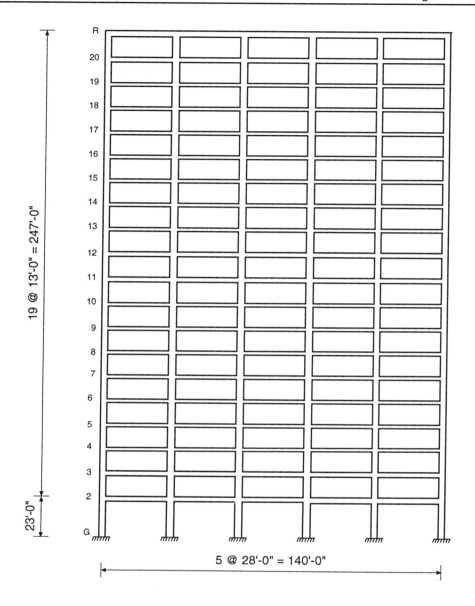

Figure 5-2 Elevation of Example Building

5.2   Design for Seismic Zone 4

5.2.1   General

This section will illustrate the design and detailing of members for Seismic Zone 4. Requirements for detailing in Seismic Zone 3 are identical to those for Zone 4. Also, the analysis for both zones is similar, except that the Seismic Zone 4 design lateral forces are 33% higher than the corresponding Seismic Zone 3 forces. Therefore, the design of the structure in Seismic Zone 3 is not separately addressed here.

5.2.2   Frame Analysis

In the majority of designs, the base shear caused by seismic forces and the vertical distribution are determined using UBC Eqs. (28-1) and (28-8) (UBC 1628). These equations are:

$$V \quad = \frac{ZICW}{R_w} \qquad (28\text{-}1)$$

$$F_x \quad = \frac{(V - F_t)w_x h_x}{\sum\limits_{i=1}^{n} w_i h_i} \qquad (28\text{-}8)$$

Calculation of the base shear and the distribution of the shear using these two equations may not be appropriate for buildings with severe setbacks, unusual configurations, considerable variations in story drift, or of significant height. For this reason, a building with a total height greater than 240 ft in Seismic Zone 3 or 4 (also non-standard occupancy buildings with heights exceeding 240 ft in Seismic Zone 2) must use a dynamic analysis procedure to determine the base shear and its distribution along the building height (UBC 1627.8.3).

The dynamic force procedure is performed using the steps outlined in Table 5-2 and discussed below.

Step I — *Determine mode shapes and corresponding periods of the structure.*

The vibration mode shapes are determined considering undamped, free vibration. The corresponding differential equation is:

$$[m] \{\ddot{u}\} + [k] \{u\} = \{0\}$$

where

$[m]$ = matrix of lumped masses at the floor levels
$[k]$ = stiffness matrix
$\{u\}$ = lateral displacement vector
$\{\ddot{u}\}$ = vector of corresponding accelerations

The solution of this equation is:

$$\{u\} \quad = \{A\} \sin (\omega t)$$

Substituting the solution into the differential equation:

$$-\omega^2 [m] \{A\} + [k] \{A\} = [0]$$

The amplitude, $\{A\}$, of the motion is not equal to zero, so it follows that the determinant of the matrix $-\omega^2[m]$ + $[k]$ must be equal to zero. Since $[m]$ and $[k]$ represent geometric properties of the structure, the only unknown quantity is the circular frequency, ω. Once the circular frequency is determined, the period is calculated as $T = 2\pi/\omega$.

Table 5-2 Outline of Dynamic Force Procedure

Step I — Determine mode shapes and corresponding periods of the structure.
Step II — Compute L_m and M_m for each mode shape. $$L_m = \sum_{i=1}^{n} w_i \phi_{im} \qquad\qquad M_m = \sum_{i=1}^{n} w_i \phi_{im}^2$$
Step III — Determine the modal seismic design coefficient, S^r_{am}, for each mode, using $$S^r_{am} = \text{spectral acceleration}/R_w$$
Step IV — Calculate the base shear, V, using modal analysis results: $$V_m = L_m^2 \times S^r_{am}/M_m$$ $$V = (V_1{}^2 + V_2{}^2 ..... V_n{}^2)^{1/2}$$
Step V — Calculate 80% of base shear using Eq. (28-1) from UBC 1628 and compare this to value from modal analysis. If modal analysis result is greater, go to Step VII.
Step VI — Scale up modal results.
Step VII — Distribute base shear throughout height of structure for each mode.
Step VIII — Perform lateral analysis for each mode.
Step IX — Combine modal results for each mode using root-mean-square combination.

The solution to the above eigenvalue problem is quite cumbersome except for very small, simple structures, beyond which the use of a computer program is required. Sample calculations showing the determination of the periods for a three-story building are given in Fig. 5-3. The values for the first five periods and mode shapes for the building shown in Figs. 5-1 and 5-2 are shown in Table 5-3. These quantities were obtained from the dynamic analysis module in *SAP90* [4.1].

Step II — *Compute L_m and M_m for each mode shape.*

The factors L_m and M_m are dependent on the weight and modal shape and are calculated as follows:

$$L_m = \sum_{i=1}^{n} w_i \phi_{im}$$

$$M_m = \sum_{i=1}^{n} w_i \phi_{im}^2$$

where w_i = weight for level i

ϕ_{im} = displacement at level i for mode m

Calculations for L_m and M_m are shown in Table 5-4.

Table 5-3 Modal Shapes and Periods for Example Building (Seismic Zone 4)

Mode, m	1	2	3	4	5
Modal Period (sec.)					
	1.752	0.419	0.187	0.114	0.081
Floor Level, i	Mode Shapes, ϕ_{im}				
R	1.00000	1.00000	1.00000	1.00000	1.00000
20	0.94502	0.78965	0.64218	0.46692	0.26538
19	0.88829	0.56747	0.25155	-0.11216	-0.48355
18	0.83043	0.34075	-0.13050	-0.60140	-0.93885
17	0.77128	0.11594	-0.45982	-0.87606	-0.90184
16	0.71115	-0.09870	-0.69577	-0.87094	-0.41197
15	0.65041	-0.29275	-0.80925	-0.61858	0.23517
14	0.58922	-0.46281	-0.79836	-0.19158	0.78541
13	0.52795	-0.60283	-0.66796	0.29332	0.99963
12	0.46713	-0.70772	-0.43783	0.70275	0.78020
11	0.40723	-0.77416	-0.14127	0.92490	0.22376
10	0.34883	-0.80054	0.18022	0.89773	-0.42554
9	0.29271	-0.78735	0.48051	0.62945	-0.87878
8	0.23993	-0.74015	0.71119	0.21497	-0.93751
7	0.19062	-0.66283	0.85442	-0.25308	-0.60479
6	0.14571	-0.56175	0.89423	-0.65651	-0.02479
5	0.10610	-0.45089	0.83985	-0.88899	0.51953
4	0.07142	-0.33415	0.70747	-0.93760	0.88272
3	0.04249	-0.21988	0.51989	-0.80626	0.95731
2	0.01999	-0.11578	0.30122	-0.52156	0.70481

• **Given:**

h_s = 10 ft

w = 390 kips/floor

E = 4000 ksi

I_{col} = 9000 in.^{4} each column

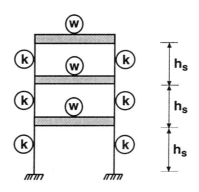

• **Determine mass matrix**

m = w/g = 390/386.4

= 1.0 kip-sec.^{2}/in.

$$[m] = \begin{bmatrix} 1 & 0 & 0 \\ 0 & 1 & 0 \\ 0 & 0 & 1 \end{bmatrix}$$

• **Determine stiffness matrix**

$12EI/h_s^3 = 12 \times 4000 \times 9000/(12 \times 10)^3 = 250$ kips/in.

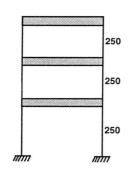
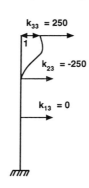
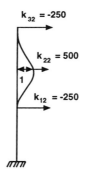
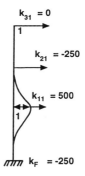

k_{ij} = force corresponding to displacement of coordinate i resulting from a unit displacement of coordinate j

$$[k] = 250 \begin{bmatrix} 2 & -1 & 0 \\ -1 & 2 & -1 \\ 0 & -1 & 1 \end{bmatrix}$$

Figure 5-3 Sample Period Calculation

• **Find determinant for matrix [k] - ω^2[m]**

$$[k] - \omega^2[m] = \begin{bmatrix} 500 - \omega^2 & -250 & 0 \\ -250 & 500 - \omega^2 & -250 \\ 0 & -250 & 250 - \omega^2 \end{bmatrix}$$

Setting the determinant of the above matrix equal to zero yields the following frequencies:

$\omega_1 = 7.036$ radians/sec.

$\omega_2 = 19.685$ radians/sec.

$\omega_3 = 28.491$ radians/sec.

The period is equal to $2\pi / \omega$:

$T_1 = 0.893$ sec.

$T_2 = 0.319$ sec.

$T_3 = 0.221$ sec.

• **Find modal shapes**

First mode:

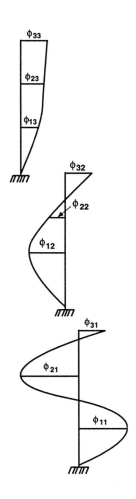

$$\begin{bmatrix} 500 - (7.036)^2 & -250 & 0 \\ -250 & 500 - (7.036)^2 & -250 \\ 0 & -250 & 250 - (7.036)^2 \end{bmatrix} \begin{bmatrix} \phi_{33} \\ \phi_{23} \\ \phi_{13} \end{bmatrix} = \begin{bmatrix} 0 \\ 0 \\ 0 \end{bmatrix}$$

$\phi_{33} = 1.0, \quad \phi_{23} = 0.802, \quad \phi_{13} = 0.445$

Second mode:

$$\begin{bmatrix} 500 - (19.685)^2 & -250 & 0 \\ -250 & 500 - (19.685)^2 & -250 \\ 0 & -250 & 250 - (19.685)^2 \end{bmatrix} \begin{bmatrix} \phi_{32} \\ \phi_{22} \\ \phi_{12} \end{bmatrix} = \begin{bmatrix} 0 \\ 0 \\ 0 \end{bmatrix}$$

$\phi_{32} = 1.0, \quad \phi_{22} = -0.55, \quad \phi_{12} = -1.22$

Third mode:

$$\begin{bmatrix} 500 - (28.491)^2 & -250 & 0 \\ -250 & 500 - (28.491)^2 & -250 \\ 0 & -250 & 250 - (28.491)^2 \end{bmatrix} \begin{bmatrix} \phi_{31} \\ \phi_{21} \\ \phi_{11} \end{bmatrix} = \begin{bmatrix} 0 \\ 0 \\ 0 \end{bmatrix}$$

$\phi_{31} = 1.0, \quad \phi_{21} = -2.25, \quad \phi_{11} = 1.802$

Figure 5-3 (continued)

Table 5-4 Calculation of L_m and M_m for Example Building (Seismic Zone 4)

Floor Level, i	Mode 1, T_1 = 1.752 sec.				Mode 2, T_2 = 0.419 sec.			Mode 3, T_3 = 0.187 sec.		
	w_i	ϕ_{i1}	$w_i \phi_{i1}$	$w_i \phi_{i1}^2$	ϕ_{i2}	$w_i \phi_{i2}$	$w_i \phi_{i2}^2$	ϕ_{i3}	$w_i \phi_{i3}$	$w_i \phi_{i3}^2$
R	3246	1.00000	3246	3246	1.00000	3246	3246	1.00000	3246	3246
20	3582	0.94502	3385	3199	0.78965	2829	2234	0.64218	2300	1477
19	3582	0.88829	3182	2826	0.56747	2033	1153	0.25155	901	227
18	3582	0.83043	2975	2470	0.34075	1221	416	-0.13050	-467	61
17	3582	0.77128	2763	2131	0.11594	415	48	-0.45982	-1647	757
16	3582	0.71115	2547	1812	-0.09870	-354	35	-0.69577	-2492	1734
15	3646	0.65041	2371	1542	-0.29275	-1067	312	-0.80925	-2951	2388
14	3646	0.58922	2148	1266	-0.46281	-1687	781	-0.79836	-2911	2324
13	3646	0.52795	1925	1016	-0.60283	-2198	1325	-0.66796	-2435	1627
12	3646	0.46713	1703	796	-0.70772	-2580	1826	-0.43783	-1596	699
11	3646	0.40723	1485	605	-0.77416	-2823	2185	-0.14127	-515	73
10	3646	0.34883	1272	444	-0.80054	-2919	2337	0.18022	657	118
9	3646	0.29271	1067	312	-0.78735	-2871	2260	0.48051	1752	842
8	3710	0.23993	890	214	-0.74015	-2746	2032	0.71119	2639	1876
7	3710	0.19062	707	135	-0.66283	-2459	1630	0.85442	3170	2708
6	3710	0.14571	541	79	-0.56175	-2084	1171	0.89423	3318	2967
5	3710	0.10610	394	42	-0.45089	-1673	754	0.83985	3116	2617
4	3710	0.07142	265	19	-0.33415	-1240	414	0.70747	2625	1857
3	3710	0.04249	158	7	-0.21988	-816	179	0.51989	1929	1003
2	4325	0.01999	86	2	-0.11578	-501	58	0.30122	1303	392
Σ	73,263		33,110	22,161		-18,274	24,397		11,940	28,993

Mode 1	Mode 2	Mode 3
L_1 = 33,110	L_2 = -18,274	L_3 = 11,940
M_1 = 22,161	M_2 = 24,397	M_3 = 28,993

Step III — *Determine modal seismic design coefficients, S^r_{am}, for each mode.*

The seismic design coefficient is determined as the spectral acceleration divided by R_w. The corresponding equation is:

$$S^r_{am} = \text{spectral acceleration}/R_w$$

$$= \frac{(Z \times I \times 1.25 \times S)}{R_w T_m^{2/3}}$$

For Z = 0.4, I = 1.0, R_w = 12, and S = 1.2, this equation reduces to:

$$S^r_{am} = \frac{0.05}{T_m^{2/3}}$$

Therefore, for Mode 1: $S^r_{a1} = \dfrac{0.05}{(1.752)^{2/3}} = 0.034$

Mode 2: $S^r_{a2} = \dfrac{0.05}{(0.419)^{2/3}} \doteq 0.089$

Mode 3: $S^r_{a3} = \dfrac{0.05}{(0.187)^{2/3}} = 0.153$

Step IV — *Calculate base shear using modal analysis.*

$$V_m = \frac{L_m^2}{M_m} S^r_{am}$$

$V_1 \text{ (Mode 1)} = \dfrac{(33,110)^2 \times 0.034}{22,161} = 1682 \text{ kips}$

$V_2 \text{ (Mode 2)} = \dfrac{(-18,274)^2 \times 0.089}{24,397} = 1218 \text{ kips}$

$V_3 \text{ (Mode 3)} = \dfrac{(11,940)^2 \times 0.153}{28,993} = 752 \text{ kips}$

$V = [(1682)^2 + (1218)^2 + (752)^2]^{1/2} = 2209 \text{ kips}$

In the above calculations, only the first three of many possible modes were considered in determining V. The number of significant modes to be considered is required to be the number of modes that will include at least 90% of the participating mass of the structure (UBC 1629.5.1). The participating mass for each mode is determined as:

$$\text{Participating Mass (PM)} = \frac{L_m^2}{M_m \times \displaystyle\sum_{i=1}^{n} w_i}$$

$\text{PM (Mode 1)} = \dfrac{(33,110)^2}{22,161 \times 73,263} = 0.675$

$\text{PM (Mode 2)} = \dfrac{(-18,274)^2}{24,397 \times 73,263} = 0.187$

$$\text{PM (Mode 3)} \quad = \frac{(11,940)^2}{28,993 \times 73,263} = 0.067$$

$$\Sigma\text{PM} \quad = 0.675 + 0.187 + 0.067 = 0.929 > 0.900$$

Therefore, using only the first three modes is sufficient.

Step V — *Calculate base shear using UBC Eq. (28-1).*

The base shear is equal to:

$$V \quad = \frac{ZICW}{R_w}$$

where
$$Z \quad = 0.4$$

$$R_w \quad = 12$$

$$I \quad = 1.0$$

$$S \quad = 1.2$$

$$T \quad = 0.02 \times h_n{}^{3/4} = 0.02 \times (270)^{3/4} = 1.33 \text{ sec.}$$

$$C \quad = 1.25 \times S/(T)^{2/3} = 1.25 \times 1.2/(1.33)^{2/3} = 1.24$$

$$W \quad = 73,263 \text{ kips}$$

$$V \quad = \frac{0.4 \times 1.0 \times 1.24 \times 73,263}{12} = 3028 \text{ kips}$$

Eighty percent of the base shear obtained from UBC Eq. (28-1) = (0.8 × 3028) = 2423 kips.   *1629.5.3*

Base shear from modal analysis = 2209 kips.

Since 2423 kips > 2209 kips, the modal forces must be scaled up (UBC 1629.5.3).

$$\text{Scaling factor} = 2423/2209 = 1.097$$

Step VI — *Scale up modal results.*

Adjusted values are:

$$V_1 \quad = 1.097 \times 1682 = 1845 \text{ kips}$$

$$V_2 \quad = 1.097 \times 1218 = 1336 \text{ kips}$$

$$V_3 = 1.097 \times 752 = 825 \text{ kips}$$

$$V = [(1845)^2 + (1336)^2 + (825)^2]^{1/2} = 2423 \text{ kips}$$

The scaled up modal base shear is equal to 80% of the base shear calculated using UBC Eq. (28-1).

Step VII — *Distribute base shear throughout the height of the structure.*

The scaled up modal base shears, V_m, are distributed over the height of the structure according to the following formula:

$$F_{xm} = C_{vxm} V_m$$

where

F_{xm} = lateral force at floor level x, corresponding to the mth mode

$$C_{vxm} = \frac{w_x \phi_{xm}}{\displaystyle\sum_{i=1}^{n} w_i \phi_{im}}$$

The distribution of the modal base shear for each mode is shown in Table 5-5.

Step VIII — *Lateral Analysis.*

Three-dimensional analysis of the structure was performed for each set of modal forces using *SAP90*. To account for accidental torsion, the mass at each level was assumed to be displaced from the center of mass a distance equal to 5% of the building dimension perpendicular to the direction of the force (UBC 1628.5). Also, a rigid diaphragm was assigned on each floor level. The results from these analyses for beams in an exterior frame, beams in an interior frame, columns in an interior frame, and shearwall parallel to the direction of the forces are given in Figs. 5-4 through 5-8.

Step IX — *Combine Results*

The member forces and moments for the three modes were combined using the root-mean-square method (UBC 1629.5.2). These resultant values are given in Figs. 5-4 through 5-8.

In a dual system, an additional safeguard against total collapse is provided by requiring that the moment-resisting space frames be capable of resisting 25% of the base shear without the benefit of the shearwalls (UBC 1627.6.5). Thus, the structure was reanalyzed using 25% of the base shear corresponding to each of the three modes without the shearwalls present (UBC 1629.5.6). The torsional effects were also included in the analysis. The resultants from these analyses are given in Figs. 5-4 through 5-7 in bold italic text.

The lateral displacements at each floor level of the dual system for each mode are shown in Table 5-6. Also given in this table are the combined displacements determined from the root-mean-square method.

Table 5-5 Distribution of Base Shear for Example Building (Seismic Zone 4)

Floor Level i	w_i	Mode 1, T_1 = 1.752 sec. V_1 = 1845 kips				Mode 2, T_2 = 0.419 sec. V_2 = 1336 kips				Mode 3, T_3 = 0.187 sec. V_3 = 825 kips			
		ϕ_{i1}	$w_i\phi_{i1}$	$\dfrac{w_i\phi_{i1}}{\Sigma w_i\phi_{i1}}$	$V_1 \times$ col. 4	ϕ_{i2}	$w_i\phi_{i2}$	$\dfrac{w_i\phi_{i2}}{\Sigma w_i\phi_{i2}}$	$V_2 \times$ col. 8	ϕ_{i3}	$w_i\phi_{i3}$	$\dfrac{w_i\phi_{i3}}{\Sigma w_i\phi_{i3}}$	$V_3 \times$ col. 12
		1	2	3	4	5	6	7	8	9	10	11	12	13
R	3246	1.00000	3246	0.098	181	1.00000	3246	-0.178	-237	1.00000	3246	0.272	224
20	3582	0.94502	3385	0.102	189	0.78965	2829	-0.155	-207	0.64218	2300	0.193	159
19	3582	0.88829	3182	0.096	177	0.56747	2033	-0.111	-149	0.25155	901	0.075	62
18	3582	0.83043	2975	0.090	166	0.34075	1221	-0.067	-89	-0.13050	-467	-0.039	-32
17	3582	0.77128	2763	0.083	154	0.11594	415	-0.023	-30	-0.45982	-1647	-0.138	-114
16	3582	0.71115	2547	0.077	142	-0.09870	-354	0.019	26	-0.69577	-2492	-0.209	-172
15	3646	0.65041	2371	0.072	132	-0.29275	-1067	0.058	78	-0.80925	-2951	-0.247	-204
14	3646	0.58922	2148	0.065	120	-0.46281	-1687	0.092	123	-0.79836	-2911	-0.244	-201
13	3646	0.52795	1925	0.058	107	-0.60283	-2198	0.120	161	-0.66796	-2435	-0.204	-168
12	3646	0.46713	1703	0.051	95	-0.70772	-2580	0.141	189	-0.43783	-1596	-0.134	-110
11	3646	0.40723	1485	0.045	83	-0.77416	-2823	0.154	206	-0.14127	-515	-0.043	-36
10	3646	0.34883	1272	0.038	71	-0.80054	-2919	0.160	213	0.18022	657	0.055	45
9	3646	0.29271	1067	0.032	59	-0.78735	-2871	0.157	210	0.48051	1752	0.147	121
8	3710	0.23993	890	0.027	50	-0.74015	-2746	0.150	201	0.71119	2639	0.221	182
7	3710	0.19062	707	0.021	39	-0.66283	-2459	0.135	180	0.85442	3170	0.265	219
6	3710	0.14571	541	0.016	30	-0.56175	-2084	0.114	152	0.89423	3318	0.278	229
5	3710	0.10610	394	0.012	22	-0.45089	-1673	0.092	122	0.83985	3116	0.261	215
4	3710	0.07142	265	0.008	15	-0.33415	-1240	0.068	91	0.70747	2625	0.220	181
3	3710	0.04249	158	0.005	9	-0.21988	-816	0.045	60	0.51989	1929	0.162	133
2	4325	0.01999	86	0.003	5	-0.11578	-501	0.027	37	0.30122	1303	0.109	90
Σ			33,110		1845		-18,274		1336		11,940		825

MODE 1 Bending Moments (ft-kips)

R				
R	-121.8	-109.0	-128.4	-123.5	-158.4
20	-162.9	-155.6	-182.5	-180.2	-217.5
19	-169.3	-160.7	-185.8	-183.1	-219.6
18	-173.8	-165.2	-190.0	-187.3	-223.5
17	-179.2	-170.4	-194.5	-191.9	-227.4
16	-183.6	-174.6	-197.9	-195.4	-230.1
15	-187.0	-177.9	-200.3	-197.9	-231.5
14	-189.8	-180.6	-202.0	-199.6	-232.1
13	-191.2	-182.0	-202.2	-200.0	-231.1
12	-191.0	-181.9	-200.7	-198.6	-228.1
11	-188.9	-179.9	-197.4	-195.4	-223.1
10	-184.6	-175.9	-191.8	-190.0	-215.6
9	-177.1	-168.8	-183.0	-181.4	-204.8
8	-167.5	-159.6	-172.1	-170.6	-191.6
7	-156.3	-149.9	-159.4	-158.2	-176.5
6	-142.8	-136.8	-146.0	-145.0	-160.6
5	-128.7	-123.8	-131.9	-131.1	-144.2
4	-111.5	-107.2	-113.3	-112.7	-123.3
3	-91.1	-87.7	-92.1	-91.7	-99.6
2	-66.5	-63.1	-64.8	-60.3	-69.4

MODE 2 Bending Moments (ft-kips)

R				
R	36.3	32.8	35.3	34.5	40.5
20	48.9	46.8	51.0	50.6	56.9
19	51.3	48.8	52.6	52.1	58.2
18	51.7	49.3	52.9	52.4	58.4
17	50.4	48.1	51.5	51.1	56.8
16	46.7	44.5	47.8	47.4	52.8
15	41.2	39.3	42.4	42.0	47.0
14	34.6	32.9	35.9	35.6	40.2
13	26.5	25.2	28.1	27.7	32.0
12	17.3	16.5	19.2	18.9	22.6
11	7.5	7.1	9.7	9.4	12.7
10	-2.5	-2.5	0.1	-0.2	2.6
9	-11.6	-11.2	-8.9	-9.1	-6.7
8	-19.6	-18.8	-16.6	-16.8	-14.9
7	-26.2	-25.1	-23.1	-23.2	-21.8
6	-30.7	-29.5	-27.7	-27.8	-26.7
5	-32.7	-31.6	-30.1	-30.1	-29.2
4	-32.9	-31.7	-30.3	-30.3	-29.7
3	-30.7	-29.7	-28.5	-28.5	-28.1
2	-26.1	-24.8	-23.5	-23.6	-23.4

MODE 3 Bending Moments (ft-kips)

R				
R	-12.9	-11.7	-11.6	-11.5	-12.4
20	-17.6	-16.9	-17.3	-17.2	-18.1
19	-18.4	-17.5	-17.7	-17.6	-18.5
18	-16.8	-16.1	-16.3	-16.2	-17.0
17	-13.2	-12.6	-12.8	-12.8	-13.5
16	-7.9	-7.5	-7.8	-7.8	-8.4
15	-1.7	-1.6	-2.0	-2.0	-2.5
14	4.3	4.1	3.6	3.6	3.2
13	9.7	9.3	8.7	8.7	8.4
12	13.8	13.2	12.6	12.6	12.3
11	16.0	15.3	14.7	14.7	14.5
10	16.1	15.4	14.8	14.8	14.6
9	13.7	13.1	12.6	12.6	12.5
8	9.6	9.2	8.9	8.9	8.8
7	4.8	4.6	4.4	4.4	4.3
6	-0.5	-0.5	-0.4	-0.4	-0.5
5	-4.8	-4.6	-4.5	-4.5	-4.5
4	-8.3	-8.0	-7.8	-7.8	-7.8
3	-10.5	-10.2	-9.9	-9.9	-9.9
2	-11.2	-10.6	-10.2	-10.2	-10.2

RESULTANT Bending Moments (ft-kips)

R									
R	127.7	*27.5*	114.4	*24.4*	133.7	*37.2*	128.7	*35.4*	164.0	*37.3*
20	171.0	*47.3*	163.4	*45.0*	190.3	*61.2*	188.0	*60.4*	225.5	*63.2*
19	177.9	*69.5*	168.9	*65.8*	193.9	*82.3*	191.2	*81.4*	227.9	*84.2*
18	182.1	*87.6*	173.1	*83.1*	197.9	*100.4*	195.2	*99.5*	231.6	*102.3*
17	186.6	*100.9*	177.5	*95.6*	201.6	*113.9*	199.0	*113.0*	234.8	*116.0*
16	189.6	*110.5*	180.3	*104.9*	203.7	*124.1*	201.2	*123.1*	236.2	*126.2*
15	191.5	*119.3*	182.2	*113.2*	204.7	*132.9*	202.3	*131.9*	236.2	*135.0*
14	193.0	*128.8*	183.6	*122.3*	205.2	*141.9*	202.8	*141.0*	235.6	*144.0*
13	193.3	*139.5*	184.0	*132.6*	204.3	*151.5*	202.1	*150.6*	233.5	*153.4*
12	192.3	*150.7*	183.1	*143.2*	202.0	*161.3*	199.9	*160.4*	229.5	*163.0*
11	189.7	*161.3*	180.7	*153.6*	198.2	*170.4*	196.2	*169.6*	223.9	*172.0*
10	185.3	*171.1*	176.6	*163.0*	192.4	*178.6*	190.6	*177.9*	216.1	*179.9*
9	178.0	*180.2*	169.7	*171.7*	183.6	*186.1*	182.1	*185.3*	205.3	*187.2*
8	168.9	*189.6*	161.0	*180.7*	173.1	*193.4*	171.1	*192.7*	192.4	*194.3*
7	158.6	*200.1*	152.1	*190.7*	161.1	*201.0*	160.0	*200.5*	177.9	*201.7*
6	146.1	*212.5*	139.9	*203.8*	148.6	*212.3*	147.6	*212.0*	162.8	*213.0*
5	132.9	*224.0*	127.9	*215.8*	135.4	*222.8*	134.6	*222.4*	147.2	*223.3*
4	116.5	*234.0*	112.1	*225.3*	117.5	*229.3*	117.0	*229.4*	127.1	*229.9*
3	96.7	*239.0*	93.2	*230.7*	96.9	*232.4*	96.5	*232.3*	104.0	*232.6*
2	72.3	*229.8*	68.6	*217.7*	69.7	*212.8*	65.6	*213.2*	73.9	*213.8*

Note: Resultant is obtained using root-mean-square values of the three modes.

Bold Italic text denotes results with 25% of design base shear applied to frames.

Figure 5-4 Bending Moments in Beams in an Exterior Frame (Seismic Zone 4)

MODE 1 Bending Moments (ft-kips)

R				
R	-65.9	-66.1	-30.1	-29.3	-58.3
20	-72.4	-72.4	-36.6	-36.3	-65.8
19	-73.9	-73.9	-38.0	-37.5	-66.8
18	-75.3	-75.4	-39.2	-38.8	-68.0
17	-76.8	-76.9	-40.5	-40.1	-69.2
16	-78.0	-78.0	-41.6	-41.1	-69.9
15	-78.7	-78.9	-42.4	-41.9	-70.3
14	-79.2	-79.3	-43.0	-42.5	-70.5
13	-79.1	-79.2	-43.3	-42.8	-70.1
12	-78.3	-78.5	-43.3	-42.8	-69.1
11	-76.8	-76.9	-42.8	-42.3	-67.6
10	-74.5	-74.6	-41.9	-41.4	-65.3
9	-70.9	-71.0	-40.2	-39.7	-61.9
8	-66.6	-66.8	-38.1	-37.6	-57.9
7	-61.5	-61.7	-35.6	-35.1	-53.3
6	-55.1	-55.2	-32.3	-31.9	-47.8
5	-48.7	-48.9	-29.0	-28.7	-42.3
4	-41.6	-41.7	-25.4	-25.1	-36.2
3	-33.4	-33.5	-21.1	-20.9	-29.2
2	-24.4	-24.5	-16.4	-16.1	-21.4

MODE 2 Bending Moments (ft-kips)

R				
R	18.4	18.4	10.5	10.3	16.1
20	20.5	20.5	12.8	12.8	18.7
19	21.2	21.2	13.7	13.6	19.3
18	21.1	21.1	13.9	13.9	19.4
17	20.4	20.5	13.7	13.6	18.8
16	18.9	18.9	12.8	12.7	17.5
15	16.8	16.8	11.4	11.3	15.6
14	14.2	14.2	9.6	9.6	13.4
13	11.1	11.1	7.5	7.4	10.7
12	7.6	7.7	5.0	5.0	7.6
11	3.9	3.9	2.4	2.3	4.3
10	0.2	0.2	-0.4	-0.4	0.9
9	-3.3	-3.3	-2.9	-2.8	-2.1
8	-6.2	-6.2	-5.1	-5.0	-4.8
7	-8.7	-8.7	-7.0	-6.9	-7.1
6	-10.3	-10.3	-8.2	-8.1	-8.7
5	-10.9	-10.9	-8.7	-8.7	-9.3
4	-10.9	-11.0	-8.9	-8.8	-9.5
3	-10.1	-10.2	-8.5	-8.4	-9.0
2	-8.8	-8.9	-7.7	-7.6	-7.9

MODE 3 Bending Moments (ft-kips)

R				
R	-5.9	-6.0	-4.3	-4.3	-5.4
20	-6.8	-6.9	-5.3	-5.4	-6.6
19	-7.0	-7.1	-5.7	-5.8	-6.9
18	-6.4	-6.5	-5.3	-5.4	-6.3
17	-5.1	-5.1	-4.2	-4.2	-5.0
16	-3.1	-3.1	-2.5	-2.6	-3.1
15	-0.9	-0.9	-0.6	-0.6	-1.0
14	1.3	1.3	1.3	1.3	1.2
13	3.3	3.3	3.1	3.1	3.1
12	4.7	4.8	4.4	4.4	4.6
11	5.6	5.6	5.1	5.1	5.4
10	5.6	5.7	5.1	5.1	5.4
9	4.8	4.9	4.3	4.4	4.7
8	3.4	3.5	3.0	3.0	3.3
7	1.8	1.8	1.4	1.4	1.6
6	-0.1	-0.1	-0.3	-0.3	-0.2
5	-1.5	-1.6	-1.7	-1.7	-1.7
4	-2.7	-2.7	-2.8	-2.8	-2.8
3	-3.4	-3.5	-3.5	-3.5	-3.6
2	-3.8	-3.9	-3.9	-3.9	-3.9

RESULTANT Bending Moments (ft-kips)

R									
R	68.7	*11.9*	68.9	*11.5*	32.2	*14.1*	31.4	*14.0*	60.7	*14.2*
20	75.6	*17.0*	75.6	*16.8*	39.1	*19.8*	38.9	*19.7*	68.7	*20.0*
19	77.2	*23.8*	77.2	*23.5*	40.8	*26.5*	40.3	*26.5*	69.9	*26.7*
18	78.5	*29.5*	78.6	*29.1*	41.9	*32.1*	41.6	*32.1*	71.0	*32.4*
17	79.6	*33.4*	79.7	*32.9*	43.0	*36.1*	42.6	*36.1*	71.9	*36.4*
16	80.3	*36.2*	80.3	*35.7*	43.6	*39.1*	43.1	*39.0*	72.1	*39.2*
15	80.5	*38.7*	80.7	*38.2*	43.9	*41.5*	43.4	*41.5*	72.0	*41.8*
14	80.5	*41.3*	80.6	*40.7*	44.1	*44.0*	43.6	*44.0*	71.8	*44.3*
13	79.9	*44.2*	80.0	*43.7*	44.1	*46.9*	43.5	*46.9*	71.0	*47.1*
12	78.8	*47.3*	79.0	*46.7*	43.8	*49.7*	43.3	*49.7*	69.7	*50.0*
11	77.1	*50.3*	77.2	*49.7*	43.2	*52.5*	42.7	*52.5*	68.0	*52.6*
10	74.7	*53.0*	74.8	*52.3*	42.2	*54.8*	41.7	*54.8*	65.5	*55.0*
9	71.1	*55.5*	71.2	*54.8*	40.5	*57.0*	40.0	*57.0*	62.1	*57.2*
8	67.0	*58.0*	67.2	*57.3*	38.6	*59.2*	38.0	*59.2*	58.2	*59.3*
7	62.1	*60.8*	62.3	*60.0*	36.3	*61.7*	35.8	*61.7*	53.8	*61.8*
6	56.1	*63.6*	56.2	*62.9*	33.3	*64.2*	32.9	*64.2*	48.6	*64.3*
5	49.9	*66.1*	50.1	*65.4*	30.3	*66.6*	30.0	*66.6*	43.3	*66.6*
4	43.1	*68.6*	43.2	*67.9*	27.1	*68.6*	26.7	*68.6*	37.5	*68.7*
3	35.1	*70.1*	35.2	*69.4*	23.1	*69.8*	22.8	*69.8*	30.8	*69.8*
2	26.2	*69.8*	26.4	*68.8*	18.5	*68.4*	18.2	*68.4*	23.1	*68.4*

Note: Resultant is obtained using root-mean-square values of the three modes.

Bold Italic text denotes results with 25% of design base shear applied to frames.

Figure 5-5 Bending Moments in Beams in an Interior Frame (Seismic Zone 4)

MODE 1 Axial Loads (kips)

	Col. A3	Col. C3
R	4.8	-1.0
20	10.1	-2.1
19	15.6	-3.2
18	21.2	-4.4
17	26.9	-5.6
16	32.6	-6.9
15	38.4	-8.2
14	44.3	-9.5
13	50.1	-10.9
12	55.8	-12.3
11	61.4	-13.6
10	66.9	-15.0
9	72.1	-16.3
8	76.9	-17.6
7	81.4	-18.8
6	85.4	-19.9
5	89.0	-20.8
4	92.0	-21.7
3	94.4	-22.4
2	96.2	-22.8

MODE 2 Axial Loads (kips)

	Col. A3	Col. C3
R	-1.3	0.4
20	-2.8	0.8
19	-4.4	1.2
18	-5.9	1.6
17	-7.4	2.0
16	-8.8	2.4
15	-10.0	2.7
14	-11.1	3.0
13	-11.9	3.2
12	-12.5	3.3
11	-12.8	3.3
10	-12.8	3.3
9	-12.6	3.2
8	-12.2	2.9
7	-11.6	2.7
6	-10.9	2.4
5	-10.2	2.1
4	-9.4	1.8
3	-8.7	1.6
2	-8.1	1.3

MODE 3 Axial Loads (kips)

	Col. A3	Col. C3
R	0.4	-0.2
20	0.9	-0.3
19	1.3	-0.5
18	1.8	-0.6
17	2.1	-0.7
16	2.4	-0.7
15	2.4	-0.7
14	2.3	-0.7
13	2.1	-0.6
12	1.8	-0.5
11	1.4	-0.4
10	1.1	-0.2
9	0.7	-0.1
8	0.5	0.0
7	0.4	0.0
6	0.4	0.0
5	0.5	0.0
4	0.7	-0.1
3	0.9	-0.2
2	1.2	-0.2

RESULTANT Axial Loads (kips)

	Col. A3		Col. C3	
R	5.0	*0.9*	1.1	*0.0*
20	10.5	*2.3*	2.3	*0.0*
19	16.3	*4.1*	3.5	*0.1*
18	22.1	*6.4*	4.7	*0.1*
17	28.0	*8.9*	6.0	*0.1*
16	33.9	*11.5*	7.3	*0.1*
15	39.8	*14.3*	8.7	*0.2*
14	45.7	*17.1*	10.0	*0.2*
13	51.5	*20.1*	11.4	*0.2*
12	57.2	*23.1*	12.7	*0.2*
11	62.7	*26.2*	14.0	*0.2*
10	68.1	*29.4*	15.4	*0.2*
9	73.2	*32.9*	16.6	*0.2*
8	77.9	*36.3*	17.8	*0.3*
7	82.2	*40.0*	19.0	*0.3*
6	86.1	*43.7*	20.0	*0.3*
5	89.6	*47.6*	20.9	*0.3*
4	92.5	*51.6*	21.8	*0.3*
3	94.8	*55.7*	22.5	*0.3*
2	96.5	*59.8*	22.8	*0.3*

Note: Resultant is obtained using root-mean-square values of the three modes.

Bold Italic text denotes results with 25% of design base shear applied to frames.

Figure 5-6 Axial Loads in Columns in an Interior Frame (Seismic Zone 4)

	MODE 1 Moments (ft-kips)		MODE 2 Moments (ft-kips)		MODE 3 Moments (ft-kips)		RESULTANT Moments (ft-kips)			
	Col. A3	Col. C3	Col. A3	Col. C3	Col. A3	Col. C3	Col. A3		Col. C3	
R	65.8	86.6	-18.5	-26.2	6.0	9.7	68.6	*11.9*	91.0	*28.1*
	31.9	47.6	-7.2	-12.9	1.4	3.7	32.7	*13.7*	49.5	*2.4*
20	40.4	54.1	-13.3	-18.6	5.5	8.4	42.9	*30.4*	57.8	*41.8*
	33.7	48.5	-9.6	-15.4	4.0	7.0	35.3	*9.1*	51.4	*8.4*
19	40.1	55.6	-11.6	-17.5	3.1	5.8	41.9	*31.0*	58.6	*46.9*
	33.7	49.1	-11.1	-17.1	5.8	8.7	36.0	*10.2*	52.7	*20.1*
18	41.5	57.4	-10.1	-16.2	0.7	3.1	42.7	*29.9*	59.7	*48.0*
	34.8	50.9	-12.6	-18.7	7.0	9.7	37.7	*16.2*	55.1	*29.8*
17	41.9	58.1	-7.9	-13.8	-1.9	-0.4	42.7	*29.6*	59.7	*48.8*
	37.3	53.7	-14.6	-20.4	7.4	9.3	40.7	*21.2*	58.2	*36.4*
16	40.6	57.1	-4.4	-9.9	-4.3	-3.6	41.1	*30.8*	58.1	*50.6*
	37.8	54.3	-14.0	-19.4	6.4	7.6	40.8	*24.0*	58.2	*40.3*
15	40.9	57.6	-2.8	-7.6	-5.5	-6.0	41.4	*33.2*	58.4	*53.7*
	39.3	56.0	-14.2	-18.9	5.0	5.1	42.1	*25.7*	59.3	*42.3*
14	39.9	56.7	-0.1	-4.1	-6.3	-7.7	40.4	*36.2*	57.4	*57.0*
	40.7	57.6	-13.9	-17.8	3.1	2.2	43.1	*24.4*	60.3	*43.7*
13	38.3	55.1	2.8	-0.3	-6.4	-8.5	38.9	*38.2*	55.8	*59.8*
	42.0	58.8	-13.0	-15.9	0.7	-1.0	44.0	*23.2*	60.9	*45.8*
12	36.2	52.9	5.3	3.4	-5.5	-8.1	37.0	*39.2*	53.6	*61.8*
	43.1	59.7	-11.5	-13.3	-1.6	-4.0	44.6	*23.7*	61.3	*48.7*
11	33.6	50.5	7.5	6.7	-4.0	-6.7	34.7	*39.6*	51.4	*63.5*
	44.1	60.4	-9.5	-10.2	-3.8	-6.5	45.3	*25.7*	61.6	*52.1*
10	30.3	46.1	9.3	9.6	-1.9	-4.2	31.8	*40.2*	47.3	*65.0*
	46.2	61.7	-6.3	-5.8	-5.8	-8.3	47.0	*28.6*	62.5	*55.3*
9	24.7	39.8	9.6	10.8	1.0	-0.8	26.5	*41.4*	41.2	*67.3*
	44.1	58.8	-4.1	-2.8	-5.9	-7.9	44.7	*30.8*	59.4	*57.5*
8	22.5	36.7	10.3	12.7	2.4	1.5	24.9	*43.4*	38.9	*70.6*
	44.3	58.4	-1.4	1.1	-5.8	-7.1	44.7	*32.3*	58.8	*59.3*
7	17.2	30.1	10.1	13.1	4.1	4.1	20.4	*44.6*	33.1	*74.0*
	45.2	56.9	2.7	5.5	-4.9	-5.3	45.5	*35.3*	57.4	*61.6*
6	9.8	22.8	7.6	11.4	4.9	5.8	13.3	*44.9*	26.1	*76.2*
	43.2	55.0	3.7	7.3	-3.7	-3.4	43.5	*33.5*	55.6	*61.9*
5	5.5	16.1	7.2	10.9	5.3	6.8	10.5	*46.0*	20.6	*79.0*
	42.6	52.8	6.2	10.0	-1.8	-0.6	43.1	*35.3*	53.7	*65.0*
4	-1.0	8.5	4.7	8.5	4.5	6.3	6.6	*42.1*	13.6	*77.4*
	41.1	50.2	8.0	11.8	0.2	1.8	41.9	*38.1*	51.6	*69.7*
3	-7.8	0.1	2.1	5.9	3.3	5.4	8.7	*36.4*	8.0	*72.4*
	38.9	47.4	8.2	12.9	0.7	3.0	39.8	*160.4*	49.2	*80.7*
2	-14.5	-10.5	0.6	2.7	3.2	5.0	14.9	*28.0*	11.9	*57.1*
	61.4	63.4	29.8	31.3	16.8	19.0	70.3	*242.8*	73.2	*301.2*

Note: Resultant is obtained using root-mean-square values of the three modes.

Bold Italic text denotes results with 25% of design base shear applied to frames.

Figure 5-7 Bending Moments in Columns in an Interior Frame (Seismic Zone 4)

	Shearwall B2-C2, Mode 1			
Level	Axial Load (kips)	Bending Moment (ft-kips)		Shear Force (kips)
		Top	Bottom	
R	8.6	749	-1563	62.6
20	17.4	2340	-2248	-7.1
19	26.2	3034	-2230	-54.1
18	34.9	3128	-1879	-96.2
17	44.1	2690	-937	-135.0
16	52.9	1757	489	-172.7
15	61.8	340	2348	-206.9
14	70.9	-1520	4622	-238.5
13	79.6	-3796	7275	-267.7
12	88.4	-6461	10,289	-294.4
11	96.6	-9492	13,637	-319.0
10	104.6	-12,870	17,301	-340.9
9	112.0	-16,571	21,297	-363.7
8	119.3	-20,612	25,574	-381.7
7	125.5	-24,945	30,140	-399.8
6	131.2	-29,580	34,960	-414.0
5	136.0	-34,468	40,040	-428.7
4	140.3	-39,619	45,374	-442.8
3	143.5	-44,976	50,966	-455.4
2	147.8	-50,730	61,739	-481.1

	Shearwall B2-C2, Mode 2			
Level	Axial Load (kips)	Bending Moment (ft-kips)		Shear Force (kips)
		Top	Bottom	
R	-2.3	-203	-181	29.4
20	-4.5	-31	-1070	84.6
19	-6.7	859	-2439	121.7
18	-9.1	2228	-4097	143.7
17	-11.0	3894	-5872	152.1
16	-12.9	5687	-7621	149.1
15	-14.8	7453	-9169	132.0
14	-16.3	9029	-10,393	104.9
13	-17.5	10,279	-11,178	69.0
12	-18.5	11,098	-11,444	26.6
11	-19.0	11,400	-11,140	-20.1
10	-19.1	11,132	-10,252	-67.9
9	-19.0	10,277	-8761	-116.6
8	-18.6	8814	-6696	-163.0
7	-18.1	6774	-4113	-204.7
6	-17.4	4207	-1066	-241.7
5	-16.6	1162	2359	-270.9
4	-15.8	-2261	6086	-294.2
3	-15.2	-5996	10,033	-310.6
2	-14.5	-9959	17,412	-324.1

	Shearwall B2-C2, Mode 3			
Level	Axial Load (kips)	Bending Moment (ft-kips)		Shear Force (kips)
		Top	Bottom	
R	0.6	62	538	-46.3
20	1.2	-476	1591	-85.8
19	1.8	-1529	2852	-101.6
18	2.2	-2794	4027	-94.9
17	2.6	-3983	4878	-68.9
16	2.9	-4850	5230	-29.3
15	2.9	-5223	4981	18.8
14	3.0	-4990	4133	65.9
13	2.9	-4160	2793	105.3
12	2.5	-2830	1126	131.0
11	2.0	-1172	-640	139.5
10	1.7	592	-2264	128.5
9	1.4	2224	-3532	100.9
8	1.1	3502	-4248	57.2
7	0.9	4233	-4300	5.3
6	0.9	4299	-3656	-49.6
5	1.1	3667	-2355	-100.9
4	1.2	2380	-497	-144.9
3	1.3	521	1786	-177.4
2	1.5	-1757	6352	-199.7

	Shearwall B2-C2, Resultant			
Level	Axial Load (kips)	Bending Moment (ft-kips)		Shear Force (kips)
		Top	Bottom	
R	8.9	778	1663	83.2
20	18.0	2388	2955	120.7
19	27.1	3504	4417	167.5
18	36.1	4749	6044	197.3
17	45.5	6186	7691	214.7
16	54.5	7678	9256	230.0
15	63.6	9107	10,696	246.1
14	72.8	10,428	12,102	268.8
13	81.6	11,721	13,626	295.8
12	90.3	13,150	15,430	323.3
11	98.5	14,881	17,620	348.7
10	106.3	17,027	20,237	370.6
9	113.6	19,626	23,298	395.0
8	120.7	22,689	26,775	419.0
7	126.8	26,193	30,722	449.2
6	132.4	30,185	35,167	481.9
5	137.0	34,682	40,179	517.1
4	141.2	39,755	45,783	551.0
3	144.3	45,377	51,975	579.1
2	148.5	51,728	64,461	613.5

Note: Resultant is obtained using root-mean-square values of the three modes.

Figure 5-8 Axial Loads, Bending Moments and Shear Forces in a Shearwall (Seismic Zone 4)

Table 5-6 Lateral Displacements of Example Building (Seismic Zone 4)

Floor Level	Mode 1 Displacement (in.)	Mode 2 Displacement (in.)	Mode 3 Displacement (in.)	Combined Displacement (in.)
R	1.70	-0.13	0.03	1.71
20	1.61	-0.11	0.02	1.61
19	1.51	-0.08	0.01	1.51
18	1.41	-0.05	0.00	1.41
17	1.31	-0.02	-0.02	1.31
16	1.21	0.01	-0.02	1.21
15	1.11	0.04	-0.03	1.11
14	1.00	0.06	-0.03	1.00
13	0.90	0.08	-0.02	0.90
12	0.80	0.10	-0.01	0.81
11	0.69	0.10	-0.01	0.70
10	0.59	0.11	0.01	0.60
9	0.49	0.11	0.02	0.50
8	0.41	0.10	0.02	0.42
7	0.33	0.09	0.03	0.34
6	0.25	0.08	0.03	0.26
5	0.18	0.06	0.03	0.19
4	0.12	0.05	0.02	0.13
3	0.07	0.03	0.02	0.08
2	0.03	0.02	0.01	0.04

5.2.3   Design of Structural Members

The objective is to determine the required flexural and shear reinforcement for a beam of an exterior frame, a beam of an interior frame, interior and exterior columns of an interior frame, a shearwall, and two beam-column connections.

The beams along column line 1 carry an unfactored dead load of 2.3 kips/ft and an unfactored live load of 0.6 kips/ft (including live load reduction).

The factored load due to gravity is:

$$w_u = (1.4 \times 2.3) + (1.7 \times 0.6) = 4.2 \text{ kips/ft}$$

Using the coefficients in UBC 1908.3.3, the factored moments due to gravity are:

- End Span—Beam A1-B1

    Exterior negative moment $= w_u \ell_n^2/16 = 4.2 \times (28 - 36/12)^2/16 = 164.1$ ft-kips

    Positive moment $= w_u \ell_n^2/14 = 4.2 \times (28 - 36/12)^2/14 = 187.5$ ft-kips

    Interior negative moment $= w_u \ell_n^2/10 = 4.2 \times (28 - 36/12)^2/10 = 262.5$ ft-kips

• Interior Span—Beam B1-C1

$$\text{Negative moment} \quad = w_u \ell_n^2/11 = 4.2 \times (28 - 36/12)^2/11 = 238.6 \text{ ft-kips}$$

$$\text{Positive moment} \quad = w_u \ell_n^2/16 = 4.2 \times (28 - 36/12)^2/16 = 164.1 \text{ ft-kips}$$

Table 5-7 shows the factored moments for different combinations of gravity and lateral loads on a beam on the third floor of an exterior frame. The gravity load moments are obtained from the calculations above and the lateral load moments are taken from Fig. 5-4. Note that in this case, the seismic moments that govern are the ones generated from the application of 25% of the base shear to the frames.

Table 5-7 Summary of Factored Bending Moments for a Beam on the 3rd Floor of an Exterior Frame (Seismic Zone 4)

(1)	$U = 1.4D + 1.7L$		Eq. (9-1), UBC 1909.2.1
(2)	$U = 1.4(D + L \pm E)$		Eq. (9-2), UBC 1921.2.7
(3)	$U = 0.9D \pm 1.4E$		Eq. (9-3), UBC 1921.2.7

Load Combination	Bending Moments (ft-kips)					
	Beam A1-B1			Beam B1-C1		
	Exterior Negative	Positive	Interior Negative	Negative	Positive	Negative
1	-164.1	187.5	-262.5	-238.6	164.1	-238.6
2: sidesway right	176.1	187.0	-576.8	94.6	158.5	-556.1
sidesway left	-493.1	175.6	69.2	-556.1	158.5	94.6
3: sidesway right	253.8	98.2	-452.4	207.7	80.8	-443.0
sidesway left	-415.4	86.7	193.6	-443.0	80.8	207.7

The beams along column line 3 carry an unfactored dead load of 3.4 kips/ft and an unfactored live load of 0.84 kips/ft.

The factored load due to gravity is:

$$w_u = (1.4 \times 3.4) + (1.7 \times 0.84) = 6.2 \text{ kips/ft}$$

Using the coefficients in UBC 1908.3.3, the factored moments due to gravity are:

• End Span—Beam A3-B3

$$\text{Exterior negative moment} = w_u \ell_n^2/16 = 6.2 \times (28 - 36/24 - 40/24)^2/16 = 239.0 \text{ ft-kips}$$

$$\text{Positive moment} \quad = w_u \ell_n^2/14 = 6.2 \times (28 - 36/24 - 40/24)^2/14 = 273.1 \text{ ft-kips}$$

$$\text{Interior negative moment} = w_u \ell_n^2/10 = 6.2 \times (28 - 36/24 - 40/24)^2/10 = 382.4 \text{ ft-kips}$$

*   Interior Span—Beam B3-C3

    Negative moment $= w_u \ell_n^2 / 11 = 6.2 \times (28 - 36/24 - 40/24)^2/11 = 347.6$ ft-kips

    Positive moment $= w_u \ell_n^2 / 16 = 6.2 \times (28 - 36/24 - 40/24)^2/16 = 239.0$ ft-kips

Table 5-8 shows the factored moments for different combinations of gravity and lateral loads on a beam on the 15th floor of an interior frame. The gravity load moments are obtained from the calculations above. The lateral load moments are taken from Fig. 5-5. In this case, the seismic moments due to the seismic loads on the dual system govern.

Table 5-8 Summary of Factored Bending Moments for a Beam on the 15th Floor of an Interior Frame (Seismic Zone 4)

(1) U $= 1.4D + 1.7L$	Eq. (9-1), UBC 1909.2.1	
(2) U $= 1.4(D + L \pm E)$	Eq. (9-2), UBC 1921.2.7	
(3) U $= 0.9D \pm 1.4E$	Eq. (9-3), UBC 1921.2.7	

Load Combination	Bending Moments (ft-kips)					
	Beam A3-B3			Beam B3-C3		
	Exterior Negative	Positive	Interior Negative	Negative	Positive	Negative
1	-239.0	273.1	-382.4	-347.6	239.0	-347.6
2: sidesway right	-116.2	261.4	-479.1	-271.3	228.9	-393.5
sidesway left	-341.6	261.4	-253.1	-394.2	228.9	-272.0
3: sidesway right	-5.3	134.7	-301.7	-110.1	118.0	-232.3
sidesway left	-230.7	134.7	-75.8	-233.0	118.0	-110.8

The different combinations of axial loads and bending moments corresponding to the appropriate load combinations are listed in Table 5-9 for an interior and an exterior column of an interior frame between grade and the 2nd floor. The seismic loads and moments are taken from Figs. 5-6 and 5-7.

UBC 1631.1 requires that buildings located in Seismic Zones 2, 3 or 4 must be designed for the effects of earthquake forces acting in a direction other than the principal axes if a column of a structure forms part of two or more intersecting lateral force-resisting systems. An exception to this is permitted if the axial load in the column due to seismic forces acting in either direction is less than 20% of the column axial load strength.

The maximum column axial load from unfactored earthquake forces is 96.5 kips (see Fig. 5-6). The design axial load strength, $\phi P_{n(max)}$, of a 36 × 36 in. column with an $f'_c = 6000$ psi and minimum reinforcement (1%) is approximately 4100 kips. Twenty percent of this capacity is 820 kips. Therefore, orthogonal effects due to seismic forces can be neglected for the columns.

Table 5-10 lists the different combinations of axial loads and bending moments for a shearwall. Seismic loads and moments are given in Fig. 5-8.

Table 5-9 Summary of Factored Axial Loads and Bending Moments for Columns between Grade and Level 2 of an Interior Frame (Seismic Zone 4)

(1) U = 1.4D + 1.7L   Eq. (9-1), UBC 1909.2.1
(2) U = 1.4(D + L ± E)   Eq. (9-2), UBC 1921.2.7
(3) U = 0.9D ± 1.4E   Eq. (9-3), UBC 1921.2.7

Load Combination	Column A3			Column C3		
	Axial Load (kips)	Bending Moment (ft-kips)		Axial Load (kips)	Bending Moment (ft-kips)	
		Top	Bottom		Top	Bottom
1	2905	-119.2	119.2	4118	-40.0	40.0
2: sidesway right	2716	-75.2	-225.5	3990	47.0	-388.8
sidesway left	2986	-153.6	454.3	4054	-112.8	454.6
3: sidesway right	1533	-19.8	-281.0	2264	79.9	-421.7
sidesway left	1804	-98.2	398.9	2328	-79.9	421.7

Table 5-10 Summary of Factored Axial Loads and Bending Moments for Shearwall B2-C2 between Grade and Level 2 (Seismic Zone 4)

(1) U = 1.4D + 1.7L   Eq. (9-1), UBC 1909.2.1
(2) U = 1.4(D + L ± E)   Eq. (9-2), UBC 1921.2.7
(3) U = 0.9D ± 1.4E   Eq. (9-3), UBC 1921.2.7

Load Combination	Shearwall B2-C2			Boundary Element
	Axial Load (kips)	Bending Moment (ft-kips)		Axial Load (kips)
		Top	Bottom	
1	8613	—	—	4307
2: sidesway right	8213	-72,419	90,245	884
sidesway left	8629	72,419	-90,245	7538
3: sidesway right	4627	-72,419	90,245	-909
sidesway left	5043	72,419	-90,245	5745

5.2.3.1   Story Drift Limitation and P-Δ Effects

UBC 1628.8.2 specifies that for buildings with a fundamental period of 0.7 seconds or greater, the calculated story drift (including torsional effects) must not exceed $0.03/R_w$ nor 0.004 times the story height.

For $h_s = 13$ ft and $R_w = 12$:

$$\frac{0.03h_s}{R_w} = \frac{0.03 \times 13 \times 12}{12} = 0.39 \text{ in. (governs)}$$

$$0.004h_s = 0.004 \times 13 \times 12 = 0.62 \text{ in.}$$

Table 5-11 shows that for all stories, the lateral drifts obtained from the prescribed lateral forces are less than the limiting value.

Table 5-11 Lateral Story Drifts of Example Building (Seismic Zone 4)

Floor Level	Mode 1 Story Drift (in.)	Mode 2 Story Drift (in.)	Mode 3 Story Drift (in.)	Resultant Story Drift (in.)
R	0.09	-0.02	0.01	0.09
20	0.10	-0.03	0.01	0.10
19	0.10	-0.03	0.01	0.10
18	0.10	-0.03	-0.02	0.11
17	0.10	-0.03	0.00	0.10
16	0.10	-0.03	0.01	0.10
15	0.11	-0.02	0.00	0.11
14	0.10	-0.02	-0.01	0.10
13	0.10	-0.02	-0.01	0.10
12	0.11	0.00	0.00	0.11
11	0.10	-0.01	-0.02	0.10
10	0.10	0.00	-0.01	0.10
9	0.08	0.01	0.00	0.08
8	0.08	0.01	-0.01	0.08
7	0.08	0.01	0.00	0.08
6	0.07	0.02	0.00	0.07
5	0.06	0.01	0.01	0.06
4	0.05	0.02	0.00	0.05
3	0.04	0.01	0.01	0.05
2	0.03	0.02	0.01	0.03

According to UBC 1628.9, P-Δ effects need not be considered when the story drift ratio is less than $0.02h_s/R_w$ = 0.26 in. It can be seen from Table 5-11 that for all stories, the lateral drifts obtained from the prescribed lateral forces are less than the limiting value. Thus, P-Δ effects can be neglected.

5.2.3.2 Proportioning and Detailing a Flexural Member of an Exterior Frame—Beam A1-B1

Members are to be designed as flexural elements if they meet the following requirements (UBC 1921.3):

- Factored axial compressive force on the member does not exceed $A_g f'_c/10$.

- Clear span of the member is not less than four times its effective depth.

- Width-to-depth ratio of the member is not less than 0.3.

- Width of the member is not (1) less than 10 in. and (2) more than the width of the supporting member (measured on a plane perpendicular to the longitudinal axis of the flexural member) plus distances on either side of the supporting member not exceeding three-quarters of the depth of the flexural member.

Since the beam under consideration has negligible axial force, the first requirement is satisfied. A check of the other three requirements is shown below.

$$\text{Clear span} = 25 \text{ ft} > 4d = \frac{4 \times 33.4}{12} = 11.1 \text{ ft} \quad \text{O.K.}$$

$$\frac{\text{Width}}{\text{Depth}} = \frac{27 \text{ in.}}{36 \text{ in.}} = 0.75 > 0.30 \quad \text{O.K.}$$

Beam width is 27 in. which is greater than 10 in. Beam width is less than width of supporting column + (1.5 × depth of beam) = 36 + (1.5 × 36) = 90 in. O.K.

5.2.3.2.1 Required Flexural Reinforcement

The maximum negative moment for this beam at the interior support is 576.8 ft-kips (see Table 5-7). The required reinforcement, ignoring the effects of any compression reinforcement, is $A_s = 3.99$ in.[2] Use 4-#9 bars at this location ($\phi M_n = 577.6$ ft-kips).

Check limitations on the reinforcement ratio:

$$\rho \quad = A_s/bd = 4.0/(27 \times 33.4) = 0.0044$$

$$\rho_{max} \quad = 0.0250 > 0.0044 > 200/f_y = 0.0033 \quad \text{O.K.} \hspace{3cm} \textit{1921.3.2.1}$$

The maximum negative moment in the beam at an exterior support is 493.1 ft-kips. Use 4-#9 bars at this support as well.

The positive moment strength at the joint face must be equal to at least 50% of the corresponding negative moment strength at that joint (UBC 1921.3.2.2). The negative moment strength at an interior or exterior joint face is the same; the positive moment reinforcement must not give a design moment strength less than 577.6/2 = 288.8 ft-kips. The maximum positive moments in the beam at the interior and exterior supports are 207.7 ft-kips and 253.8 ft-kips, respectively (see Table 5-7). Thus, the required number of reinforcing bars is 2-#9 at both supports, giving a moment strength of 294.7 ft-kips. However, to meet the minimum reinforcement requirements given in UBC 1921.3.2.1, use 3-#9 bars ($\rho = 0.0033$) at these locations ($\phi M_n = 437.6$ ft-kips).

The maximum positive design moment at the midspan of the member is 187.5 ft-kips. According to UBC 1921.3.2.2, the positive moment strength at any section along the member length must not be less than 25% of the maximum moment strength provided at the face of either joint. Thus, the minimum design value of the positive moment is 577.6/4 = 144.4 ft-kips which is less than 187.5 ft-kips. Use 3-#9 bars to meet minimum reinforcement requirements.

Similarly, the negative moment strength at any section along the member length must not be less than 25% of the maximum moment strength provided at the face of either joint (577.6/4 = 144.4 ft-kips). Three #9 bars will be required, giving a moment strength of 437.6 ft-kips.

5.2.3.2.2 Required Shear Reinforcement

According to UBC 1921.3.4, the beams must be designed for shear forces that are based on the probable moment strength at each joint face using a strength reduction factor of 1.0 and a stress in the tensile flexural reinforcement equal to $1.25f_y$, combined with the shear caused by unfactored gravity loads. It should be noted that ACI 318-89 (Revised 1992) uses the same procedure, but with factored gravity loads.

Figure 5-9 shows the probable flexural strengths M_{pr} at the joint faces for this beam for sidesway to the right and to the left. As noted above, these flexural strengths are determined using a steel stress of $1.25\,f_y$ and a strength reduction factor of 1.0. Figure 5-9 also shows the unfactored gravity loads on the beam, as well as the shear forces at the faces of the columns. In general, the shear forces V_e are determined as follows:

$$V_e = \frac{M^{\pm}_{pre} + M^{\mp}_{pri}}{\ell_n} \pm \frac{w\ell_n}{2}$$

where $M^{\pm}_{pre}$ = probable flexural strength at the exterior support

$M^{\mp}_{pri}$ = probable flexural strength at the interior support

ℓ_n = clear span length

w   = unfactored gravity load

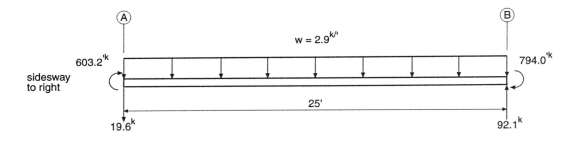

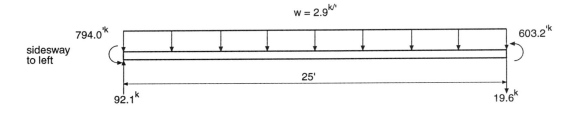

Figure 5-9 Design Shear Forces for Beam A1-B1 (Seismic Zone 4)

The shear strength of the concrete is not considered in Seismic Zones 3 and 4 if the factored axial force including earthquake effects is less than $A_g f'_c/20$ and the shear caused by seismic forces represents more than half of the total shear (UBC 1921.3.4.2). The shear caused by the seismic forces is $(603.2 + 794.0)/25 = 55.9$ kips which is larger than 50% of 92.1 kips; thus, V_c is taken as zero. The maximum shear force that shear reinforcement must be designed for is:

$$\phi V_s = V_u - \phi V_c$$

$$V_s = 92.1/0.85 - 0 = 108.4 \text{ kips}$$

Shear strength contributed by shear reinforcement cannot be taken greater than $0.75 \times 8\sqrt{f'_c} \times b_w d$ for light-weight concrete (UBC 1911.2 and 1911.5.6.8):

$$V_{s,max} = 0.75 \times 8\sqrt{4000} \times (27 \times 33.4)/1000 = 342.2 \text{ kips} > 108.4 \text{ kips} \quad \text{O.K.}$$

In Seismic Zones 3 and 4, the following detailing requirements must be satisfied in addition to the shear strength requirements. Hoops must be provided:

* Over a length equal to twice the member depth, starting from the face of the supporting member toward midspan, at both ends of the flexural member (UBC 1921.3.3.1).

* Over lengths equal to twice the member depth where flexural yielding may occur in connection with inelastic lateral displacements of the frame (UBC 1921.3.3.1).

Where hoops are not required, stirrups with 135-degree seismic hooks shall be spaced at no more than d/2 throughout the length of the member (UBC 1921.3.3.4).

The maximum allowable hoop spacing within a distance 2h = 72 in. from the face of the supports is (UBC 1921.3.3.2):

$$s_{max} = \frac{d}{4} = \frac{33.4}{4} = 8.4 \text{ in. (governs)}$$

$$= 8 \times \text{(diameter of smallest longitudinal bar)} = 8 \times 1.128 = 9.0 \text{ in.}$$

$$= 24 \times \text{(diameter of hoop bars)} = 24 \times 0.5 = 12.0 \text{ in.}$$

$$= 12.0 \text{ in.}$$

The required spacing of #4 closed stirrups (hoops) for the factored shear force of 108.4 kips is:

$$s = \frac{A_v f_y d}{V_s} = \frac{0.40 \times 60 \times 33.4}{108.4} = 7.4 \text{ in.}$$

Therefore, hoops must be spaced at 7 in. on center with the first hoop located at 2 in. from the face of the column. Provide 11 hoops at 7 in. spacing, with the last hoop at 72 in. from the support face. At this location, the shear force is:

$$V_u = 92.1 - (2.9 \times 72/12) = 74.7 \text{ kips}$$

Therefore,

$$V_s = \frac{74.7}{0.85} = 87.9 \text{ kips}$$

The required spacing of #4 stirrups at this location is:

$$s = \frac{0.40 \times 60 \times 33.4}{87.9} = 9.1 \text{ in.}$$

A 9-in. spacing, starting at 72 in. from the column face to the start of the longitudinal bar splices at midspan, will be sufficient.

Note that lap splices in longitudinal reinforcement are permitted only if hoop or spiral reinforcement is provided over the lap length (UBC 1921.3.2.3). The maximum spacing of the transverse reinforcement enclosing the lapped bars must not exceed d/4 or 4 in. Lap splices must not be used within a distance of twice the member depth from the joint face, nor at locations where analysis indicates flexural yielding caused by inelastic lateral displacements of the frame to be likely.

5.2.3.2.3 Reinforcing Bar Cutoff Points and Splices

The negative reinforcement at the supports consists of 4-#9 bars. The location where one of the four can be terminated away from an interior support will be determined. The loading used to find the cutoff points is 0.9 times the dead load in combination with the probable moment strengths, M_{pr}, at the member ends (conservatively using $f_s = 1.25f_y$). The design bending moment strength, ϕM_n, provided by 3-#9 reinforcing bars in the beam is 437.6 ft-kips. Therefore, the three reinforcing bars can be terminated only after the factored moment, M_u, has been reduced to 437.6 ft-kips.

With $\phi = 1.0$ and $f_s = 75$ ksi, $M^+_{pr} = 603.2$ ft-kips at the exterior end of the beam, and $M^-_{pr} = 794.0$ ft-kips at the interior end (see Fig. 5-9). The dead load is equal to $0.9 \times 2.3 = 2.1$ kips/ft. The distance from the face of the support to where the moment under the loading considered equals 437.6 ft-kips is readily obtained by summing moments about section a-a in Fig. 5-10:

$$\frac{1}{2}(2.1x^2) + 794.0 - 82.1x = 437.6$$

Solving for x gives a distance of 4.6 ft. Therefore, the negative reinforcing bars can be terminated at a distance of $4.6 + d = 7.4$ ft from the face of the interior support (noting that $d = 33.4$ in. $> 12d_b = 12 \times 1.128 = 13.5$ in.; see UBC 1912.10.3).

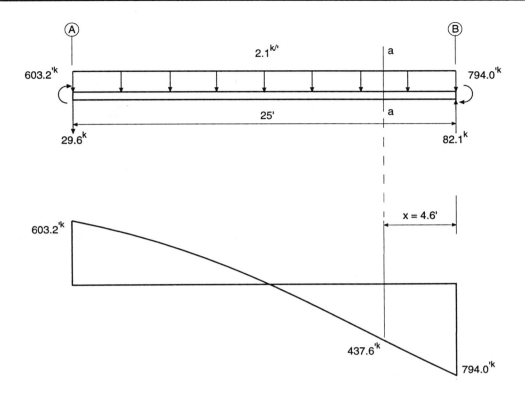

*Figure 5-10 Moment Diagram for Cutoff Location of Negative Bars at Interior Support
of Beam A1-B1 (Seismic Zone 4)*

The #9 bars being terminated must be properly developed at the support. For lightweight concrete used in Seismic Zone 4, the development length reinforcement in tension when the depth of concrete cast in one lift beneath the bar exceeds 12 in. is given by:

$$\ell_d = 1.25 \times 3.5 \times f_y \times d_b/65\sqrt{f'_c} \qquad\qquad 1921.5.4.2$$

$$= 1.25 \times 3.5 \times 60,000 \times 1.128/(65\sqrt{4000})$$

$$= 72.0 \text{ in.} < (7.4 \times 12) = 88.8 \text{ in.} \quad \text{O.K.}$$

Lap splices of flexural reinforcement must not be placed within a distance of 2h from the faces of the supports, nor within regions of potential plastic hinging (UBC 1921.3.2.3). All lap splices must be confined by hoops or spirals with a maximum spacing of d/4 or 4 in. over the length of the lap.

The required lap splice length is the product of the basic development length and applicable modification factors. The basic development length for #9 bars is:

$$\ell_{db} = 0.04A_b f_y/\sqrt{f'_c} \qquad\qquad 1912.2.2$$

$$= 0.04 \times 1.0 \times 60,000/\sqrt{4000} = 38.0 \text{ in.}$$

The development length modification factors are shown in Table 5-12.

Table 5-12 Development Length Factors

Factor I

Criteria	Cover	Transverse Reinforcement Requirements	Clear Bar Spacing	Confinement provided UBC 1912.2.3.5	Applicable Factor
1	≥ 1.5 in. clear interior or ≥ 2.0 in. clear exterior exposure	Meeting UBC 1907.10.5 or UBC 1911.5.4 and 1911.5.5.3	≥ 3d_b	YES	0.75
				NO	1.00
			≥ 5d_b & side cover ≥ 2.5 d_b	YES	0.60
				NO	0.80
2	≥ 1.5 in. clear interior or ≥ 2.0 in. clear exterior exposure	$A_{tr} \ge \dfrac{d_b s N}{40}$	—	YES	0.75
				NO	1.00
			≥ 5d_b & side cover ≥ 2.5 d_b	YES	0.60
				NO	0.80
3	≥ 2 d_b	—	≥ 3d_b	YES	0.75
				NO	1.00
			≥ 5d_b & side cover ≥ 2.5 d_b	YES	0.60
				NO	0.80
4	Either ≤ d_b	—	Or < 2d_b	YES	1.50
				NO	2.00
5	All other conditions		—	YES	1.05
				NO	1.40
			≥ 5d_b & side cover ≥ 2.5 d_b	YES	0.84
				NO	1.12

Bar Size	#4	#5	#6	#7	#8	#9	#10	#11
Minimum Value of Factor I*	1.88	1.51	1.28	1.09	0.95	0.85	0.75	0.68

* Based on $0.03 \, d_b f_y / \sqrt{f'_c}$

Factor II

Bar with more than 12 in. of fresh concrete cast in one lift below it	1.3
All others	1.0

Factor III

Bars Embedded in Lightweight Concrete	1.3 or $6.7 \sqrt{f'_c} \, \ell_d$ but > 1.0
Bars Embedded in Normal Weight Concrete	1.0

Factor IV

Epoxy Coated Bars	Cover < 3d_b	1.5
	Clear Spacing < 6d_b	1.5
	All other	1.2
Uncoated Bars		1.0

Factor V

$$= \frac{A_s \text{ (required)} *}{A_s \text{ (provided)}}$$

* This factor cannot be used in Seismic Zones 3 & 4

ℓ_d = (Factor I) × (Factor II) × (Factor III) × (Factor IV) × (Factor V) × ℓ_{db}

The required development length is the product of the applicable factors and the basic development length. The bottom reinforcing bars will be subject to stresses greater than $0.5f_y$ and will require a Class B splice.

The information required for determining the development length factors is:

Minimum bottom cover to bars being spliced $= 2.0$ in. $(1.77\ d_b)$

Minimum side cover $= 2.0$ in. $(1.77\ d_b)$

Clear spacing of bottom bars being spliced $= 9.8$ in. $(8.7\ d_b)$

Transverse reinforcement meets requirements of UBC 1911.5.4 and 1912.2.3.5.

This gives a value of 0.75 for Factor I, but the limiting value for #9 bars of 0.85 will control (see Table 5-12). The other factors are all equal to 1.0, with the exception of Factor III for lightweight concrete (Factor III = 1.3).

Class B splice length = $1.3 \times (0.85 \times 1.0 \times 1.3 \times 1.0 \times 1.0 \times 38.0) = 54.6$ in.

The top reinforcing bars are subject to stresses less than $0.5f_y$ at midspan and a Class A splice will be used.

The information required for determining the development length factors is:

Minimum top cover to bars being spliced $= 2.0$ in. $(1.77\ d_b)$

Minimum side cover   large (slab)

Clear spacing of top bars being spliced $= 6.2$ in. $(5.5\ d_b)$

Transverse reinforcement meets requirements of UBC 1911.5.4 and 1912.2.3.5.

This gives a value of 0.60 for Factor I, but the limiting value for #9 bars of 0.85 will control.   The other factors are all equal to 1.0, with the exception of Factor III for lightweight concrete and Factor II for top bars, which are both equal to 1.3.

Class A splice length = $1.0 \times (0.85 \times 1.3 \times 1.3 \times 1.0 \times 1.0 \times 38.0) = 54.6$ in.

Figure 5-11 shows the reinforcement details for beam A1-B1.

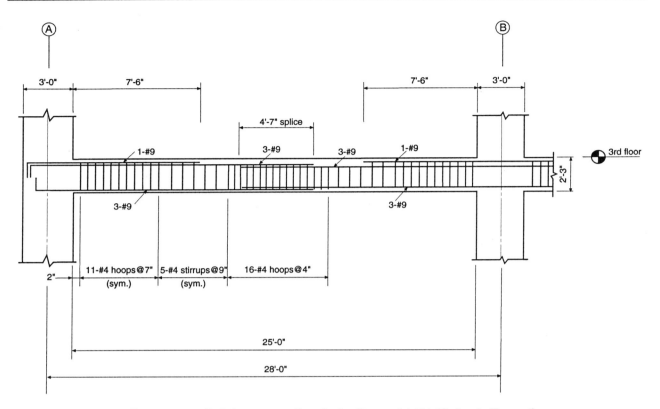

Figure 5-11 Reinforcement Details for Beam A1-B1 (Seismic Zone 4)

5.2.3.3   Proportioning and Detailing a Flexural Member of an Interior Frame—Beam A3-B3

Members are to be designed as flexural elements if they meet the following requirements (UBC 1921.3):

- Factored axial compressive force on the member does not exceed $A_g f'_c/10$.

- Clear span of the member is not less than four times its effective depth.

- Width-to-depth ratio of the member is not less than 0.3.

- Width of the member is not (1) less than 10 in. and (2) more than the width of the supporting member (measured on a plane perpendicular to the longitudinal axis of the flexural member) plus distances on either side of the supporting member not exceeding three quarters of the depth of the flexural member.

Since the beam under consideration has negligible axial force, the first requirement is satisfied. A check of the other three requirements is shown below.

$$\text{Clear span} = 24.83 \text{ ft} > 4d = \frac{4 \times 21.44}{12} = 7.2 \text{ ft} \quad \text{O.K.}$$

$$\frac{\text{Width}}{\text{Depth}} = \frac{24 \text{ in.}}{24 \text{ in.}} = 1.0 > 0.3 \quad \text{O.K.}$$

Beam width is 24 in. which is greater than 10 in. Beam width is less than width of supporting column + (1.5 × depth of beam) = 36 + (1.5 × 24) = 72 in.   O.K.

5.2.3.3.1 Required Flexural Reinforcement

The maximum negative moment for this beam at the interior support is 479.1 ft-kips (see Table 5-8). The required reinforcement, ignoring the effects of any compression reinforcement, is A_s = 5.48 in.[2] Use 6-#9 bars at this location (ϕM_n = 519.1 ft-kips).

Check limitations on the reinforcement ratio:

$$\rho \qquad = A_s/bd = 6.0/(24 \times 21.44) = 0.0117$$

$$\rho_{max} = 0.0250 > 0.0117 > 200/f_y = 0.0033 \quad \text{O.K.} \qquad \qquad 1921.3.2.1$$

The maximum negative moment in the beam at an exterior support is 341.6 ft-kips. Use 4-#9 bars, giving a moment strength of 359.4 ft-kips.

The positive moment strength at the joint face must be equal to at least 50% of the corresponding negative moment strength at that joint (UBC 1921.3.3.2). The negative moment strengths at the interior and the exterior joint faces are 519.1 ft-kips and 359.1 ft-kips, respectively. There are no positive moments at the exterior end of the beam or at the interior joint face. Thus, the governing positive moment strengths are 519.1/2 = 259.6 ft-kips at the interior joint and 359.1/2 = 179.6 ft-kips at the exterior joint. Use 4-#8 bars at both ends, giving a moment strength of 288.3 ft-kips.

The maximum positive moment at the midspan of the member is 273.1 ft-kips. According to UBC 1921.3.2.2, the positive moment strength at any section along the member length must not be less than 25% of the maximum moment strength provided at the face of either joint. Thus, the minimum design value of the positive moment is 519.1/4 = 129.8 ft-kips which is less than 273.1 ft-kips. Use 4-#8 bars, giving a moment strength of 288.3 ft-kips.

Similarly, the negative moment strength at any section along the member length must not be less than 25% of the maximum moment strength provided at the face of either joint (519.1/4 = 129.8 ft-kips). Two #9 bars will be required, giving a moment strength of 186.3 ft-kips.

5.2.3.3.2 Required Shear Reinforcement

According to UBC 1921.3.4, the beams must be designed for shear forces that are based on the probable moment strength at each joint face using a strength reduction factor of 1.0 and a stress in the tensile flexural reinforcement equal to $1.25f_y$, combined with the shear caused by unfactored gravity loads. It should be noted that ACI 318-89 (Revised 1992) uses the same procedure, but with factored gravity loads.

Figure 5-12 shows the probable flexural strengths M_{pr} at the joint faces for this beam for sidesway to the right and to the left. As noted above, these flexural strengths are determined using a steel stress of $1.25 f_y$ and a strength reduction factor of 1.0. Figure 5-12 also shows the unfactored gravity loads on the beam, as well as the shear forces at the faces of the supports.

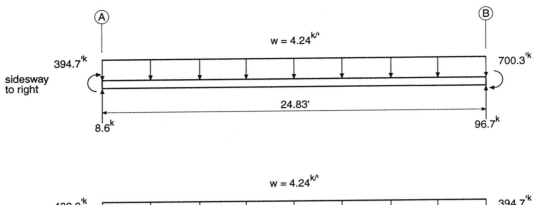

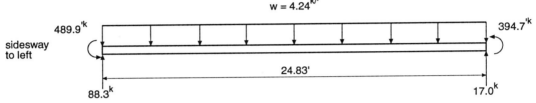

Figure 5-12 Design Shear Forces for Beam A3-B3 (Seismic Zone 4)

If the earthquake-induced shear force represents more than one-half of the total shear and the factored axial compressive force including earthquake effects is less than $A_g f'_c/20$, the quantity V_c is taken equal to zero (UBC 1921.3.4.2). At the faces of the columns, the earthquake induced shear force is less than one-half of the total shear. However, away from the column faces, the earthquake forces soon become more than one-half the total design shear, so that V_c is taken equal to zero in this design.

$$\phi V_s = V_u - \phi V_c$$

$$V_s = 96.7/0.85 - 0 = 113.8 \text{ kips}$$

Shear strength contributed by shear reinforcement cannot be taken greater than $0.75 \times 8 \sqrt{f'_c} \times b_w d$ noting that 0.75 accounts for the use of the lightweight concrete (UBC 1911.2 and 1911.5.6.8):

$$V_{s,max} = 0.75 \times 8 \sqrt{4000} \times (24 \times 21.44)/1000 = 195.3 \text{ kips} > 113.8 \text{ kips} \quad \text{O.K.}$$

In Seismic Zones 3 and 4, the following detailing requirements must be satisfied in addition to the shear strength requirements. Hoops must be provided:

- Over a length equal to twice the member depth, starting from the face of the supporting member toward midspan, at both ends of the flexural member (UBC 1921.3.3.1).

- Over lengths equal to twice the member depth where flexural yielding may occur in connection with inelastic lateral displacements of the frame (UBC 1921.3.3.1).

Where hoops are not required, stirrups with 135-degree seismic hooks shall be spaced at no more than $d/2$ throughout the length of the member (UBC 1921.3.3.4).

The maximum allowable hoop spacing within a distance $2h = 48$ in. from the face of the supports is:

$$s_{max} = \frac{d}{4} = \frac{21.44}{4} = 5.4 \text{ in. (governs)}$$

$$= 8 \times \text{(diameter of smallest longitudinal bar)} = 8 \times 1.0 = 8.0 \text{ in.}$$

$$= 24 \times \text{(diameter of hoop bars)} = 24 \times 0.5 = 12.0 \text{ in.}$$

$$= 12.0 \text{ in.}$$

The required spacing of #4 closed stirrups (hoops) for the factored shear of 113.8 kips is:

$$s = \frac{A_v f_y d}{V_s} = \frac{0.40 \times 60 \times 21.44}{113.8} = 4.5 \text{ in.}$$

Therefore, hoops should be spaced at 4.5 in. on center with the first being located 2 in. from the face of the support. Provide 12 hoops at 4.5 in. spacing, with the last hoop at 51.5 in. from the support face. At this location, the shear force is:

$$V_u = 96.7 - (4.24 \times 51.5/12) = 78.5 \text{ kips}$$

$$V_s = 78.5/0.85 = 92.4 \text{ kips}$$

The required spacing of #4 stirrups at this location is:

$$s = \frac{0.40 \times 60 \times 21.44}{92.4} = 5.6 \text{ in.}$$

A 5.5 in. spacing, starting at 51.5 in. from the column face to the start of the longitudinal bar splices at midspan, will be sufficient.

Note that lap splices in longitudinal reinforcement are permitted only if hoop or spiral reinforcement is provided over the lap length (UBC 1921.3.2.3). The maximum spacing of the transverse reinforcement enclosing the lapped bars must not exceed $d/4$ or 4 in. Lap splices must not be used within a distance of twice the member depth from the joint face, nor at locations where analysis indicates flexural yielding caused by inelastic lateral displacements of the frame to be likely.

5.2.3.3.3 Reinforcing Bar Cutoff Points and Splices

The negative reinforcement consists of 6-#9 bars at the interior support. The location where three of the six can be terminated away from the interior support will be determined. The loading used to find the cutoff points is 0.9 times the dead load in combination with the probable moment strengths, M_{pr}, at the member ends (conservatively using $f_s = 1.25f_y$). The design moment strength, ϕM_n provided by 3-#9 reinforcing bars in the beam is 274.5 ft-kips. Therefore, the three reinforcing bars can be terminated only after the factored moment, M_u, has been reduced to 274.5 ft-kips.

With $\phi = 1.0$ and $f_s = 75$ ksi, $M^+_{pr} = 394.7$ ft-kips at the exterior end of the beam, and $M^-_{pr} = 700.3$ ft-kips at the interior end. The dead load is equal to $0.9 \times 3.4 = 3.1$ kips/ft. The distance from the face of the support to where the moment under the loading considered equals 274.5 ft-kips is readily obtained by summing moments about section a-a in Fig. 5-13:

$$\frac{1}{2}(3.1x^2) + 700.3 - 82.6x = 274.5$$

Solving for x gives a distance of 5.8 ft. Therefore, the negative reinforcing bars can be terminated at a distance of $5.8 + d = 7.6$ ft from the face of the interior support (noting that $d = 21.44$ in. $> 12\,d_b = 12 \times 1.128 = 13.5$ in.; see UBC 1912.10.3).

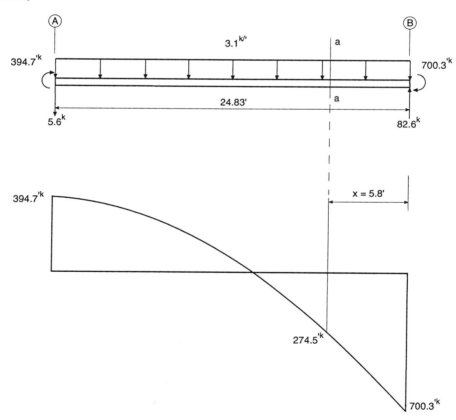

Figure 5-13 Moment Diagram for Cutoff Location of Negative Bars at Interior Support of Beam A3-B3
(Seismic Zone 4)

The #9 bars being terminated must be properly developed at the support. For lightweight concrete used in Seismic Zone 4, the development length of reinforcement in tension when the depth of concrete cast in one lift beneath the bar exceeds 12 in., is given by:

$$\ell_d = 1.25 \times 3.5 \times f_y \times d_b/65\sqrt{f'_c} \qquad\qquad 1921.5.4.2$$

$$= 1.25 \times 3.5 \times 60{,}000 \times 1.128/(65\sqrt{4000})$$

$$= 72.0 \text{ in.} < (7.6 \times 12) = 91.2 \text{ in.} \quad \text{O.K.}$$

Using the same procedure outlined above, the cutoff point for one of the four bars at the exterior end of the span is 4.9 ft from the exterior column face.

Lap splices of flexural reinforcement must not be placed within a distance of 2h from the faces of the support, nor within regions of potential plastic hinging (UBC 1921.3.2.3). All lap splices must be confined by hoops or spirals with a maximum spacing of d/4 or 4 in. over the length of the lap.

The development length on which the lap splice length is based is computed as the product of the basic development length and applicable modification factors. The basic development length for #9 bars is:

$$\ell_{db} = 0.04 A_b f_y / \sqrt{f'_c}$$

$$= 0.04 \times 1.0 \times 60{,}000/\sqrt{4000}$$

$$= 38.0 \text{ in.}$$

For #8 bars:

$$\ell_{db} = 0.04 A_b f_y / \sqrt{f'_c}$$

$$= 0.04 \times 0.79 \times 60{,}000/\sqrt{4000}$$

$$= 30.0 \text{ in.}$$

The development length modification factors are shown in Table 5-12.

The bottom reinforcing bars will be subject to stresses greater than $0.5f_y$ and will require a Class B splice.

The information required for determining the development length factors is:

Minimum bottom cover to bars being spliced	$= 2.0$ in. $(2.0\ d_b)$
Minimum side cover	$= 2.0$ in. $(2.0\ d_b)$
Clear spacing of bottom bars being spliced	$= 5.3$ in. $(5.3\ d_b)$

Transverse reinforcement meets requirements of UBC 1911.5.4 and 1912.2.3.5.

This gives a value of 0.75 for Factor I, but the limiting value for #8 bars of 0.95 will control (see Table 5-12). The other factors are all equal to 1.0, with the exception of Factor III for lightweight concrete (Factor III = 1.3).

Class B splice length = 1.3 × (0.95 × 1.0 × 1.3 ×1.0 × 1.0 × 30.0) = 48.2 in.

The top reinforcing bars are subject to stresses less than $0.5f_y$ at midspan and a Class A splice will be used.

The information required for determining the development length factors is:

Minimum top cover to bars being spliced   = 2.0 in. (1.77 d_b)

Minimum side cover   large (slab)

Clear spacing of top bars being spliced   = 6.42 in. (5.69 d_b)

Transverse reinforcement meets requirements of UBC 1911.5.4 and 1912.2.3.5.

This gives a value of 0.60 for Factor I, but the limiting value for #9 bars of 0.85 will control.   The other factors are all equal to 1.0, with the exception of Factor III for lightweight concrete and Factor II for top bars, which are both equal to 1.3.

Class A splice length = 1.0 × (0.85 × 1.3 × 1.3 ×1.0 × 1.0 × 38.0) = 54.6 in.

Figure 5-14 shows the reinforcement details for beam A3-B3.

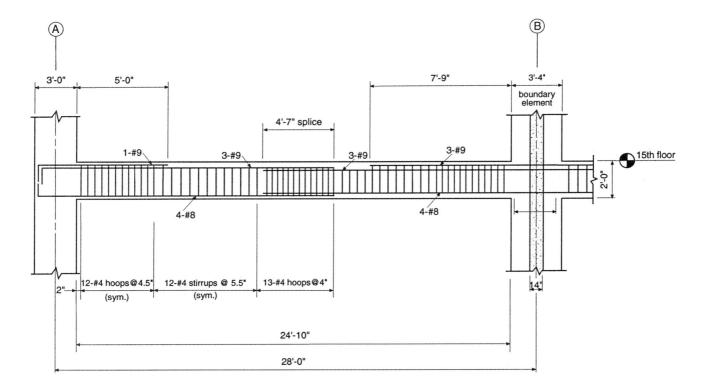

Figure 5-14 Reinforcement Details for Beam A3-B3 (Seismic Zone 4)

5.2.3.4   Proportioning and Detailing an Interior Column

5.2.3.4.1 General Code Requirements

This section will present the design of column C3 between the ground level and the 2nd floor. The maximum factored axial force on this column at this level is 4118 kips (see Table 5-9). The requirements in UBC 1921.4 apply to all structural members resisting earthquake forces and factored axial forces exceeding $A_g f'_c/10$. The maximum axial force of 4118 kips exceeds $A_g f'_c/10 = (36 \times 36 \times 6)/10 = 778$ kips; therefore, the two following criteria must be satisfied:

- Shortest column dimension shall not be less than 12 in. This criterion is met since the columns are 36 in. square.

- Ratio of the shortest cross-sectional dimension to the perpendicular dimension shall not be less than 0.4. This criterion is met since the columns are square.

Based on the loading combinations in Table 5-9, a 36 in. square column with 12-#11 bars will be adequate (see Fig. 5-15). The reinforcement ratio for columns must be greater than 1% but less than 6% (UBC 1921.4.3.1):

$$0.01 < \rho = (12 \times 1.56)/(36 \times 36) = 0.0144 < 0.06 \quad \text{O.K.}$$

5.2.3.4.2 Relative Flexural Strengths of Columns and Girders

UBC 1921.4.2.2 requires that the sum of the flexural strengths of columns at a joint must be greater than or equal to 6/5 times the sum of the flexural strengths of girders framing into that joint.

The beam designed in Section 5.2.3.3 was for the 15th floor of the structure. In this section the column being designed is between the ground and the 2nd floor. The reinforcement for the beams framing into this column is assumed to be the same as that used for the beams at the 15th floor since the gravity load moments are equal and the seismic load moments are comparable. The negative flexural strength of the beam at the face of the column is 519.1 ft-kips (see Section 5.2.3.3.1). The positive moment strength of the beam on the opposite face of the column is 288.3 ft-kips.

The maximum factored axial load on the upper column framing into the top of the joint is 4118 kips (see Table 5-9). The corresponding moment strength using $f'_c = 6$ ksi, $f_y = 60$ ksi and 12-#11 reinforcing bars is 1426.7 ft-kips (see Fig. 5-15). Assuming the same strength above and below the second floor level:

$$\Sigma M_e = 1426.7 + 1426.7 = 2853.4 \text{ ft-kips}$$

$$\Sigma M_g = 519.1 + 288.3 = 807.4 \text{ ft-kips}$$

$$2853.4 \text{ ft-kips} > \frac{6}{5} \times 807.4 \text{ ft-kips} = 968.9 \text{ ft-kips} \quad \text{O.K.}$$

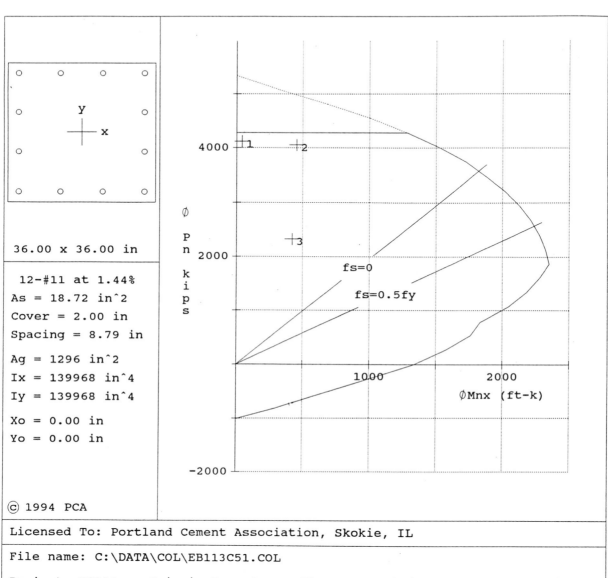

36.00 x 36.00 in

12-#11 at 1.44%
As = 18.72 in^2
Cover = 2.00 in
Spacing = 8.79 in

Ag = 1296 in^2
Ix = 139968 in^4
Iy = 139968 in^4

Xo = 0.00 in
Yo = 0.00 in

© 1994 PCA

Licensed To: Portland Cement Association, Skokie, IL

File name: C:\DATA\COL\EB113C51.COL

Project: EB113 -- Seismic Zone 4   f'c = 6.00 ksi   Ec = 4696 ksi

Column Id: Column C3   fc = 5.10 ksi   eu = 0.003 in/in

Engineer: Block stress profile, Beta1 = 0.75

Date: Time: fy = 60 ksi   Es = 29000 ksi

Code: ACI 318-89   Confinement: Tied

Units: English   phi(c) = 0.70, phi(b) = 0.90, a = 0.80

X-axis slenderness is not considered.

Figure 5-15 Design Strength Interaction Diagram for Column C3 (Seismic Zone 4)

5.2.3.4.3 Confinement Reinforcement

Special transverse reinforcement for confinement is required over a distance ℓ_o at the column ends where ℓ_o equals the maximum of:

$$\frac{1}{6}(\text{clear height}) = \frac{(21 \times 12)}{6} = 42.0 \text{ in. (governs)} \qquad\qquad 1921.4.4.4$$

Depth of member = 36.0 in.

18.0 in.

According to UBC 1921.4.4.2, the maximum allowable spacing of rectangular hoops within the 42 in. length is 4 in. The columns in the structure are subject to double curvature, and therefore the point of contraflexure is within the middle half of the clear height, and special confinement will not be required for the full length of the column (UBC 1921.4.4.1).

The required cross-sectional area of confinement reinforcement in the form of hoops is:

$$A_{sh} \geq \begin{cases} 0.09 sh_c \dfrac{f'_c}{f_{yh}} \\[12pt] 0.3 sh_c \left[\dfrac{A_c}{A_{ch}} - 1\right] \dfrac{f'_c}{f_{yh}} \end{cases} \qquad\qquad 1921.4.4.1$$

where

s =   spacing of transverse reinforcement (in.)

h_c =   cross-sectional dimension of column core, measured center-to-center of confining reinforcement (in.)

A_{ch} =   cross-sectional area of a structural member measured out-to-out of transverse reinforcement (in.^{2})

f_{yh} =   specified yield strength of transverse reinforcement (psi)

Using a hoop spacing of 4 in., f_{yh} = 60,000 psi, and tentatively assuming #5 bar hoops, the required cross-sectional area is:

$$A_{sh} \geq \begin{cases} 0.09 \times 4 \times 32.4 \times 6000 / 60,000 = 1.17 \text{ in.}^2 \text{ (governs)} \\[10pt] 0.3 \times 4 \times 32.4 \times [1296 / 1089 - 1] \times 6000 / 60,000 = 0.74 \text{ in.}^2 \end{cases}$$

Using #5 hoops with 2-#5 crossties provides $A_{sh} = 4 \times 0.31 = 1.24 \text{ in.}^2$ Therefore, use 4 in. spacing for the hoops at the column ends.

5.2.3.4.4 Transverse Reinforcement for Shear

Similar to the design of shear reinforcement for beams, the design of columns for shear is not based on the factored shear forces obtained from a lateral load analysis, but rather on the probable moment strengths provided at the column ends. UBC 1921.4.5 requires that the design shear force shall be determined from the consideration of the maximum forces that can be developed at the faces of the joints. These forces shall be determined using the probable moment strengths calculated without strength reduction factors and assuming that the stress in the tensile reinforcement is equal to $1.25f_y$. The member shear need not exceed that determined from joint strengths based on the probable moment strength of the transverse members framing into the joint.

The largest probable moment strength that may develop in the column can conservatively be assumed to correspond to the balanced point of the column interaction diagram. The balanced point nominal moment strength for this column with $f_s = 1.25f_y$ is equal to $2464.0/\phi = 2464.0/0.70 = 3520.0$ ft-kips (see Fig. 5-16).

In Section 5.2.3.3.2, the probable moment strengths of the beams were determined for $\phi = 1.0$ and $f_s = 1.25f_y$. The probable negative moment strength at the face of the interior column is 700.3 ft-kips. The probable positive moment strength at the same location is 394.7 ft-kips. The largest moment that can develop from the beams is $700.3 + 394.7 = 1095$ ft-kips. This moment is less than the sum of the probable moment strengths at the column ends ($2 \times 3520.0 = 7040.0$ ft-kips). Therefore, the columns need only be designed to resist the maximum shear that can be transferred through the beams. The design factored shear for the column is:

$$V_u = \frac{1095.0/2 + 1095.0/2}{21} = 52.1 \text{ kips}$$

Since the factored axial forces are greater than $A_g f'_c/20 = (36 \times 36 \times 6)/20 = 389$ kips, the shear strength of the concrete may be used (see Table 5-9 and UBC 1921.4.5.2):

$$V_c = 2\left[1 + \frac{N_u}{2000A_g}\right]\sqrt{f'_c}\, b_w d \qquad\qquad 1911.3.1.2$$

Conservatively use the minimum axial load from Table 5-9 (i.e., $N_u = 2264$ kips); therefore,

$$V_c = 2\left[1 + \frac{2,264,000}{2000 \times 1296}\right]\sqrt{6000} \times 36 \times 33.2/1000 = 346.9 \text{ kips}$$

$$\phi V_c = 0.85 \times 346.9 = 294.9 \text{ kips} > 52.1 \text{ kips}$$

Thus, use the 4 in. spacing over the distance $\ell_o = 42$ in. near the column ends that was determined by the requirements for confinement.

The remainder of the column length must contain hoop reinforcement with center-to-center spacing not to exceed either six times the diameter of the longitudinal column bars or 6 in. (UBC 1921.4.4.6). In this case, the 6 in. controls.

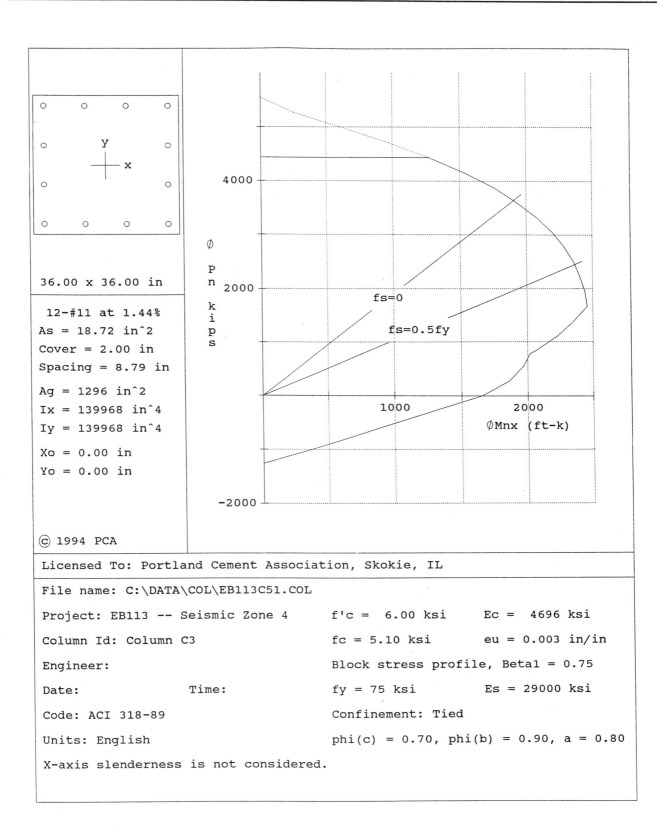

36.00 x 36.00 in

12-#11 at 1.44%
As = 18.72 in^2
Cover = 2.00 in
Spacing = 8.79 in

Ag = 1296 in^2
Ix = 139968 in^4
Iy = 139968 in^4

Xo = 0.00 in
Yo = 0.00 in

© 1994 PCA

Licensed To: Portland Cement Association, Skokie, IL

File name: C:\DATA\COL\EB113C51.COL

Project: EB113 -- Seismic Zone 4   f'c = 6.00 ksi   Ec = 4696 ksi

Column Id: Column C3   fc = 5.10 ksi   eu = 0.003 in/in

Engineer: Block stress profile, Beta1 = 0.75

Date: Time: fy = 75 ksi   Es = 29000 ksi

Code: ACI 318-89   Confinement: Tied

Units: English   phi(c) = 0.70, phi(b) = 0.90, a = 0.80

X-axis slenderness is not considered.

Figure 5-16 Design Strength Interaction Diagram for Column C3 with f_y = 75 ksi

Note that lap splices in longitudinal reinforcement are permitted only if hoop or spiral reinforcement is provided over the lap length (UBC 1921.4.3.2). The maximum spacing of the transverse reinforcement enclosing the lapped bars must not exceed 4 in. Lap splices are permitted only within the center half of the column height.

5.2.3.4.5 Minimum Length of Lap Splices of Column Vertical Bars

Lap splices will be proportioned as Class B tension splices. The basic development length for #11 bars is:

$$\ell_{db} = \frac{0.04 A_b f_y}{\sqrt{f'_c}} = \frac{0.04 \times 1.56 \times 60,000}{\sqrt{6000}} = 48.3 \text{ in.} \qquad\qquad 1912.2.2$$

The information required for determining the development length factors is:

Minimum cover to bars being spliced   = 2.125 in. (1.51 d_b)

Clear bar spacing   = 8.7 in. (6.17 d_b)

Transverse reinforcement meets requirements of UBC 1907.10.5 and 1912.2.3.5.

This gives a value of 0.75 for Factor I (see Table 5-12). The other factors are all equal to 1.0, so that:

Class B splice length = 1.3 ℓ_d = 1.3 × (0.75 × 1.0 × 1.0 ×1.0 × 1.0 × 48.3) = 47.1 in.

Figure 5-17 shows the reinforcement details for column C3.

5.2.3.5   Proportioning and Detailing an Exterior Column

5.2.3.5.1 General Requirements

This section will present the design of column A3 between the ground level and the 2nd floor. The maximum factored axial force on this column at this level is 2986 kips (see Table 5-9). The requirements in UBC 1921.4 apply to all structural members resisting earthquake forces and factored axial forces exceeding $A_g f'_c/10$. The maximum axial force of 2986 kips exceeds $A_g f'_c/10 = (36 \times 36 \times 6)/10 = 778$ kips; therefore, the two following criteria must be satisfied:

- Shortest column dimension shall not be less than 12 in. This criterion is met since the columns are 36 in. square.

- Ratio of the shortest cross sectional dimension to the perpendicular dimension shall not be less than 0.4. This criterion is met since the columns are square.

Based on the loading combinations in Table 5-9, a 36 in. square column with 12-#10 bars will be used (see Fig. 5-18). The reinforcement ratio for columns must be greater than 1% but less than 6% (UBC 1921.4.3.1):

$$0.01 < \rho = (12 \times 1.27)/(36 \times 36) = 0.0118 < 0.06 \quad \text{O.K.}$$

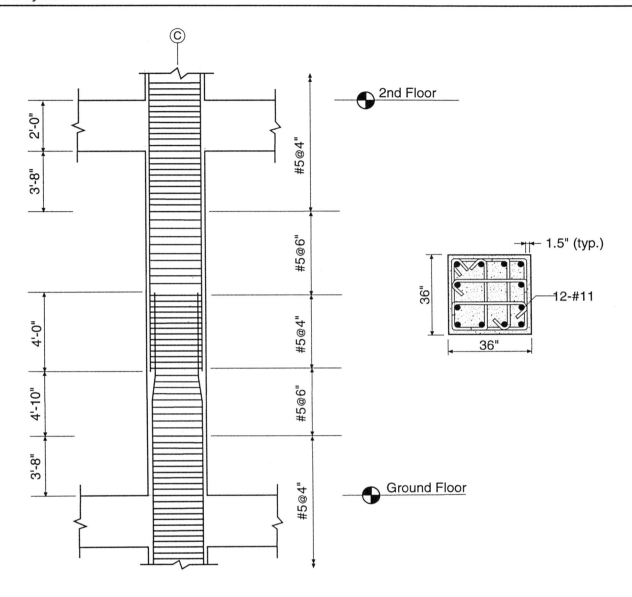

Figure 5-17 Reinforcement Details for Column C3 (Seismic Zone 4)

5.2.3.5.2 Relative Flexural Strengths of Columns and Girders

UBC 1921.4.2.2 requires that the sum of the flexural strengths of columns at a joint must be greater than or equal to 6/5 times the sum of the flexural strengths of girders framing into that joint.

Assuming that the negative reinforcement in the 2nd floor beam is the same as in the 15th floor beam, the negative design flexural strength of the beam at the exterior column is 359.4 ft-kips (see Section 5.2.3.3.1)

The maximum factored axial load on the upper column framing into the top of the joint is 2905 kips (see Table 5-9). The design flexural strength using $f'_c = 6$ ksi, $f_y = 60$ ksi and 12-#10 reinforcing bars is 2049.0 ft-kips (see Fig. 5-18). Assuming the same moment capacity above and below the joint:

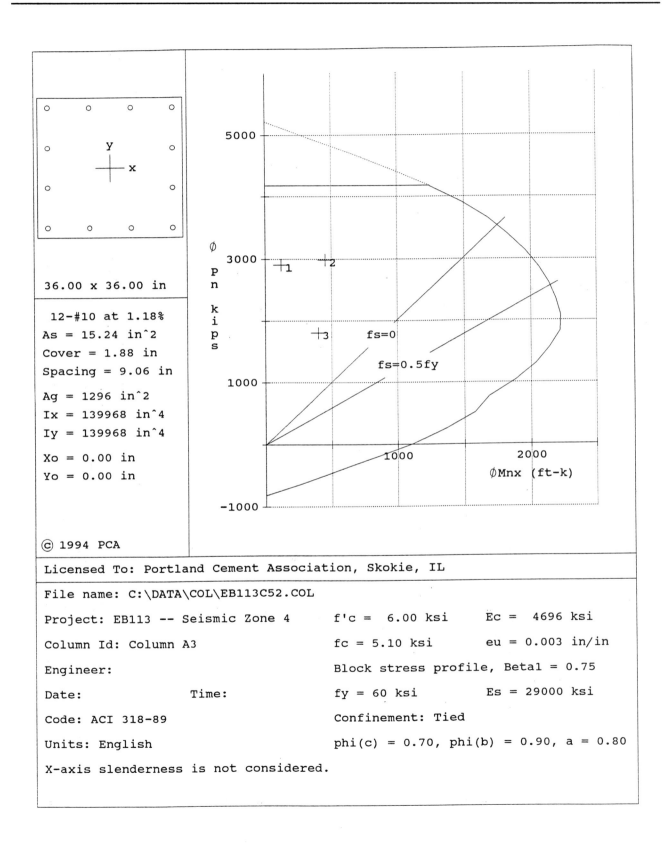

36.00 x 36.00 in

12-#10 at 1.18%
As = 15.24 in^2
Cover = 1.88 in
Spacing = 9.06 in

Ag = 1296 in^2
Ix = 139968 in^4
Iy = 139968 in^4

Xo = 0.00 in
Yo = 0.00 in

© 1994 PCA

Licensed To: Portland Cement Association, Skokie, IL

File name: C:\DATA\COL\EB113C52.COL

Project: EB113 -- Seismic Zone 4   f'c = 6.00 ksi   Ec = 4696 ksi

Column Id: Column A3   fc = 5.10 ksi   eu = 0.003 in/in

Engineer: Block stress profile, Beta1 = 0.75

Date: Time: fy = 60 ksi   Es = 29000 ksi

Code: ACI 318-89   Confinement: Tied

Units: English   phi(c) = 0.70, phi(b) = 0.90, a = 0.80

X-axis slenderness is not considered.

Figure 5-18 Design Strength Interaction Diagram for Column A3 (Seismic Zone 4)

$$\Sigma M_e = 2049.0 + 2049.0 = 4098.0 \text{ ft-kips}$$

$$\Sigma M_g = 359.4 \text{ ft-kips}$$

$$4098.0 \text{ ft-kips} > \frac{6}{5} \times 359.4 = 431.3 \text{ ft-kips} \quad \text{O.K.}$$

5.2.3.5.3 Confinement Reinforcement

Special transverse reinforcement for confinement is required over a distance ℓ_o at the column ends where ℓ_o equals the maximum of:

$$\frac{1}{6}(\text{clear height}) = \frac{(21 \times 12)}{6} = 42.0 \text{ in. (governs)} \qquad\qquad 1921.4.4.4$$

Depth of member = 36.0 in.

18.0 in.

According to UBC 1921.4.4.2, the maximum allowable spacing of rectangular hoops within the 42 in. length is 4 in. The columns in the structure are subject to double curvature, and therefore the point of contraflexure is within the middle half of the clear height, and special confinement will not be required for the full length of the column (UBC 1921.4.4.1).

The required cross-sectional area of confinement reinforcement in the form of hoops is:

$$A_{sh} \geq \begin{cases} 0.09 s h_c \dfrac{f'_c}{f_{yh}} \\[2em] 0.3 s h_c \left[\dfrac{A_c}{A_{ch}} - 1 \right] \dfrac{f'_c}{f_{yh}} \end{cases} \qquad\qquad 1921.4.4.1$$

where

s	=	spacing of transverse reinforcement (in.)
h_c	=	cross-sectional dimension of column core, measured center-to-center of confining reinforcement (in.)
A_{ch}	=	cross-sectional area of a structural member measured out-to-out of transverse reinforcement (in.^{2})
f_{yh}	=	specified yield strength of transverse reinforcement (psi)

Using a hoop spacing of 4 in., $f_{yh} = 60,000$ psi, and tentatively assuming #5 bar hoops, the required cross-sectional area is:

$$A_{sh} \geq \begin{cases} 0.09 \times 4 \times 32.4 \times 6000\,/\,60{,}000 \,=\, 1.17 \text{ in.}^2 \quad \text{(governs)} \\ 0.3 \times 4 \times 32.4 \times [1296\,/\,1089 - 1] \times 6000\,/\,60{,}000 \,=\, 0.74 \text{ in.}^2 \end{cases}$$

Using #5 hoops with 2-#5 crossties provides $A_{sh} = 4 \times 0.31 = 1.24$ in.^{2} Therefore, use 4 in. spacing for the hoops at the column ends.

5.2.3.5.4 Transverse Reinforcement for Shear

As noted in Sect. 5.2.3.4.4, the design shear force for the column is based on the probable moment strengths of the ends of the column (UBC 1921.4.5).

The largest probable moment strength that may develop in the column can conservatively be assumed to correspond to the balanced point of the column interaction diagram. The balanced point nominal moment strength for this column with $f_s = 1.25 f_y$ is equal to $2325.0\,/\,\phi = 2325.0\,/\,0.70 = 3321.4$ ft-kips (see Fig. 5-19).

In Section 5.2.3.3.2, the probable moment strengths of the beams were determined for $\phi = 1.0$ and $f_s = 1.25 f_y$. The probable moment strength at the interior face of the exterior column is 489.9 ft-kips. This moment is less than the sum of the probable moment strengths at the column ends ($2 \times 3321.4 = 6642.8$ ft-kips). Therefore, the columns need only be designed to resist the maximum shear that can be transferred through the beams. The design factored shear for the column is:

$$V_u = \frac{489.9/2 + 489.9/2}{21} = 23.3 \text{ kips}$$

Since the factored axial forces are greater than $A_g f'_c / 20 = (36 \times 36 \times 6)/20 = 389$ kips, the shear strength of the concrete may be used (see Table 5-9 and UBC 1921.4.5.2):

$$V_c = 2\left[1 + \frac{N_u}{2000 A_g}\right]\sqrt{f'_c}\, b_w d \qquad\qquad \textit{1911.3.1.2}$$

Using the minimum axial load from Table 3-9:

$$V_c = 2\left[1 + \frac{1{,}533{,}000}{2000 \times 1296}\right]\sqrt{6000} \times 36 \times 33.2\,/\,1000 = 294.7 \text{ kips}$$

$$\phi V_c = 0.85 \times 294.7 = 250.5 \text{ kips} > 23.3 \text{ kips}$$

Thus, use the 4 in. spacing over the distance $\ell_o = 42$ in. near the column ends that was determined by the requirements for confinement.

The remainder of the column length must contain hoop reinforcement with center-to-center spacing not to exceed either six times the diameter of the longitudinal column bars or 6 in (UBC 1921.4.4.6). In this case, 6 in. controls.

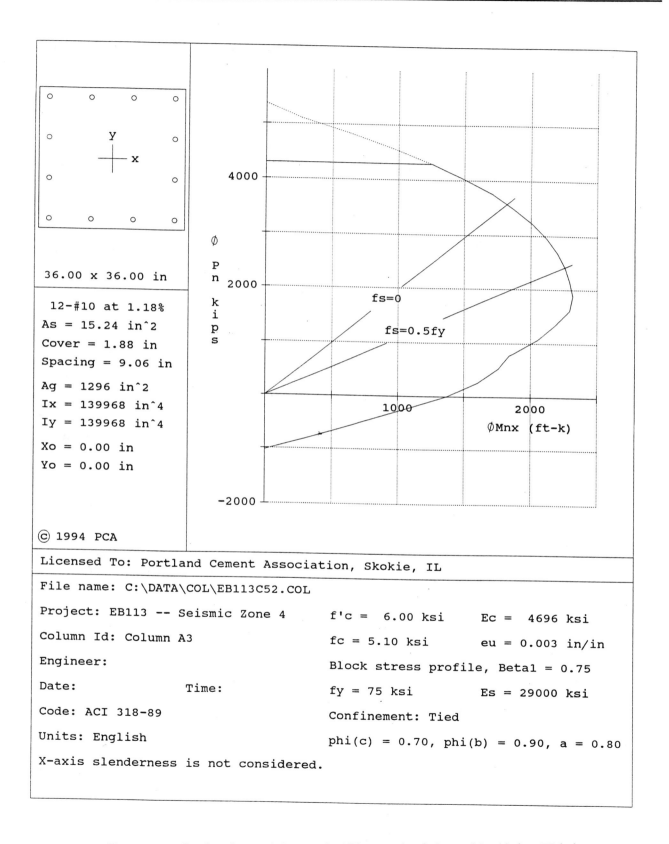

36.00 x 36.00 in

12-#10 at 1.18%
As = 15.24 in^2
Cover = 1.88 in
Spacing = 9.06 in

Ag = 1296 in^2
Ix = 139968 in^4
Iy = 139968 in^4

Xo = 0.00 in
Yo = 0.00 in

© 1994 PCA

Licensed To: Portland Cement Association, Skokie, IL

File name: C:\DATA\COL\EB113C52.COL

Project: EB113 -- Seismic Zone 4   f'c = 6.00 ksi   Ec = 4696 ksi

Column Id: Column A3   fc = 5.10 ksi   eu = 0.003 in/in

Engineer: Block stress profile, Beta1 = 0.75

Date: Time: fy = 75 ksi   Es = 29000 ksi

Code: ACI 318-89   Confinement: Tied

Units: English   phi(c) = 0.70, phi(b) = 0.90, a = 0.80

X-axis slenderness is not considered.

Figure 5-19 Design Strength Interaction Diagram for Column A3 with f_y = 75 ksi

Note that lap splices in longitudinal reinforcement are permitted only if hoop or spiral reinforcement is provided over the lap length (UBC 1921.4.3.2). The maximum spacing of the transverse reinforcement enclosing the lapped bars must not exceed 4 in. Lap splices are permitted only within the center half of the column height.

5.2.3.5.5 Minimum Length of Lap Splices of Column Vertical Bars

Lap splices will be proportioned as Class B tension splices. The basic development length for #10 bars is:

$$\ell_{db} = \frac{0.04 A_b f_y}{\sqrt{f'_c}} = \frac{0.04 \times 1.27 \times 60,000}{\sqrt{6000}} = 39.4 \text{ in.} \qquad\qquad 1912.2.2$$

The information required for determining the development length factors is:

Minimum cover to bars being spliced	$= 2.125$ in. $(1.67\ d_b)$
Clear bar spacing	$= 8.9$ in. $(7.0\ d_b)$

Transverse reinforcement meets requirements of UBC 1907.10.5 and 1912.2.3.5.

This gives a value of 0.75 for Factor I (see Table 5-12). The other factors are all equal to 1.0, so that

$$\text{Class B splice length} = 1.3\,\ell_d = 1.3 \times (0.75 \times 1.0 \times 1.0 \times 1.0 \times 1.0 \times 39.4) = 38.4 \text{ in.}$$

Figure 5-20 shows the reinforcement details for column A3.

5.2.3.6   Proportioning and Detailing of an Interior Beam-Column Connection

This section will check the shear strength and detailing requirements for interior beam-column connection C3 on the third floor level of the example building. The column considered is an interior column of an interior frame along either principal direction of the building, and has beams framing in on all four sides: two transverse beams and two longitudinal beams.

5.2.3.6.1 Transverse Reinforcement for Confinement

UBC 1921.5.2.1 requires that transverse hoop reinforcement as specified in UBC 1921.4.4 shall be provided within a joint, unless the joint is confined by structural members as specified in UBC 1921.5.2.2.

The joint in question has beams framing in on all four faces. However, the width of each beam (24 in.) is smaller than three-quarters of the column face width (3/4 × 36 in. = 27 in.). Thus, the joint is not confined by framing members, according to the criteria in UBC 1921.5.2.2. The special column-end transverse reinforcement shown for the first story column C3 in Fig. 5-17 (#5 closed hoops with two crossties in each direction at a 4 in. spacing) must continue through the beam-column joint being designed.

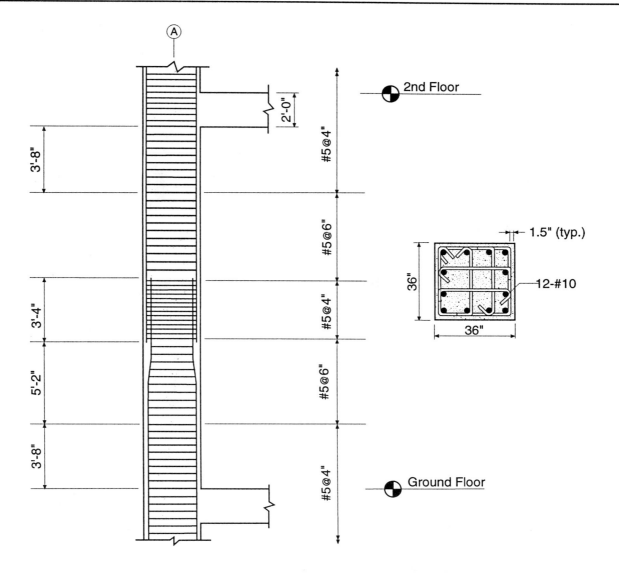

Figure 5-20 Reinforcement Details for Column A3 (Seismic Zone 4)

5.2.3.6.2 Shear Strength of Joint

Figure 5-21 shows the beam-column joint at the 3rd floor level of the structure. The shear strength is checked along column line C of the building. Because of symmetry, the computations in the orthogonal direction would be identical. The shear at section x-x is obtained by subtracting the column horizontal shear from the sum of the tensile force in the top beam reinforcement on one side and the concrete compressive force near the top of the beam on the opposite face of the column. It is assumed that the 3rd floor beams are identical to the 15th floor beams designed in Sect. 5.2.3.3.

The column horizontal shear, V_h, can be obtained by assuming that the beams in the adjoining floors are deformed so that plastic hinges form at their junctions with the column, with M^-_{pr} (beam) = 700.3 ft-kips and M^+_{pr} (beam) = 394.7 ft-kips. Therefore, the sum of the column moments above and below the joint will equal 700.3 + 394.7 = 1095.0 ft-kips.

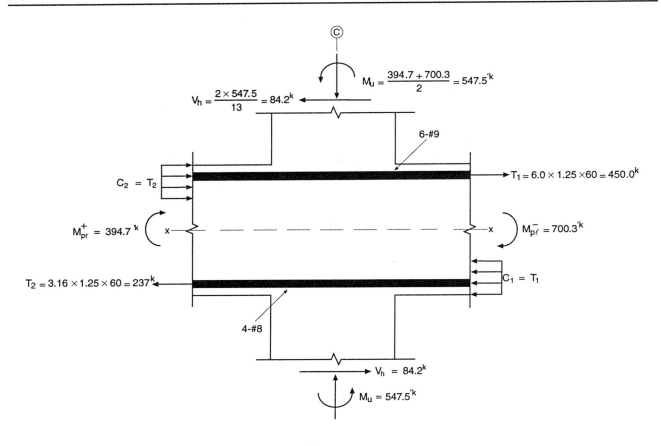

Figure 5-21 Shear Analysis of Interior Beam-Column Joint C3 (Seismic Zone 4)

It is assumed that beam end moments are resisted equally by the columns above and below the joint; the horizontal shear at the column ends is then equal to:

$$V_h = \frac{2 \times \left(\dfrac{700.3 + 394.7}{2}\right)}{13} = 84.2 \text{ kips}$$

According to UBC 1921.5.1.1, the tensile force in the top beam reinforcement (6-#9 bars) = $6.0 \times 1.25 \times 60 = 450$ kips. The compressive force on the opposite side of the column is equal to the tensile force of 4-#8 bars which is $3.16 \times 1.25 \times 60 = 237$ kips.

The net shear at section x-x of the joint is $T_1 + C_2 - V_h = 450 + 237 - 84.2 = 602.8$ kips.

For a joint not confined on any of the four faces, the nominal shear strength is (UBC 1921.5.3.1):

$$\phi V_c = \phi 12 \sqrt{f'_c} \times A_j / 1000$$

$$= 0.85 \times 12 \sqrt{6000} \times 36 \times 36 / 1000 = 1024.0 \text{ kips} > 602.8 \text{ kips} \quad \text{O.K.}$$

5.2.3.7 Proportioning and Detailing of an Exterior Beam-Column Connection

This section will check the shear strength and detailing requirements for exterior beam-column connection A3 on the 3rd floor level of the example building. The column considered is an interior column of an exterior frame in one direction of the building, and an exterior column of an interior frame in the orthogonal direction. The column has beams framing in on three sides: two transverse beams and one longitudinal beam.

5.2.3.7.1 Transverse Reinforcement for Confinement

UBC 1921.5.2.1 requires that transverse hoop reinforcement as specified in UBC 1921.4.4 shall be provided within a joint, unless the joint is confined by structural members as specified in UBC 1921.5.2.2.

The joint in question has beams framing in on the three faces only. However, it is not confined by these framing members, according to the criteria in UBC 1921.5.2.2. The special column-end transverse reinforcement shown for the first story column A3 in Fig. 5-20 (#5 closed hoops with two crossties in each direction at a 4 in. spacing) must continue through the beam-column joint being designed.

5.2.3.7.2 Shear Strength of Joint

Figure 5-22 shows the beam-column joint at the 3rd floor level of the structure. The shear strength is checked along column line 3 of the building. The shear at section x-x is obtained by subtracting the column horizontal shear from the tensile force in the top beam reinforcement. It will be assumed that the 3rd floor beams are identical to the 15th floor beams designed in Sect. 5.2.3.3.

The column horizontal shear, V_h, can be obtained by assuming that the beam in the adjoining floor is deformed so that a plastic hinge forms at the junctions with the column, with M^-_{pr} (beam) = 489.9 ft-kips.

It is assumed that the beam end moments are resisted equally by the column above and below the joint. The horizontal shear at the column ends is then equal to:

$$V_h = \frac{2 \times \left(\dfrac{489.9}{2}\right)}{13} = 37.7 \text{ kips}$$

The tensile force in the top beam reinforcement (4-#9 bars) = $4.0 \times 1.25 \times 60 = 300$ kips. The net shear at section x-x of the joint is $T_1 - V_h = 300 - 37.7 = 262.3$ kips.

Although the joint has beams framing in on three faces, it is confined only on two opposite faces where the transverse beam width of 27 in. is equal to three-quarters of the column width (3/4 × 36 in. = 27 in.). It is not confined on the third face, since the beam width of 24 in. is less than three-quarters of the column width (UBC 1921.5.3.1).

For a joint confined on two opposite faces:

$$\phi V_c = \phi 15 \sqrt{f'_c} \times A_j /1000$$

$$= 0.85 \times 15 \sqrt{6000} \times 36 \times 36/1000 = 1280.0 \text{ kips} > 262.3 \text{ kips} \quad \text{O.K.}$$

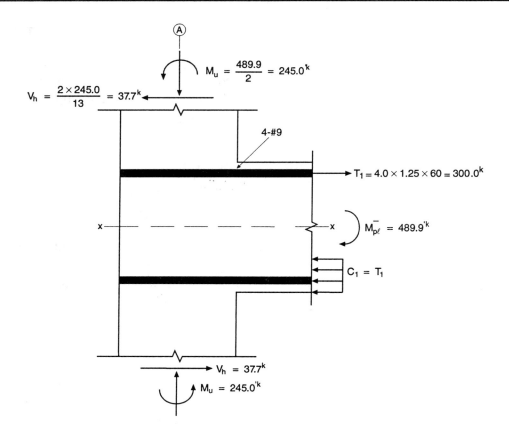

Figure 5-22 Shear Analysis of Exterior Beam-Column Joint A3 (Seismic Zone 4)

5.2.3.8 Proportioning and Detailing a Shearwall

A significant change in the design of shearwalls occurred with the adoption of the 1994 edition of the Uniform Building Code. In UBC-91 Sect. 2625(f)3, the boundary elements of the shearwall, whenever required, must be designed to carry all of the factored gravity loads tributary to the wall plus the vertical force required to resist the overturning moment at the base of the wall due to the factored seismic loads. In other words, the boundary elements are to be designed completely ignoring the contribution of the web. As a result of this provision, large boundary elements with congested reinforcement are usually obtained.

To alleviate this problem, the provisions in UBC-94 Sect. 1921.6.5 give a design procedure which utilizes the entire cross-section of the wall in resisting the factored axial loads and moments. Since the current edition of the ACI Code uses the provisions in UBC-91, and for comparison purposes, both design methods will be presented in the following two sections. In Sect. 5.2.3.8.1, the section numbers given are from UBC-91 and ACI 318-89 (Revised 1992).

5.2.3.8.1 Design of Shearwall B2-C2 Based on UBC-91 Provisions

The shearwalls are part of the lateral load resisting system and must meet the applicable special detailing requirements of UBC 2625.

According to UBC 2625(f)3A, a shearwall is required to have boundary elements if the maximum compressive stress in the shearwall exceeds $0.2\,f'_c$. A boundary element is a column-like member provided at the end of a shearwall; it is contiguous with the wall section. If the compressive stress at the edges of a shearwall exceeds $0.2\,f'_c$ and boundary elements are not provided, the entire shearwall must have transverse reinforcement meeting the requirements specified for members subjected to bending and axial load (UBC 2625(e)4A through C). It is usually more economical to provide boundary elements than to meet the special transverse reinforcement requirement.

The maximum stress in the shearwall occurs at the ends of the shearwall and is given by:

$$f_c = \frac{P_u}{A_g} \pm \frac{M_u \, (\ell_w / 2)}{I_g}$$

The stress is computed using factored loads and gross section properties.

Assuming that the length of the wall is 28.0 ft + 40.0 in. = 31.33 ft, the section properties of the wall are:

$$A_g = 31.33 \times \frac{16}{12} = 41.78 \text{ ft}^2$$

$$I_g = \frac{1}{12} \times \frac{16}{12} \times 31.33^3 = 3418.0 \text{ ft}^4$$

Using the factored loads from Table 5-10, the maximum fiber stress is:

$$f_c = \frac{8629}{41.78} + \frac{90,245 \, (31.33/2)}{3418.0}$$

$$= 206.5 + 413.6 = 620.1 \text{ ksf} = 4.3 \text{ ksi}$$

$$4.3 \text{ ksi} > 0.2 \; f'_c = 0.2 \times 6 = 1.2 \text{ ksi}$$

Therefore, provide boundary elements per UBC 2625(f)3.

- ### *Design of Boundary Elements*

Boundary elements must be designed to carry all factored gravity loads on the wall plus the vertical force required to resist the overturning moment calculated from factored forces related to earthquake effects (UBC 2625(f)3C; ACI 21.6.5.3).

The maximum factored axial compressive force on a boundary element at the base of the wall is 7538 kips (see Table 5-10). This value was determined as follows:

$$P_u = 1.4 \, (P_D + P_L + P_E\}$$

$$P_{D \text{ (boundary element)}} = P_{D \text{ (shearwall)}}/2$$

$$= \frac{1}{2}\left\{(0.116 \times 1568) + (19 \times 0.106 \times 1568) + \left(20 \times \frac{24^2}{144} \times 0.115 \times 42\right)\right.$$

$$+ \left(19 \times \frac{40^2}{144} \times 0.150 \times 13\right) + \left(\frac{40^2}{144} \times 0.150 \times 23\right) + \left(\frac{296}{144} \times 0.150 \times 13\right)$$

$$\left[(12 \times 5) + (14 \times 7) + (16 \times 7)\right] + \left.\left(\frac{16 \times 296}{144} \times 0.150 \times 23\right)\right\} = 2685.9 \text{ kips}$$

$$P_{L \text{ (boundary element)}} = P_{L \text{ (shearwall)}}/2$$

$$= \frac{1}{2} \times \left[0.03 + 19(0.02)\right] \times 1568 = 321.5 \text{ kips}$$

$$P_{E \text{ (boundary element)}} = \frac{M_{E \text{ (shearwall)}}}{\ell} + \frac{P_E}{2} = \frac{64,461}{28} + \frac{148.5}{2} = 2376.4 \text{ kips}$$

$$P_{u \text{ (boundary element)}} = 1.4 \times (2685.9 + 321.5 + 2376.4) = 7538.0 \text{ kips}$$

The axial load strength of the boundary element acting as a short column is:

$$P_u = \phi P_n = 0.8 \, \phi \left[0.85 f'_c \left(A_g - A_{st}\right) + f_y A_{st}\right]$$

UBC 2610(d)5B

ACI 10.3.5.2

Solving for the required area of steel:

$$7538 = 0.8 \times 0.7 \times \left[0.85 \times 6.0 \times \left(40^2 - A_{st}\right) + 60 \, A_{st}\right]$$

$$A_{st} = 96.6 \text{ in.}^2$$

Use 62-#11 bars (A_{st} = 96.7 in.^{2})

The reinforcement ratio, ρ_g, must be greater than 1% and less than 6%:

$$\rho_g = \frac{A_{st}}{A_g} = \frac{96.7}{40^2} = 0.060 \quad \text{O.K.}$$

UBC 2625(e)3A

ACI 21.4.3.1

Although 62-#11 bars gives a reinforcement ratio equal to the maximum allowed, the 62 bars will not adequately fit within the 40 × 40 in. section. The clear spacing between bars on the face of the section that has 17 bars is [40 - 2(1.5 + 0.625) - (17 × 1.41)]/16 = 0.74 in. which is less than the minimum allowed value of 1.5 d_b = 1.5 × 1.41 = 2.12 in. (UBC 2607(g) and ACI 7.6.3). Therefore, increase the size of the boundary element to 44 × 44 in. Note that at this time, the structure should be reanalyzed and the axial forces on the boundary elements should be recomputed based on the increase in the size of the boundary elements. However, it is assumed for our purposes here that the reactions on the wall do not change significantly. Consequently, the required A_{st} is:

$$7538 = 0.8 \times 0.7 \times [0.85 \times 6.0 \times (44^2 - A_{st}) + 60A_{st}]$$

$$A_{st} = 65.3 \text{ in.}^2$$

Use 42-#11 bars (ρ_g = 3.4%; clear space between bars = 2.08 in. ≅ 2.12 in.).

The maximum factored tensile force on a boundary element is 909 kips (see Table 5-10). The amount of steel required to carry the tensile force is:

$$A_s = \frac{P_u}{\phi f_y} = \frac{909}{0.9 \times 60} = 16.8 \text{ in.}^2 < 65.5 \text{ in.}^2 \quad \text{O.K.}$$

The required area of confinement reinforcement in the form of rectangular hoops is:

$$A_{sh} \geq \begin{cases} 0.09 s h_c \dfrac{f'_c}{f_{yh}} \\[2ex] 0.3 s h_c \left[\dfrac{A_g}{A_{ch}} - 1 \right] \dfrac{f'_c}{f_y} \end{cases} \qquad \begin{array}{l} \textit{UBC 2625(e)4A} \\ \textit{ACI 21.4.4.1} \end{array}$$

where

    s   = spacing of transverse reinforcement, in.

    h_c = cross-sectional dimension of column core measured center-to-center of transverse reinforcement, in.

    f_{yh} = specified yield strength of transverse reinforcement, psi

    A_g = gross area of section, in.^{2}

    A_{ch} = cross-sectional area of a column measured out-to-out of transverse reinforcement, in.^{2}

Using the maximum allowed hoop spacing of 4 in. and assuming #5 bar hoops with 1.5 in. clear cover all around, the required cross-sectional area is:

$$A_{sh} \geq \begin{cases} 0.09 \times 4 \times 40.4 \times 6000 / 60,000 = 1.45 \text{ in.}^2 \text{ (governs)} \\[2ex] 0.3 \times 4 \times 40.4 \times [1936 / 1681 - 1] \times 6000 / 60,000 = 0.74 \text{ in.}^2 \end{cases}$$

#5 hoops with crossties around every other longitudinal bar provides A_{sh} = 6 × 0.31 = 1.86 in.^{2} > 1.45 in.^{2}

• Design of Wall

At least two curtains of reinforcement must be used in the wall between the boundary elements if the in-plane factored shear force V_u for the wall exceeds $2A_{cv}\sqrt{f'_c}$ where A_{cv} is the net area of concrete section bounded by the wall thickness and length in the direction of shear force (UBC 2625(f)2B; ACI 21.6.2.2). From Fig. 5-8, the shear force at the base of the wall is 613.5 kips. Using Eq. (9-2), the factored shear V_u is (UBC 2625(c)4):

$$V_u = 1.4 \times 613.5 = 858.9 \text{ kips}$$

$$858.9 \text{ kips} < 2A_{cv}\sqrt{f'_c} = 2 \times 16 \times (31.67 \times 12) \sqrt{6000}/1000 = 941.9 \text{ kips}$$

Although not required, use two curtains of reinforcement in the wall.

The reinforcement ratio along the longitudinal and the transverse direction must not be less than 0.0025; also, the spacing along these directions must not exceed 18 in. (UBC 2625(f)2A; ACI 21.6.2).

The upper limit on the shear strength of a shearwall is given by:

$$\phi V_n = \phi 8 A_{cv}\sqrt{f'_c} \qquad\qquad\qquad \text{UBC 2625(h)3D}$$
$$\text{ACI 21.6.4.6}$$

$$= 0.6 \times 8 \times 16 \times (31.67 \times 12) \sqrt{6000}/1000 = 2260.8 \text{ kips} > 858.9 \text{ kips} \quad \text{O.K.}$$

Typically, the shear strength reduction factor is 0.85. However, this factor must be taken as 0.6 for any structural member if its nominal shear strength is less than the shear corresponding to the development of the nominal flexural strength of the member (UBC 2609(d)4; ACI 9.3.4.1). In this case a conservative value of 0.6 is used in the design of the wall.

The nominal shear strength is given by:

$$V_n = A_{cv} \times \left[\alpha_c \sqrt{f'_c} + \rho_n f_y \right] \qquad\qquad \text{UBC 2625(h)3}$$
$$\text{ACI 21.6.4}$$

where α_c varies linearly from 3.0 for $h_w / \ell_w = 1.5$ to 2.0 for $h_w / \ell_w = 2.0$

or $\alpha_c = 2.0$ if $h_w / \ell_w \geq 2.0$

The α_c factor adjusts for a higher allowable shear strength for walls with low ratios of height to horizontal length. In this case, $\alpha_c = 2.0$ since the ratio $h_w / \ell_w = 270/31.67 = 8.5$.

Using 2-#5 horizontal bars at 14 in. gives the shear strength of the wall as follows:

$$\rho_n = \frac{2 \times 0.31}{16 \times 14} = 0.0028$$

$$\phi V_n = 0.6 \times (31.67 \times 12 \times 16) \times [2\sqrt{6000} + (0.0028 \times 60,000)]/1000$$

$$= 1178.1 \text{ kips} > V_u = 858.9 \text{ kips} \quad \text{O.K.}$$

The vertical reinforcement ratio ρ_v shall not be less than the horizontal reinforcement ratio, ρ_n, when the ratio h_w / ℓ_w is less than 2.0 (UBC 2625(h)3C; ACI 21.6.4.5). Since $h_w / \ell_w = 8.5 > 2.0$, the minimum reinforcement ratio $\rho_v = 0.0025$ will be used. The required area of vertical steel is:

$$A_{sv} = 0.0025 \times 16 \times 12 = 0.48 \text{ in.}^2/\text{ft}$$

Using 2-#5 bars:

$$s_{req} = \frac{12}{0.48} \times 0.62 = 15.5 \text{ in.} < 18 \text{ in.}$$

Use #5 vertical bars on each face at 14 in. o.c.

The adequacy of the shearwall section at the base under combined factored axial loads and bending moments in the plane of the wall must be checked. An interaction diagram for the shearwall with the reinforcement previously determined is shown in Fig. 5-23. It can be seen from the figure that the wall is adequate for the load combinations given in Table 5-10.

• *Splice Lengths for Reinforcement*

The splice length must be determined for the vertical bars in the web portion and the boundary element of the shearwall. For the boundary element, use a Class B splice, which has a length of $1.3 \, \ell_d$. The basic development length for #11 bars is:

$$\ell_{db} = 0.04 \, A_b f_y / \sqrt{f'_c} \qquad\qquad \text{UBC 2612(c)2}$$
$$\text{ACI 12.2.2}$$

$$= 0.04 \times 1.56 \times 60{,}000/ \sqrt{6000} = 48.3 \text{ in.}$$

The information required to find the applicable factors is:

   Minimum cover to bars being spliced   = 2.13 in. (1.51 d_b)

   Clear spacing between bars being spliced = 2.08 in. (1.48 d_b)

   Transverse reinforcement meets requirements of UBC 2611(f)4 & 5 and UBC 2612(c)3E.

   Minimum clear cover meets requirement of UBC 2607(h).

This gives a value of 1.5 for Factor I (see Table 5-12). The other factors are all equal to 1.0, giving:

   Class B splice length = $1.3 \times (1.5 \times 1.0 \times 1.0 \times 1.0 \times 1.0 \times 48.3) = 94.2$ in.

The lap splices for the vertical reinforcement in the web will be a Class B splice. The basic development length for #5 bars is:

$$\ell_{db} = 0.04 \, A_b f_y / \sqrt{f'_c}$$

$$= 0.04 \times 0.31 \times 60{,}000/ \sqrt{6000} = 9.6 \text{ in.}$$

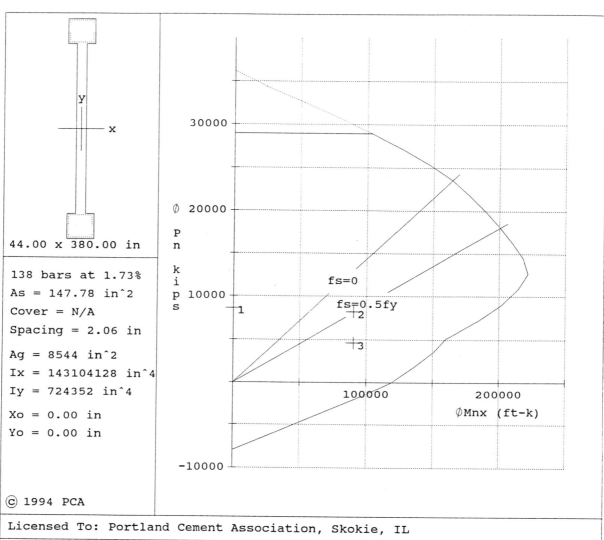

44.00 x 380.00 in

138 bars at 1.73%
As = 147.78 in^2
Cover = N/A
Spacing = 2.06 in

Ag = 8544 in^2
Ix = 143104128 in^4
Iy = 724352 in^4

Xo = 0.00 in
Yo = 0.00 in

© 1994 PCA

Licensed To: Portland Cement Association, Skokie, IL

File name: C:\DATA\COL\EB113C53.COL

Project: EB113 -- Seismic Zone 4   f'c = 6.00 ksi   Ec = 4696 ksi

Column Id: Shearwall B2 - C2   fc = 5.10 ksi   eu = 0.003 in/in

Engineer: Block stress profile, Beta1 = 0.75

Date: Time: fy = 60 ksi   Es = 29000 ksi

Code: ACI 318-89   Confinement: Tied

Units: English   phi(c) = 0.70, phi(b) = 0.90, a = 0.80

X-axis slenderness is not considered.

Figure 5-23 *Design Strength Interaction Diagram for Shearwall B2-C2, UBC-91 (Seismic Zone 4)*

The information required to find the applicable factors is:

Minimum clear cover   = 2.13 in. ($3.4\,d_b$)

Side cover   large

Clear bar spacing   = 10.5 in. ($16.8 d_b$)

Reinforcement is an inner layer of wall reinforcement.

A maximum value of 1.51 will control for Factor I. The other factors are equal to 1.0, giving:

Class B splice length = $1.3 \times (1.51 \times 1.0 \times 1.0 \times 1.0 \times 1.0 \times 9.6) = 18.9$ in.

The development length for the #5 horizontal bars will be calculated assuming that no hooks are used in the boundary element. According to UBC 2625(g)4B and ACI 21.5.4.2, the basic development length is:

$$\ell_{dh} = f_y d_b / 65\sqrt{f'_c}$$

$$= (60,000 \times 0.625)/(65\sqrt{6000}) = 7.5 \text{ in.} > 8 d_b = 5.0 \text{ in.}$$

For a straight bar, the required development length $\ell_d = 2.5 \times 7.5 = 18.8$ in.

This length can be accommodated within the confined core of the boundary element. Figure 5-24 shows the reinforcement details for this shearwall.

5.2.3.8.2 Design of Shearwall B2-C2 Based on UBC-94 Provisions

The design procedure for shearwalls based on the provisions given in UBC-94 can be summarized in three steps:

1) Design the shearwall for flexure and axial load by considering the entire cross-section of the wall to act as a short column; this will yield the required flexural reinforcement (UBC 1921.6.5.1). Check the shear resistance of the wall provided by the web (UBC 1921.6.4).

2) Provide the boundary zone detail requirements given in UBC 1921.6.5.6 at each end of the wall unless $P_u \leq 0.10\,A_g f'_c$ (for geometrically symmetrical walls) and either $M_u/V_u \ell_w \leq 1.0$ or $V_u \leq 3\,\ell_w h\sqrt{f'_c}$ and $M_u/V_u \ell_w \leq 3.0$ (UBC 1921.6.5.4). In any case, shearwalls with $P_u > 0.35\,P_o$ are not permitted to resist earthquake-induced forces (UBC 1921.6.5.3).

3) For shearwalls where boundary elements with specially detailed reinforcement have to be provided, determine the length of the boundary zone by using one of two methods. In the conservative method, the boundary zones are provided over a maximum distance of $0.25\,\ell_w$ at each of the walls (UBC 1921.6.5.4). Alternatively, the boundary zone length may be based on the compressive strain levels at the edges of the wall when the wall is subjected to displacements resulting from the ground motions specified in UBC 1629.2, using cracked section properties of the wall (UBC 1921.6.5.5). Confinement is to be provided wherever the compressive strain exceeds 0.003.

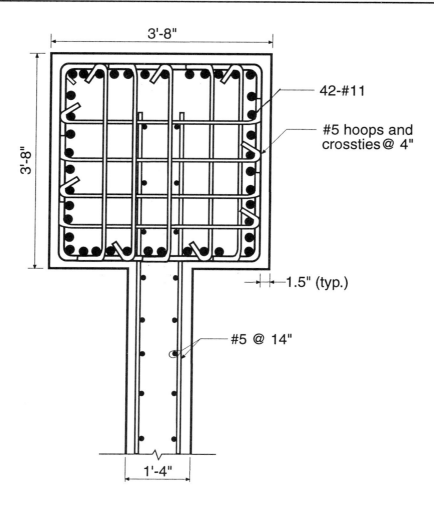

42-#11

#5 hoops and
crossties @ 4"

1.5" (typ.)

#5 @ 14"

Figure 5-24 Reinforcement Details for Shearwall B2-C2, UBC-91 (Seismic Zone 4)

• **Design of Boundary Zone**

Assuming a 1 ft-4 in. × 31 ft-4 in. wall, determine if boundary zones are required. From Table 5-10, the maximum factored axial load is 8629 kips. Check the criterion in UBC 1921.6.5.4:

$$0.10 \, A_g \, f'_c = 0.10 \, (16 \times 31.33 \times 12) \times 6.0 = 3609 \text{ kips} < P_u = 8629 \text{ kips}$$

Therefore, boundary zones are required. Assume 40 × 40 in. boundary elements reinforced with 24-#11 bars. Since the provisions in UBC 1921.6.4 for the nominal shear strength of shearwalls is the same as in UBC-91 Sect. 2625(f)2B, use 2-#5 @ 14 in. vertical bars in the web (see Sect. 5.2.3.8.1). Based on this reinforcement, the nominal axial load capacity of the wall at zero eccentricity is:

$$P_o \quad = 0.85 \, f'_c \, (A_g - A_{st}) + f_y A_{st}$$

$$= 0.85 \times 6 \, (7936 - 91.62) + (60 \times 91.62) = 45{,}504 \text{ kips}$$

Since $P_u = 8629$ kips $= 0.19\ P_o$, conservatively provide boundary zones at each end at a distance of $0.17\ \ell_w = 0.17 \times 31.33 = 5.3$ ft (UBC 1921.6.5.4). The special reinforcement details given in UBC 1921.6.5.6 are to be provided in this area at each end of the wall.

Alternatively, determine the length of the boundary zone using the provisions in UBC 1921.6.5.5. Before proceeding with the strain compatibility analysis, determine if the wall can be considered to contribute to the strength of the structure for resisting earthquake-induced forces. Since $P_u = 8629$ kips $< 0.35\ P_o = 15,926$ kips, the wall can be used to resist its portion of the seismic loads.

When performing the strain compatibility analysis for the wall, it is important to note that the compressive strain in the concrete is not allowed to be larger than ε_{max} where

$$\varepsilon_{max} = \frac{3R_w}{8}\ (0.004) \leq 0.015$$

In this case,

$$\varepsilon_{max} = \frac{3 \times 12}{8}\ (0.004) = 0.018 > 0.015, \text{ use } \varepsilon_{max} = 0.015$$

For shearwalls in which the flexural limit state is governed by yielding at the base of the wall, the total curvature demand ϕ_t is given in UBC Eq. (21-9):

$$\phi_t = \frac{\Delta_i}{(h_w - \ell_p / 2)\ell_p} + \phi_y$$

where Δ_i = inelastic deflection at the top of the wall

$$= \Delta_t - \Delta_y$$

   Δ_t = total deflection at the top of the wall equal to $3R_w/8$ times the elastic displacement using cracked section properties, or may be taken as $2(3R_w/8)\ \Delta_E$.

   Δ_y = displacement at top of wall corresponding to the extreme fiber compressive strain of 0.003 at the critical section, or may be taken as $(M_n' / M_E)\Delta_E$ where M_E is the moment at the critical section when the top of the wall is displaced Δ_E and M_n' is the nominal flexural strength of the critical section at an axial load $P_u' = 1.2D + 0.5L + E$.

   Δ_E = elastic displacement at the top of the wall using gross section properties and code-specified seismic forces

   ℓ_p = height of the plastic hinge above the critical section, which may be taken as $0.5\ \ell_w$.

   ϕ_y = $0.003/c_u'$

   c_u' = neutral axis depth at P_u' and M_n'

The first step in the analysis will be determining the location of the neutral axis c'_u. Using the design loads listed in Sect. 5.1.2 (including live load reduction) and the seismic load given in Fig. 5-8, the load P'_u acting at the base of the wall is:

$$P'_u = (1.2 \times 5371.8) + (0.5 \times 642.9) + 148.5 = 6916.1 \text{ kips}$$

In lieu of hand computations, the PCA computer program *PCACOL* [2.2] was used to obtain the neutral axis depth c'_u. For 40 × 40 in. boundary elements reinforced with 24-#11 bars and a 16 in. web with 2-#5 @ 14 in., $c'_u = 55.6$ in. The nominal moment strength M'_n corresponding to $P'_u = 6916.1$ kips is 174,020 ft-kips (see Fig. 5-25). Figure 5-26 shows the interaction diagram with the governing load combinations given in Table 5-10. As can be seen from this figure, the wall is adequate to carry the factored loads.

From Table 5-6, $\Delta_E = 1.71$ in. Therefore,

$$\Delta_t = 2 \times \left(\frac{3 \times 12}{8}\right) \times 1.71 = 15.4 \text{ in.}$$

Also, using $M_E = 64,461$ ft-kips from Fig. 5-8:

$$\Delta_y = \left(\frac{M'_n}{M_E}\right) \Delta_E$$

$$= \left(\frac{174,020}{64,461}\right) \times 1.71 = 4.6 \text{ in.}$$

The inelastic deflection at the top of the wall is:

$$\Delta_i = \Delta_t - \Delta_y = 15.4 - 4.6 = 10.8 \text{ in.}$$

Assuming that $\ell_p = 0.5 \, \ell_w = 0.5 \times 31.33 \times 12 = 188$ in., the total curvature demand is:

$$\phi_t = \frac{10.8}{[(270 \times 12) - (188/2)](188)} + \frac{0.003}{55.6}$$

$$= 1.83 \times 10^{-5} + 5.40 \times 10^{-5} = 7.23 \times 10^{-5}$$

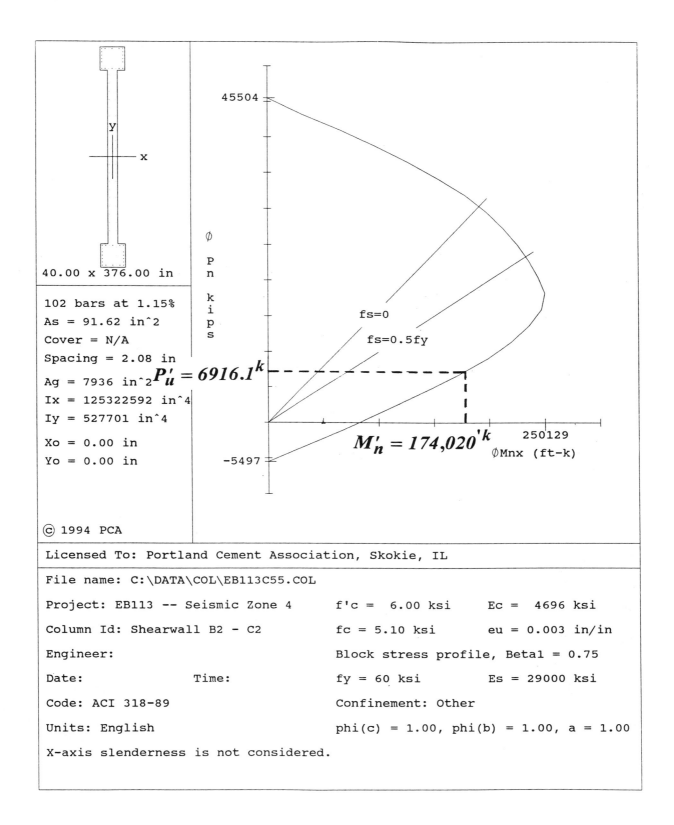

Figure 5-25 P_n - M_n Interaction Diagram for Shearwall B2-C2, UBC-94 (Seismic Zone 4)

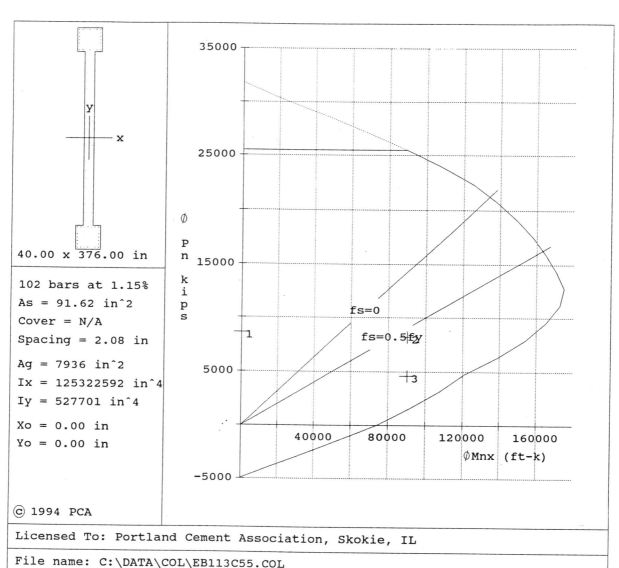

40.00 x 376.00 in

102 bars at 1.15%
As = 91.62 in^2
Cover = N/A
Spacing = 2.08 in

Ag = 7936 in^2
Ix = 125322592 in^4
Iy = 527701 in^4

Xo = 0.00 in
Yo = 0.00 in

© 1994 PCA

Licensed To: Portland Cement Association, Skokie, IL

File name: C:\DATA\COL\EB113C55.COL

Project: EB113 -- Seismic Zone 4   f'c = 6.00 ksi   Ec = 4696 ksi

Column Id: Shearwall B2 - C2   fc = 5.10 ksi   eu = 0.003 in/in

Engineer: Block stress profile, Beta1 = 0.75

Date: Time: fy = 60 ksi   Es = 29000 ksi

Code: ACI 318-89   Confinement: Tied

Units: English   phi(c) = 0.70, phi(b) = 0.90, a = 0.80

X-axis slenderness is not considered.

Figure 5-26 Design Strength Interaction Diagram for Shearwall B2-C2, UBC-94 (Seismic Zone 4)

Therefore, assuming the compressive strains to vary linearly over the depth c'_u, the maximum compressive strain ε_u is:

$$\varepsilon_u = c'_u \, \phi_t$$

$$= 55.6 \times 7.23 \times 10^{-5} = 0.0040 > 0.0030 < \varepsilon_{max} = 0.0150 \quad \text{O.K.}$$

Confinement is required over

$$\left(55.6 - \frac{0.0030}{0.0040} \times 55.6 \right) = 13.9 \text{ in.} < 40.0 \text{ in.}$$

For practical purposes, the entire 40×40 in. boundary element will be confined according to the provisions in UBC 1921.6.5.6. Check minimum dimensions of the boundary zone:

$$\text{Minimum thickness} = \frac{\ell_u}{10} = \frac{(23 \times 12) - 24}{10} = 25.2 \text{ in.} < 40.0 \text{ in.} \quad \text{O.K.}$$

$$\text{Minimum length} = \frac{\ell_w}{10} = \frac{31.33 \times 12}{10} = 37.6 \text{ in.} < 40.0 \text{ in.} \quad \text{O.K.}$$

The minimum area of confinement reinforcement A_{sh} is given in UBC Eq. (21-10):

$$A_{sh} = 0.09 s h_c \, f'_c / f_{yh}$$

Using the maximum allowable spacing of 6 in. and #5 hoops:

$$A_{sh} = 0.09 \times 6 \times 36.375 \times \frac{6}{60} = 1.96 \text{ in.}^2$$

#5 hoops with crossties around every longitudinal bar provides $A_{sh} = 7 \times 0.31 = 2.17$ in.^{2} Note that over the lap splice length of the vertical bars in the boundary zone, the hoops must be spaced at 4 in. Also, the minimum area of vertical reinforcement in the boundary zone is $0.005 \times 40^2 = 8.0$ in.^{2} which is less than the area provided by 24-#11 bars.

- **Design of Wall**

As noted above, the nominal shear strength provisions for shearwalls in UBC 1921.6.4 are the same as those given in UBC-91. Therefore, calculations similar to those shown in Sect. 5.2.3.8.1 yield 2-#5 horizontal bars at 14 in. and 2-#5 vertical bars at 14 in.

- **Splice Lengths for Reinforcement**

For the boundary elements, use a Class B splice. The information required to find the applicable factors is:

Minimum cover to bars being spliced   = 2.13 in. (1.51 d_b)

Clear space between bars being spliced   = 4.31 in. (3.06 d_b)

Transverse reinforcement meets requirements of UBC 1911.5.4 and 1911.5.5.3, UBC 1907.10.5, and UBC 1912.2.3.5.

Minimum clear cover meets requirements of UBC 1907.7.1.

This gives a value of 0.75 for Factor I (see Table 5-12). The other factors are all equal to 1.0, giving:

Class B splice length = 1.3 × (0.75 × 1.0 × 1.0 × 1.0 × 1.0 × 48.3) = 47.1 in.

The splice lengths for the web reinforcement is the same given in Sect. 5.2.3.8.1. Figure 5-27 shows the reinforcement details for this shearwall. UBC 1921.6.5.6 gives other detailing requirements for the shearwall.

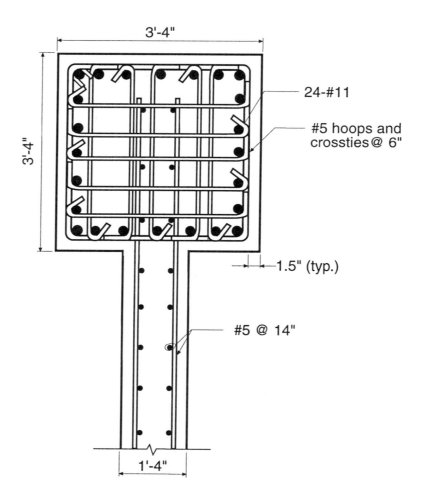

Figure 5-27 Reinforcement Details for Shearwall B2-C2, UBC-94 (Seismic Zone 4)

5.2.3.8.3 Comparison of Results for Shearwall B2-C2 using UBC-91 and UBC-94

Table 5-13 shows a comparison of the designs of the shearwall using the 1991 and 1994 UBC provisions. It is clear from the table that in this design example, the UBC-94 provisions yield a much more economical design, without sacrificing safety.

Table 5-13 Comparison of Shearwall Designs by UBC-91 and UBC-94 Provisions

| Code | Dimensions of Boundary Elements (in.) | Longitudinal Reinforcement in Boundary Elements | Confinement Reinforcement in Boundary Elements | Web | | |
				Thickness (in.)	Vertical Reinforcement	Horizontal Reinforcement
UBC-91	44 × 44	42-#11	#5 hoops @ 4 in. with 5 crossties in one direction and 4 in the other	16	2-#5 @ 14"	2-#5 @ 14"
UBC-94	40 × 40	24-#11	#5 hoops @ 6 in.* with 5 crossties in each direction	16	2-#5 @ 14"	2-#5 @ 14"

* Provide 4 in. spacing over the lap splice length of the vertical bars in the boundary zone.

5.3 Design for Seismic Zone 2B

5.3.1 General

For structures in Seismic Zones 4 and 3 and for non-standard occupancy structures in Zone 2 with a total height greater than 240 ft, a dynamic lateral force procedure must be used to determine the base shear and its distribution along the building height. Standard occupancy structures of any height in zones of moderate seismicity (including Zone 2B) are exempt from this requirement. Instead, the static lateral force procedures of UBC 1628 may be used. The calculation for the base shear, V, for this structure is shown at the bottom of Table 5-14. For this example, the importance factor, I, and the site soil factor, S, have been taken equal to 1.0 and 1.2, respectively. The seismic forces resulting from the distribution of the base shear in accordance with UBC Eqs. (28-6), (28-7), and (28-8) are also listed in Table 5-14.

As discussed in the introduction of this chapter, the resistance to lateral loads in a dual system is provided by both the specially detailed moment-resisting frames and the shearwalls. An additional safeguard is provided by requiring the moment-resisting frames to be capable of resisting 25% of the design base shear without the benefit of the shearwalls.

A three-dimensional analysis of the structure was carried out using *SAP90*. To account for uncertainties in the location of the loads, UBC 1628.5 requires that the mass at each level be assumed to be displaced a distance equal to 5% of the building dimension at that level perpendicular to the direction of the force under consideration. As before, this eccentricity was included in the analysis.

The results from the lateral analysis for beams of an exterior frame, beams of an interior frame, columns of an interior frame, and a shearwall are shown in Figs. 5-28 through 5-30 and Table 5-15. These figures also show results from the lateral analysis of the beam-column frames under 25% of the design base shear (shown in bold italic text).

Table 5-14 Seismic Forces for Zone 2B

Floor Level	Weight, w_x (kips)	Height, h_x (ft)	$w_x h_x$ (ft-kips)	Lateral Force (kips)
R	3075	270	830,250	313.2
20	3318	257	852,726	146.4
19	3318	244	809,592	139.0
18	3318	231	766,458	131.6
17	3318	218	723,324	124.2
16	3318	205	680,190	116.8
15	3318	192	637,056	109.4
14	3318	179	593,922	102.0
13	3318	166	550,788	94.6
12	3318	153	507,654	87.2
11	3318	140	464,520	79.8
10	3318	127	421,386	72.3
9	3318	114	378,252	64.9
8	3318	101	335,118	57.5
7	3318	88	291,984	50.1
6	3318	75	248,850	42.7
5	3318	62	205,716	35.3
4	3318	49	162,582	27.9
3	3318	36	119,448	20.5
2	3662	23	84,226	14.5
Σ	66,461		9,664,042	1829.8

Building height h_n = 270 ft
Building weight W = 66,461 kips
Fundamental period T = 0.02 $(h_n)^{3/4}$ = 1.33 sec.
Seismic zone factor Z = 0.20
Importance factor I = 1.0
Response coefficient R_w = 9
Soil factor S = 1.2
Coefficient C = 1.25 $S/T^{2/3}$ = 1.24
Base shear V = $ZICW/R_w$ = 1829.8 kips
Top level force F_t = 0.07TV = 170.6 kips

Bending Moments (ft-kips)

R									
R	-67.6	*-18.9*	-65.4	*-17.9*	-63.0	*-26.7*	-62.5	*-26.5*	-80.9	*-24.5*
20	-76.6	*-25.8*	-75.7	*-25.4*	-75.0	*-34.9*	-74.9	*-34.8*	-93.6	*-33.0*
19	-77.1	*-34.4*	-75.9	*-33.8*	-74.9	*-42.9*	-74.6	*-42.8*	-93.0	*-41.0*
18	-77.9	*-43.2*	-76.8	*-42.4*	-75.8	*-51.4*	-75.5	*-51.3*	-93.7	*-49.5*
17	-78.8	*-51.8*	-77.6	*-50.9*	-76.6	*-59.7*	-76.3	*-59.6*	-94.2	*-57.8*
16	-79.4	*-60.1*	-78.2	*-59.1*	-77.2	*-67.6*	-77.0	*-67.6*	-94.3	*-65.8*
15	-79.7	*-67.9*	-78.6	*-66.8*	-77.5	*-75.1*	-77.3	*-75.1*	-94.1	*-73.3*
14	-79.6	*-75.3*	-78.5	*-74.1*	-77.4	*-82.1*	-77.2	*-82.0*	-93.4	*-80.4*
13	-79.1	*-82.2*	-78.0	*-80.9*	-76.9	*-88.4*	-76.7	*-88.4*	-92.1	*-86.9*
12	-78.0	*-88.6*	-76.9	*-87.2*	-75.8	*-94.3*	-75.6	*-94.3*	-90.2	*-92.8*
11	-76.2	*-94.5*	-75.1	*-93.0*	-74.1	*-99.7*	-73.9	*-99.7*	-87.5	*-98.3*
10	-73.7	*-99.9*	-72.7	*-98.3*	-71.6	-104.5	-71.5	-104.5	-84.1	-103.2
9	-70.4	*-104.7*	-69.4	*-103.1*	-68.5	*-108.7*	-68.3	*-108.7*	-79.7	*-107.5*
8	-66.3	*-109.0*	-65.3	*-107.4*	-64.4	*-112.3*	-64.3	*-112.3*	-74.5	*-111.2*
7	-61.2	*-112.6*	-60.3	*-110.9*	-59.4	*-115.0*	-59.4	*-115.0*	-68.2	*-114.1*
6	-54.8	*-115.3*	-54.1	*-113.7*	-53.4	*-117.2*	-53.4	*-117.2*	-60.8	*-116.4*
5	-48.7	*-116.3*	-48.1	*-114.9*	-47.5	*-117.9*	-47.5	*-117.9*	-53.7	*-117.2*
4	-41.7	*-115.5*	-41.2	*-114.1*	-40.7	*-116.2*	-40.7	*-116.2*	-45.7	*-115.7*
3	-33.7	*-110.8*	-33.3	*-109.5*	-33.0	*-110.9*	-33.0	*-110.9*	-36.7	*-110.5*
2	-24.7	*-98.7*	-24.3	*-96.9*	-23.9	*-96.9*	-23.9	*-96.9*	-26.2	*-96.7*

Shear Forces (kips)

R									
R	4.8	*1.3*	-4.8	*-1.3*	4.5	*1.9*	-4.5	*-1.9*	5.8	*1.8*
20	5.4	*1.8*	-5.4	*-1.8*	5.4	*2.5*	-5.4	*-2.5*	6.7	*2.4*
19	5.5	*2.4*	-5.5	*-2.4*	5.3	*3.1*	-5.3	*-3.1*	6.7	*2.9*
18	5.5	*3.1*	-5.5	*-3.1*	5.4	*3.7*	-5.4	*-3.7*	6.7	*3.6*
17	5.6	*3.7*	-5.6	*-3.7*	5.5	*4.3*	-5.5	*-4.3*	6.7	*4.1*
16	5.6	*4.3*	-5.6	*-4.3*	5.5	*4.8*	-5.5	*-4.8*	6.7	*4.7*
15	5.7	*4.8*	-5.7	*-4.8*	5.5	*5.4*	-5.5	*-5.4*	6.7	*5.2*
14	5.6	*5.3*	-5.6	*-5.3*	5.5	*5.9*	-5.5	*-5.9*	6.7	*5.7*
13	5.6	*5.8*	-5.6	*-5.8*	5.5	*6.3*	-5.5	*-6.3*	6.6	*6.2*
12	5.5	*6.3*	-5.5	*-6.3*	5.4	*6.7*	-5.4	*-6.7*	6.4	*6.6*
11	5.4	*6.7*	-5.4	*-6.7*	5.3	*7.1*	-5.3	*-7.1*	6.3	*7.0*
10	5.2	*7.1*	-5.2	*-7.1*	5.1	*7.5*	-5.1	*-7.5*	6.0	*7.4*
9	5.0	*7.4*	-5.0	*-7.4*	4.9	*7.8*	-4.9	*-7.8*	5.7	*7.7*
8	4.7	*7.7*	-4.7	*-7.7*	4.6	*8.0*	-4.6	*-8.0*	5.3	*8.0*
7	4.3	*8.0*	-4.3	*-8.0*	4.2	*8.2*	-4.2	*-8.2*	4.9	*8.2*
6	3.9	*8.2*	-3.9	*-8.2*	3.8	*8.4*	-3.8	*-8.4*	4.4	*8.3*
5	3.5	*8.3*	-3.5	*-8.3*	3.4	*8.4*	-3.4	*-8.4*	3.8	*8.4*
4	3.0	*8.2*	-3.0	*-8.2*	2.9	*8.3*	-2.9	*-8.3*	3.3	*8.3*
3	2.4	*7.9*	-2.4	*-7.9*	2.4	*7.9*	-2.4	*-7.9*	2.6	*7.9*
2	1.8	*7.0*	-1.8	*-7.0*	1.7	*6.9*	-1.7	*-6.9*	1.9	*6.9*

Bold Italic text denotes results with 25% of design base shear applied to frames.

Figure 5-28 Bending Moments and Shear Forces in Beams in an Exterior Frame (Seismic Zone 2B)

Bending Moments (ft-kips)

R	-85.3	*-18.0*	-84.7	*-17.1*	-37.7	*-24.8*	-37.7	*-24.7*	-75.3	*-23.0*
20	-94.3	*-24.4*	-94.1	*-24.0*	-47.3	*-32.3*	-47.3	*-32.3*	-84.6	*-30.8*
19	-94.2	*-32.3*	-93.8	*-31.7*	-46.7	*-39.6*	-46.5	*-39.6*	-83.6	*-38.0*
18	-95.1	*-40.4*	-94.7	*-39.7*	-47.6	*-47.6*	-47.4	*-47.6*	-84.4	*-45.9*
17	-95.9	*-48.4*	-95.6	*-47.6*	-48.4	*-55.3*	-48.2	*-55.3*	-84.9	*-53.7*
16	-96.4	*-56.1*	-96.1	*-55.1*	-49.1	*-62.6*	-48.8	*-62.6*	-85.2	*-61.1*
15	-96.6	*-63.4*	-96.3	*-62.3*	-49.5	*-69.6*	-49.3	*-69.6*	-85.1	*-68.1*
14	-96.3	*-70.2*	-96.0	*-69.0*	-49.7	*-76.0*	-49.5	*-76.0*	-84.5	*-74.6*
13	-95.5	*-76.6*	-95.1	*-75.3*	-49.7	*-82.0*	-49.4	*-82.0*	-83.5	*-80.6*
12	-93.9	*-82.5*	-93.6	*-81.2*	-49.2	*-87.5*	-49.0	*-87.5*	-81.8	*-86.2*
11	-91.6	*-88.0*	-91.3	*-86.6*	-48.4	*-92.5*	-48.2	*-92.5*	-79.5	*-91.3*
10	-88.4	*-92.9*	-88.1	*-91.5*	-47.1	*-97.0*	-46.9	*-97.0*	-76.5	*-95.8*
9	-84.2	*-97.4*	-83.9	*-95.9*	-45.3	*-100.9*	-45.1	*-100.9*	-72.7	*-99.8*
8	-78.9	*-101.4*	-78.7	*-99.8*	-43.0	*-104.2*	-42.8	*-104.2*	-68.0	*-103.3*
7	-72.5	*-104.7*	-72.3	*-103.1*	-40.0	*-106.8*	-39.8	*-106.8*	-62.3	*-106.0*
6	-64.7	*-107.2*	-64.6	*-105.7*	-36.4	*-108.8*	-36.2	*-108.8*	-55.7	*-108.1*
5	-57.1	*-108.1*	-57.0	*-106.8*	-32.7	*-109.5*	-32.5	*-109.5*	-49.2	*-108.9*
4	-48.6	*-107.3*	-48.5	*-106.0*	-28.4	*-107.9*	-28.3	*-107.9*	-41.9	*-107.5*
3	-38.8	*-102.8*	-38.7	*-101.7*	-23.5	*-103.0*	-23.4	*-103.0*	-33.7	*-102.6*
2	-27.8	*-91.6*	-27.7	*-89.9*	-17.8	*-89.9*	-17.7	*-89.9*	-24.2	*-89.7*

Shear Forces (kips)

R	6.1	*1.3*	-6.1	*-1.3*	2.7	*1.8*	-2.7	*-1.8*	5.4	*1.7*
20	6.7	*1.7*	-6.7	*-1.7*	3.4	*2.3*	-3.4	*-2.3*	6.0	*2.2*
19	6.7	*2.3*	-6.7	*-2.3*	3.3	*2.8*	-3.3	*-2.8*	6.0	*2.7*
18	6.8	*2.9*	-6.8	*-2.9*	3.4	*3.4*	-3.4	*-3.4*	6.0	*3.3*
17	6.8	*3.4*	-6.8	*-3.4*	3.5	*3.9*	-3.5	*-3.9*	6.1	*3.8*
16	6.9	*4.0*	-6.9	*-4.0*	3.5	*4.5*	-3.5	*-4.5*	6.1	*4.4*
15	6.9	*4.5*	-6.9	*-4.5*	3.5	*5.0*	-3.5	*-5.0*	6.1	*4.9*
14	6.9	*5.0*	-6.9	*-5.0*	3.6	*5.4*	-3.6	*-5.4*	6.0	*5.3*
13	6.8	*5.4*	-6.8	*-5.4*	3.5	*5.9*	-3.5	*-5.9*	6.0	*5.8*
12	6.7	*5.9*	-6.7	*-5.9*	3.5	*6.3*	-3.5	*-6.3*	5.9	*6.2*
11	6.5	*6.2*	-6.5	*-6.2*	3.5	*6.6*	-3.5	*-6.6*	5.7	*6.5*
10	6.3	*6.6*	-6.3	*-6.6*	3.4	*6.9*	-3.4	*-6.9*	5.5	*6.8*
9	6.0	*6.9*	-6.0	*-6.9*	3.2	*7.2*	-3.2	*-7.2*	5.2	*7.1*
8	5.6	*7.2*	-5.6	*-7.2*	3.1	*7.4*	-3.1	*-7.4*	4.9	*7.4*
7	5.2	*7.4*	-5.2	*-7.4*	2.9	*7.6*	-2.9	*-7.6*	4.5	*7.6*
6	4.6	*7.6*	-4.6	*-7.6*	2.6	*7.8*	-2.6	*-7.8*	4.0	*7.7*
5	4.1	*7.7*	-4.1	*-7.7*	2.3	*7.8*	-2.3	*-7.8*	3.5	*7.8*
4	3.5	*7.6*	-3.5	*-7.6*	2.0	*7.7*	-2.0	*-7.7*	3.0	*7.7*
3	2.8	*7.3*	-2.8	*-7.3*	1.7	*7.4*	-1.7	*-7.4*	2.4	*7.3*
2	2.0	*6.5*	-2.0	*-6.5*	1.3	*6.4*	-1.3	*-6.4*	1.7	*6.4*

Bold Italic text denotes results with 25% of design base shear applied to frames.

Figure 5-29 Bending Moments and Shear Forces in Beams in an Interior Frame (Seismic Zone 2B)

Bending Moments (ft-kips)

Floor	Col. A3		Col. C3	
R	85.1	*17.9*	111.6	*47.1*
	49.1	*-6.5*	72.8	*15.1*
20	45.1	*30.9*	58.3	*47.8*
	45.6	*-2.8*	62.1	*17.8*
19	48.6	*35.1*	67.3	*59.3*
	45.6	*1.1*	63.6	*26.3*
18	49.4	*39.3*	67.4	*66.7*
	46.6	*5.8*	64.8	*34.6*
17	49.2	*42.6*	67.6	*73.7*
	47.5	*10.6*	65.9	*43.0*
16	48.8	*45.5*	67.4	*80.0*
	48.5	*15.3*	67.1	*51.1*
15	48.1	*48.0*	66.6	*85.8*
	49.4	*19.9*	68.0	*58.9*
14	46.8	*50.3*	65.4	*91.1*
	50.2	*24.3*	68.8	*66.1*
13	45.2	*52.3*	63.5	*95.8*
	50.9	*28.4*	69.2	*72.9*
12	42.9	*54.0*	61.1	*100.0*
	51.4	*32.4*	69.3	*79.2*
11	40.2	*55.5*	57.9	*103.7*
	51.5	*36.2*	69.0	*85.1*
10	36.8	*56.7*	53.9	*106.8*
	51.4	*39.8*	68.3	*90.5*
9	32.7	*57.7*	49.1	*109.3*
	50.8	*43.3*	66.9	*95.4*
8	28.0	*58.1*	43.5	*111.2*
	50.0	*46.4*	65.3	*100.3*
7	22.5	*58.3*	36.6	*111.5*
	49.6	*51.7*	62.4	*102.8*
6	15.1	*55.4*	29.3	*113.4*
	47.1	*54.4*	60.0	*109.9*
5	10.1	*53.8*	21.7	*107.8*
	45.6	*61.2*	56.9	*115.7*
4	2.9	*46.1*	13.3	*99.0*
	43.5	*73.8*	53.2	*124.0*
3	-4.7	*29.0*	4.1	*81.0*
	40.8	*88.2*	50.0	*147.3*
2	-13.0	*3.4*	-8.1	*31.4*
	56.3	*276.3*	59.2	*290.2*

Shear Force (kips)

Floor	Col. A3		Col. C3	
R	-10.3	*-0.9*	-14.2	*-4.8*
20	-7.0	*-2.2*	-9.3	*-5.0*
19	-7.2	*-2.8*	-10.1	*-6.6*
18	-7.4	*-3.5*	-10.2	*-7.8*
17	-7.4	*-4.1*	-10.3	*-9.0*
16	-7.5	*-4.7*	-10.3	*-10.1*
15	-7.5	*-5.2*	-10.4	*-11.1*
14	-7.5	*-5.7*	-10.3	*-12.1*
13	-7.4	*-6.2*	-10.2	*-13.0*
12	-7.3	*-6.7*	-10.0	*-13.8*
11	-7.1	*-7.1*	-9.8	*-14.5*
10	-6.8	*-7.4*	-9.4	*-15.2*
9	-6.4	*-7.8*	-8.9	*-15.8*
8	-6.0	*-8.0*	-8.4	*-16.3*
7	-5.6	*-8.5*	-7.6	*-16.5*
6	-4.8	*-8.5*	-6.9	*-17.2*
5	-4.3	*-8.8*	-6.0	*-17.2*
4	-3.6	*-9.2*	-5.1	*-17.2*
3	-2.8	*-9.0*	-4.2	*-17.6*
2	-1.9	*-12.2*	-2.2	*-14.0*

Axial Loads (kips)

Floor	Col. A3		Col. C3	
R	6.0	*1.2*	-1.3	*0.3*
20	12.6	*2.9*	-2.7	*0.7*
19	19.3	*5.2*	-4.2	*1.0*
18	26.0	*8.0*	-5.7	*1.3*
17	32.8	*11.4*	-7.2	*1.7*
16	39.6	*15.3*	-8.8	*2.0*
15	46.4	*19.8*	-10.4	*2.4*
14	53.2	*24.7*	-12.0	*2.7*
13	59.9	*30.1*	-13.6	*3.0*
12	66.6	*36.0*	-15.2	*3.3*
11	73.1	*42.2*	-16.9	*3.6*
10	79.3	*48.7*	-18.5	*3.9*
9	85.3	*55.6*	-20.0	*4.2*
8	90.9	*62.8*	-21.5	*4.5*
7	96.0	*70.2*	-22.9	*4.7*
6	100.6	*77.8*	-24.1	*4.9*
5	104.6	*85.4*	-25.2	*5.1*
4	108.1	*93.1*	-26.2	*5.2*
3	110.8	*100.4*	-27.0	*5.4*
2	112.8	*106.8*	-27.5	*5.4*

Bold Italic text denotes results with 25% of design base shear applied to frames.

Figure 5-30 Bending Moments, Shear Forces and Axial Loads in Columns in an Interior Frame (Seismic Zone 2B)

Table 5-15 Axial Loads, Bending Moments, and Shear Forces in a Shearwall (Seismic Zone 2B)

Level	Axial Load (kips)	Bending Moment (ft-kips)		Shear Force (kips)
		Top	Bottom	
R	10.2	968	-875	-7.1
20	20.9	1884	-977	-70.0
19	31.6	1982	-618	-104.9
18	42.1	1633	183	-139.7
17	52.7	838	1409	-172.8
16	63.3	-384	3039	-204.3
15	73.9	-2015	5053	-233.7
14	84.4	-4036	7438	-261.6
13	94.7	-6430	10,171	-287.9
12	104.7	-9184	13,246	-312.4
11	114.4	-12,287	16,653	-335.8
10	123.8	-15,729	20,373	-357.3
9	132.5	-19,496	24,407	-377.7
8	140.6	-23,588	28,741	-396.5
7	148.1	-27,992	33,372	-414.1
6	154.7	-32,708	38,305	-430.5
5	160.4	-37,722	43,500	-444.5
4	165.1	-43,006	48,957	-457.6
3	168.8	-48,571	54,662	-468.6
2	171.2	-54,387	65,539	-482.9

5.3.2   Design of Structural Members

The objective is to determine the required flexural and shear reinforcement for a typical beam in an interior and an exterior frame, an interior and exterior column in an interior frame, and a shearwall.

The beams along column line 1 carry an unfactored dead load of 2.0 kips/ft, and an unfactored live load of 0.6 kips/ft.

The factored load due to gravity is:

$$w_u = (1.4 \times 2.0) + (1.7 \times 0.6) = 3.8 \text{ kips/ft}$$

Using the coefficients in UBC 1908.3.3, the factored moments due to gravity are:

• End Span—Beam A1-B1

Exterior negative moment $\quad = w_u \ell_n^2/16 = 3.8 \times (28 - 32/12)^2/16 = 152.4$ ft-kips

Positive moment $\quad = w_u \ell_n^2/14 = 3.8 \times (28 - 32/12)^2/14 = 174.2$ ft-kips

Interior negative moment $\quad = w_u \ell_n^2/10 = 3.8 \times (28 - 32/12)^2/10 = 243.9$ ft-kips

- Interior Span—Beam B1-C1

    Negative moment $= w_u \ell_n^2/11 = 3.8 \times (28 - 32/12)^2/11 = 221.7$ ft-kips

    Positive moment $= w_u \ell_n^2/16 = 3.8 \times (28 - 32/12)^2/16 = 152.4$ ft-kips

Table 5-16 shows the factored moments for different combinations of gravity and lateral loads on a beam on the 5th floor of an exterior frame. The gravity load moments are obtained from the calculations above and the lateral load moments are taken from Fig. 5-28.

Table 5-16 Summary of Factored Bending Moments for a Beam on the 5th Floor of an Exterior Frame (Seismic Zone 2B)

(1) U $= 1.4D + 1.7L$ Eq. (9-1), UBC 1909.2.1
(2) U $= 0.75[1.4D + 1.7L \pm 1.87E]$ Eq. (9-2), UBC 1909.2.3
(3) U $= 0.9D \pm 1.43E$ Eq. (9-3), UBC 1909.2.3

Load Combination	Bending Moment (ft-kips)					
	Beam A1-B1			Beam B1-C1		
	Exterior Negative	Positive	Interior Negative	Negative	Positive	Negative
1	-152.4	174.2	-243.9	-221.7	152.4	-221.7
2: sidesway right sidesway left	48.8 -277.4	131.6 129.8	-344.1 -21.8	-1.0 -331.6	114.3 114.3	-331.6 -1.0
3: sidesway right sidesway left	94.1 -271.3	83.5 81.6	-279.9 48.8	63.6 -273.6	72.2 72.2	-273.6 63.6

The beams along column line 3 carry an unfactored dead load of 3.5 kips/ft, and an unfactored live load of 0.84 kips/ft.

The factored load due to gravity is:

$w_u = (1.4 \times 3.5) + (1.7 \times 0.84) = 6.3$ kips/ft

- End Span—Beam A3-B3

    Exterior negative moment $= w_u \ell_n^2/16 = 6.3 \times (28 - 32/12)^2/16 = 252.7$ ft-kips

    Positive moment $= w_u \ell_n^2/14 = 6.3 \times (28 - 32/12)^2/14 = 288.8$ ft-kips

    Interior negative moment $= w_u \ell_n^2/10 = 6.3 \times (28 - 32/12)^2/10 = 404.3$ ft-kips

- Interior Span—Beam B3-C3

Negative moment $\qquad = w_u \ell_n{}^2/11 = 6.3 \times (28 - 32/12)^2/11 = 367.6$ ft-kips

Positive moment $\qquad = w_u \ell_n{}^2/16 = 6.3 \times (28 - 32/12)^2/16 = 252.7$ ft-kips

Table 5-17 shows the factored moments for different combinations of gravity and lateral loads on a beam on the 5th floor of an interior frame. The gravity load moments are obtained from the calculations above and the lateral load moments are taken from Fig. 5-29.

The different combinations of axial loads and bending moments corresponding to the appropriate load combinations are listed in Table 5-18 for an interior and an exterior column of an interior frame between grade and the 2nd floor. The seismic loads and moments are taken from Fig. 5-30.

Table 5-19 lists the different combinations of axial loads and bending moments for a shearwall. Seismic loads and moments are given in Table 5-15.

Table 5-17 Summary of Factored Bending Moments for a Beam on the 5th Floor of an Interior Frame (Seismic Zone 2B)

(1) $\quad U = 1.4D + 1.7L$	Eq. (9-1), UBC 1909.2.1
(2) $\quad U = 0.75\,[1.4D + 1.7L \pm 1.87E]$	Eq. (9-2), UBC 1909.2.3
(3) $\quad U = 0.9D \pm 1.43E$	Eq. (9-3), UBC 1909.2.3

Load Combination	Bending Moments (ft-kips)					
	Beam A3-B3			Beam B3-C3		
	Exterior Negative	Positive	Interior Negative	Negative	Positive	Negative
1	-252.7	288.8	-404.3	-367.6	252.7	-367.6
2: sidesway right sidesway left	-37.9 -341.1	217.4 215.8	-453.0 -153.4	-122.1 -429.3	216.6 216.6	-429.3 -122.1
3: sidesway right sidesway left	28.2 -280.9	145.2 143.5	-354.9 -49.4	-27.2 -340.4	126.4 126.4	-340.4 -27.2

Table 5-18 Summary of Factored Axial Loads and Bending Moments for Columns between Grade and Level 2 of an Interior Frame (Seismic Zone 2B)

(1) U = 1.4D + 1.7L   Eq. (9-1), UBC 1909.2.1
(2) U = 0.75 [1.4D + 1.7L ± 1.87E]   Eq. (9-2), UBC 1909.2.3
(3) U = 0.9D ± 1.43E   Eq. (9-3), UBC 1909.2.3

| Load Combination | Column A3 | | | Column C3 | | |
| | Axial Load (kips) | Bending Moment (ft-kips) | | Axial Load (kips) | Bending Moment (ft-kips) | |
		Top	Bottom		Top	Bottom
1	2592	-127.0	127.0	4089	-41.7	41.7
2: sidesway right	1786	-90.5	-292.3	3059	12.8	-375.7
sidesway left	2102	-100.0	482.8	3074	-75.3	438.3
3: sidesway right	1308	-58.3	-331.9	2270	44.9	-415.0
sidesway left	1631	-68.0	458.3	2285	-44.9	415.0

Table 5-19 Summary of Factored Axial Loads, Bending Moments and Shear Forces for a Shearwall between Grade and Level 2 (Seismic Zone 2B)

(1) U = 1.4D + 1.7L   Eq. (9-1), UBC 1909.2.1
(2) U = 0.75 [1.4D + 1.7L ± 1.87E]   Eq. (9-2), UBC 1909.2.3
(3) U = 0.9D ± 1.43E   Eq. (9-3), UBC 1909.2.3

| Load Combination | Shearwall B2-C2 | | | |
| | Axial Load (kips) | Bending Moment (ft-kips) | | Shear Force (kips) |
		Top	Bottom	
1	8211	—	—	—
2: sidesway right	5918	-76,278	91,755	-676.1
sidesway left	6398	76,278	-91,755	676.1
3: sidesway right	4331	-77,773	93,721	-690.6
sidesway left	4821	77,773	-93,721	690.6

5.3.2.1   P-Δ Effects

The member forces and the story drifts induced by P-Δ effects need to be considered in the evaluation of overall structural frame stability. P-Δ effects can be neglected when the ratio of the secondary moment to the primary moment is less than or equal to 0.1 (UBC 1628.9). Based on the results shown in Table 5-20, the P-Δ effects need not be considered.

Table 5-20 P-Δ Check under Seismic Zone 2B Forces

Floor Level	Gravity Loads * Σw_i (kips)	Story Drift Δ_i (in.)	Story Shear V_i (kips)	Story Height h_{si} (in.)	$\Sigma w_i \, \Delta_i / V_i \, h_{si}$
R	4092	0.16	313.2	156	0.013
20	8224	0.16	459.6	156	0.018
19	12,356	0.16	598.6	156	0.021
18	16,488	0.16	730.2	156	0.023
17	20,620	0.16	854.4	156	0.025
16	24,752	0.17	971.2	156	0.028
15	28,884	0.16	1080.6	156	0.027
14	33,016	0.16	1182.6	156	0.029
13	37,148	0.16	1277.2	156	0.030
12	41,280	0.16	1364.4	156	0.031
11	45,412	0.15	1444.2	156	0.030
10	49,544	0.15	1516.5	156	0.031
9	53,676	0.13	1581.4	156	0.028
8	57,808	0.13	1638.9	156	0.029
7	61,940	0.12	1689.0	156	0.028
6	66,072	0.10	1731.7	156	0.025
5	70,204	0.09	1767.0	156	0.023
4	74,336	0.07	1794.9	156	0.019
3	78,468	0.06	1815.4	156	0.017
2	83,288	0.05	1829.8	276	0.008

* Includes floor weight, superimposed dead load, and reduced roof and floor live loads

5.3.2.2   Story Drift Limitation

For buildings having a fundamental period greater than 0.7 seconds, the calculated story drift (including torsional effects) must not exceed $0.03/R_w$ nor 0.004 times the story height (UBC 1628.8.2).

For $h_s = 13$ ft and $R_w = 9$:

$$\frac{0.03 h_s}{R_w} = \frac{0.03 \times 13 \times 12}{9} = 0.52 \text{ in.} \text{(governs)}$$

$$0.004 h_s = 0.004 \times 13 \times 12 = 0.62 \text{ in.}$$

For $h_s = 23$ ft and $R_w = 9$:

$$\frac{0.03 h_s}{R_w} = \frac{0.03 \times 23 \times 12}{9} = 0.92 \text{ in.} \text{(governs)}$$

$$0.004 h_s = 0.004 \times 23 \times 12 = 1.10 \text{ in.}$$

It can be seen from Table 5-20 that for all stories, the lateral drifts obtained from the prescribed lateral forces are less than the limiting value.

5.3.2.3 Proportioning and Detailing a Flexural Member of an Exterior Frame—Beam A1-B1

5.3.2.3.1 Required Flexural Reinforcement

Members designed for Seismic Zone 2 earthquake forces must meet the requirements in UBC 1921.8 for frames in regions of moderate seismic risk. According to these requirements, a member must be designed as a beam if the factored axial load on the member is less than or equal to $A_g f'_c/10$. The beams in the example building carry negligible axial loads, and thus qualify as flexural members.

The maximum negative moment for the beam at an interior support is 344.1 ft-kips (see Table 5-16). Ignoring the effects of any compression reinforcement, the required area of reinforcing steel is $A_s = 4.67$ in.^{2} Use 6-#8 bars ($\phi M_n = 348.5$ ft-kips).

Check limitations on reinforcement ratio:

$$\rho = A_s/bd = (6.0 \times 0.79)/(32 \times 17.65) = 0.0084$$

$$\rho_{max} = 0.0214 > 0.0084 > 200/f_y = 0.0033 \quad \text{O.K.}$$

1910.3.3
1910.5.1

The maximum negative moment at the exterior support is 277.4 ft-kips. Use 5-#8 bars at this location ($\phi M_n = 294.3$ ft-kips).

According to UBC 1921.8.4.1, the positive moment strength at a joint face must be equal to at least 33% of the corresponding negative moment strength at that joint. In this case, the positive moment strength at the interior support must be at least 348.5/3 = 116.2 ft-kips, and at the exterior support it must be at least 294.3/3 = 98.1 ft-kips. The maximum positive factored moments from Table 5-16 of 63.6 ft-kips for the interior support and 94.1 ft-kips for the exterior support are less than the minimum moments. Using 3-#8 bars gives a design positive moment strength of 181.2 ft-kips. Check reinforcement ratio:

$$\rho = 3.0 \times 0.79/(32 \times 17.65) = 0.0042$$

$$0.0213 > 0.0042 > 0.0033 \quad \text{O.K.}$$

The maximum positive factored moment at midspan is 174.2 ft-kips. The positive moment strength at any section along the member length must not be less than 20% of the maximum moment strength provided at the face of either joint (UBC 1921.8.4.1). Thus, the minimum value is 348.5/5 = 69.7 ft-kips. Use 3-#8 bars giving a design moment strength of 181.2 ft-kips.

Similarly, the negative moment strength at any section along the member must not be less than 20% of the maximum moment strength provided at the face of either joint. Use 3-#8 bars giving a design moment strength of 181.2 ft-kips.

5.3.2.3.2 Required Shear Reinforcement

According to UBC 1921.8.3, the design shear strength of beams resisting earthquake effects must not be less than the sum of the shear associated with the development of nominal moment strength of the member, M_n, at each restrained end of the clear span and the calculated shear for gravity loads. Note that ACI 318-89 (Revised 1992) uses the same procedure but with factored gravity loads. Figure 5-31 shows the shear forces at the faces of the supports for sidesway to the right and to the left. The shear capacity of the concrete is:

$$V_c = 0.75 \times 2\sqrt{f_c'} \times bd \qquad\qquad\qquad 1911.3.1.1$$

$$= \frac{0.75 \times 2\sqrt{4000} \times 32 \times 17.65}{1000} = 53.6 \text{ kips} \qquad\qquad 1911.2.1.2$$

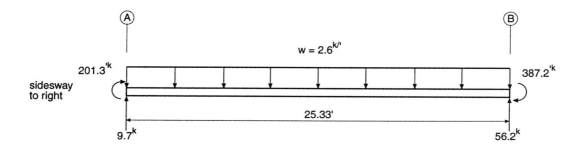

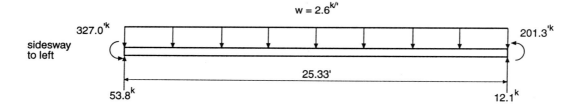

Figure 5-31 Design Shear Forces for Beam A1-B1 (Seismic Zone 2B)

The required strength to be provided by shear reinforcement becomes:

$$\phi V_s = V_u - \phi V_c = 56.2 - (0.85 \times 53.6) = 10.6 \text{ kips}$$

$$V_s = \frac{10.6}{0.85} = 12.5 \text{ kips}$$

The required spacing of #3 stirrups is:

$$s = \frac{A_v f_y d}{V_s} = \frac{0.22 \times 60 \times 17.65}{12.5} = 18.6 \text{ in.}$$ *1911.5.6.2*

The maximum allowable stirrup spacing within a distance of $2h = 2 \times 20 = 40$ in. from the faces of supports is (UBC 1921.8.4.2):

$$s_{max} = \frac{d}{4} = \frac{17.65}{4} = 4.4 \text{ in. (governs)}$$

$$= 8 \times \text{(diameter of smallest longitudinal bar)} = 8 \times 1.0 = 8.0 \text{ in.}$$

$$= 24 \times \text{(diameter of stirrup bars)} = 24 \times 0.375 = 9.0 \text{ in.}$$

$$= 12.0 \text{ in.}$$

Use #3 closed stirrups at a 4.0 in. spacing.

Beyond a distance of $2h$ from the supports, the maximum spacing of stirrups becomes (UBC 1921.8.4.3):

$$s_{max} = \frac{d}{2} = 8.8 \text{ in.}$$

The minimum area of shear reinforcement A_v is:

$$A_v = \frac{50 \times b \times s}{f_y} = \frac{50 \times 32 \times 8.0}{60,000} = 0.21 \text{ in.}^2$$ *1911.5.5.3*

Thus, #3 stirrups at 8 in. will be adequate.

The shear strength of the member with this spacing is:

$$\phi V_n = 0.85 \times (V_s + V_c)$$

$$= 0.85 \times \left(\frac{0.22 \times 60 \times 17.65}{8} + 53.6 \right)$$

$$= 0.85 \times (29.1 + 53.6) = 70.3 \text{ kips}$$

Two-thirds of this strength is:

$$\frac{2}{3}\phi V_n = \frac{2 \times 70.3}{3} = 46.9 \text{ kips}$$

The location where the factored shear is equal to 46.9 kips is 3.6 ft from the face of the interior support. The corresponding location is 2.7 ft from the exterior support. Therefore, any reinforcing bars terminated in a tension zone beyond 3.6 ft from the column faces will meet the requirements in UBC 1912.10.5.1.

5.3.2.3.3 Reinforcing Bar Cutoff Points and Splices

The location where three of the 6-#8 bars can be terminated away from an interior support will be determined. The loading used to find the cutoff point is 0.9 times the dead load in combination with the nominal moment strength, M_n, at the member ends (using $f_s = f_y = 60$ ksi and $\phi = 1.0$). The design moment strength, ϕM_n, provided by 3-#8 reinforcing bars is 181.2 ft-kips.

With $\phi = 1.0$ and $f_s = f_y = 60$ ksi, $M^+_n = 181.2/0.9 = 201.3$ ft-kips at the exterior end of the beam, and $M^-_n = 348.5/0.9 = 387.2$ ft-kips at the interior end. The dead load is equal to $0.9 \times 2.0 = 1.8$ kips/ft. The distance from the face of the interior support to where the moment under the loading considered equals 181.2 ft-kips is readily obtained by summing moments about section a-a in Fig. 5-32:

$$\frac{1}{2}(1.8x^2) + 387.2 - 46x = 181.2$$

Solution of this equation gives $x = 5.0$ ft. Three of the 6-#8 top bars near the interior support may be discontinued at $(x + d) = (5.0 + 17.65/12) = 6.5$ ft from the face of the support.

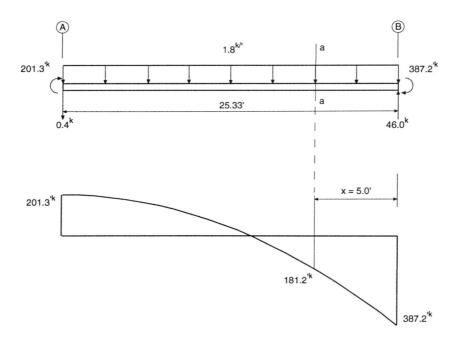

Figure 5-32 Moment Diagram for Cutoff Location of Negative Bars at Interior Support of Beam A1-B1 (Seismic Zone 2B)

Using similar calculations, two of the 5-#8 top bars near the exterior support may be discontinued at 5.1 ft from the face of the support.

Reinforcing bars must be properly developed at the supports. The development length is the product of the basic development length and the factors previously presented in Table 5-12. The basic development length for a #8 bar is 30 in.

The information required for determining the development length factors is:

Minimum top or bottom cover to bars being developed	$= 1.88$ in. $(1.88\ d_b)$
Minimum side cover	large (slab)
Clear spacing between bars being developed	$= 4.45$ in. $(4.45\ d_b)$

Transverse reinforcement meets requirements of UBC 1911.5.4 and 1911.5.5.3.

Factor I has a value of 1.0. Factors II and II are 1.3. All other factors are 1.0; thus,

Top bar $\ell_d = 1.0 \times (1.3 \times 1.3 \times 1.0 \times 1.0 \times 30.0)/12 = 4.2$ ft < 6.5 ft and 5.1 ft

The information required to determine the splice length of the #8 bars is:

	Top Bars	Bottom Bars
Stress level	$< 0.5\ f_y$	$> 0.5\ f_y$
Splice type	A	B
Minimum top or bottom cover to bars	1.88 in. $(1.88\ d_b)$	1.88 in. $(1.88\ d_b)$
Minimum side cover	large (slab)	1.88 in. $(1.88\ d_b)$
Clear spacing between bars being spliced	9.9 in. $(9.9\ d_b)$	12.63 in. $(12.63\ d_b)$

Transverse reinforcement meets requirements of UBC 1911.5.4 and 1912.2.3.5 for both top and bottom bars.

This gives a value of 0.95 and 1.0 for Factor I for top and bottom splices, respectively. Thus,

Top bar splice length (Class A) $= 1.0 \times (0.95 \times 1.3 \times 1.3 \times 1.0 \times 1.0 \times 30.0) = 48.2$ in.

Bottom bar splice length (Class B) $= 1.3 \times (1.0 \times 1.0 \times 1.3 \times 1.0 \times 1.0 \times 30.0) = 50.7$ in.

Figure 5-33 shows the reinforcement details of Beam A1-B1.

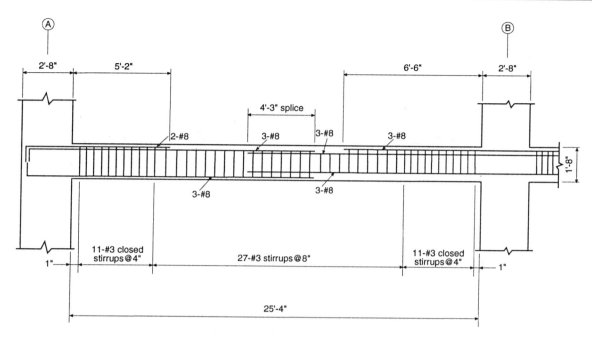

Figure 5-33 Reinforcement Details for Beam A1-B1 (Seismic 2B)

5.3.2.4 Proportioning and Detailing a Flexural Member of an Interior Frame—Beam A3-B3

5.3.2.4.1 Required Flexural Reinforcement

The maximum negative moment for the beam at an interior support is 453.0 ft-kips (see Table 5-17). The required reinforcing steel, ignoring the effects of any compression reinforcement, is $A_s = 6.33$ in.[2] Use 7-#9 bars ($\phi M_n = 495.0$ ft-kips).

The maximum negative moment at an exterior support is 341.1 ft-kips. Use 5-#9 bars ($\phi M_n = 366.0$ ft-kips).

According to UBC 1921.8.4.1, the positive moment strength at the joint face must be equal to at least 33% of the corresponding negative moment strength at that joint. In this case, the positive moment strength must be at least $495.0/3 = 165.0$ ft-kips at the interior support and $366.0/3 = 122.0$ ft-kips at the exterior support. This is larger than the 28.2 ft-kips at the exterior support; there is no positive moment at the interior support (see Table 5-17). Three #8 bars will be required, giving a positive moment strength of 181.2 ft-kips at both ends of the beam.

The maximum positive factored moment at midspan is 288.8 ft-kips. The positive moment strength at any section along the member length must not be less than 20% of the maximum moment strength provided at the face of either joint. Thus, the minimum value for both spans is $495.0/5 = 99.0$ ft-kips which is less than 288.8 ft-kips. Using 5-#8 bars gives a design moment strength of 294.3 ft-kips.

Similarly, the negative moment strength at any section along the member length must not be less than 20% of the maximum moment strength provided at the face of either joint. Thus, the minimum value is $495.0/5 = 99.0$ ft-kips. Use 2-#9 bars, which gives a design moment strength, $\phi M_n = 153.9$ ft-kips.

5.3.2.4.2 Required Shear Reinforcement

According to UBC 1921.8.3, the design shear strength of beams resisting earthquake effects must not be less than the sum of the shear associated with the development of the nominal moment strength of the member, M_n, at each restrained end and the shear calculated for unfactored gravity loads. It should be noted that ACI 318-89 uses the same procedure but with factored gravity loads. Figure 5-34 shows the shear forces at the faces of the supports for sidesway to the right and to the left. The shear capacity of the concrete is:

$$V_c = 0.75 \times 2\sqrt{f'_c} \times bd \qquad \text{1911.3.1.1}$$

$$= \frac{0.75 \times 2\sqrt{4000} \times 32 \times 17.65}{1000} = 53.6 \text{ kips} \qquad \text{1911.2.1.2}$$

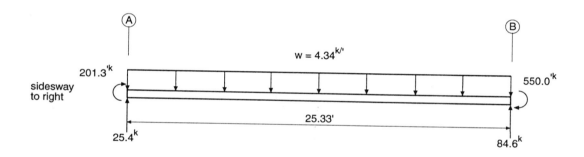

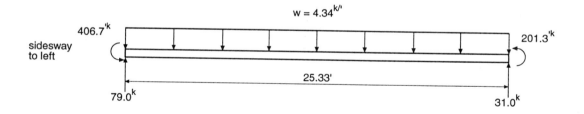

Figure 5-34 Design Shear Forces for Beam A3-B3 (Seismic Zone 2B)

The required strength to be provided by shear reinforcement becomes:

$$\phi V_s = V_u - \phi V_c = 84.6 - (0.85 \times 53.6) = 39.0 \text{ kips}$$

$$V_s = 39.0/0.85 = 45.9 \text{ kips}$$

The required spacing of #3 stirrups is:

$$s = \frac{A_v f_y d}{V_s} = \frac{0.22 \times 60 \times 17.65}{45.9} = 5.1 \text{ in.} \qquad \text{1911.5.6.2}$$

The maximum allowable stirrup spacing within a distance of 2h = 2 × 20 = 40 in. from the faces of supports is (UBC 1921.8.4.2):

$$s_{max} = \frac{d}{4} = \frac{17.65}{4} = 4.4 \text{ in. (governs)}$$

$$= 8 \times (\text{diameter of smallest longitudinal bar}) = 8 \times 1.0 = 8.0 \text{ in.}$$

$$= 24 \times (\text{diameter of stirrup bars}) = 24 \times 0.375 = 9.0 \text{ in.}$$

$$= 12.0 \text{ in.}$$

Use #3 closed stirrups at a 4.0 in. spacing.

Beyond a distance of 2h from the supports, the maximum spacing of stirrups becomes (UBC 1921.8.4.3):

$$s_{max} = \frac{d}{2} = \frac{17.65}{2} = 8.8 \text{ in.}$$

The minimum area of shear reinforcement is:

$$A_v = \frac{50 \times b \times s}{f_y} = \frac{50 \times 32 \times 8}{60,000} = 0.21 \text{ in.}^2 \qquad\qquad 1911.5.5.3$$

Thus, #3 closed stirrups at 8 in. will be adequate.

The shear strength of the member with this spacing is:

$$\phi V_n = 0.85 \times (V_s + V_c)$$

$$= 0.85 \times \left(\frac{0.22 \times 60 \times 17.65}{8} + 53.6 \right)$$

$$= 70.3 \text{ kips}$$

Two-thirds of this strength is:

$$\frac{2}{3} \phi V_n = \frac{2 \times 70.3}{3} = 46.9 \text{ kips}$$

The location where the factored shear is equal to 46.9 kips is 8.7 ft from the face of the interior support. The corresponding location is 7.4 ft from the exterior support. Therefore, any reinforcing bars terminated in a tension zone beyond 8.7 ft from the interior column face and 7.4 ft from the exterior column face will meet the requirements in UBC 1912.10.5.1.

5.3.2.4.3 Reinforcing Bar Cutoff Points and Splices

The negative reinforcement at interior supports consists of 7-#9 bars. The location where five of the #9 bars can be terminated away from an interior support will be determined. The loading used to find the cutoff point is 0.9 times the dead load in combination with the nominal moment strength, M_n, at the member ends (using $f_s = f_y = 60$ ksi and $\phi = 1.0$). The design bending moment strength, ϕM_n, provided by 2-#9 reinforcing bars is 153.9 ft-kips.

With $\phi = 1.0$ and $f_s = f_y = 60$ ksi, $M^+_n = 181.2/0.9 = 201.3$ ft-kips at the exterior end of the beam, and $M^-_n = 495.0/0.9 = 550.0$ ft-kips at the interior end. The dead load is equal to $0.9 \times 3.5 = 3.15$ kips/ft. The distance from the face of the right support to where the moment under the loading considered equals 153.9 ft-kips is readily obtained by summing moments about section a-a in Fig. 5-35:

$$\frac{1}{2}(3.15x^2) + 550.0 - 69.6x = 153.9$$

Solution of this equation gives $x = 6.7$ ft. Five #9 top bars near the interior support may be discontinued at $(x + d) = (6.7 + 17.65/12) = 8.2$ ft from the face of the support. However, discontinue the bars at 8.7 ft from the face of the support so that the requirement in UBC 1912.10.5.1 is satisfied.

The determination of the cutoff point for the negative reinforcing bars at the exterior support is made in a similar manner. Three of the 5-#9 bars can be terminated at 5.9 ft from the face of the exterior support. Use at least 7.4 ft bar lengths to satisfy UBC 1912.10.5.1.

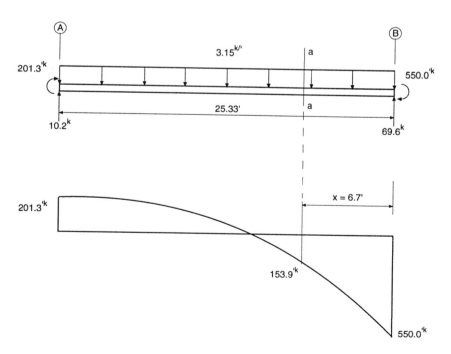

Figure 5-35 Moment Diagram for Cutoff Location of Negative Bars at Interior Support of Beam A3-B3 (Seismic Zone 2B)

Five #8 bars are required at midspan for positive moment strength. Similarly, three #8 bars are required at the supports. The location where two of the five bars at midspan can be terminated will be determined. The design bending moment strength, ϕM_n, provided by 3-#8 reinforcing bars is 181.2 ft-kips.

With $\phi = 1.0$ and $f_s = f_y = 60$ ksi, the nominal moment strength of 5-#8 bars at midspan is equal to 294.3/0.9 = 327.0 ft-kips. The load on the beam will be taken equal to the total factored gravity load which is 6.3 kips/ft. The factored moment at the exterior end of the beam will be taken equal to 252.7 ft-kips as determined from the loading combinations shown in Table 5-17. The distance from the midspan of the beam to where the moment under the loading considered equals 181.2 ft-kips is readily obtained by summing moments about section a-a in Fig. 5-36:

$$\frac{1}{2}(6.3x^2) - 327.0 + 5.9x = -181.2$$

Solution of this equation gives x = 5.9 ft. Two of the five #8 bars can be discontinued at (5.9 + 17.65/12) = 7.4 ft from the midspan section of the beam.

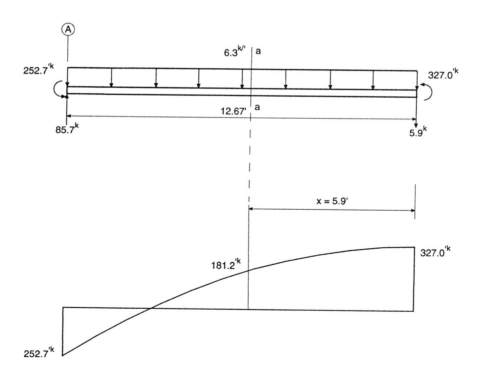

Figure 5-36 Moment Diagram for Cutoff Location of Positive Bars at Midspan of Beam A3-B3
(Seismic Zone 2B)

Reinforcing bars must be properly developed. The development length is the product of the basic development length and the factors previously presented in Table 5-12. The basic development length for a #9 bar is 38.0 in. and for a #8 bar it is 30.0 in.

The information required for determining the development length factors is:

	Top Bars	Bottom Bars
Minimum top or bottom cover to bars being developed	1.88 in. (1.67 d_b)	1.88 in. (1.88 d_b)
Minimum side cover	large (slab)	1.88 in. (1.88 d_b)
Clear spacing between bars being developed	3.39 in. (3.0 d_b)	5.81 in. (5.81 d_b)

Transverse reinforcement meets requirements of UBC 1911.5.4 and 1911.5.5.3 for both top and bottom bars.

Factor I has a value of 1.0 for both the top and the bottom bars. Using the other applicable factors gives:

Top bar $\ell_d = 1.0 \times (1.3 \times 1.3 \times 1.0 \times 1.0 \times 38.0)/12 = 5.4$ ft < 8.7 ft

Bottom bar $\ell_d = 1.0 \times (1.0 \times 1.3 \times 1.0 \times 1.0 \times 30.0)/12 = 3.3$ ft < 7.4 ft

The information required for determining the splice length is:

	Top Bars	Bottom Bars
Stress level	$< 0.5 f_y$	$> 0.5 f_y$
Splice type	A	B
Minimum top or bottom cover to bars being spliced	1.88 in. (1.67 d_b)	1.88 in. (1.88 d_b)
Minimum side cover	large (slab)	1.88 in. (1.88 d_b)
Clear spacing between bars being spliced	26.0 in. (23.0 d_b)	12.63 in. (12.63 d_b)

Transverse reinforcement meets requirements of UBC 1911.5.4 and 1911.5.5.3 for both top and bottom bars.

This gives a value of 0.85 and 1.0 for Factor I for the top and the bottom bar splices, respectively.

Top bar splice length (Class A) $= 1.0 \times (0.85 \times 1.3 \times 1.3 \times 1.0 \times 1.0 \times 38.0) = 54.6$ in.

Bottom bar splice length (Class B) $= 1.3 \times (1.0 \times 1.0 \times 1.3 \times 1.0 \times 1.0 \times 30.0) = 50.7$ in.

Figure 5-37 shows the reinforcement details for Beam A3-B3.

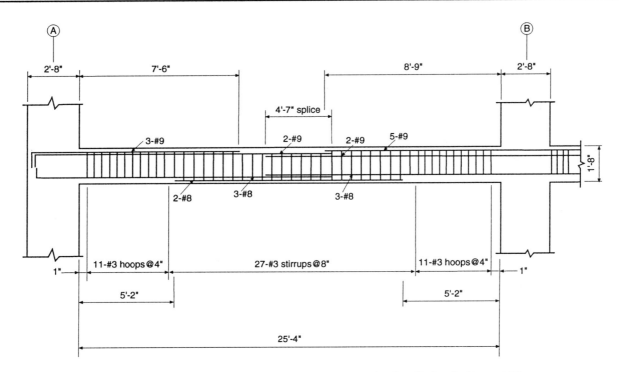

Figure 5-37 Reinforcement Details for Beam A3-B3 (Seismic Zone 2B)

5.3.2.5   Proportioning and Detailing an Exterior Column of an Interior Frame

5.3.2.5.1 General Code Requirements

This section will present the design of column A3 supporting the second floor of the structure. The maximum factored axial load on this column is 2592 kips (see Table 5-18). This factored axial load is larger than $A_g f'_c/10$ so that the section must be designed as a column (UBC 1921.8.2):

$$P_u \quad = 2592 \text{ kips} > \frac{A_g f'_c}{10} \quad = \frac{32 \times 32 \times 6}{10} = 614 \text{ kips}$$

Check the reinforcement ratio for the 32 in. square column reinforced with 12-#9 bars:

$$0.08 \quad > \rho_g \quad = \frac{A_{st}}{A_g} \quad = \frac{12 \times 1.0}{32 \times 32} = 0.0117 > 0.01 \quad \text{O.K.}$$

Note that this section is adequate for the load combinations given in Table 5-18 (see Fig. 5-38).

5.3.2.5.2 Confinement Reinforcement

Special transverse reinforcement for confinement is required over a distance ℓ_o at column ends, where ℓ_o equals the maximum of:

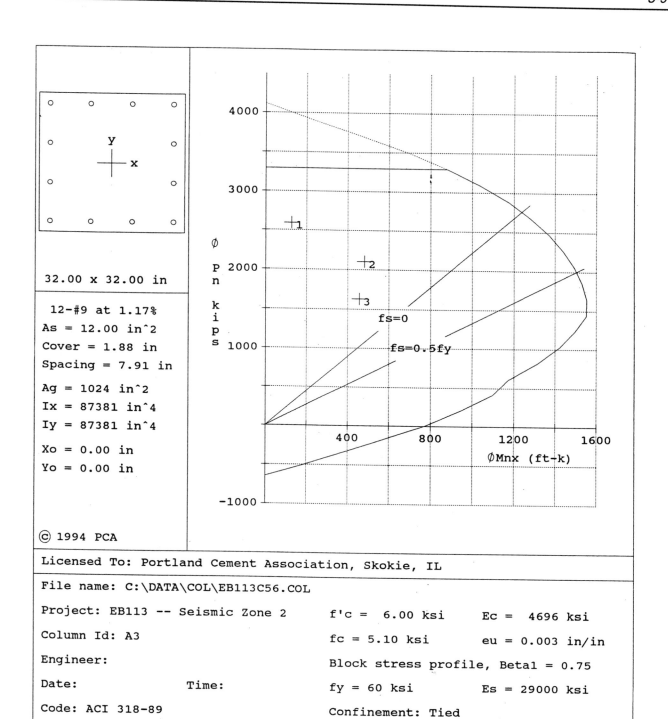

32.00 x 32.00 in

12-#9 at 1.17%
As = 12.00 in^2
Cover = 1.88 in
Spacing = 7.91 in

Ag = 1024 in^2
Ix = 87381 in^4
Iy = 87381 in^4

Xo = 0.00 in
Yo = 0.00 in

© 1994 PCA

Licensed To: Portland Cement Association, Skokie, IL

File name: C:\DATA\COL\EB113C56.COL

Project: EB113 -- Seismic Zone 2   f'c = 6.00 ksi   Ec = 4696 ksi

Column Id: A3   fc = 5.10 ksi   eu = 0.003 in/in

Engineer: Block stress profile, Beta1 = 0.75

Date: Time: fy = 60 ksi   Es = 29000 ksi

Code: ACI 318-89   Confinement: Tied

Units: English   phi(c) = 0.70, phi(b) = 0.90, a = 0.80

X-axis slenderness is not considered.

Figure 5-38 Design Strength Interaction Diagram for Column A3 (Seismic Zone 2B)

$$\frac{1}{6} \text{ (clear height)} = \frac{(21.33 \times 12)}{6} = 42.7 \text{ in. (governs)} \qquad\qquad 1921.8.5.1$$

Depth of member = 32 in.

18 in.

The maximum allowable spacing of #3 ties over the previously determined length ℓ_o is:

$$s_o \quad = 8 \times \text{(diameter of smallest longitudinal bar)} = 8 \times 1.128 = 9.0 \text{ in. (governs)} \qquad 1921.8.5.1$$

$$= 24 \times \text{(diameter of tie bars)} = 24 \times 0.375 = 9.0 \text{ in.}$$

$$= \frac{h}{2} = \frac{32}{2} = 16 \text{ in.}$$

$$= 12.0 \text{ in.}$$

The maximum spacing of ties outside of the distance ℓ_o is $2s_o = 2 \times 9 = 18$ in. (UBC 1921.8.5.4). These ties must be arranged such that every corner and alternate longitudinal bar shall have support provided by the corner of a tie or crosstie.

5.3.2.5.3 Transverse Reinforcement for Shear

According to UBC 1921.8.3, the design shear in columns resisting earthquake effects in moderate seismic zones is not based on the factored shears obtained from a lateral load analysis, but rather on the nominal moment strengths that can be developed at the faces of the joints. The moment strength for the column is conservatively taken to correspond to the balanced axial load from the interaction diagram (see Fig. 5-38): $M_n = 1564/0.7 = 2234$ ft-kips. The nominal moment strength of the beam framing into the joint is $366.0/0.9 = 406.7$ ft-kips (see Sect. 5.3.2.4.1). Therefore, the beam moment strength will control and the design shear is:

$$V_u \quad = \frac{406.7/2 + 406.7/2}{21.33} = 19.1 \text{ kips}$$

Since the member is subject to axial compression, the shear strength provided by the concrete, V_c, is calculated as:

$$V_c \quad = 2\left[1 + \frac{N_u}{2000A_g}\right]\sqrt{f'_c}\, b_w d \qquad\qquad 1911.3.1.2$$

Using the minimum value of axial compression for the column (see Table 5-18), V_c is:

$$V_c \quad = 2 \times \left[1 + \frac{1,308,000}{2000 \times 1024}\right] \times \sqrt{6000} \times 32 \times 29.56/1000 = 240.1 \text{ kips}$$

$$\phi V_c = 0.85 \times 240.1 = 204.1 \text{ kips} > V_u = 19.1 \text{ kips}$$

The value of ϕV_c exceeds $2V_u$; therefore, use maximum tie spacing of 18 in.

Use #3 ties with #3 crossties spaced at 9 in. within a distance of 45 in. from the column ends, and spaced at a maximum of 18 in. over the remainder of the column height.

No specific requirements for column splices are provided for columns in Seismic Zone 2. For this design, the requirements for Seismic Zone 4 will be used rather than the alternative of using those for regions of low or no seismicity. In Seismic Zone 4, UBC 1921.4.3.2 requires tension splices to be within the center of the member length.

The basic development length for #9 bars is 31.0 in.

The information required for determining the development length factors is:

Minimum clear cover to bars being spliced	= 1.88 in. (1.67 d_b)
Clear spacing between bars being spliced	= 6.3 in. (5.6 d_b)
Transverse reinforcement meets requirements of UBC 1907.10.5.	

This gives a value of 1.0 for Factor I. The other factors are all equal to 1.0, giving:

Class B splice length = $1.3 \times (1.0 \times 1.0 \times 1.0 \times 1.0 \times 1.0 \times 31.0) = 40.3$ in.

Reinforcement details for column A3 are shown in Fig. 5-39.

5.3.2.6 Proportioning and Detailing an Interior Column of an Interior Frame

5.3.2.6.1 General Code Requirements

This section will present the design of column C3 supporting the second floor of the structure. The maximum factored axial load on this column is 4089 kips (see Table 5-18). This factored axial load is larger than $A_g f'_c/10$ so that the section must be designed as a column (UBC 1921.8.2):

$$P_u = 4089 \text{ kips} > \frac{A_g f'_c}{10} = \frac{32 \times 32 \times 6}{10} = 614 \text{ kips}$$

Check the reinforcement ratio for the 32-in. square column reinforced with 28-#11 bars:

$$0.08 > \rho_g = \frac{A_{st}}{A_g} = \frac{28 \times 1.56}{32 \times 32} = 0.0427 > 0.01 \quad \text{O.K.}$$

Note that this section is adequate for the load combinations given in Table 5-18 (see Fig. 5-40).

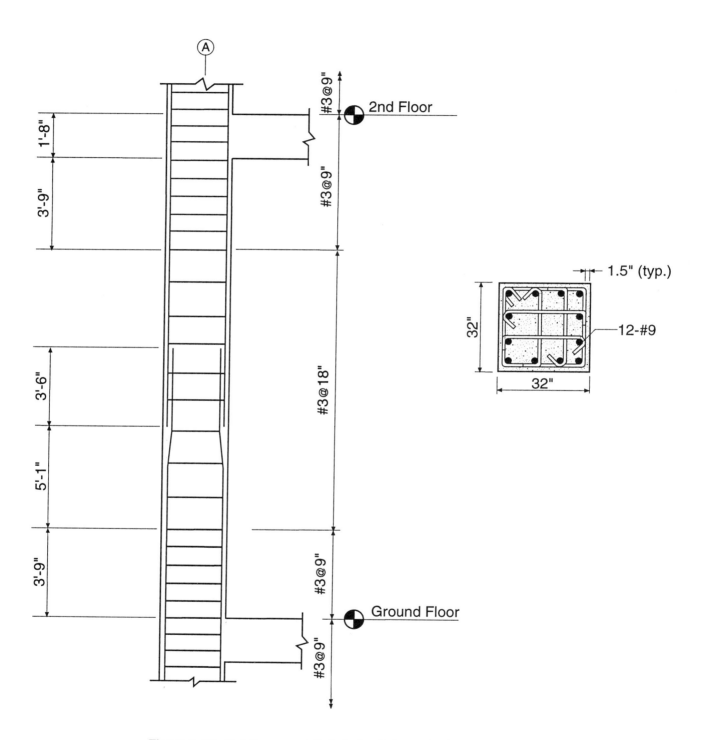

Figure 5-39 Reinforcement Details for Column A3 (Seismic Zone 2B)

5.3.2.6.2 Confinement Reinforcement

Special transverse reinforcement for confinement is required over a distance ℓ_o at column ends, where ℓ_o equals the maximum of:

$$\frac{1}{6} \text{ (clear height)} = \frac{(21.33 \times 12)}{6} = 42.7 \text{ in. (governs)} \qquad \textit{1921.8.5.1}$$

Depth of member = 32 in.

18 in.

The maximum allowable spacing of #4 ties over the previously determined length ℓ_o is:

$$s_o = 8 \times \text{(diameter of smallest longitudinal bar)} = 8 \times 1.41 = 11.3 \text{ in. (governs)} \qquad \textit{1921.8.5.1}$$

$$= 24 \times \text{(diameter of tie bars)} = 24 \times 0.5 = 12.0 \text{ in.}$$

$$= \frac{h}{2} = \frac{32}{2} = 16 \text{ in.}$$

$$= 12.0 \text{ in.}$$

The maximum spacing of ties outside of the distance ℓ_o is $2s_o = 2 \times 11.3 = 22.6$ in. (UBC 1921.8.5.4). These ties must be arranged such that every corner and alternate longitudinal bar shall have support provided by the corner of a tie or crosstie.

5.3.2.6.3 Transverse Reinforcement for Shear

According to UBC 1921.8.3, the design shear in columns resisting earthquake effects in moderate seismic zones is not based on the factored shears obtained from a lateral load analysis, but rather on the nominal moment strengths that can be developed at the faces of the joints. The moment strength for the column is conservatively taken to correspond to the balanced axial load from the interaction diagram (see Fig. 5-40): $M_n = 2503/0.7 = 3576$ ft-kips. The nominal moment strengths of the beams framing into the joint are 201.3 ft-kips and 550.0 ft-kips (see Sect. 5.3.2.4.1). Therefore, the beam flexural strength will control and the design shear is:

$$V_u = \frac{2 \times (201.3/2 + 550.0/2)}{21.33} = 35.2 \text{ kips}$$

Since the member is subject to axial compression, the shear strength provided by the concrete, V_c, is calculated as:

$$V_c = 2\left[1 + \frac{N_u}{2000 A_g}\right] \sqrt{f'_c}\, b_w d \qquad \textit{1911.3.1.2}$$

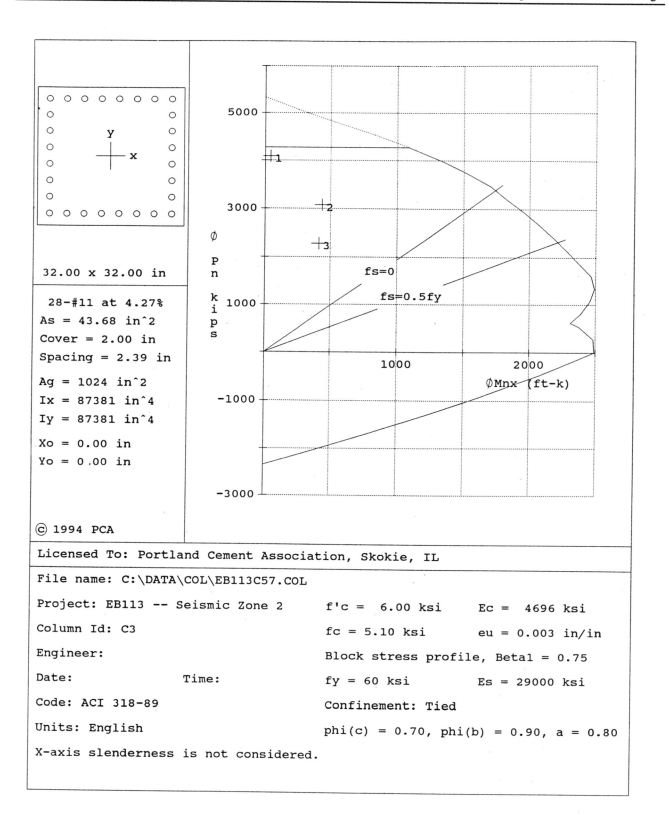

Licensed To: Portland Cement Association, Skokie, IL

File name: C:\DATA\COL\EB113C57.COL

Project: EB113 -- Seismic Zone 2   f'c = 6.00 ksi   Ec = 4696 ksi

Column Id: C3   fc = 5.10 ksi   eu = 0.003 in/in

Engineer:

Date: Time: Block stress profile, Beta1 = 0.75

                                          fy = 60 ksi   Es = 29000 ksi

Code: ACI 318-89   Confinement: Tied

Units: English   phi(c) = 0.70, phi(b) = 0.90, a = 0.80

X-axis slenderness is not considered.

Figure 5-40 Design Strength Interaction Diagram for Column C3 (Seismic Zone 2B)

Using the minimum value of axial compression for the column (see Table 5-18), V_c is:

$$V_c = 2 \times \left[1 + \frac{2,270,000}{2000 \times 1024} \right] \times \sqrt{6000} \times 32 \times 29.56/1000 = 309.0 \text{ kips}$$

$$\phi V_c = 0.85 \times 309.0 = 262.6 \text{ kips} > V_u = 35.2 \text{ kips}$$

The value of ϕV_c exceeds $2V_u$; therefore, use maximum tie spacing of 22 in.

Use #4 ties with #4 crossties spaced at 10 in. within a distance of 50 in. from the column ends and spaced at 20 in. over the remainder of the column height.

No specific requirements for column splices are provided for columns in Seismic Zone 2. For this design, the requirements for Seismic Zone 4 will be used rather than the alternative of using those for regions of low or no seismicity. In Seismic Zone 4, UBC 1921.4.3.2 requires tension splices to be within the center of the member length.

The basic development length for #11 bars is 48.3 in.

The information required for determining the splice length is:

Minimum clear cover to bars being spliced	= 2.0 in. (1.42 d_b)
Clear spacing between bars being spliced	= 2.39 in. (1.69 d_b)

Transverse reinforcement meets requirements of UBC 1907.10.5.

This gives a value of 2.0 for Factor I. The other factors are all equal to 1.0, giving:

Class B splice length = $1.3 \times (2.0 \times 1.0 \times 1.0 \times 1.0 \times 1.0 \times 48.3) = 125.6$ in.

The reinforcement details for column C3 are given in Fig. 5-41.

5.3.2.7 Proportioning and Detailing of an Interior Beam-Column Connection

This section will check the shear strength and detailing requirements for interior beam-column connection C3 on the third floor level of the example building. The column considered is an interior column of an interior frame along either principal direction of the building, and has beams framing in on all four sides: two transverse beams and two longitudinal beams. The beams are all as wide as the column.

It should be noted that UBC-94 contains no specific requirements for the design of beam-column joints in regions of moderate and low seismicity. This section contains a suggested design procedure.

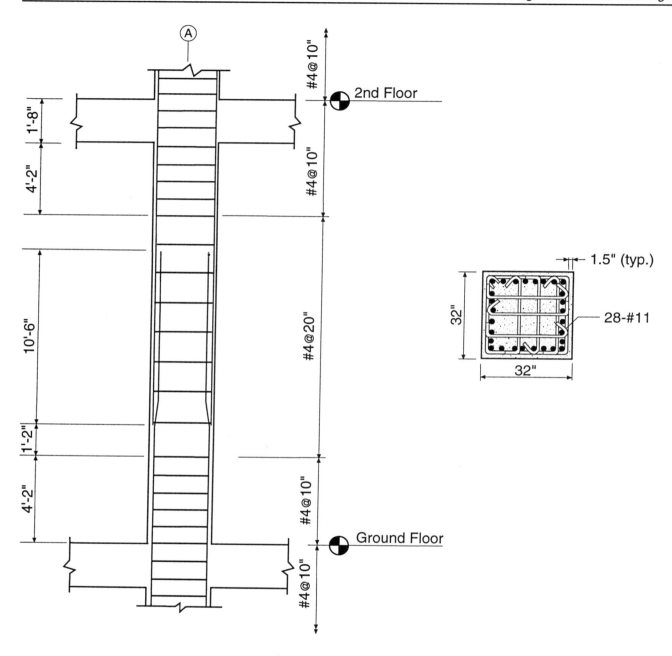

Figure 5-41 Reinforcement Details for Column C3 (Seismic Zone 2B)

5.3.2.7.1 Transverse Reinforcement for Confinement

The columns above and below the joint were not designed.   Assuming that the transverse reinforcement in the critical end regions is the same as for the column designed in Section 5.3.2.6, #4 ties with three crossties in each direction at a 10 in. spacing will be continued through the joint. Although the joint in question is confined on all four faces by beams that are as wide as the column, according to the criteria of UBC 1921.5.2, any relaxation in the moderate amount of transverse reinforcement provided is felt to be unjustified.

5.3.2.7.2 Shear Strength of Joint

Figure 5-42 shows the beam-column joint being designed at the 3rd floor level of the structure. The shear strength is checked along column line C of the building. Because of symmetry, the computations in the orthogonal direction would be identical. The shear across section x-x is obtained by subtracting the column horizontal shear from the sum of the tensile force in the top beam reinforcement on one side and the concrete compressive force near the top of the beam on the opposite face of the column.

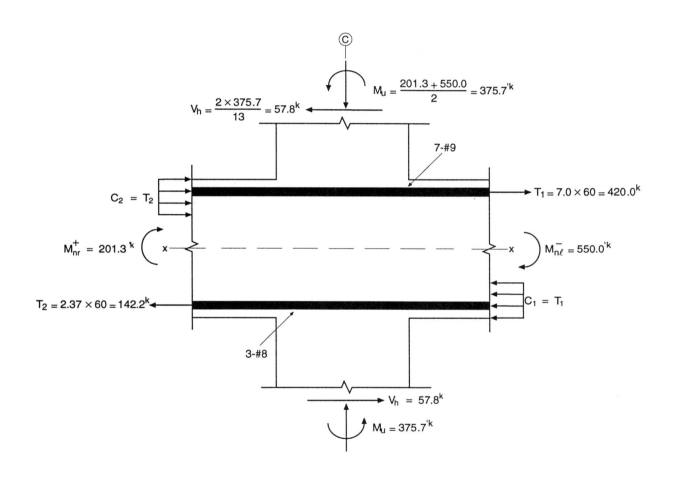

$$M_u = \frac{201.3 + 550.0}{2} = 375.7^{'k}$$

$$V_h = \frac{2 \times 375.7}{13} = 57.8^k$$

7-#9

$$T_1 = 7.0 \times 60 = 420.0^k$$

$$C_2 = T_2$$

$$M^+_{nr} = 201.3^k \qquad M^-_{n\ell} = 550.0^{'k}$$

$$T_2 = 2.37 \times 60 = 142.2^k$$

$$C_1 = T_1$$

3-#8

$$V_h = 57.8^k$$

$$M_u = 375.7^{'k}$$

Figure 5-42 Shear Analysis of Interior Beam-Column Joint C3 (Seismic Zone 2B)

The column horizontal shear, V_h, can be obtained by assuming that the beams in the adjoining floors are deformed so that plastic hinges form at their junctions with the column, with M^-_n (beam) = 550.0 ft-kips and M^+_n (beam) = 201.3 ft-kips. Therefore, the sum of the column moments above and below the joint will equal 550.0 + 201.3 = 751.3 ft-kips.

It is assumed that beam end moments are resisted equally by the columns above and below the joint. The horizontal shear at the column ends is then equal to:

$$V_h = \frac{2 \times (201.3/2 + 550.0/2)}{13} = 57.8 \text{ kips}$$

The tensile force in the top beam reinforcement (7-#9 bars) = 7.0 × 60 = 420.0 kips. The compression on the opposite side of the column is equal to the tensile force of 3-#8 bars which is 2.37 × 60 = 142.2 kips.

The net shear at section x-x of the joint is $T_1 + C_2 - V_h$ = 420.0 +142.2 - 57.8 = 504.4 kips.

For a lightweight concrete joint confined on all four faces:

$$\phi V_c = 0.75 \times \phi 20 \sqrt{f'_c} \times A_j /1000 \qquad\qquad 1921.5.3.1$$

$$= 0.75 \times 0.85 \times 20 \sqrt{6000} \times 32 \times 32/1000 = 1011.3 \text{ kips} > 504.4 \text{ kips} \quad \text{O.K.}$$

5.3.2.8   Proportioning and Detailing an Exterior Beam-Column Connection

This section will check the shear strength and detailing requirements for exterior beam-column connection A3 on the third floor level of the example building. The column considered is an interior column of an exterior frame in one direction of the building, and an exterior column of an interior frame in the orthogonal direction. The column has beams framing in on three sides: two transverse beams and one longitudinal beam.

It should be noted again that UBC-94 contains no specific requirements for the design of beam-column joints in regions of moderate and low seismicity. This section contains a suggested design procedure.

5.3.2.8.1 Transverse Reinforcement for Confinement

The columns above and below the joint were not designed. Assuming that the transverse reinforcement in the critical end regions is the same as for the column designed in Section 5.3.2.5, #3 ties with two crossties in each direction at a 9 in. spacing will be continued through the joint. It may be noted as a matter of interest that the joint in question is unconfined according to the criteria of UBC 1921.5.2.2.

5.3.2.8.2 Shear Strength of Joint

Figure 5-43 shows the beam-column joint being designed at the 3rd floor level of the structure. The shear strength is checked along column line 3 of the building. The shear across section x-x is obtained by subtracting the column horizontal shear from the tensile force in the top beam reinforcement.

The column horizontal shear, V_h, can be obtained by assuming that the beam in the adjoining floor is deformed so that a plastic hinge forms at the junctions with the column, with M^-_n (beam) = 406.7 ft-kips.

It is assumed that the beam end moments are resisted equally by the columns above and below the joint. The horizontal shear at the column ends is then equal to:

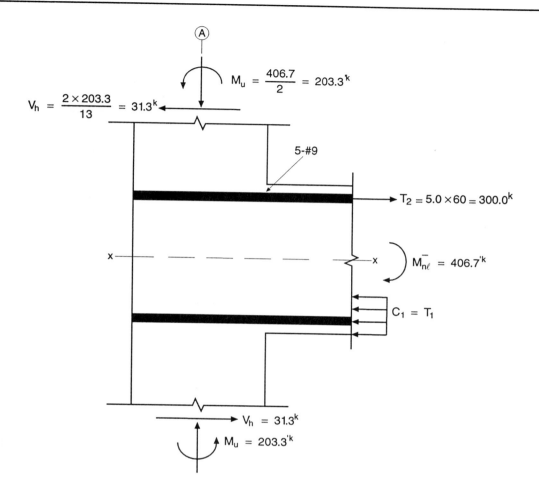

Figure 5-43 Shear Analysis of Exterior Beam-Column Joint A3 (Seismic Zone 2B)

$$V_h = \frac{2 \times (406.7 / 2)}{13} = 31.3 \text{ kips}$$

The tensile force in the top beam reinforcement (5-#9 bars) = $5.0 \times 60 = 300.0$ kips. The net shear at section x-x of the joint is $T_1 - V_h = 300.0 - 31.3 = 268.7$ kips.

For a lightweight concrete joint confined on three faces:

$$\phi V_c = 0.75 \times \phi 15 \sqrt{f_c'} \times A_j / 1000$$

$$= 0.75 \times 0.85 \times 15 \sqrt{6000} \times 32 \times 32/1000 = 758.5 \text{ kips} > 268.7 \text{ kips} \quad \text{O.K.}$$

5.3.2.9   Proportioning and Detailing a Shearwall

5.3.2.9.1 General

No special proportioning and detailing requirements for structural walls (shearwalls) are required in regions of moderate seismic risk (Zones 2A and 2B). It is anticipated that structural walls proportioned to resist earthquake effects and designed according to UBC Chapter 19 (excluding UBC 1921) will possess sufficient toughness at the low drift levels which they would be likely to experience in regions of moderate seismicity.

5.3.2.9.2 Design of Shearwall B2-C2

Boundary elements are not required in Seismic Zone 2B. The ends of the shearwall for this example are thickened not to act as boundary elements but to provide increased flexural strength to the shearwall, and to maintain a regular column layout.

The design shear strength of a shearwall is limited to:

$$\phi V_n = \phi \, 10 \sqrt{f'_c} \times hd \qquad\qquad\qquad \textit{1911.10.3}$$

where d equals 80% of the length of the wall (UBC 1911.10.4). In this case, the maximum shear force is 690.6 kips (see Table 5-19). Therefore,

$$\phi V_n = 0.85 \times 10 \sqrt{6000} \times 12 \times [0.8 \times (28 \times 12 + 32)]/1000 = 2326.0 \text{ kips} > 690.6 \text{ kips O.K.}$$

The shear strength provided by the concrete is:

$$\phi V_c = \phi \, 2 \sqrt{f'_c} \times hd$$

$$= 0.85 \times 2 \sqrt{6000} \times 12 \times [0.8 \times (28 \times 12 + 32)]/1000 \qquad\qquad \textit{1911.10.5}$$

$$= 465.2 \text{ kips} < V_u = 690.6 \text{ kips}$$

Therefore, shear reinforcement is required. The maximum spacing of horizontal bars is:

$$s_{max} = \frac{\ell_w}{5} = \frac{(28 \times 12 + 32)}{5} = 73.6 \text{ in.} \qquad\qquad \textit{1911.10.9.3}$$

$$= 3h = 3 \times 12 = 36.0 \text{ in.}$$

$$= 18.0 \text{ in. (governs)}$$

The design shear strength of the wall with #5 reinforcing bars on both faces at 18 in. spacing is:

$$\phi V_n = \phi (V_s + V_c)$$

$$\phi V_n = \phi A_{vh} f_y d/s + \phi V_c$$

$$= 0.85 \times 0.62 \times 60 \times [0.8 \times (28 \times 12 + 32)]/18 + 465.2$$

$$= 517.2 + 465.2 = 982.4 \text{ kips} > V_u = 690.6 \text{ kips}$$

The ratio of horizontal shear reinforcement area to gross concrete area of the vertical cross section must not be less than 0.0025 (UBC 1911.10.9.2):

$$A_{vh} = 0.0025 \times 12 \times 12 = 0.36 \text{ in.}^2/\text{ft}$$

Area of horizontal reinforcement provided $= 0.62 \times 12/18 = 0.41$ in.^{2}/ft > 0.36 in.^{2}/ft   O.K.

The ratio of vertical reinforcement area to gross area of horizontal section must not be less than:

$$\rho_n = 0.0025 + 0.5 \times (2.5 - h_w/\ell_w) \times (\rho_h - 0.0025) \text{ or } 0.0025 \qquad \textit{1911.10.9.4}$$

where $\rho_h = 0.41/(12 \times 12) = 0.0028$

$$\rho_n = 0.0025 + 0.5 [2.5 - (270/(28 + 32/12)] \times (0.0028 - 0.0025)$$

$$= 0.0016 < 0.0025$$

Use $\rho_n = 0.0025$

The required vertical reinforcing steel area is:

$$A_n = 0.0025 \times 12 \times 12 = 0.36 \text{ in.}^2/\text{ft}$$

The spacing for 2 layers of #5 bars is $0.62 \times 12/0.36 = 20.7$ in.

The maximum spacing for the vertical #5 bars is:

$$s_{max} = \frac{\ell_w}{3} = \frac{(28 \times 12 + 32)}{3} = 122.7 \text{ in.} \qquad \textit{1911.10.9.5}$$

$$= 3h = 3 \times 12 = 36.0 \text{ in.}$$

$$= 18.0 \text{ in. (governs)}$$

Use two layers of #5 vertical bars at 18 in. spacing.

An interaction diagram for the shearwall is shown in Fig. 5-44. The interaction diagram shows that the shearwall has adequate strength to withstand the load combinations given in Table 5-19.

See Fig. 5-45 for reinforcement details of the shearwall.

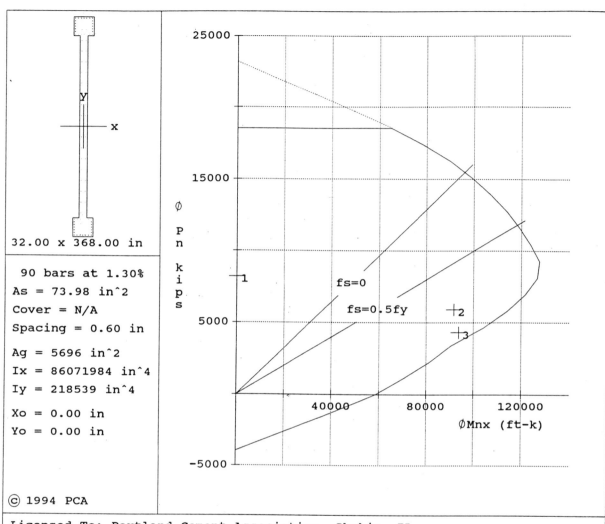

32.00 x 368.00 in

90 bars at 1.30%
As = 73.98 in^2
Cover = N/A
Spacing = 0.60 in

Ag = 5696 in^2
Ix = 86071984 in^4
Iy = 218539 in^4

Xo = 0.00 in
Yo = 0.00 in

© 1994 PCA

Licensed To: Portland Cement Association, Skokie, IL

File name: C:\DATA\COL\EB113C58.COL

Project: EB113 -- Seismic Zone 2   f'c = 6.00 ksi   Ec = 4696 ksi

Column Id: Shearwall B2 - C2   fc = 5.10 ksi   eu = 0.003 in/in

Engineer: Block stress profile, Beta1 = 0.75

Date: Time: fy = 60 ksi   Es = 29000 ksi

Code: ACI 318-89   Confinement: Tied

Units: English   phi(c) = 0.70, phi(b) = 0.90, a = 0.80

X-axis slenderness is not considered.

Figure 5-44 *Design Strength Interaction Diagram for Shearwall B2-C2 (Seismic Zone 2B)*

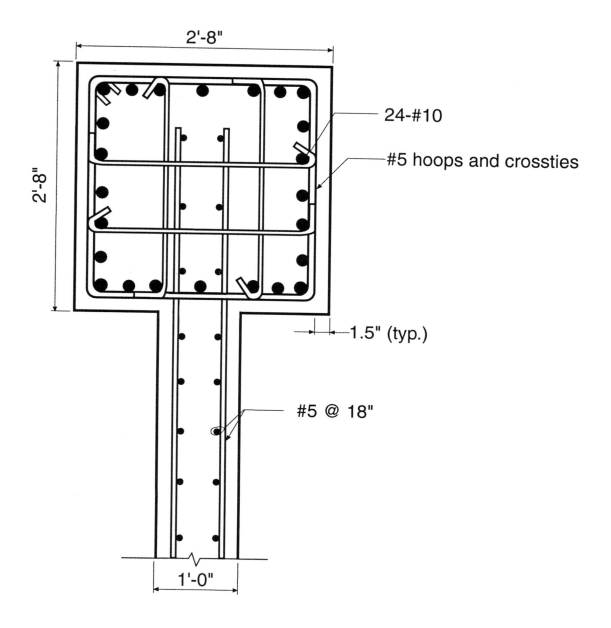

2'-8"

2'-8"

24-#10

#5 hoops and crossties

1.5" (typ.)

#5 @ 18"

1'-0"

Figure 5-45 Reinforcement Details for Shearwall B2-C2 (Seismic Zone 2B)

Chapter 6

Frame-Shearwall System

6.1   Introduction

6.1.1   General

The computation of design loads for a 10-story reinforced concrete shearwall-frame building under the requirements of the Uniform Building Code (UBC), 1994 edition, is illustrated. These computations are performed for UBC Seismic Zone 1 and 70 mph (Exposure B) wind forces. Typical members are designed and detailed in each case.

Shearwalls used in conjunction with beam-column frames to carry lateral loads in zones of high seismicity constitute structures that must be designed as either Building Frame Systems or as Dual Systems. The concept of the Dual System loses its relevance in zones of low seismicity where the beam-column frame or the slab-column frame without any special detailing cannot really act as a backup to the shearwall subsystem. The concept of the Building Frame System also loses its appeal since there is little to be gained from assigning the entire lateral resistance to the shearwall subsystem, in the absence of any special detailing requirements for the beam-column or slab-column frames. In zones of low seismicity, it is usual practice to design the shearwalls and the frames in a shearwall-frame structure to resist lateral loads in proportion to their relative rigidities, considering interaction between the two subsystems at all levels. Such a system is referred to here as simply a Frame-Shearwall System.

6.1.2   Design Criteria

A typical floor plan and elevation of the building are shown in Figs. 6-1 and 6-2, respectively. The columns, beams, and slabs have constant cross-sections throughout the height of the building. Though the uniformity and symmetry used in this example have been adopted primarily for simplicity, these are generally considered to be sound engineering design concepts that should be utilized wherever practicable for seismic design. Even though

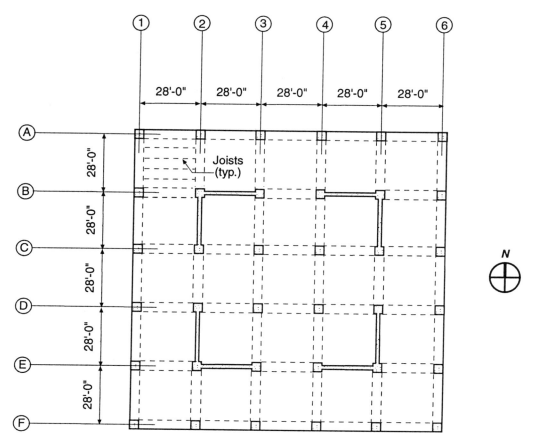

Figure 6-1 Typical Floor Plan of Example Building

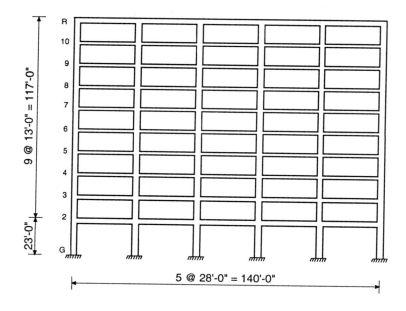

Figure 6-2 Elevation of Example Building

the member dimensions in this example are within the practical range, the structure itself is a hypothetical one, and has been chosen mainly for illustrative purposes. Other pertinent design data are as follows:

Service Loads:
    Live load: 50 psf
    Superimposed dead load: 32 psf

Material Properties:
    Concrete: f'_c = 4 ksi
                       w_c = 150 pcf
    Reinforcement: f_y = 60 ksi

Beams: 18 × 22 in.

Joists and topping: 86 psf

Columns: 18 × 18 in.

Shearwalls: 12 in. thick

6.2   Design for Seismic Zone 1

6.2.1   Frame Analysis

On the basis of the given data and the dimensions of the building, the weights of the floors are listed in Table 6-1. This example uses an importance factor I of 1.0 and a site soil coefficient S of 1.2. In regions of low seismic risk (Seismic Zone 1), there are no special proportioning and detailing requirements; all reinforced concrete members are designed according to Chapter 19 excluding Section 1921. Without any special detailing for the beam-column framing, the value of R_w is taken as 8, the same as that for a Building Frame System. The design lateral forces resulting from the distribution of the base shear computed in accordance with UBC Eqs. (28-6), (28-7), and (28-8) are shown in Table 6-1.

Analysis of the structure under the forces determined above was performed using *SAP90* [4.1]. The building was analyzed considering accidental torsional effects, as required in UBC 1628.5.

For buildings having a fundamental period greater than 0.7 seconds, the calculated story drift (including torsional effects) must not exceed $0.03/R_w$ nor 0.004 times the story height (UBC 1628.8.2).

For typical floors:

$$\frac{0.03h_s}{R_w} = \frac{0.03 \times 13 \times 12}{8} = 0.59 \text{ in. (governs)}$$

$$0.004h_s = 0.004 \times 13 \times 12 = 0.62 \text{ in.}$$

For the lower level:

$$\frac{0.03h_s}{R_w} = \frac{0.03 \times 23 \times 12}{8} = 1.04 \text{ in.} \quad (\text{governs})$$

$$0.004h_s = 0.004 \times 23 \times 12 = 1.10 \text{ in.}$$

Table 6-1 Seismic Forces for Zone 1

Floor Level	Weight, w_x (kips)	Height, h_x (ft)	$w_x h_x$ (ft-kips)	Lateral Force (kips)
R	3165	140	443,100	115.2
10	3451	127	438,277	82.7
9	3451	114	393,414	74.2
8	3451	101	348,551	65.7
7	3451	88	303,688	57.3
6	3451	75	258,825	48.8
5	3451	62	213,962	40.4
4	3451	49	169,099	31.9
3	3451	36	124,236	23.4
2	3671	23	84,433	15.9
Σ	34,444		2,777,585	555.6

Building height	h_n	= 140 ft	
Building weight	W	= 34,444 kips	
Fundamental period	T	= 0.020 $(h_n)^{3/4}$ = 0.81 sec.	
Seismic zone factor	Z	= 0.075	
Importance factor	I	= 1.0	
Response coefficient	R_w	= 8	
Soil factor	S	= 1.2	
Coefficient	C	= 1.25 $S/T^{2/3}$ = 1.72	
Base shear	V	= $ZICW/R_w$ = 555.6 kips	
Top level force	F_t	= 0.07TV = 31.7 kips	

Table 6-2 shows the lateral displacements and drifts obtained from the analyses described above. From this table it is clear that the lateral drifts obtained using the prescribed lateral forces for all stories are less than the limiting values.

In Seismic Zone 1, P-Δ effects are potentially more significant than in higher seismic zones, because the stiffness of the lateral load resisting systems in the higher zones is required to be greater than that in the lower zones. Therefore, it is prudent to check the P-Δ effects prior to any member design for Seismic Zone 1. UBC 1628.9 allows P-Δ effects to be neglected where the ratio of the secondary moment to the primary moment is less than or equal to 0.1. The results of the P-Δ check are shown in Table 6-3. Based on the calculations shown, the P-Δ effects for this frame need not be considered.

The results of the structural analysis under earthquake forces, including torsion, for an interior frame along column line 3 and shearwall B2-C2 are shown in Fig. 6-3 and Table 6-4, respectively.

Table 6-2 Lateral Displacements and Drifts Due to Seismic Zone 1 Forces

Floor Level	Total Lateral Displacement (in.)	Interstory Drift (in.)
R	0.28	0.03
10	0.25	0.03
9	0.22	0.03
8	0.19	0.04
7	0.15	0.03
6	0.12	0.03
5	0.09	0.03
4	0.06	0.03
3	0.03	0.01
2	0.02	0.02

Table 6-3 P-Δ Check under Seismic Zone 1 Forces

Floor Level	Gravity Loads* Σw_i (kips)	Story Drift Δ_i (in.)	Story Shear V_i (kips)	Story Height h_{si} (in.)	$\Sigma w_i \Delta_i / V_i h_{si}$
R	3766	0.03	115.2	156	0.006
10	7618	0.03	197.9	156	0.007
9	11,470	0.03	272.1	156	0.008
8	15,322	0.04	337.8	156	0.012
7	19,174	0.03	395.1	156	0.009
6	23,026	0.03	443.9	156	0.010
5	26,878	0.03	484.3	156	0.011
4	30,730	0.03	516.2	156	0.011
3	34,582	0.01	539.6	156	0.004
2	38,873	0.02	555.6	276	0.005

* Includes floor weight, superimposed dead load, and reduced roof and floor live loads

Table 6-4 Reactions on Shearwall B2-C2 Due to Seismic Zone 1 Forces

Level	Shearwall B2-C2			
	Axial Load (kips)	Bending Moment (ft-kips)		Shear Force (kips)
		Top	Bottom	
R	1.7	145	154	-23.3
10	3.6	3	609	-47.3
9	5.4	-452	1322	-67.0
8	7.1	-1164	2266	-84.8
7	8.6	-2118	3423	-100.5
6	10.3	-3278	4759	-113.9
5	11.9	-4626	6257	-125.5
4	13.2	-6139	7893	-134.9
3	14.1	-7797	9641	-142.0
2	14.8	-9571	13,011	-149.7

R	-9.5	-10.7	-3.8	-2.8	-7.3
10	-13.1	-13.7	-6.4	-5.5	-10.0
9	-12.6	-13.3	-6.2	-5.3	-9.7
8	-12.7	-13.4	-6.5	-5.5	-9.7
7	-12.4	-13.1	-6.4	-5.5	-9.4
6	-11.9	-12.6	-6.4	-5.5	-9.0
5	-11.0	-11.6	-6.1	-5.3	-8.3
4	-9.8	-10.3	-5.7	-4.9	-7.4
3	-8.2	-8.6	-5.0	-4.3	-6.2
2	-5.6	-6.1	-3.7	-3.1	-4.1

Bending Moments in Beams (ft-kips)

R	-0.7	-0.7	-0.2	-0.2	-0.5
10	-1.0	-1.0	-0.4	-0.4	-0.7
9	-0.9	-0.9	-0.4	-0.4	-0.7
8	-0.9	-0.9	-0.4	-0.4	-0.7
7	-0.9	-0.9	-0.4	-0.4	-0.7
6	-0.9	-0.9	-0.4	-0.4	-0.6
5	-0.8	-0.8	-0.4	-0.4	-0.6
4	-0.7	-0.7	-0.4	-0.4	-0.5
3	-0.6	-0.6	-0.3	-0.3	-0.4
2	-0.4	-0.4	-0.2	-0.2	-0.3

Shear Forces in Beams (kips)

R	Col. A3	Col. C3
R	9.4 / 7.2	9.5 / 8.1
10	5.9 / 6.2	7.1 / 7.2
9	6.4 / 6.4	7.4 / 7.4
8	6.3 / 6.3	7.3 / 7.4
7	6.0 / 6.2	7.1 / 7.3
6	5.6 / 5.9	6.7 / 7.1
5	5.1 / 5.5	6.2 / 6.6
4	4.2 / 4.8	5.3 / 5.9
3	3.4 / 4.5	4.4 / 5.4
2	1.1 / 2.4	1.4 / 2.6

Bending Moments in Columns (ft-kips)

	Col. A3	Col. C3
R	-1.3	-1.4
10	-1.0	-1.1
9	-1.0	-1.1
8	-1.0	-1.1
7	-0.9	-1.1
6	-0.9	-1.1
5	-0.8	-1.0
4	-0.7	-0.9
3	-0.6	-0.8
2	-0.2	-0.2

Shear Forces in Columns (kips)

	Col. A3	Col. C3
R	0.7	-0.3
10	1.7	-0.6
9	2.6	-0.9
8	3.5	-1.3
7	4.4	-1.6
6	5.3	-1.9
5	6.1	-2.3
4	6.8	-2.6
3	7.4	-2.8
2	7.8	-3.0

Axial Forces in Columns (kips)

Figure 6-3 Results of Analysis of Interior Frame along Column Line 3 under Seismic Zone 1 Forces

6.2.2   Design of Structural Members

The objective is to determine the required flexural and shear reinforcement for a typical beam and a typical column of an interior frame, and for a typical shearwall. The structural members will be designed for dead and live loads, in combination with UBC Seismic Zone 1 forces.

The gravity loads acting on the beams along column line 3 are as follows:

    Tributary area = 784 ft^2

Dead Loads:

Beams and slab	= 99.4 psf
Superimposed	= 32 psf
Total	= 131.4 psf
Load per foot	= 131.4 × 784/28 = 3679 plf

Live Load:

Reduction factor	= r(A - 150) = 0.08 × (784 - 150) = 50.7% > 40%
Load per foot	= (1.0 - 0.4) × 784 × 50/28 = 840 plf

The factored load due to gravity on a beam of an interior frame is given by:

$$w_u = (1.4 \times 3.68) + (1.7 \times 0.84) = 6.58 \text{ kips/ft}$$

Using the coefficients in UBC 1908.3.3, the factored moments due to gravity are:

• End Span

Exterior negative moment $= w_u \ell_n^2/16 = 6.58 \times (28 - 18/12)^2/16 = 288.8$ ft-kips

Positive moment $= w_u \ell_n^2/14 = 6.58 \times (28 - 18/12)^2/14 = 330.1$ ft-kips

Interior negative moment $= w_u \ell_n^2/10 = 6.58 \times (28 - 18/12)^2/10 = 462.1$ ft-kips

• Interior Span

Negative moment $= w_u \ell_n^2/11 = 6.58 \times (28 - 18/12)^2/11 = 420.1$ ft-kips

Positive moment $= w_u \ell_n^2/16 = 6.58 \times (28 - 18/12)^2/16 = 288.8$ ft-kips

Table 6-5 shows the bending moments for different combinations of gravity and seismic loads on a beam on the tenth floor in an interior frame. The gravity load moments are obtained from the calculations above and the seismic moments are taken from Fig. 6-3.

The different combinations of axial loads and bending moments corresponding to the appropriate load combinations are listed in Table 6-6 for an exterior column of an interior frame between the third and the fourth floors. The effects of seismic forces are taken from Fig. 6-3.

Shearwall B2-C2 is 29.5 ft long from end to end. Boundary elements are not required for shearwalls in Seismic Zone 1. The shearwalls in the example building have columns at the ends, basically to maintain regularity of the plan layout. The end columns are 18 in. by 18 in., and the wall is 12 in. thick. Design forces for shearwall B2-C2 resulting from the different load combinations are listed in Table 6-7.

Table 6-5 Summary of Factored Bending Moments for a Beam on the 10th Floor of an Interior Frame (Seismic Zone 1)

(1) U = 1.4D + 1.7L   Eq. (9-1), UBC 1909.2.1
(2) U = 0.75 [1.4D + 1.7L ± 1.87E]   Eq. (9-2), UBC 1909.2.3
(3) U = 0.9D ± 1.43E   Eq. (9-3), UBC 1909.2.3

| Load Combination | Bending Moment (ft-kips) | | | | | |
| | Beam A3-B3 | | | Beam B3-C3 | | |
	Exterior Negative	Positive	Interior Negative	Negative	Positive	Negative
1	-288.8	330.1	-462.1	-420.1	288.8	-420.1
2: sidesway right	-192.8	247.2	-365.8	-306.1	216.6	-322.8
sidesway left	-235.0	248.0	-327.4	-324.1	216.6	-307.4
3: sidesway right	-126.6	165.7	-252.2	-202.3	145.4	-219.3
sidesway left	-164.1	166.6	-213.0	-220.6	145.4	-203.6

Table 6-6 Summary of Factored Axial Loads and Bending Moments for an Exterior Column in an Interior Frame between the 3rd and the 4th Floors (Seismic Zone 1)

(1) U = 1.4D + 1.7L   Eq. (9-1), UBC 1909.2.1
(2) U = 0.75 [1.4D + 1.7L ± 1.87E]   Eq. (9-2), UBC 1909.2.3
(3) U = 0.9D ± 1.43E   Eq. (9-3), UBC 1909.2.3

| Load Combination | Column A3 | | |
| | Axial Load (kips) | Bending Moment (ft-kips) | |
		Top	Bottom
1	785	-144.4	144.4
2: sidesway right	579	-102.4	101.6
sidesway left	598	-114.2	115.0
3: sidesway right	413	-66.7	65.9
sidesway left	432	-78.7	79.6

Table 6-7 Summary of Factored Axial Loads, Bending Moments and Shear Forces for Shearwall B2-C2 between Grade and the 2nd Floor (Seismic Zone 1)

(1) U = 1.4D + 1.7L   Eq. (9-1), UBC 1909.2.1
(2) U = 0.75 [1.4D + 1.7L ± 1.87E]   Eq. (9-2), UBC 1909.2.3
(3) U = 0.9D ± 1.43E   Eq. (9-3), UBC 1909.2.3

| Load Combination | Shearwall B2-C2 | | | |
| | Axial Load (kips) | Bending Moments (ft-kips) | | Shear Force (kips) |
		Top	Bottom	
1	4356	—	—	—
2: sidesway right	3246	-13,423	18,215	-209.6
sidesway left	3288	13,423	-18,215	-209.6
3: sidesway right	2420	-13,687	18,606	-214.1
sidesway left	2462	13,687	-18,606	-214.1

6.2.2.1 General Code Requirements

In Seismic Zone 1, structural members are designed in accordance with all the applicable provisions of UBC 1901 through 1918; the ductility related detailing requirements of UBC 1921 are not applicable in Zone 1. The UBC requirements concerning orthogonal effects also are not applicable.

6.2.2.2 Proportioning and Detailing a Flexural Member of an Interior Frame

6.2.2.2.1 Required Flexural Reinforcement

The maximum negative moment at an interior support is 462.1 ft-kips (see Table 6-5). The required area of steel, ignoring the effects of any compression reinforcement, is $A_s = 6.22$ in.^{2} Use 5-#10 bars at this location ($\phi M_n = 469.7$ ft-kips).

Check limitations on the reinforcement ratio:

$$\rho \quad = A_s/bd = 6.35/(18 \times 19.56) = 0.0180$$

$$\rho_{max} = 0.75\,\rho_{bal} = 0.0213 > 0.0180 > 200/f_y = 0.0033 \quad \text{O.K.}$$

1910.3.3

1910.5.1

The maximum negative moment in the beam at the exterior support is 288.8 ft-kips. Use 4-#9 bars, giving a design moment strength of 316.7 ft-kips.

The maximum positive moment at midspan is 330.1 ft-kips. Use 5-#9 bars ($\phi M_n = 384.8$ ft-kips).

6.2.2.2.2 Required Shear Reinforcement

The loading to be used in determining shear reinforcement requirements in Seismic Zone 1 differs from that used in higher seismic zones. In Seismic Zones 3 and 4, the probable moment strengths of relevant sections are used for the end moments. In Seismic Zone 2, the nominal moment strengths are used. In Seismic Zone 1, the shear reinforcement requirements are determined from the applied loading only.

The shear forces caused by gravity and seismic loads at a distance d from the face of an interior support are:

$$V_D \quad = 1.15 \times (w_D \times \ell_n)/2 - (w_D \times d)$$

$$= 1.15 \times (3.68 \times 26.5/2) - (3.68 \times 19.56/12) = 50.1 \text{ kips}$$

$$V_L \quad = 1.15 \times (w_L \times \ell_n)/2 - (w_L \times d)$$

$$= 1.15 \times (0.84 \times 26.5/2) - (0.84 \times 19.56/12) = 11.4 \text{ kips}$$

$$V_E \quad = 1.0 \text{ kip (see Fig. 6-3)}$$

Load combinations:

$\qquad$ (1) $V_u \quad = 1.4V_D + 1.7V_L = 89.5$ kips (governs)

$\qquad$ (2) $V_u \quad = 0.75 [1.4V_D + 1.7V_L + 1.87V_E] = 68.5$ kips

$\qquad$ (3) $V_u \quad = 0.9\ V_D + 1.43V_E = 46.5$ kips

The shear force that is carried by the concrete is:

$$V_c \quad = 2\sqrt{f_c'} \times bd$$

$$= 2\sqrt{4000} \times 18 \times 19.56/1000 = 44.5 \text{ kips} \qquad \textit{1911.3.1.1}$$

The shear force that must be carried by the stirrups is:

$$V_s \quad = V_u/\phi - V_c = (89.5/0.85) - 44.5 = 60.8 \text{ kips} \qquad \begin{matrix}\textit{1911.1.1} \\ \textit{1911.5.4.3}\end{matrix}$$

$$< 4\sqrt{f_c'} \times bd = 89.0 \text{ kips}$$

Using #4 stirrups, the required spacing near the interior support is:

$$s \quad = A_v \times f_y \times d/V_s \qquad \textit{1911.5.6.2}$$

$$= 0.4 \times 60 \times 19.56/60.8 = 7.7 \text{ in.}$$

The maximum allowable stirrup spacing s_{max} is:

$$s_{max} \quad = A_s \times f_y/(50 \times b_w) = 0.4 \times 60,000/(50 \times 18) = 26.7 \text{ in.} \qquad \textit{1911.5.5.3}$$

$$= \frac{d}{2} = \frac{19.56}{2} = 9.8 \text{ in. (governs)}$$

$$= 24.0 \text{ in.}$$

Therefore, use #4 stirrups at 7 in.

Stirrups can be discontinued when $V_u < \phi V_c/2 = 18.9$ kips. This will occur at approximately 11.5 ft from the column face. Therefore, stirrups will be spaced at 7 in. throughout the entire length of the beam.

The design shear strength of the beam with #4 stirrups at a 7 in. spacing is:

$$\phi V_n = \phi (V_s + V_c) = 0.85(67.1 + 44.5) = 94.8 \text{ kips}$$

Two-thirds of this strength is:

$$\frac{2}{3}(\phi V_n) = \frac{2}{3} \times 94.8 = 63.2 \text{ kips}$$

The location where the factored shear is equal to 63.2 kips is 4.6 ft from the face of the interior support. Therefore, any reinforcing bars terminated beyond 4.6 ft from the column face will meet the requirements of UBC 1912.10.5.1, since $2/3 (\phi V_n) > V_u$.

6.2.2.2.3 Reinforcing Bar Cutoff Points and Splices

The negative reinforcement at the interior support consists of 5-#10 bars. The location where three of the five bars can be terminated away from an interior support will be determined. The design moment strength of 2-#10 bars is 209.3 ft-kips. The distance from the face of the right support to where the factored moment is reduced to 209.3 ft-kips must be determined. Based on the various loading conditions listed in Table 6-5, the location where the bars can be cut off is obtained by summing moments about section a-a in Fig. 6-4 and equating these to 209.3 ft-kips. Three of the 5-#10 top bars near the interior support can be discontinued at $(x + d) = (3.0 + 19.56/12) = 4.6$ ft from the face of the interior support. Similarly, 2-#9 bars at the exterior support can be discontinued at 3.2 ft away from that support.

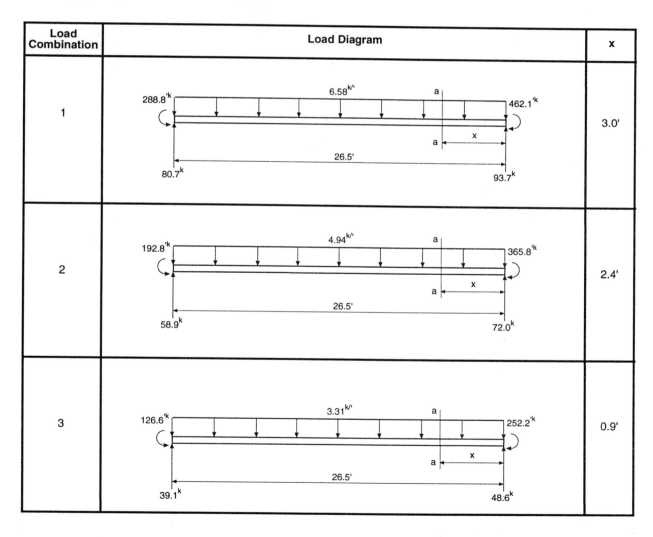

Figure 6-4 Load Diagrams for Cutoff Location of Negative Bars at Interior Support of Beam A3-B3
(Seismic Zone 1)

Five #9 bars are required at midspan for positive moment strength. The location where two of the five bars at midspan can be terminated will be determined. The design moment strength, ϕM_n, provided by 2-#9 bars is 167.2 ft-kips. The moments at the interior and the exterior columns are equal to 462.1 ft-kips and 288.8 ft-kips, respectively, when the gravity load is equal to 6.58 kips/ft. The location where the bars can be cut off is obtained by summing moments about section a-a in Fig. 6-5:

$$\frac{6.58x^2}{2} + 462.1 - 93.7x = -167.2$$

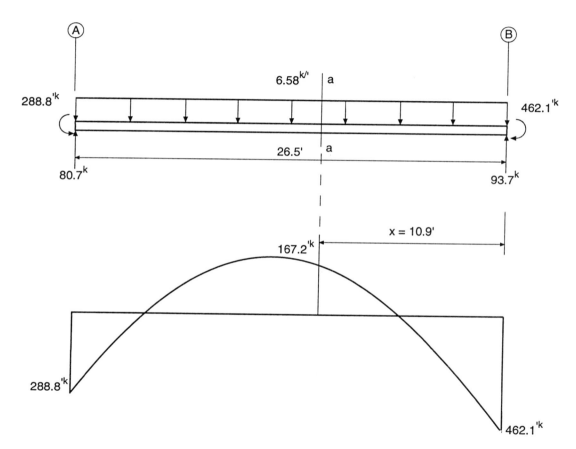

Figure 6-5 Moment Diagram for Cutoff Location of Positive Bars in Beam A3-B3 (Seismic Zone 1)

Solution of this equation gives x = 10.9 ft. Two of the 5-#9 bars can be terminated at [(26.5/2) - 10.9 + (19.56/12)] = 4.0 ft from the midspan section of the beam. This length must be compared to the one corresponding to the required development length ℓ_d (see below).

The reinforcing bars must be properly developed at the supports. The development length is the product of the basic development length and the factors previously presented in Table 3-13. The basic development length for a #9 bar is 38.0 in. and for a #10 bar it is 48.2 in.

The information required for determining the development length factors is:

	#9 Top Bars	#10 Top Bars	#9 Bottom Bars
Minimum cover to bars being developed	2.0 in. (1.77 d_b)	2.0 in. (1.58 d_b)	2.0 in. (1.77 d_b)
Minimum side cover	large (slab)	large (slab)	2.0 in. (1.77 d_b)
Clear spacing between bars being developed	3.2 in. (2.80 d_b)	1.9 in. (1.51 d_b)	2.1 in. (1.85 d_b)

Transverse reinforcement meets requirements of UBC 1911.5.4 and 1911.5.5.3 for both top and bottom bars.

Factor I has a value of 1.4 for the #9 top bars and 2.0 for the other bars. Using the other applicable factors gives:

#9 top bar $\ell_d = 1.4 \times (1.3 \times 1.0 \times 1.0 \times 1.0 \times 38.0)/12 = 5.8$ ft

#10 top bar $\ell_d = 2.0 \times (1.3 \times 1.0 \times 1.0 \times 1.0 \times 48.2)/12 = 10.4$ ft

#9 bottom bar $\ell_d = 2.0 \times (1.0 \times 1.0 \times 1.0 \times 1.0 \times 38.0)/12 = 6.3$ ft

Based on the #10 top bars, the information required for determining the splice length is:

	Top Bars	Bottom Bars
Stress level	< 0.5 f_y (midspan)	< 0.5 f_y (support)
Splice type	A	A
Minimum top or bottom cover to bars being spliced	2.0 in. (1.58 d_b)	2.0 in. (1.77 d_b)
Minimum side cover	large (slab)	2.0 in. (1.77 d_b)
Clear spacing between bars being spliced	11.5 in. (9.0 d_b)	11.7 in. (10.4 d_b)

Transverse reinforcement meets requirements of UBC 1911.5.4 and 1911.5.5.3.

This gives a value of 0.80 for Factor I for the top splice and 1.0 for the bottom splice; therefore,

Top bar splice length (Class A) = $1.0 \times (0.80 \times 1.3 \times 1.0 \times 1.0 \times 1.0 \times 48.2) = 50.1$ in.

Bottom bar splice length (Class A) = $1.0 \times (1.0 \times 1.0 \times 1.0 \times 1.0 \times 1.0 \times 38.0) = 38.0$ in.

Figure 6-6 shows the reinforcement details for this beam.

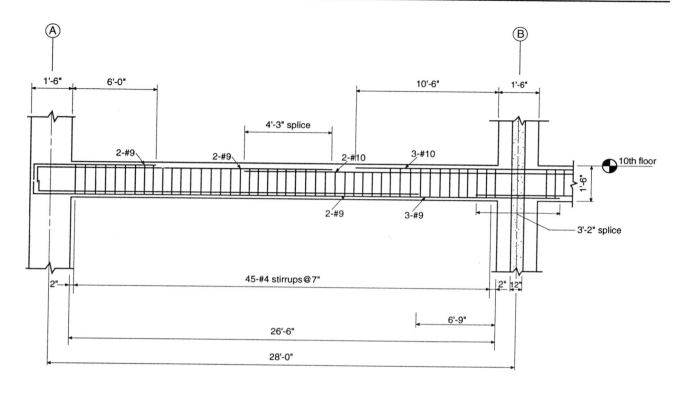

Figure 6-6 Reinforcement Details for Beam A3-B3 (Seismic Zone 1)

6.2.2.3   Proportioning and Detailing an Exterior Column

6.2.2.3.1 General Requirements

This section will present the design of an exterior column of an interior frame supporting the fourth floor of the structure. The maximum axial load on this column at this level is 785 kips (see Table 6-6).

An 18-in. square tied column with 8-#9 bars will be adequate for the load combinations listed in Table 6-6. The design strength interaction diagram for this column is shown in Fig. 6-7.

6.2.2.3.2 Transverse Reinforcement Requirements

Tie reinforcement for compression members must conform to the following maximum spacing requirements, for #9 longitudinal bars and #3 ties (UBC 1907.10.5):

$$16 \times \text{longitudinal bar diameter} = 16 \times 1.128 = 18.1 \text{ in.}$$

$$48 \times \text{tie bar diameter} = 48 \times 0.375 = 18.0 \text{ in.} \quad \text{(governs)}$$

$$\text{minimum column dimension} = 18.0 \text{ in.}$$

The column shear caused by gravity loads is determined from the beam moments on the interior face of the column, divided by the clear column height. The unfactored column shear caused by the dead load is:

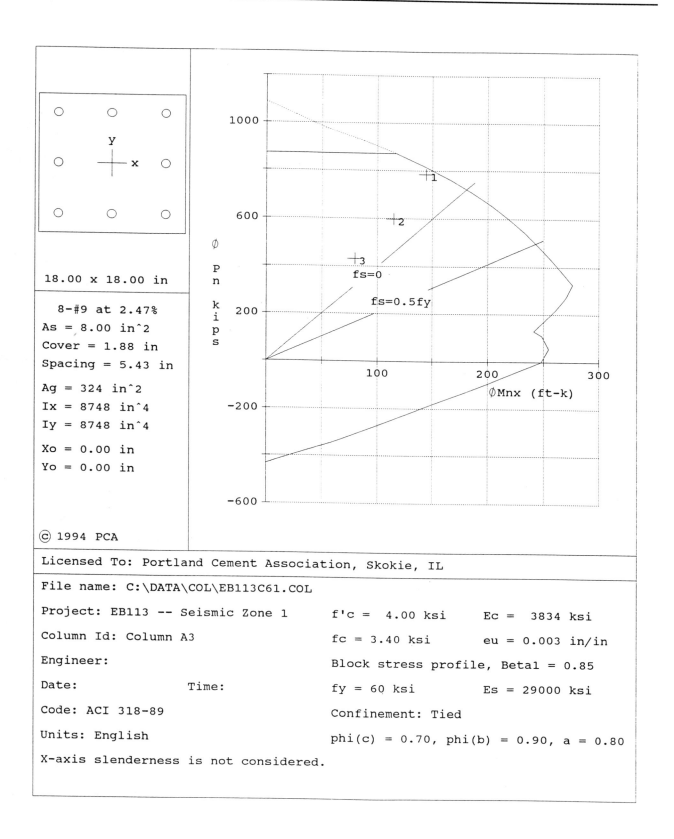

Figure 6-7 Design Strength Interaction Diagram for Column A3 (Seismic Zone 1)

$$V_D = \frac{w_D \ell_n^2}{16 \ell_c}$$

$$= \frac{3.68 \times (28 - 18/12)^2}{16 \times 11.17} = 14.5 \text{ kips}$$

The unfactored column shear caused by live loads is:

$$V_L = \frac{w_L \ell_n^2}{16 \ell_c}$$

$$= \frac{0.84 \times (28 - 18/12)^2}{16 \times 11.17} = 3.3 \text{ kips}$$

From Fig. 6-3:

$$V_E = 0.7 \text{ kips}$$

The load combinations for the column are:

(1) $V_u = 1.4 V_D + 1.7 V_L = 25.9 \text{ kips}$ (governs)

(2) $V_u = 0.75 [1.4 V_D + 1.7 V_L + 1.87 V_E] = 20.4 \text{ kips}$

(3) $V_u = 0.9 V_D + 1.43 V_E = 14.1 \text{ kips}$

For members subject to axial compression, the shear strength provided by the concrete is:

$$V_c = 2 \left(1 + \frac{N_u}{2000 A_g} \right) \sqrt{f_c'} \ b_w d \qquad\qquad 1911.3.1.2$$

From Table 6-6, the minimum factored axial load on the column is 413 kips. Therefore,

$$\phi V_c = 0.85 \times 2 \left(1 + \frac{413,000}{2000 \times 324} \right) \sqrt{4000} \times 18 \times 15.6/1000$$

$$= 49.4 \text{ kips}$$

Since $\phi V_c = 49.4$ kips is less than $2 V_u = 51.8$ kips, minimum shear reinforcement must be provided throughout the length of the column (UBC 1911.5.5.3). Assuming #3 ties, the maximum allowable spacing is:

$$s = \frac{A_v f_y}{50 b_w} = \frac{0.22 \times 60,000}{50 \times 18} = 14.7 \text{ in.}$$

Use #3 ties spaced at 14 in.

From a review of the load combinations in Table 6-6 and the design strength interaction diagram given in Fig. 6-7 for this column, it can be seen that the reinforcing bars will be subjected to compressive stresses only. Therefore, a compression splice will be used. The length of a compression splice for #9 bars is (UBC 1912.16.1):

$$0.0005f_yd_b = 0.0005 \times 60,000 \times 1.128 = 33.8 \text{ in.}$$

Use a 34 in. lap splice.

Figure 6-8 shows the reinforcement details for this column.

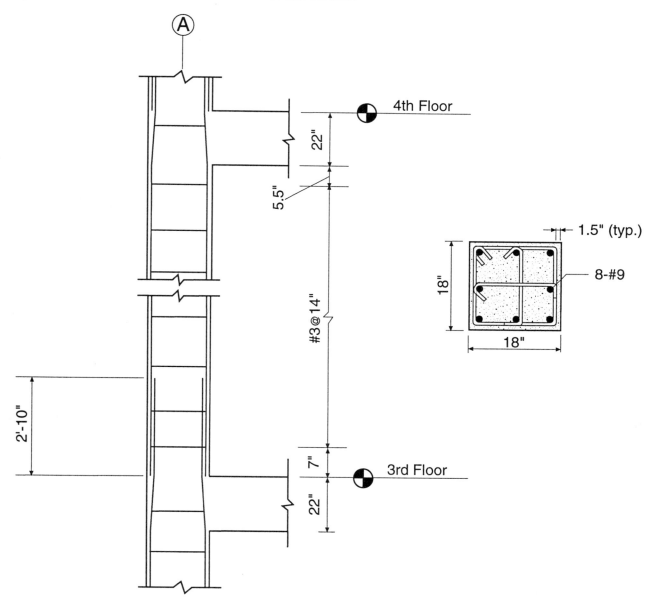

Figure 6-8 Reinforcement Details for Column A3 (Seismic Zone 1)

6.2.2.4   Proportioning and Detailing a Shearwall

6.2.2.4.1 Design of Shearwall B2-C2

The design shear strength of a shearwall is limited to:

$$\phi V_n = \phi 10\sqrt{f_c'} \times hd \qquad\qquad\qquad 1911.10.3$$

where d equals 80% of the length of the wall (UBC 1911.10.4). In this case, ϕV_n is:

$$\phi V_n = 0.85 \times 10\sqrt{4000} \times 12 \times (0.8 \times 29.5 \times 12)/1000 = 1826.9 \text{ kips}$$

Using Table 6-7, the maximum factored shear at the base of the wall is:

$$V_u \quad = 214.1 \text{ kips} < 1826.9 \text{ kips} \quad \text{O.K.}$$

The shear strength provided by the concrete is:

$$\phi V_c = \phi 2\sqrt{f_c'} \times hd \qquad\qquad\qquad 1911.10.5$$

$$= 0.85 \times 2\sqrt{4000} \times 12 \times (0.8 \times 29.5 \times 12)/1000$$

$$= 365.4 \text{ kips} > V_u = 214.1 \text{ kips}$$

Therefore, use maximum allowable spacing of horizontal shear reinforcement. The maximum spacing for #5 reinforcing bars is:

$$s_{max} \quad = \ell_w/5 = (29.5 \times 12)/5 = 70.8 \text{ in.} \qquad\qquad 1911.10.9.3$$

$$= 3h = 3 \times 12 = 36.0 \text{ in.}$$

$$= 18.0 \text{ in. (governs)}$$

The ratio of horizontal shear reinforcement area to gross concrete area of the vertical cross section must not be less than 0.0025 (UBC 1911.10.9.2).

With a #5 horizontal bar on each face:

$$\rho_h \quad = 0.62/(12 \times 18) = 0.0029 > 0.0025 \quad \text{O.K.}$$

Therefore, use two layers of #5 horizontal bars spaced at 18 in.

The ratio of vertical reinforcement area to gross area of horizontal section must not be less than:

$$\rho_n \quad = 0.0025 + 0.5\,(2.5 - h_w/\ell_w)\,(\rho_h - 0.0025) \qquad\qquad 1911.10.9.4$$

$$= 0.0025 + 0.5\,[2.5 - (140/29.5)] \times (0.0029 - 0.0025)$$

$$= 0.0021 < 0.0025$$

The required vertical steel area is:

$$A_n = 0.0025 \times 12 \times 12 = 0.36 \text{ in.}^2/\text{ft}$$

Assuming two layers of #5 bars, the required spacing is $(0.62 \times 12)/0.36 = 20.7$ in.

The maximum spacing for the vertical #5 bars is:

$$s_{max} = \ell_w/3 = (29.5 \times 12)/3 = 118.0 \text{ in.}$$

1911.10.9.5

$$= 3h = 3 \times 12 = 36.0 \text{ in.}$$

$$= 18.0 \text{ in. (governs)}$$

Use two layers of #5 vertical bars spaced at 18 in.

A design strength interaction diagram for the shearwall is shown in Fig. 6-9. Note that the two ends of the wall have been reinforced with 12-#5 bars. The interaction diagram shows that the shearwall has adequate strength to carry the load combinations listed in Table 6-7.

Figure 6-10 shows the reinforcement details for shearwall B2-C2.

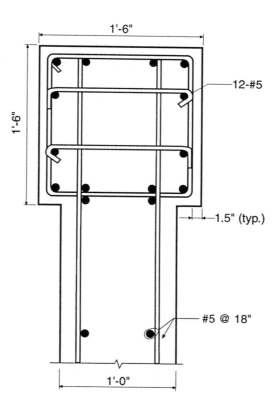

Figure 6-10 Reinforcement Details for Shearwall B2-C2 (Seismic Zone 1)

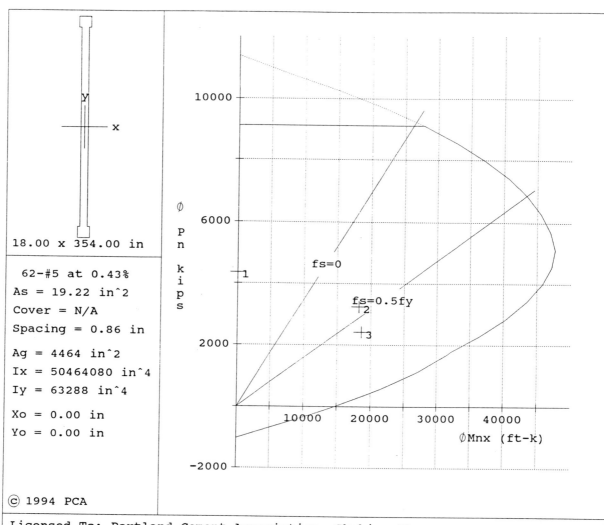

18.00 x 354.00 in

62-#5 at 0.43%
As = 19.22 in^2
Cover = N/A
Spacing = 0.86 in

Ag = 4464 in^2
Ix = 50464080 in^4
Iy = 63288 in^4

Xo = 0.00 in
Yo = 0.00 in

© 1994 PCA

Licensed To: Portland Cement Association, Skokie, IL

File name: C:\DATA\COL\EB113C62.COL

Project: EB113 -- Seismic Zone 1   f'c = 4.00 ksi   Ec =   3834 ksi

Column Id: Shearwall B2-C2   fc = 3.40 ksi   eu = 0.003 in/in

Engineer: Block stress profile, Beta1 = 0.85

Date: Time: fy = 60 ksi   Es = 29000 ksi

Code: ACI 318-89   Confinement: Tied

Units: English   phi(c) = 0.70, phi(b) = 0.90, a = 0.80

X-axis slenderness is not considered.

Figure 6-9 Design Strength Interaction Diagram for Shearwall B2-C2 (Seismic Zone 1)

6.3 Design for Wind

This section presents the design of the example building for wind forces corresponding to a fastest mile wind speed of 70 mph (Exposure B) and an importance factor I = 1.0, computed in accordance with UBC 1613. The wind forces are shown in Table 6-8. The same member sizes and material strengths used in the previous design (Seismic Zone 1) will be utilized. The results of the analysis for the building subjected to the wind forces are shown in Fig. 6-11 and Table 6-9.

Table 6-8 Wind Forces on Example Building

Floor Level	Height (ft)	Tributary Height (ft)	Windward			Leeward			Total Force (kips)
			C_e	Pressure (psf)	Force (kips)	C_e	Pressure (psf)	Force (kips)	
R	140	6.5	1.26	12.7	11.6	1.26	7.9	7.2	18.8
10	127	13.0	1.22	12.3	22.4	1.26	7.9	14.4	36.8
9	114	13.0	1.18	11.9	21.7	1.26	7.9	14.4	36.1
8	101	13.0	1.13	11.4	20.8	1.26	7.9	14.4	35.2
7	88	13.0	1.08	10.9	19.8	1.26	7.9	14.4	34.2
6	75	13.0	1.02	10.3	18.8	1.26	7.9	14.4	33.2
5	62	13.0	0.96	9.7	17.7	1.26	7.9	14.4	32.1
4	49	13.0	0.90	9.1	16.6	1.26	7.9	14.4	31.0
3	36	13.0	0.80	8.1	14.7	1.26	7.9	14.4	29.1
2	23	18.0	0.70	7.1	17.9	1.26	7.9	19.9	37.8

6.3.1 Design Forces

The objective is to determine the required flexural and shear reinforcement for a typical beam and a typical column of an interior frame, and for a shearwall.

The gravity loads will be the same as those determined in the Seismic Zone 1 example, since the member sizes are the same. In Section 6.2.2, the factored moments due to gravity loads for a beam of an interior frame were determined to be:

• End Span

    Exterior negative moment   = 288.8 ft-kips

    Positive moment   = 330.1 ft-kips

    Interior negative moment   = 462.1 ft-kips

• Interior Span

    Negative moment   = 420.1 ft-kips

    Positive moment   = 288.8 ft-kips

R	-3.7	-4.1	-1.3	-0.9	-2.8
10	-5.1	-5.4	-2.4	-2.0	-3.9
9	-5.0	-5.3	-2.4	-2.1	-3.9
8	-5.1	-5.4	-2.6	-2.2	-3.9
7	-5.1	-5.4	-2.6	-2.3	-3.9
6	-5.0	-5.3	-2.7	-2.3	-3.8
5	-4.7	-5.0	-2.6	-2.3	-3.6
4	-4.3	-4.6	-2.6	-2.2	-3.2
3	-3.7	-3.9	-2.3	-2.0	-2.8
2	-2.7	-2.9	-1.9	-1.5	-1.9

Bending Moments in Beams (ft-kips)

R	-0.3	-0.3	-0.1	-0.1	-0.2
10	-0.4	-0.4	-0.2	-0.2	-0.3
9	-0.4	-0.4	-0.2	-0.2	-0.3
8	-0.4	-0.4	-0.2	-0.2	-0.3
7	-0.4	-0.4	-0.2	-0.2	-0.3
6	-0.4	-0.4	-0.2	-0.2	-0.3
5	-0.4	-0.4	-0.2	-0.2	-0.3
4	-0.3	-0.3	-0.2	-0.2	-0.2
3	-0.3	-0.3	-0.2	-0.2	-0.2
2	-0.2	-0.2	-0.1	-0.1	-0.1

Shear Forces in Beams (kips)

	Col. A3	Col. C3
R	3.7	3.5
	2.8	2.9
10	2.4	2.8
	2.4	2.8
9	2.6	2.9
	2.6	2.9
8	2.6	3.0
	2.6	3.0
7	2.5	3.0
	2.6	3.0
6	2.4	2.9
	2.5	3.0
5	2.2	2.7
	2.4	2.9
4	1.9	2.4
	2.1	2.7
3	1.6	2.0
	2.0	2.4
2	0.6	0.9
	1.3	1.5

Bending Moments in Columns (ft-kips)

	Col. A3	Col. C3
R	-0.5	-0.5
10	-0.4	-0.4
9	-0.4	-0.5
8	-0.4	-0.5
7	-0.4	-0.5
6	-0.4	-0.5
5	-0.4	-0.4
4	-0.3	-0.4
3	-0.3	-0.3
2	-0.1	-0.1

Shear Forces in Columns (kips)

	Col. A3	Col. C3
R	0.3	-0.1
10	0.7	-0.2
9	1.0	-0.4
8	1.4	-0.5
7	1.8	-0.6
6	2.1	-0.8
5	2.5	-0.9
4	2.8	-1.0
3	3.1	-1.2
2	3.3	-1.3

Axial Forces in Columns (kips)

Figure 6-11 Results of Analysis of Interior Frame along Column Line 3 under Wind forces

Table 6-9 Reactions on Shearwall B2-C2 Due to Wind Forces

| Level | Shearwall B2-C2 | | | |
| | Axial Load (kips) | Bending Moment (ft-kips) | | Shear Force (kips) |
		Top	Bottom	
R	0.7	58	-35	-1.8
10	1.4	97	65	-12.5
9	2.1	-4	290	-22.2
8	2.9	-227	640	-31.8
7	3.5	-578	1109	-41.0
6	4.3	-1049	1698	-50.1
5	4.7	-1643	2406	-58.8
4	5.4	-2351	3229	-67.6
3	5.9	-3186	4168	-75.6
2	6.1	-4137	6154	-87.7

Table 6-10 shows the factored moments for different combinations of gravity and wind loads for a beam on the tenth floor of an interior frame. The gravity load moments are given above and the wind load moments are taken from Fig. 6-11.

Table 6-10 Summary of Factored Bending Moments for a Beam on the 10th Floor of an Interior Frame (Seismic Zone 0)

(1)	$U = 1.4D + 1.7L$	Eq. (9-1), UBC 1909.2.1
(2)	$U = 0.75 [1.4D + 1.7L \pm 1.7W]$	Eq. (9-2), UBC 1909.2.2
(3)	$U = 0.9D \pm 1.3W$	Eq. (9-3), UBC 1909.2.2

| Load Combination | | Bending Moment (ft-kips) | | | | | |
| | | Beam A3-B3 | | | Beam B3-C3 | | |
		Exterior Negative	Positive	Interior Negative	Negative	Positive	Negative
1		-288.8	330.1	-462.1	-420.1	288.8	-420.1
2:	sidesway right	-210.1	247.3	-353.5	-312.0	216.6	-320.1
	sidesway left	-223.1	247.8	-339.7	-318.1	216.6	-310.1
3:	sidesway right	-138.7	165.9	-239.6	-208.3	145.4	-216.5
	sidesway left	-152.0	166.4	-225.5	-214.5	145.4	-206.3

The different load combinations are listed in Table 6-11 for an exterior column of an interior frame between the third and the fourth floors. The effects of the wind forces are taken from Fig. 6-11. The gravity loads are the same as those used in the Seismic Zone 1 design. Similarly, the load combinations for shearwall B2-C2 are given in Table 6-12.

Table 6-11 Summary of Factored Axial Loads and Bending Moments for an Exterior Column in an Interior Frame between the 3rd and the 4th Floors (Seismic Zone 0)

(1) U = 1.4D + 1.7L	Eq. (9-1), UBC 1909.2.1
(2) U = 0.75(1.4D + 1.7L ± 1.7W)	Eq. (9-2), UBC 1909.2.2
(3) U = 0.9D ± 1.3W	Eq. (9-3), UBC 1909.2.2

Load Combination	Column A3		
	Axial Load (kips)	Bending Moment (ft-kips)	
		Top	Bottom
1	785	-144.4	144.4
2: sidesway right	585	-105.9	105.6
sidesway left	592	-110.7	111.0
3: sidesway right	419	-70.3	70.0
sidesway left	426	-75.2	75.5

Table 6-12 Summary of Factored Axial Loads, Bending Moments and Shear Forces for Shearwall B2-C2 between Grade and the 2nd Floor (Seismic Zone 0)

(1) U = 1.4D + 1.7L	Eq. (9-1), UBC 1909.2.1
(2) U = 0.75 [1.4D + 1.7L ± 1.7W]	Eq. (9-2), UBC 1909.2.2
(3) U = 0.9D ± 1.3W	Eq. (9-3), UBC 1909.2.2

Load Combination	Shearwall B2-C2			
	Axial Load (kips)	Bending Moment (ft-kips)		Shear Force (kips)
		Top	Bottom	
1	4356	—	—	—
2: sidesway right	3259	5275	-7846	111.8
sidesway left	3275	-5275	7846	-111.8
3: sidesway right	2433	5378	-8000	114.0
sidesway left	2449	-5378	8000	-114.0

6.3.2   Design of Structural Members

When subjected to Seismic Zone 1 loads, the design of the structural members was controlled by either gravity loads or minimum reinforcement requirements. A review of the loading combinations for the structure subjected to wind loads shows that the governing load combinations involve gravity loads only. Therefore, since the member sizes and the concrete strength are the same, the required reinforcement for the members in this example is exactly the same as that determined in the previous example.

Chapter 7

Bearing Wall System

7.1 Introduction

7.1.1 General

The computation of the design loads for a reinforced concrete building utilizing the Bearing Wall System under the requirements of the Uniform Building Code (UBC), 1994 edition, is illustrated. The Bearing Wall System is composed of shearwalls that resist both gravity and lateral loads. The computations are performed for UBC Seismic Zones 4 and 2B, and also for a 70 mph wind, Exposure B.

7.1.2 Design Criteria

A typical plan and elevation of the example building are shown in Figs. 7-1 and 7-2, respectively. The building consists of 12 walls (8 T-shaped and 4 L-shaped). It is assumed that lateral forces are carried by the wall segments that are parallel to the direction of analysis. The wall thickness used in Seismic Zone 4 is 8 in. In Seismic Zone 2B, the walls are 6 in. thick.

The floor system is composed of one-way joists (5 ft modules with 16 in. rib depth, 7 in. rib width, and a 4.5 in. thick slab), spanning in the longitudinal direction of the structure. The floor in the corridor is an 8 in. thick solid slab. Other pertinent design data are as follows:

Service Loads:
    Live load: 50 psf
    Superimposed dead load: 20 psf

Material Properties:
    Concrete: f_c' = 6 ksi in Seismic Zone 4
                             = 4 ksi elsewhere

$$w_c \quad = 150 \text{ pcf}$$

Reinforcement: $\quad f_y \quad = 60 \text{ ksi}$

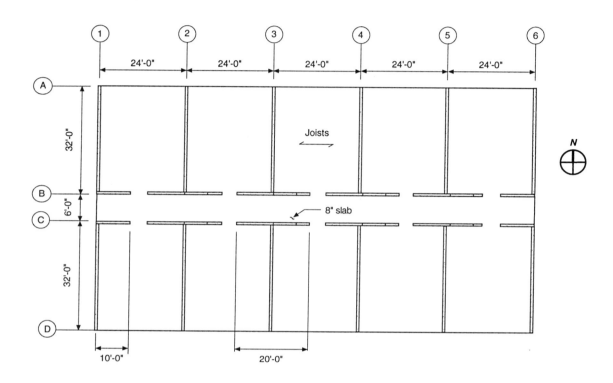

Figure 7-1 Typical Floor Plan of Example Building

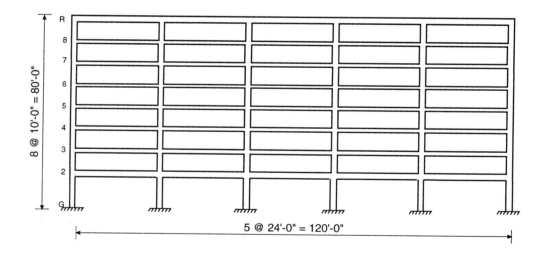

Figure 7-2 Elevation of Example Building

7.2 Design for Seismic Zone 4

On the basis of the given data and the dimensions of the building, the weights of the floors are listed in Table 7-1. It should be noted that the building has the same lateral load resisting system in both principal directions. Since the estimated period is independent of the plan dimensions of the building, the seismic forces will be the same in both the longitudinal and the transverse directions of the building. However, since the building is not symmetrical, the lateral shears produced by torsion will most likely not be equal in the two directions.

The calculations of the base shear V for the transverse and the longitudinal directions are shown at the bottom of Table 7-1. For this example, the importance factor I and the site soil coefficient S are taken equal to 1.0 and 1.2, respectively. The value of $R_w = 6$ is used in both directions. The seismic forces resulting from the distribution of the base shear in accordance with UBC Eqs. (28-6), (28-7), and (28-8) are also listed in Table 7-1.

Table 7-1 Seismic Forces for Zone 4

Floor Level	Weight, w_x (kips)	Height, h_x (ft)	$w_x h_x$ (ft-kips)	Lateral Force (kips)
R	1137	80	90,960	313.1
8	1429	70	100,030	344.3
7	1429	60	85,740	295.1
6	1429	50	71,450	245.9
5	1429	40	57,160	196.8
4	1429	30	42,870	147.6
3	1429	20	28,580	98.4
2	1429	10	14,290	49.2
Σ	11,140		491,080	1690.4

Building height	h_n	$= 80\ ft$
Building weight	W	$= 11,140\ kips$
Fundamental period	T	$= 0.020\ (h_n)^{3/4} = 0.53\ sec.$
Seismic zone factor	Z	$= 0.4$
Importance factor	I	$= 1.0$
Response coefficient	R_w	$= 6$
Soil factor	S	$= 1.2$
Coefficient	C	$= 1.25\ S/T^{2/3} = 1.72$
Base shear	V	$= ZICW/R_w = 1690.4\ kips$
Top level force	F_t	$= 0\ (T < 0.7\ sec.)$

The design story shear must be distributed to the various elements of the lateral force resisting system in proportion to their rigidities, considering the rigidity of the diaphragm. To account for uncertainties in the locations of the loads, the mass at each level must be assumed to be displaced from the calculated center of mass in each direction a distance equal to 5% of the building dimension at that level perpendicular to the direction of the force under consideration (UBC 1628.5).

The three-dimensional analysis of the structure was performed using *SAP90* [4.1]. The lateral loads in both the transverse and longitudinal directions were applied at a distance equal to 5% of the appropriate building dimension from the center of mass to allow for accidental torsion. Also, rigid diaphragms were specified at each floor level. The results of the analyses for interior transverse and longitudinal walls are summarized in Table 7-2.

Table 7-2 Results of Analyses for Interior Transverse and Longitudinal Walls under Seismic Zone 4 Forces

Transverse Direction (Stem Segment—Shearwall 2B)				
Level	Axial Load (kips)	Bending Moment (ft-kips)		Shear Force (kips)
		Top	Bottom	
R	6	96	212	-31
8	22	41	551	-59
7	46	-169	1012	-84
6	77	-511	1562	-105
5	115	-958	2174	-122
4	157	-1505	2837	-133
3	202	-2118	3516	-140
2	242	-2877	4295	-142

Shearwall Longitudinal Direction (Flange Segment—Shearwall 2B)				
Level	Axial Load (kips)	Bending Moment (ft-kips)		Shear Force (kips)
		Top	Bottom	
R	5	339	-68	27
8	9	641	-22	62
7	12	771	153	92
6	17	829	356	118
5	22	783	611	139
4	27	619	941	156
3	31	235	1438	167
2	32	-565	2245	168

For structures having a fundamental period less than 0.7 seconds, the calculated story drift must not exceed $0.04/R_w$ nor 0.005 times the story height (UBC 1628.8.2).

For $h_s = 10$ ft and $R_w = 6$:

$$\frac{0.04 h_s}{R_w} = \frac{0.04 \times 10 \times 12}{6} = 0.8 \text{ in.}$$

$$0.005 h_s = 0.005 \times 10 \times 12 = 0.6 \text{ in.} \quad \text{(governs)}$$

As shown in Table 7-3, the story drift limits are not exceeded at any floor level.

In Seismic Zone 4, P-Δ effects need not be considered as long as the story drift does not exceed $0.02 h_s / R_w$ (UBC 1628.9):

$$\frac{0.02 h_s}{R_w} = \frac{0.02 \times 10 \times 12}{6} = 0.4 \text{ in.}$$

The value is greater than any of the interstory drifts in Table 7-3, so that P-Δ effects need not be considered.

Table 7-3 Lateral Displacements and Drifts Due to Seismic Zone 4 Forces

Floor Level	Longitudinal Direction		Transverse Direction	
	Displacement (in.)	Drift (in.)	Displacement (in.)	Drift (in.)
R	0.10	0.02	0.09	0.01
8	0.08	0.01	0.08	0.01
7	0.07	0.02	0.07	0.01
6	0.05	0.01	0.06	0.01
5	0.04	0.01	0.05	0.01
4	0.03	0.01	0.04	0.02
3	0.02	0.01	0.02	0.01
2	0.01	0.01	0.01	0.01

7.2.1   Design of Shearwall 2B

Table 7-4 shows gravity loads for the flange and stem segments of the shearwall. The axial loads, bending moments, and shear forces corresponding to the appropriate load combinations are listed in Tables 7-5 and 7-6.

Table 7-4 Gravity Loads on Shearwall 2B (Seismic Zone 4)

Level	Tributary Area (ft^2)		Cumulative Dead Load (kips)		Cumulative Reduced Live Load (kips)	
	Flange	Stem	Flange	Stem	Flange	Stem
R	72	768	29	112	4	23
8	72	768	57	224	7	38
7	72	768	86	336	11	54
6	72	768	115	447	14	69
5	72	768	143	559	17	84
4	72	768	172	671	20	100
3	72	768	200	783	22	115
2	72	768	229	895	25	131

A significant change in the design of shearwalls occurred with the adoption of the 1994 edition of the Uniform Building Code. In UBC-91 Sect. 2625(f)3, the boundary elements of the shearwall, whenever required, must be designed to carry all of the factored gravity loads tributary to the wall plus the vertical force required to resist the overturning moment at the base of the wall due to the factored seismic loads. In other words, the boundary elements are to be designed completely ignoring the contribution of the web. As a result of this provisions, large boundary elements with congested reinforcement are usually obtained.

To alleviate this problem, the provisions in UBC-94 Sect. 1921.6.5 give a design procedure which utilizes the entire cross-section of the wall in resisting the factored axial loads and moments. Since the current edition of the ACI Code uses the provisions in UBC-91, and for comparison purposes, both design methods will be presented in the following sections. In Sects. 7.2.1.1.2 and 7.2.1.3.2, the section numbers given are from UBC-91 and ACI 318-89 (Revised 1992).

Table 7-5 Summary of Factored Axial Loads, Bending Moments, and Shear Forces for the Stem Segment of Shearwall 2B between Grade and the 2nd Floor (Seismic Zone 4)

(1) U = 1.4D + 1.7L	Eq. (9-1), UBC 1909.2.1
(2) U = 1.4(D + L ± E)	Eq. (9-2), UBC 1921.2.7
(3) U = 0.9D ± 1.4E	Eq. (9-3), UBC 1921.2.7

Load Combination	Axial Load (kips)	Bending Moment (ft-kips)		Shear Force (kips)
		Top	Bottom	
1	1476	—	—	—
2: sidesway right	1775	4028	-6013	199
sidesway left	1098	-4028	6013	-199
3: sidesway right	1144	4028	-6013	199
sidesway left	467	-4028	6013	-199

Table 7-6 Summary of Factored Axial Loads, Bending Moments, and Shear Forces for the Flange Segment of Shearwall 2B between Grade and the 2nd Floor (Seismic Zone 4)

(1) U = 1.4D + 1.7L	Eq. (9-1), UBC 1909.2.1
(2) U = 1.4(D + L ± E)	Eq. (9-2), UBC 1921.2.7
(3) U = 0.9D ± 1.4E	Eq. (9-3), UBC 1921.2.7

Load Combination	Axial Load (kips)	Bending Moment (ft-kips)		Shear Force (kips)
		Top	Bottom	
1	363	—	—	—
2: sidesway right	400	791	-3143	235
sidesway left	311	-791	3143	-235
3: sidesway right	251	791	-3143	235
sidesway left	161	-791	3143	-235

7.2.1.1   Design of Wall Stem Segment Based on UBC-91 Provisions

7.2.1.1.1 Boundary Element Requirements

A shearwall is required to have boundary elements if the maximum extreme fiber stress in the shearwall exceeds $0.2 f_c'$. A boundary element is a column-like member provided at the ends of a shearwall; it is contiguous with the wall. If the extreme fiber stress in the shearwall exceeds $0.2 f_c'$ and boundary elements are not provided, the entire shearwall must have transverse reinforcement meeting the requirements specified for the critical end regions of columns. It is usually more economical to provide boundary elements than to meet the transverse reinforcement requirements.

The maximum stress in a shearwall occurs at the ends of the shearwall, and is given by:

$$f_c = \frac{P_u}{A_g} + \frac{M_u(\ell_w/2)}{I_g}$$

The calculations are performed using factored loads and gross section properties.

The geometric properties of the stem segment are:

$$A_g = \ell_w \times h = 32 \times 8/12 = 21.3 \text{ ft}^2$$

$$I_g = h \times \ell_w{}^3/12 = 0.67 \times 32^3/12 = 1830 \text{ ft}^4$$

Using the factored loads from Table 7-5, the maximum concrete fiber stress is:

$$f_c = \frac{1775}{21.3} + \frac{6013(32/2)}{1830}$$

$$= 83.3 + 52.6$$

$$= 135.9 \text{ ksf} = 0.9 \text{ ksi} < 0.2\,f_c' = 0.2 \times 6 = 1.2 \text{ ksi}$$

Therefore, boundary elements are not required.

7.2.1.1.2 Design and Detailing

At least two curtains of reinforcement must be provided in the wall if the in-plane factored shear force, V_u, for the wall exceeds $2A_{cv}\sqrt{f_c'}$ (UBC 2625(f)2B; ACI 21.6.2.2). The term A_{cv} refers to the gross area of the concrete section bounded by the width and the length of the wall, rather than the width and the effective depth.

$$2A_{cv}\sqrt{f_c'} = 2 \times 8 \times 32 \times 12\sqrt{6000}/1000 = 476 \text{ kips} > V_u = 199 \text{ kips (see Table 7-5)}$$

Therefore, two curtains of reinforcement are not required.

The reinforcement ratio along the longitudinal and the transverse direction must not be less than 0.0025 and the spacing must not exceed 18 in. (UBC 2625(f)2A; ACI 21.6.2.1).

The upper limit on the design shear strength is:

$$\phi V_n = \phi\, 8 A_{cv}\sqrt{f_c'} \qquad\qquad\qquad \textit{UBC 2625(h)3D}$$

$$= 0.6 \times 8 \times 8 \times 32 \times 12\sqrt{6000}/1000 = 1142 \text{ kips} \qquad\qquad \textit{ACI 21.6.4.6}$$

The shear strength reduction factor is usually 0.85. However, in Seismic Zones 3 and 4, this factor must be taken as 0.6 for any structural member if its nominal shear strength is less than the shear corresponding to the development of the nominal flexural strength of the member (UBC 2609(d)4; ACI 9.3.4). For this design, a conservative value of 0.6 is used for the design of the wall. The nominal shear strength is given by:

$$V_n = A_{cv} \times \left[\alpha_c \sqrt{f'_c} + \rho_n f_y \right]$$

UBC 2625(h)3
ACI 21.6.4

where α_c varies linearly from 3.0 for $h_w / \ell_w = 1.5$ to 2.0 for $h_w / \ell_w = 2.0$

or $\alpha_c = 2.0$ if $h_w / \ell_w \geq 2.0$

The α_c factor adjusts for a higher allowable shear strength for walls with low ratios of height to horizontal length. In this case, $\alpha_c = 2.0$ since the ratio $h_w / \ell_w = 80/32 = 2.5$.

Using #6 bars at 18 in.:

$$\rho_n = 0.44/(8 \times 18) = 0.0031 > 0.0025 \quad \text{O.K.}$$

$$\phi V_n = (0.6 \times 32 \times 12 \times 8) \times [2\sqrt{6000} + (0.0031 \times 60{,}000)] / 1000 = 628 \text{ kips}$$

From Table 7-5, the factored shear force to be resisted is:

$$V_u = 199 \text{ kips} < \phi V_n = 628 \text{ kips} \quad \text{O.K.}$$

Therefore, use one curtain of #6 bars spaced at 18 in. o.c. in the horizontal direction.

Vertical distributed reinforcement ratio ρ_v must not be less than the horizontal reinforcement ratio ρ_n when the ratio h_w / ℓ_w is less than 2.0 (UBC 2625(h)3C; ACI 21.6.4.5). Since $h_w / \ell_w = 2.5 > 2.0$, the minimum reinforcement ratio $\rho_v = 0.0025$, as specified in UBC 2625(f)2A or ACI 21.6.2.1. will be used:

$$A_{sv} = 0.0025 \times 8 \times 12 = 0.24 \text{ in.}^2/\text{ft}$$

Using #6 bars, the required bar spacing is:

$$s = \frac{0.44}{0.24} \times 12 = 22 \text{ in.} > 18 \text{ in.}$$

Use #6 bars at 18 in. in the vertical direction also.

The adequacy of the shearwall section at the base under combined factored axial loads and bending moments in the plane of the wall must be checked. A design strength interaction diagram for the shearwall with the reinforcement previously determined is shown in Fig. 7-3. Adequate strength corresponding to the various load combinations shown in Table 7-5 is provided by the reinforcement used.

Review of the load combinations and the interaction diagram for this wall segment shows that the stress in the reinforcement will exceed $0.5f_y$. Therefore, a Class B splice will be required. The basic development length for a #6 bar is:

$$\ell_{db} = 0.04 \, A_b f_y / \sqrt{f'_c}$$

$$= 0.04 \times 0.44 \times 60{,}000/\sqrt{6000} = 13.6 \text{ in.}$$

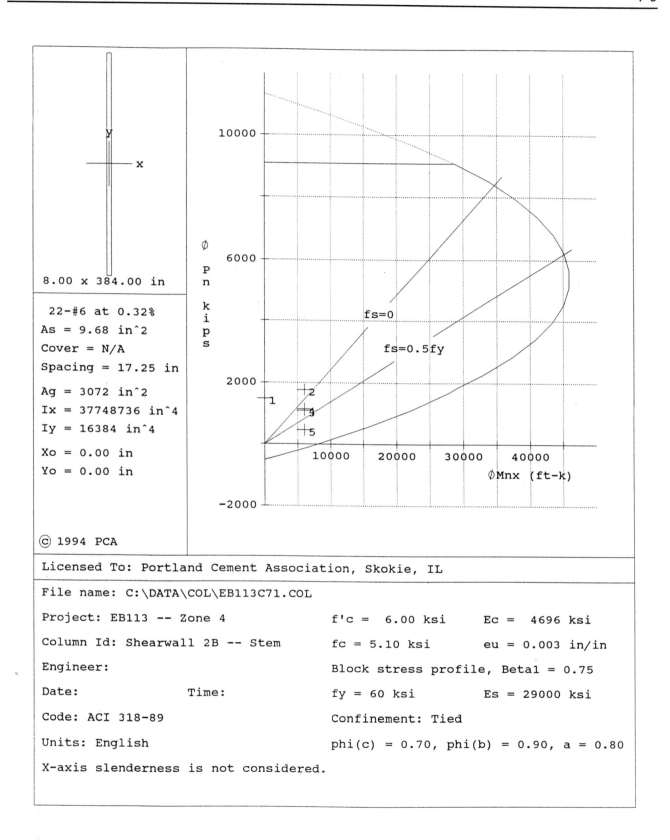

8.00 x 384.00 in

22-#6 at 0.32%
As = 9.68 in^2
Cover = N/A
Spacing = 17.25 in

Ag = 3072 in^2
Ix = 37748736 in^4
Iy = 16384 in^4

Xo = 0.00 in
Yo = 0.00 in

© 1994 PCA

Licensed To: Portland Cement Association, Skokie, IL

File name: C:\DATA\COL\EB113C71.COL

Project: EB113 -- Zone 4   f'c = 6.00 ksi   Ec = 4696 ksi

Column Id: Shearwall 2B -- Stem   fc = 5.10 ksi   eu = 0.003 in/in

Engineer: Block stress profile, Beta1 = 0.75

Date: Time: fy = 60 ksi   Es = 29000 ksi

Code: ACI 318-89   Confinement: Tied

Units: English   phi(c) = 0.70, phi(b) = 0.90, a = 0.80

X-axis slenderness is not considered.

Figure 7-3 Design Strength Interaction Diagram for Stem Segment of Shearwall 2B (Seismic Zone 4)

The information required to find the applicable factors is:

Minimum clear cover	= 2.63 in. (3.5 d_b)
Side cover	= 3.25 in. (4.3d_b)
Clear bar spacing	= 17.25 in. (23 d_b)

Reinforcement does not form an inner layer of wall reinforcement.

This gives a value of 1.28 for Factor I (see Table 5-12). The other factors are equal to 1.0, giving:

Class B splice length = $1.3 \times (1.28 \times 1.0 \times 1.0 \times 1.0 \times 1.0 \times 13.6) = 22.6$ in.

7.2.1.2   Design of Wall Stem Based on UBC-94 Provisions

7.2.1.2.1 Boundary Element Requirement

The design procedure for shearwalls based on the provisions given in UBC-94 can be summarized in three steps:

1) Design the shearwall for flexure and axial load by considering the entire cross-section of the wall to act as a short column; this will yield the required flexural reinforcement (UBC 1921.6.5.1). Check the shear resistance of the wall provided by the web (UBC 1921.6.4).

2) Provide the boundary zone detail requirements given in UBC 1921.6.5.6 at each end of the wall unless $P_u \leq$ 0.10 $A_g f'_c$ (for geometrically symmetrical walls) and either $M_u/V_u \ell_w \leq 1.0$ or $V_u \leq 3 \ell_w h \sqrt{f'_c}$ and $M_u/V_u \ell_w \leq 3.0$ (UBC 1921.6.5.4). In any case, shearwalls with $P_u > 0.35 \, P_o$ are not permitted to resist earthquake-induced forces (UBC 1921.6.5.3).

3) For shearwalls where boundary elements with specially detailed reinforcement have to be provided, determine the length of the boundary zone by using one of two methods. In the conservative method, the boundary zones are provided over a maximum distance of 0.25 ℓ_w at each of the walls (UBC 1921.6.5.4). Alternatively, the boundary zone length may be based on the compressive strain levels at the edges of the wall when the wall is subjected to displacements resulting from the ground motions specified in UBC 1629.2, using cracked section properties of the wall (UBC 1921.6.5.5). Confinement is to be provided wherever the compressive strain exceeds 0.003.

Using an 8 in. × 384 in. wall, determine if boundary zones are required. From Table 7-5, the maximum factored axial load is 1775 kips. Check the criterion in UBC 1921.6.5.4:

$$0.10 \, A_g f'_c \; = 0.10 \, (8 \times 384) \times 6.0 = 1843 \text{ kips} > P_u = 1775 \text{ kips}$$

$$M_u/V_u \ell_w \; = 6013/(199 \times 32) = 0.94 < 1.0$$

Therefore, boundary zones are not required. Since the provisions in UBC 1921.6.4 for the nominal shear strength of shearwalls are the same as in UBC-91 Sect. 2625(f) 2B, use #6 @ 18 in. vertical bars in the web (see Sect. 7.2.1.1.2). Based on this reinforcement, the nominal axial load capacity of the wall at zero eccentricity is:

$$P_o = 0.85 f'_c (A_g - A_{st}) + f_y A_{st}$$

$$= 0.85 \times 6 (3072 - 9.7) + (60 \times 9.7) = 16{,}200 \text{ kips}$$

Since $P_u = 1775$ kips $< 0.35 P_o = 5670$ kips, the wall can be used to resist its portion of the seismic loads.

7.2.1.2.2 Design and Detailing

As noted above, the nominal shear provisions in UBC 1921.6.4 are the same as those given in UBC-91. Therefore, calculations similar to those in Sect. 7.2.1.1.2 yield #6 @ 18 in. for both the horizontal and vertical reinforcement.

7.2.1.3 Design of Wall Flange Segment Based on UBC-91 Provisions

7.2.1.3.1 Boundary Element Requirement

The geometric properties of the flange segment are:

$$A_g = \ell_w \times h = 20 \times 8/12 = 13.3 \text{ ft}^2$$

$$I_g = h \ell_w^3/12 = 0.67 \times 20^3/12 = 447 \text{ ft}^4$$

Using the factored loads from Table 7-6:

$$f_c = \frac{P_u}{A_g} + \frac{M_u(\ell_w/2)}{I_g}$$

$$= \frac{400}{13.3} + \frac{3143(20/2)}{447}$$

$$= 30.1 + 70.3$$

$$= 100.4 \text{ ksf} = 0.7 \text{ ksi} < 0.2 f'_c = 1.2 \text{ ksi}$$

Therefore, boundary elements are not required.

7.2.1.3.2 Design and Detailing

At least two curtains of reinforcement must be provided in the wall if the in-plane factored shear force, V_u, for the wall exceeds $2A_{cv}\sqrt{f'_c}$ (UBC 2625(f)2B; ACI 21.6.2.2):

$$2A_{cv}\sqrt{f'_c} = 2 \times 8 \times 20 \times 12\sqrt{6000}/1000 = 297 \text{ kips} > V_u = 235 \text{ kips}$$

Therefore, two curtains of reinforcement are not required.

The reinforcement ratio along the longitudinal and transverse direction must not be less than 0.0025 and the spacing must not exceed 18 in. (UBC 2625(f)2A; ACI 21.6.2.1).

The upper limit on the design shear strength:

$$\phi V_n = \phi 8 A_{cv} \sqrt{f'_c}$$

UBC 2625(h)3D

$$= 0.6 \times 8 \times 8 \times 20 \times 12 \sqrt{6000}/1000 = 714 \text{ kips}$$

ACI 21.6.4.6

The nominal shear strength is given by:

$$V_n = A_{cv} \times \left[\alpha_c \sqrt{f'_c} + \rho_n f_y \right]$$

UBC 2625(h)3

ACI 21.6.4

where α_c varies linearly from 3.0 for $h_w / \ell_w = 1.5$ to 2.0 for $h_w / \ell_w = 2.0$

or $\alpha_c = 2.0$ if $h_w / \ell_w \geq 2.0$

In this case, $\alpha_c = 2.0$ since the ratio $h_w / \ell_w = 80/20 = 4.0$.

Using #6 bars at 18 in.:

$$\rho_n = 0.44/(8 \times 18) = 0.0031 > 0.0025 \qquad \text{O.K.}$$

$$\phi V_n = (0.6 \times 20 \times 12 \times 8) \times [2\sqrt{6000} + (0.0031 \times 60,000)]/1000 = 393 \text{ kips}$$

The factored shear force to be resisted is:

$$V_u = 235 \text{ kips} < \phi V_n = 393 \text{ kips} \quad \text{O.K.}$$

Therefore, use #6 bars spaced at 18 in. o.c. in the horizontal direction.

Vertical distributed reinforcement ratio ρ_v must not be less than the horizontal reinforcement ratio ρ_n when the ratio h_w / ℓ_w is less than 2.0 (UBC 2625(h)3C; ACI 21.6.4.5). Since $h_w / \ell_w = 4.0 > 2.0$, the minimum reinforcement ratio $\rho_v = 0.0025$, as specified in UBC 2625(f)2A or ACI 21.6.2.1 will be used:

$$A_{sv} = 0.0025 \times 8 \times 12 = 0.24 \text{ in.}^2/\text{ft}$$

Using #6 bars, the required bar spacing is:

$$s = \frac{0.44}{0.24} \times 12 = 22 \text{ in.} > 18 \text{ in.}$$

Use #6 bars at 18 in. in the vertical direction also.

The adequacy of the shearwall section at the base under combined factored axial loads and bending moments in the plane of the wall must be checked. A design strength interaction diagram for the shearwall is shown in Fig. 7-4. Adequate strength with respect to the various load combinations shown in Table 7-6 is provided for by the reinforcement used.

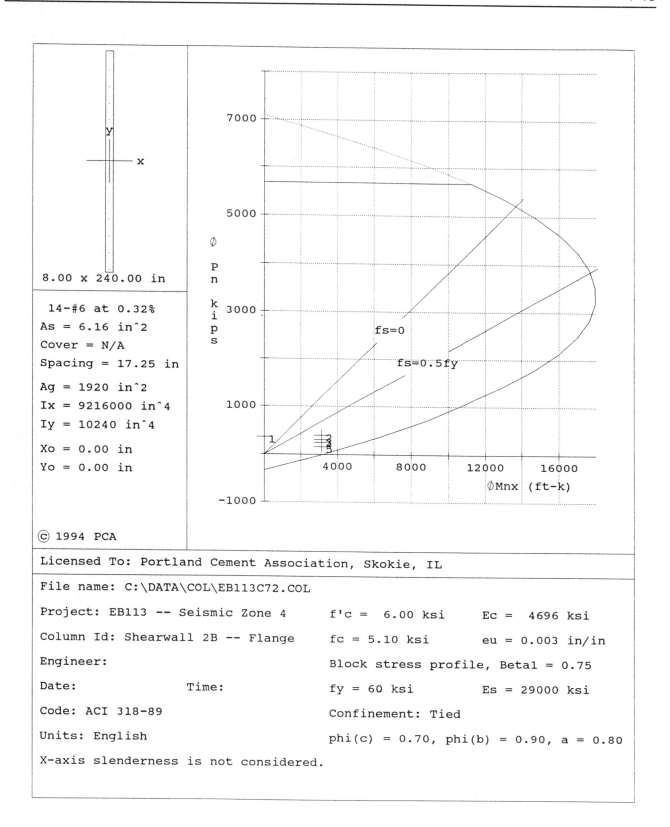

8.00 x 240.00 in

14-#6 at 0.32%
As = 6.16 in^2
Cover = N/A
Spacing = 17.25 in

Ag = 1920 in^2
Ix = 9216000 in^4
Iy = 10240 in^4

Xo = 0.00 in
Yo = 0.00 in

© 1994 PCA

Licensed To: Portland Cement Association, Skokie, IL

File name: C:\DATA\COL\EB113C72.COL

Project: EB113 -- Seismic Zone 4   f'c = 6.00 ksi   Ec = 4696 ksi

Column Id: Shearwall 2B -- Flange   fc = 5.10 ksi   eu = 0.003 in/in

Engineer: Block stress profile, Beta1 = 0.75

Date: Time: fy = 60 ksi   Es = 29000 ksi

Code: ACI 318-89   Confinement: Tied

Units: English   phi(c) = 0.70, phi(b) = 0.90, a = 0.80

X-axis slenderness is not considered.

Figure 7-4 Design Strength Interaction Diagram for Flange Segment of Shearwall 2B (Seismic Zone 4)

Review of the load combinations (Table 7-8) and the interaction diagram (Fig. 7-6) for this wall segment shows that stress in the reinforcing bars will be tensile and in excess of $0.5f_y$. Therefore, a Class B splice will be used. Use a 24 in. splice length (See Sect. 7.2.1.1.2).

Figure 7-5 shows the reinforcement details for the walls.

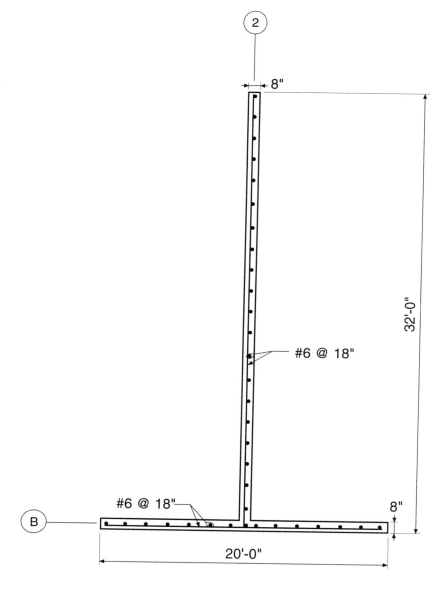

Figure 7-5 Reinforcement Details for Shearwall 2B (Seismic Zone 4)

7.3   Design for Seismic Zone 2B

The geometry of the building and walls will remain the same as in the previous section, except that the walls will be 6 in. thick. An f_c' of 4000 psi will be used for the walls.

The weights of the floors are listed in Table 7-7. The building has the same lateral force resisting system in both principal directions. Therefore, the seismic design forces are the same in both the longitudinal and the transverse directions of the building.

Table 7-7 Seismic Forces for Zone 2B

Floor Level	Weight, w_x (kips)	Height, h_x (ft)	$w_x h_x$ (ft-kips)	Lateral Force (kips)
R	1074	80	85,920	147.3
8	1293	70	90,510	155.2
7	1293	60	77,580	133.0
6	1293	50	64,650	110.9
5	1293	40	51,720	88.7
4	1293	30	38,790	66.5
3	1293	20	25,860	44.3
2	1293	10	12,930	22.2
Σ	10,125		447,960	768.2

Building height	h_n	$= 80\ ft$
Building weight	W	$= 10,125\ kips$
Fundamental period	T	$= 0.020\ (h_n)^{3/4} = 0.53\ sec.$
Seismic zone factor	Z	$= 0.2$
Importance factor	I	$= 1.0$
Response coefficient	R_w	$= 6$
Soil factor	S	$= 1.2$
Coefficient	C	$= 1.25\ S/T^{2/3} = 2.28$
Base shear	V	$= ZICW/R_w = 768.2\ kips$
Top level force	F_t	$= 0\ (T < 0.7\ sec.)$

The weight of a typical floor includes all elements located between two imaginary planes passing through the mid-height of the walls above and below the floor considered. The calculations for the base shear V for the transverse and longitudinal directions are shown at the bottom of Table 7-7. For this example, the importance factor I and the site soil factor S have been taken equal to 1.0 and 1.2, respectively. The value of R_w is equal to 6. The seismic forces resulting from the distribution of the base shear in accordance with UBC Eqs. (28-6), (28-7) and (28-8) are also listed in Table 7-7. The member forces resulting from the seismic loads, including accidental torsional effects, are determined in the same manner as used for Zone 4, as outlined in Section 7.2. The results of the analysis for an interior transverse and an interior longitudinal frame are summarized in Table 7-8.

The member forces and the story drifts induced by P-Δ effects need to be considered in the evaluation of overall structural stability. The code allows P-Δ effects to be neglected where the ratio of the secondary moment to the primary moment is less than or equal to 0.1 (UBC 1628.9). Based on the results shown in Table 7-9, the P-Δ effects may be ignored for the design example.

For structures having a fundamental period less than 0.7 seconds, the calculated story drift must not exceed 0.04/R_w nor 0.005 times the story height (UBC 1628.9).

For $h_s = 10$ ft and $R_w = 6$:

Table 7-8 Results for Analyses of Interior Transverse and Longitudinal Walls under Seismic Zone 2B Forces

Transverse Direction (Stem Segment—Shearwall 2B)				
Level	Axial Load (kips)	Bending Moment (ft-kips)		Shear Force (kips)
		Top	Bottom	
R	4	65	59	-12
8	11	70	183	-25
7	23	12	358	-37
6	38	-114	577	-46
5	55	-286	824	-54
4	75	-505	1099	-60
3	96	-760	1388	-63
2	114	-1090	1732	-64

Longitudinal Direction (Flange Segment—Shearwall 2B)				
Level	Axial Load (kips)	Bending Moment (ft-kips)		Shear Force (kips)
		Top	Bottom	
R	3	194	-65	13
8	4	353	-65	29
7	5	408	22	43
6	7	411	134	55
5	8	344	298	64
4	10	204	512	72
3	11	-55	822	77
2	12	-508	1270	78

Table 7-9 P-Δ Check for Seismic Zone 2B

Floor Level	Gravity Loads * Σw_i (kips)	Transverse Story Drift Δ_i (in.)	Longitudinal Story Drift Δ_i (in.)	Story Shear V_i (kips)	Story Height h_{si} (in.)	Transverse $\Sigma w_i \Delta_i / V_i h_{si}$	Longitudinal $\Sigma w_i \Delta_i / V_i h_{si}$
R	1326	0.01	0.01	147.3	120	0.001	0.001
8	2787	0.01	0.01	302.5	120	0.001	0.001
7	4248	0.01	0.01	435.5	120	0.001	0.001
6	5709	0.01	0.01	546.4	120	0.001	0.001
5	7170	0.01	0.01	635.1	120	0.001	0.001
4	8631	0.01	0.01	701.6	120	0.001	0.001
3	10,092	0.01	0.01	745.9	120	0.001	0.001
2	11,553	0.00	0.01	768.2	120	0.000	0.001

* Includes floor weight, superimposed dead load, and reduced roof and floor live loads

$$\frac{0.04h_s}{R_w} = \frac{0.04 \times 10 \times 12}{6} = 0.8 \text{ in.}$$

$$0.005h_s = 0.005 \times 10 \times 12 = 0.6 \text{ in. (governs)}$$

As shown in Table 7-9, the story drift limits are not exceeded at any floor level.

7.3.1   Design of Shearwall 2B

The shearwalls are part of the lateral load resisting system. In Seismic Zone 2B, the only special detailing requirements are those in UBC 1921.8. Review of this section shows that none of these detailing requirements apply to shearwalls. This points out the advantage of the Bearing Wall System in Seismic Zone 2.

The shearwall 2B is 32 ft long from end to end (in the transverse direction of the building). The wall is 6 in. thick. Table 7-10 shows gravity loads for the flange and stem segments of the shearwall.

The axial loads and bending moments corresponding to the appropriate load combinations are listed in Tables 7-11 and 7-12.

Table 7-10 Gravity Loads on Shearwall 2B (Seismic Zone 2B)

Level	Tributary Area (ft²)		Cumulative Dead Load (kips)		Cumulative Reduced Live Load (kips)	
	Flange	Stem	Flange	Stem	Flange	Stem
R	72	768	24	104	4	23
8	72	768	47	208	7	38
7	72	768	71	312	11	54
6	72	768	95	415	14	69
5	72	768	118	519	17	84
4	72	768	142	623	20	100
3	72	768	165	727	22	115
2	72	768	189	831	25	131

Table 7-11 Summary of Factored Axial Loads, Bending Moments, and Shear Forces for Stem Segment of Shearwall 2B between Grade and the 2nd Floor (Seismic Zone 2B)

(1)	$U = 1.4D + 1.7L$	Eq. (9-1), UBC 1909.2.1
(2)	$U = 0.75(1.4D + 1.7L \pm 1.87E)$	Eq. (9-2), UBC 1909.2.3
(3)	$U = 0.9D \pm 1.43E$	Eq. (9-3), UBC 1909.2.3

Load Combination	Axial Load (kips)	Bending Moment (ft-kips)		Shear Force (kips)
		Top	Bottom	
1	1386	—	—	—
2: sidesway right	1200	1526	-2425	90
sidesway left	880	-1526	2425	-90
3: sidesway right	911	1559	-2477	92
sidesway left	585	-1559	2477	-92

Table 7-12 Summary of Factored Axial Loads, Bending Moments, and Shear Forces for Flange Segment of Shearwall 2B between Grade and the 2nd Floor (Seismic Zone 2B)

(1) U = 1.4D + 1.7L	Eq. (9-1), UBC 1909.2.1	
(2) U = 0.75(1.4D + 1.7L ± 1.87E)	Eq. (9-2), UBC 1909.2.3	
(3) U = 0.9D ± 1.43E	Eq. (9-3), UBC 1909.2.3	

Load Combination	Axial Load (kips)	Bending Moment (ft-kips)		Shear Force (kips)
		Top	Bottom	
1	307	—	—	—
2: sidesway right	247	-711	-1778	109
sidesway left	213	711	1778	-109
3: sidesway right	187	-726	-1816	112
sidesway left	153	726	1816	-112

7.3.1.1   Wall Stem Segment

Boundary elements are only required in Seismic Zones 3 and 4. Therefore, the requirements of UBC 1921.6.5 need not be checked.

The design shear strength of a shearwall is limited to:

$$\phi V_n = \phi 10 \sqrt{f_c'} \times hd \qquad\qquad \textit{1911.10.3}$$

where d equals 80% of the length of the wall (UBC 1911.10.4). In this case:

$$\phi V_n = 0.85 \times 10 \sqrt{4000} \times 6 \times 0.8 \times (32 \times 12)/1000 = 991 \text{ kips} > 92 \text{ kips}$$

The shear strength provided by the concrete is:

$$\phi V_c = \phi 2 \sqrt{f_c'} \times hd \qquad\qquad \textit{1911.10.5}$$

$$= 0.85 \times 2 \sqrt{4000} \times 0.8 \times 32 \times 12 \times 6 /1000 = 198 \text{ kips} > 92 \text{ kips}$$

Therefore, minimum shear reinforcement is required. The maximum spacing of horizontal reinforcing bars is:

$$s_{max} = \ell_w/5 = (32 \times 12)/5 = 77 \text{ in.} \qquad\qquad \textit{1911.10.9.3}$$

$$= 3h = 3 \times 6 = 18 \text{ in.}$$

$$= 18 \text{ in.} \quad \text{(governs)}$$

The ratio of horizontal shear reinforcement area to gross concrete area of the vertical cross section must not be less than 0.0025 (UBC 1911.10.9.2):

$$A_{vh} = 0.0025 \times 6 \times 12 = 0.18 \text{ in.}^2/\text{ft}$$

#5 bars spaced at 18 in. gives:

$$A_{vh} = 0.31 \times 12/18 = 0.21 \text{ in.}^2/\text{ft} > 0.18 \text{ in.}^2/\text{ft} \qquad \text{O.K.}$$

Therefore, use #5 horizontal bars spaced at 18 in.

The ratio of vertical reinforcement area to gross concrete of horizontal section must not be less than:

$$\rho_n = 0.0025 + 0.5 \times (2.5 - h_w / \ell_w) \times (\rho_h - 0.0025) \qquad \text{1911.10.9.4}$$

where $\rho_h = 0.31/(18 \times 6) = 0.0029$

$$\rho_n = 0.0025 + 0.5 \times (2.5 - 80/32) \times (0.0029 - 0.0025) = 0.0025$$

The required area of vertical reinforcing steel is:

$$A_n = 0.0025 \times 6 \times 12 = 0.18 \text{ in.}^2/\text{ft}$$

The maximum spacing of vertical bars is:

$$s_{max} = \ell_w/3 = (32 \times 12)/3 = 128 \text{ in.} \qquad \text{1911.10.9.5}$$

$$= 3h = 3 \times 6 = 18 \text{ in.}$$

$$= 18 \text{ in.} \quad \text{(governs)}$$

5 bars spaced at 18 in. gives:

$$A_n = 0.31 \times 12/18 = 0.21 \text{ in.}^2/\text{ft} > 0.18 \text{ in.}^2/\text{ft} \qquad \text{O.K.}$$

Therefore, use #5 vertical bars spaced at 18 in.

A design strength interaction diagram for the shearwall is shown in Fig. 7-6. The interaction diagram shows that the shearwall has adequate strength with respect to the load combinations listed in Table 7-11.

Class B splices will be used in this case. The basic development length for #5 bars is:

$$\ell_{db} = 0.04A_b f_y / \sqrt{f_c'}$$

$$= 0.04 \times 0.31 \times 60,000 / \sqrt{4000} = 11.8 \text{ in.}$$

The information required to find the applicable factors is:

Minimum clear cover	= 2.69 in. (4.3 d_b)
Side cover	= 2.4 in. (3.8d_b)
Clear bar spacing	= 17.4 in. (27.8 d_b)

Reinforcement does not form an inner layer of wall reinforcement.

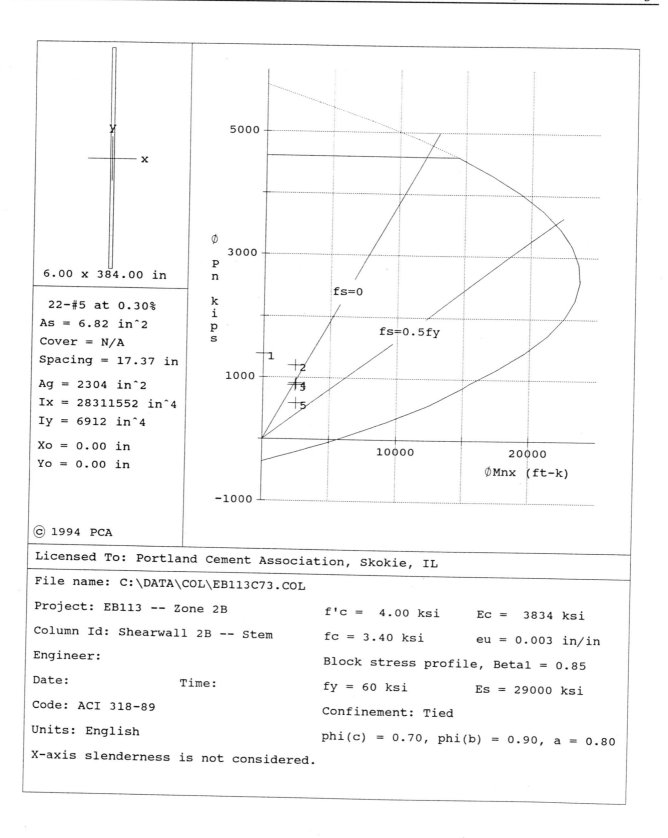

6.00 x 384.00 in

22-#5 at 0.30%
As = 6.82 in^2
Cover = N/A
Spacing = 17.37 in

Ag = 2304 in^2
Ix = 28311552 in^4
Iy = 6912 in^4

Xo = 0.00 in
Yo = 0.00 in

© 1994 PCA

Licensed To: Portland Cement Association, Skokie, IL

File name: C:\DATA\COL\EB113C73.COL

Project: EB113 -- Zone 2B

Column Id: Shearwall 2B -- Stem

Engineer:

Date: Time:

Code: ACI 318-89

Units: English

X-axis slenderness is not considered.

f'c = 4.00 ksi   Ec = 3834 ksi

fc = 3.40 ksi   eu = 0.003 in/in

Block stress profile, Beta1 = 0.85

fy = 60 ksi   Es = 29000 ksi

Confinement: Tied

phi(c) = 0.70, phi(b) = 0.90, a = 0.80

Figure 7-6 Design Strength Interaction Diagram for Stem Segment of Shearwall 2B (Seismic Zone 2B)

This gives a value of 1.51 for Factor I (see Table 5-12). The other factors are equal to 1.0, giving:

Class B splice length = $1.3 \times (1.51 \times 1.0 \times 1.0 \times 1.0 \times 1.0 \times 11.8) = 23.2$ in.

7.3.1.2   Wall Flange Segment

Boundary elements are only required in Seismic Zones 3 and 4. Therefore, the requirements of UBC 1921.6.5 need not be checked.

The design shear strengths of a shearwall is limited to:

$$\phi V_n = \phi 10 \sqrt{f'_c} \times hd \qquad\qquad\qquad 1911.10.3$$

where d equals 80% of the length of the wall (UBC 1911.10.4). In this case:

$$\phi V_n = 0.85 \times 10 \sqrt{4000} \times 6 \times 0.8 \times (20 \times 12)/1000 = 619 \text{ kips} > 112 \text{ kips} \qquad \text{O.K.}$$

The shear strength provided by the concrete is:

$$\phi V_c = \phi 2 \sqrt{f'_c} \times hd \qquad\qquad\qquad 1911.10.5$$

$$= 0.85 \times 2 \sqrt{4000} \times 0.8 \times 20 \times 12 \times 6/1000 = 124 \text{ kips} > 112 \text{ kips}$$

Therefore, only minimum shear reinforcement is required. The maximum spacing of horizontal reinforcing bars is:

$$s_{max} = \ell_w/5 = (20 \times 12)/5 = 48 \text{ in.} \qquad\qquad 1911.10.9.3$$

$$= 3h = 3 \times 6 = 18 \text{ in.}$$

$$= 18 \text{ in.} \quad \text{(governs)}$$

The ratio of horizontal shear reinforcement area to gross concrete area of the vertical cross section must not be less than 0.0025 (UBC 1911.10.9.2).

$$A_{vh} = 0.0025 \times 6 \times 12 = 0.18 \text{ in.}^2/\text{ft}$$

#5 bars spaced at 18 in. gives:

$$A_{vh} = 0.31 \times 12/18 = 0.21 \text{ in.}^2/\text{ft} > 0.18 \text{ in.}^2/\text{ft} \qquad \text{O.K.}$$

Therefore, use #5 horizontal bars spaced at 18 in.

The ratio of vertical reinforcement area to gross concrete of horizontal section shall not be less than:

$$\rho_n = 0.0025 + 0.5 \times (2.5 - h_w/\ell_w) \times (\rho_h - 0.0025) \qquad\qquad 1911.10.9.4$$

where $\rho_h = 0.31/(18 \times 6) = 0.0029$

$$\rho_n \quad = 0.0025 + 0.5 \times (2.5 - 80/20) \times (0.0029 - 0.0025)$$

$$= 0.0022 < 0.0025$$

Use $\rho_n = 0.0025$.

The required area of vertical reinforcing steel is:

$$A_n \quad = 0.0025 \times 6 \times 12 = 0.18 \text{ in.}^2/\text{ft}$$

The maximum spacing of vertical bars is:

$$s_{max} \quad = \ell_w/3 = (20 \times 12)/3 = 80 \text{ in.}$$

1911.10.9.5

$$= 3h = 3 \times 6 = 18 \text{ in.}$$

$$= 18 \text{ in.} \quad \text{(governs)}$$

#5 bars spaced at 18 in. gives:

$$A_n \quad = 0.31 \times 12/18 = 0.21 \text{ in.}^2/\text{ft} > 0.18 \text{ in.}^2/\text{ft} \quad \text{O.K.}$$

Therefore, use #5 vertical bars spaced at 18 in.

A strength interaction diagram for the shearwall is shown in Fig. 7-7. The interaction diagram shows that the shearwall has adequate strength with respect to the load combinations listed in Table 7-12.

Review of the load combinations and the interaction diagram for this wall segment shows that stress in the reinforcement will be tensile and in excess of $0.5f_y$. Therefore, a Class B splice as designed for the stem segment will be used.

Figure 7-8 shows the reinforcement details for this wall.

7.4   Design for Wind Loads

This section will present the design of the structure for wind forces corresponding to a fastest mile wind speed of 70 mph and Exposure B. The calculations of the lateral forces are shown in Tables 7-13 and 7-14. The same member sizes and material strengths used in the previous design (Seismic Zone 2B) will be utilized. The results from the analysis of the structure under the design wind forces are shown in Table 7-15.

Story drift limitations and P-Δ effect considerations specified under earthquake design regulations are not applicable to structures designed for wind loads.

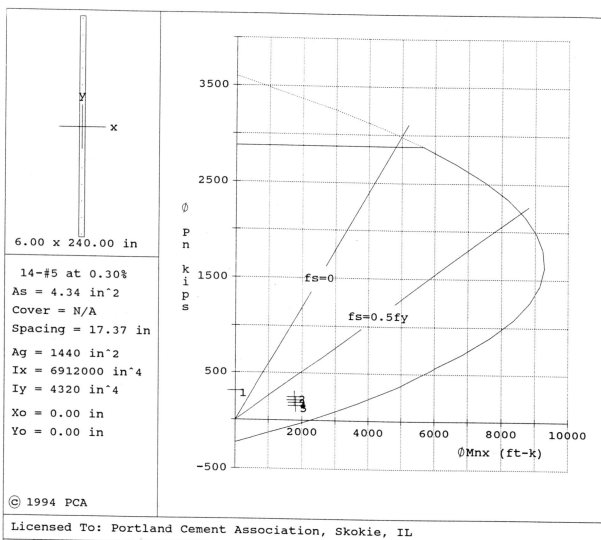

6.00 x 240.00 in

14-#5 at 0.30%
As = 4.34 in^2
Cover = N/A
Spacing = 17.37 in

Ag = 1440 in^2
Ix = 6912000 in^4
Iy = 4320 in^4

Xo = 0.00 in
Yo = 0.00 in

© 1994 PCA

Licensed To: Portland Cement Association, Skokie, IL

File name: C:\DATA\COL\EB113C74.COL

Project: EB113 -- Seismic Zone 2

Column Id: Shearwall 2B -- Flange

Engineer:

Date: Time:

Code: ACI 318-89

Units: English

X-axis slenderness is not considered.

f'c = 4.00 ksi   Ec =   3834 ksi

fc = 3.40 ksi   eu = 0.003 in/in

Block stress profile, Beta1 = 0.85

fy = 60 ksi   Es = 29000 ksi

Confinement: Tied

phi(c) = 0.70, phi(b) = 0.90, a = 0.80

Figure 7-7 Design Strength Interaction Diagram for Flange Segment of Shearwall 2B (Seismic Zone 2B)

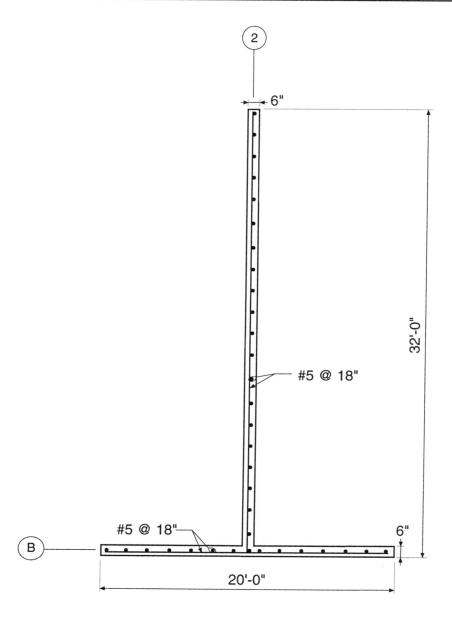

Figure 7-8 Reinforcement Details for Shearwall 2B (Seismic Zone 2B)

7.4.1   Design of Shearwall 2B

In the example for Seismic Zone 2B, shearwall 2B was designed. The design of this wall was controlled by minimum reinforcements requirements. A comparison of the design forces for shearwall 2B from wind forces (Table 7-15) and Seismic Zone 2B forces (Table 7-8), shows that Seismic Zone 2B forces are larger in all cases. Since the members are of the same size, the concrete strength is the same, and the design forces are smaller, it follows that the reinforcement used for the Seismic Zone 2B example would also be adequate for wind design. The reinforcement quantities cannot be reduced, because minimum reinforcement requirements must not be violated.

Table 7-13 Wind Forces in Transverse Direction

Floor Level	Height (ft)	Tributary Height (ft)	Windward			Leeward			Total Force (kips)
			C_e	Pressure (psf)	Force (kips)	C_e	Pressure (psf)	Force (kips)	
R	80	5	1.04	10.5	6.3	1.04	6.6	4.0	10.3
8	70	10	1.00	10.1	12.1	1.04	6.6	7.9	20.0
7	60	10	0.95	9.6	11.5	1.04	6.6	7.9	19.4
6	50	10	0.90	9.1	10.9	1.04	6.6	7.9	18.8
5	40	10	0.84	8.5	10.2	1.04	6.6	7.9	18.1
4	30	10	0.76	7.7	9.2	1.04	6.6	7.9	17.1
3	20	10	0.67	6.8	8.2	1.04	6.6	7.9	16.1
2	10	10	0.62	6.3	7.6	1.04	6.6	7.9	15.5

Table 7-14 Wind Forces in Longitudinal Direction

Floor Level	Height (ft)	Tributary Height (ft)	Windward			Leeward			Total Force (kips)
			C_e	Pressure (psf)	Force (kips)	C_e	Pressure (psf)	Force (kips)	
R	80	5	1.04	10.5	3.7	1.04	6.6	2.3	6.0
8	70	10	1.00	10.1	7.1	1.04	6.6	4.6	11.7
7	60	10	0.95	9.6	6.7	1.04	6.6	4.6	11.3
6	50	10	0.90	9.1	6.4	1.04	6.6	4.6	11.0
5	40	10	0.84	8.5	6.0	1.04	6.6	4.6	10.6
4	30	10	0.76	7.7	5.4	1.04	6.6	4.6	10.0
3	20	10	0.67	6.8	4.8	1.04	6.6	4.6	9.4
2	10	10	0.62	6.3	4.4	1.04	6.6	4.6	9.0

Table 7-15 *Results of Analyses of Interior Transverse and Longitudinal Walls under Wind Forces*

Transverse Direction (Stem Segment—Shearwall 2B)				
Level	Axial Load (kips)	Bending Moment (ft-kips)		Shear Force (kips)
		Top	Bottom	
R	1	7	4	-1
8	1	10	15	-3
7	2	7	36	-4
6	4	-6	66	-6
5	7	-27	100	-7
4	9	-54	142	-9
3	13	-90	191	-10
2	16	-143	251	-11

Longitudinal Direction (Flange Segment—Shearwall 2B)				
Level	Axial Load (kips)	Bending Moment (ft-kips)		Shear Force (kips)
		Top	Bottom	
R	0	12	-7	1
8	0	26	-9	2
7	0	33	-5	3
6	1	37	4	4
5	1	34	17	5
4	1	24	39	6
3	1	2	71	7
2	1	-41	119	8